# Introduction to Computational Materials Science

Emphasizing essential methods and universal principles, this textbook provides everything students need to understand the basics of simulating materials behavior. All the key topics are covered, from electronic structure methods to microstructural evolution, appendices provide crucial background material, and a wealth of practical resources are available online to complete the teaching package.

- Examines modeling materials across a broad range of scales, from the atomic to the mesoscale, providing students with a solid foundation for future study and research.
- Presents detailed, accessible explanations of the fundamental equations underpinning materials modeling, and includes a full chapter summarizing essential mathematical background.
- Extensive appendices, including essential background on classical and quantum mechanics, electrostatics, statistical thermodynamics and linear elasticity, provide all the background necessary to fully engage with the fundamentals of computational modeling.
- Exercises, worked examples, computer codes and discussions of practical implementations methods are all provided online to give students the hands-on experience they need.

**Richard LeSar** is the Lynn Gleason Professor of Interdisciplinary Engineering in the Department of Materials Science and Engineering, Iowa State University, and the former Chair of the Materials Science and Engineering program. He is highly experienced in teaching the modeling and simulation of materials at both undergraduate and graduate levels, and has made extensive use of these methods throughout his own research.

"Finally, an introductory textbook on computational methods that addresses the breadth of materials science. Finally, an introductory textbook that emphasizes understanding the foundations of the subject. Kudos to Professor Richard LeSar for producing such a beautifully pedagogical introductory text that covers the major methods of the field, relates them to their underlying science, and provides links to accessible simulation codes. *Introduction to Computational Materials Science* is the perfect companion to a first-course on this rapidly growing segment of our field."

*David J. Srolovitz, University of Pennsylvania*

"Professor LeSar has written an elegant book on the methods that have been found to be useful for simulating materials. Unlike most texts, he has made the effort to give clear, straightforward explanations, so that readers can implement the models for themselves. He has also covered a wider range of techniques and length-/time-scales than typical textbooks that ignore anything coarser than the atom. This text will be useful for a wide range of materials scientists and engineers."

*Anthony Rollett, Carnegie Mellon University*

"Richard LeSar has successfully summarized the computational techniques that are most commonly used in Materials Science, with many examples that bring this field to life. I have been using drafts of this book in my Computational Materials course, with very positive student response. I am delighted to see the book in print – it will become a classic!"

*Chris G. Van de Walle, University of California, Santa Barbara*

# Introduction to Computational Materials Science

## Fundamentals to Applications

RICHARD LESAR

**Iowa State University**

# CAMBRIDGE
## UNIVERSITY PRESS

University Printing House, Cambridge CB2 8BS, United Kingdom

One Liberty Plaza, 20th Floor, New York, NY 10006, USA

477 Williamstown Road, Port Melbourne, VIC 3207, Australia

314-321, 3rd Floor, Plot 3, Splendor Forum, Jasola District Centre, New Delhi-110025, India

79 Anson Road, #06-04/06, Singapore 079906

Cambridge University Press is part of the University of Cambridge.

It furthers the University's mission by disseminating knowledge in the pursuit of
education, learning and research at the highest international levels of excellence.

www.cambridge.org
Information on this title: www.cambridge.org/9780521845878

First published 2013
Reprinted 2016

*A catalogue record for this publication is available from the British Library*

ISBN 978-0-521-84587-8 Hardback

Additional resources for this publication at www.cambridge.org/lesar

# CONTENTS

## PART THREE    MESOSCOPIC METHODS

## PART FOUR   SOME FINAL WORDS

## PART FIVE   APPENDICES

# PREFACE

The goal of this book is to introduce the basic methods used in the computational modeling of materials. The text reflects many tradeoffs: breadth versus depth, pedagogy versus detail, topic versus topic. The intent was to provide a sufficient background in the theory of these methods that the student can begin to apply them to the study of materials. That said, it is not a "computation" book – details of how to implement these methods in specific computer languages are not discussed in the text itself, though they are available from an online resource, which will be described a bit later in this preface.

Modeling and simulation are becoming critical tools in the materials researcher's tool box. My hope is that this text will help attract and prepare the next generation of materials modelers, whether modeling is their principal focus or not.

## Structure of the book

This book is intended to be used by upper-level undergraduates (having taken statistical thermodynamics and at least some classical and quantum mechanics) and graduate students. Reflecting the nature of materials research, this text covers a wide range of topics. It is thus broad, but not deep. References to more detailed texts and discussions are given so that the interested reader can probe more deeply. For those without a materials science background, a brief introduction to crystallography, defects, etc. is given in Appendix B.

This text covers a wide range of methods, covering a variety of time and length scales. The text is divided into parts, each with a specific focus. Part One, for example, introduces the basic methods used in essentially all simulations. The random-walk model in Chapter 2 illustrates some important concepts used throughout the text, including the stochastic nature of simulations. Since every method in this text involves representing the materials system on a lattice or grid, Chapter 3 focuses on how to sum quantities, such as the energy, on a lattice.

In Part Two, the text focuses on modeling of systems of atoms and molecules, starting with Chapter 4, which describes the fundamental methods used to calculate the electronic structure of materials. To extend length scales beyond what can be done with such methods, analytic descriptions of the interaction potentials between atoms are discussed for a variety of materials types in Chapter 5. Molecular dynamics methods, in which Newton's equations are solved to monitor the motion of atoms, are presented in Chapter 6, enabling the modeling of the thermodynamic and dynamic properties of materials. The Monte Carlo method of Chapter 7 is thermodynamically driven, with great flexibility to be applied not only to systems of atoms but also to any models for which there is an expression for the energy in terms of a set of variables. The final chapter of this part of the text, Chapter 8, discusses how to extend these ideas to molecular systems.

One of the most important distinctions between how materials scientists view materials and how many of those in other fields do so is the materials scientist's focus on defects and defect distributions, which involve a range of length and time scales intermediate between the atomic scale and the continuum scale of everyday objects. This range of length and times is often referred to as the mesoscale. Recognizing this distinction, Part Three extends the discussion to methods used in modeling the physics of materials at the mesoscale. Chapter 9 introduces the kinetic Monte Carlo method, which evolves a system based on the rates of its underlying processes. Standard Monte Carlo is revisited in Chapter 10, with a focus on a common approach used to model microstructural evolution. Cellular automata are rule-based methods with great flexibility, and some limitations, and are discussed in Chapter 11. One of the fastest-growing, and most powerful, methods in materials computation is the phase-field method, which is based on thermodynamics. The presentation in Chapter 12 is basic, using relatively simple examples of applications of the phase-field method. In the final chapter of this part of the text, Chapter 13, a set of methods is introduced that are based on what I call mesodynamics – the application of standard dynamical methods to systems of collective variables that represent, generally, defects in a material.

The single chapter of Part Four summarizes and integrates many of the ideas brought forth in the previous text. Chapter 14 is couched in the ideas of *Integrated Computational Materials Engineering* (ICME), a new field that integrates experiment with computation to accelerate materials development. The basic ideas behind *materials informatics* are introduced and their important role in ICME is discussed.

A series of appendices provide background material to the text, covering such diverse topics as classical mechanics, electronic structure, statistical thermodynamics, rate theory, and elasticity.

The text does not present any continuum-level modeling, such as the finite element method, heat or fluid flow, etc. It is not that these topics are not interesting or that they are unimportant in materials science and engineering – there simply was not sufficient space in the text.

## Computation

Both the development of materials models as well as how one can utilize models in a computer-based calculation are covered in this text. A variety of numerical methods used in such calculations are presented. Understanding the limitations of the methods discussed in the text will require a working knowledge of those methods.

That said, this is not a computer programming text and details of the implementation of the methods and algorithms into a computer program are not discussed. Sample applications are, however, available on the web at http://www.cambridge.org/lesar, in which the implementation of the methods is discussed in some detail. These codes are based on the commercial platforms MATLAB® and Mathematica®, which are commonly available in many universities. These platforms are powerful enough to use for the examples in this book, yet straightforward enough so that programming is understandable. The graphics capabilities needed for the examples are typically built into them. Note that the sample applications may not be well-optimized computer codes – they were written to be clear not efficient. Students who want a more extensive experience are urged to create their own codes based on the algorithms in the text.

Computational modeling is best learned by doing. I encourage the use of the computer codes available at http://www.cambridge.org/lesar. Not only is doing these types of simulations essential for understanding the methods, I generally find that students enjoy running calculations more than reading about them.

## Acknowledgments

A number of people have read parts of this text over time, giving excellent advice on how to improve it. I particularly want to thank Chris Van de Walle, Tony Rollett, Simon Phillpot, Robin Grimes, Scott Beckman, Jeff Rickman, and Alan Constant for critically reading chapters of the text. Their comments helped me greatly. Any remaining errors and inadequacies in the text are, of course, my own.

This text originated in a course on computational modeling that I taught in the Department of Materials at the University of California Santa Barbara while I was a member of the technical staff at the Los Alamos National Laboratory. I want to thank Professor David Clarke, who, as then Chair of the Department of Materials at UCSB, invited me to share with him, and then teach, a course on computational modeling. I also want to thank my managers at Los Alamos, who tolerated my periodic disappearances off to Santa Barbara. The members of the Department of Materials Science and Engineering at Iowa State University also deserve thanks for putting up with a distracted Chair as I finished this project. All of these people contributed to the rather twisted path I followed to create this book and I thank them.

# 1 Introduction to materials modeling and simulation

With the development of inexpensive, yet very fast, computers and the availability of software for many applications, computational modeling and simulation of materials has moved from being entirely in the hands of specialists to being accessible to those who use modeling not as their principal activity, but as an adjunct to their primary interests. With that change in accessibility of materials modeling and simulation come exciting new opportunities for using computational modeling to greatly advance the development and refinement of materials and materials processing.

The goal of this text is not to make experts – there are entire books on subjects that are treated in a few pages here. The text is, by design, introductory and we leave out many, if not most, details about implementation. We will present the key features and possibilities of computational materials science and engineering and discuss how to use them to advance the discovery, development, and application of materials.

## 1.1 MODELING AND SIMULATION

Before we start discussing materials modeling and simulation, it is appropriate to consider those words a bit more carefully. What do we mean by a "model" or a "simulation"? How are they different? Not to be overly pedantic, but it may help our discussion if we are a bit more precise in our definitions of these terms.

A *model* is an idealization of real behavior, i.e., an approximate description based on some sort of empirical and/or physical reasoning. A model most often begins life as a set of concepts, and then is usually transcribed into a mathematical form from which one can calculate some quantity or behavior. The distinction between a *theory* and a model is that, in the creation of a model, the attempt is to create an idealization of real behavior to within some accuracy, not a fundamental description that is strictly true.

A *simulation* is a study of the response of a modeled system to external forces and constraints. We perform simulations by subjecting models to inputs and constraints that simulate real events. A key thing to remember about simulations is that they are based on models. Thus, a simulation does not represent reality, rather it is a model of reality.

The accuracy of a simulation relative to the real system it is trying to emulate can depend on many factors, some involving the simulation method itself, for example the accuracy in numerically solving sets of equations. Often, however, the biggest errors in a simulation, at least with respect to how well it describes a real system, are the inadequacies of the models upon which the simulation is based. Thus, we cannot separate simulations from the underlying models.

In this text, we deal with both models and simulations. We will discuss in some detail how to model specific materials behavior and how to create and understand the models and their limitations. We will also describe in detail many of the commonly used simulation methods, indicating some of the critical issues that must be addressed when developing accurate numerical methods.

## 1.2 WHAT IS MEANT BY COMPUTATIONAL MATERIALS SCIENCE AND ENGINEERING?

In the most general terms, Computational Materials Science and Engineering (CMSE) is the computer-based employment of modeling and simulation to understand and predict materials behavior. In practice, we generally make a distinction between computational materials science, in which the goals are to better understand and predict materials behavior, and computational materials engineering, which is focused on practical applications of materials, typically with an emphasis on products. We note that this distinction is arbitrary and not well defined, given that the basic methods are generally the same and it is the applications of those methods that have different goals. For this text, our focus is on the methods and we will not be concerned with the distinction between science and engineering.

We can use CMSE for many purposes. We could, for example, take a simple model that incorporates in some way the essential physical behavior of a system and then interrogate that model to describe the phenomenology of a process or property. The goal of such calculations is generally to seek understanding and not to describe the behavior in an accurate way. For example, a modeler could eliminate all the physical processes except one of interest, thus performing a "clean experiment" that sheds light on the role of that process in behavior – sort of an ultimate *gedanken* (thought) experiment. We could also, however, develop more detailed models and methods with the goal of predicting some property or behavior of a specific material, for example the prediction of the thermodynamic behavior of a new alloy or the mechanical properties of a doped ceramic. The models upon which such calculations would be based could be complicated or simple, depending on the actual goals of the study and the desired accuracy of the calculations. Both types of materials modeling are common and will be represented in this text.

CMSE is most powerful when it has a strong tie to experiment. At its simplest, experimental data can serve a validation of the accuracy of the models and the calculations based on them. However, when used together, CMSE can provide a deeper understanding of a materials system than possible by experiment alone by probing phenomena that experiments cannot see. Modeling can also predict behavior, whether under conditions for which we have no experimental data or as a screen for systems with so many parameters that performing all possible experiments is not feasible. Indeed, at its best, CMSE serves as an equal partner with experiment.

| Unit | Length scale | Time scale | Mechanics |
|---|---|---|---|
| Complex structure | $10^3$ m | $10^6$ s | Structural mechanics |
| Simple structure | $10^1$ m | $10^3$ s | Fracture mechanics |
| Component | $10^{-1}$ m | $10^0$ s | Continuum mechanics |
| Grain microstructure | $10^{-3}$ m | $10^{-3}$ s | Crystal plasticity |
| Dislocation microstructure | $10^{-5}$ m | $10^{-6}$ s | Micro-mechanics |
| Single dislocation | $10^{-7}$ m | $10^{-9}$ s | Dislocation dynamics |
| Atomic | $10^{-9}$ m | $10^{-12}$ s | Molecular dynamics |
| Electron orbitals | $10^{-11}$ m | $10^{-15}$ s | Quantum mechanics |

**Figure 1.1** Length and time scales in materials science adapted from [12]. On the left, we indicate the important unit structure at each scale, in the middle, the approximate length and time scales, and at the right, the approach used to simulate the material's mechanical behavior.

## 1.3 SCALES IN MATERIALS STRUCTURE AND BEHAVIOR

The modeling and simulation of materials is challenging, largely because of the extreme range of length and time scales that govern materials response. Length scales that govern a phenomenon may span from the nanometers of atoms to the meters of engineered structures. Similarly, important time scales can range from the femtoseconds of atomic vibrations to the decades of use of materials in products. Given the range of physical processes at each of these scales, it should not be surprising that no single technique will work for all scales. Thus, many methods have been developed, each focused on a specific set of physical phenomena and appropriate for a given range of lengths and times. In this text we will provide a background into some of the most important of these methods.

In Figure 1.1, we show a schematic view of the important length and time scales for just one type of materials behavior – the mechanical behavior of crystalline materials [12]. This figure is just an example – similar tables could be developed for other properties as well.

In the left column of Figure 1.1, we list the fundamental structural "unit" whose behavior dominates the materials response at the given length and time scales.[1] At the smallest scale, that "unit" represents the electrons in the solid, while at the largest scale it is some sort of complex structure (e.g., the wing of an airplane). In between are the other structures that matter for the scales listed: atoms, dislocations, grains, etc.

Consider as an example the general range of 100 microns to 10 millimeters, in which the dominant structural features in a material are the grains (in this schematic view). It is the behavior

---

[1] If these terms are unfamiliar, please see Appendix B for a brief introduction to materials.

of the ensemble of those grains that dominates the mechanical response of the material at that scale. Of course, the deformation behavior of a grain depends on the dislocations, which depend on the atoms, which depend on the bonding. Thus, the behavior at each scale is dependent on what happens at smaller scales. In a model of the deformation of a set of grains, while we may explicitly include the dislocations (and atoms and electrons), more likely we will develop a model that reflects, in some averaged way, the behavior of the dislocations (and atoms and electrons). That model will describe the mechanical behavior at the grain level (and is usually referred to as "crystal plasticity").

Each scale in Figure 1.1 reflects behavior that is dominated by its structural "unit", which is described by its own set of models. For example, consider the range of length scales from 1 Å to 100 microns ($10^{-10}$ to $10^{-4}$ m). At the smallest of these scales, the bonding between atoms dominates the behavior. This bonding arises, of course, from the underlying electronic structure and to describe it requires the use of methods that can calculate the distribution of electrons. Such methods require quantum mechanics and are described (briefly) in Chapter 4. At a somewhat larger scale, we need to consider the behavior of many atoms. While we can use electronic structure methods to describe the bonding, in general those methods are so complicated that we must approximate the bonding with some sort of empirical or approximate function. Such functions are called interatomic potentials and are discussed in Chapter 5. The interatomic potentials are thus models of the interactions between atoms. To understand the behavior of the atoms, we must simulate their behavior, which we can do with various atomistic simulation methods, such as molecular dynamics (Chapter 6) or the Monte Carlo method (Chapter 7). As discussed above, if we use a model for the interatomic interactions then we are not simulating the material, but rather a model of the material, and our results will be good only to the extent that the model represents the true interactions.

At still larger scales, there are too many atoms for us to consider, so we must find new approaches that focus on the dominant "units". These units may be dislocations, grain boundaries, or some other defect, and the simulations are based on these defects being the fundamental units. The length scale that is dominated by defects is often called the *mesoscale*.[2] We describe a number of mesoscale modeling methods in Part III of the text.

As we will see, while great strides have been made in extending the *length* scales of many of the simulation methods we will cover here, often the methods are still very restrictive in their ability to describe *time* scales of the order of what we measure in the laboratory. For example, we will see that molecular dynamics methods, used to describe dynamic motions of atoms, have a fundamental time scale in the $10^{-14}$ seconds range and atomistic simulations of more than a few nanoseconds are challenging.[3] Thus, even for problems for which an approach can describe the length scale, often we must find new approaches to also cover the time scales of interest.

---

[2] There is no definitive definition of what the "mesoscale" is. For our purposes, it represents the length and time scales between phenomena that can be described by atoms and those that can be described by continuum theories.

[3] Advances have been made in *accelerated dynamics* methods, which will be discussed in Chapter 6.

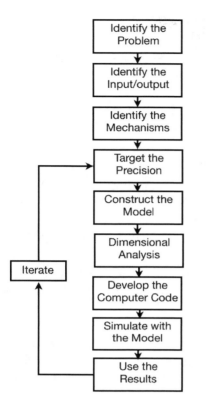

**Figure 1.2** The stages in model development, adapted from [12].

## 1.4 HOW TO DEVELOP MODELS

The first, and most important, step in any computational materials project is the creation of a model that describes the properties of a material at the length and time scales of interest. Here we want to introduce the logical steps one takes to create a model. Our discussion is based on a very useful article by Ashby, in which he describes a process for the systematic development and validation of materials models [12]. He gives a flow chart for the modeling process, which we reproduce in a somewhat streamlined fashion in Figure 1.2.

It may seem obvious, but the first step in model development is to identify the problem (given at the top of Figure 1.2). Often models go astray because the developer does not start with a clear idea of what he or she is actually trying to model. This type of misstep may arise from not understanding the problem well enough or from just not thinking through what the model has to accomplish.

Given the problem, the next step in any model development must be to specify what information the model is going to yield and what information one has at hand to use in that model – in other words, the outputs and inputs. This step is critical, and one that is often not well considered. Ignoring information that may be important can lead to either poor-quality or overly complex models (or both), creating problems for any subsequent simulations.

The next step is the identification of the physical mechanisms, which is often the most challenging part of any modeling effort. We may not have a complete picture of the fundamental phenomenology of a problem. It is in this step that modeling at a smaller scale often comes into play, serving to help identify the underlying physical processes at the larger scale. Close examination of experimental data and trends also can lead to the development of a clearer understanding of the physical mechanisms.

While identifying the problem, the inputs and outputs, and the mechanisms that define the framework for the real work of model development, it is also essential to specify the necessary quality of the model. Models should be no more complicated than needed for a specific problem. Anything more is a waste of effort. Seeking perfection in a model generally leads to high complexity, which may present difficulty when one has to "interrogate" the model, i.e., when one actually calculates something based on the model.

After creating the model, the simple act of doing a dimensional analysis often shows where the model might be incorrect. Dimensional analysis is a way to check relations among physical quantities based on their dimensions. A simple consequence of physics is that any equation must have the same dimensions on the left and right sides. Checking that the dimensions are equal is the basic step of dimensional analysis. We cannot stress enough the value of dimensional analysis as a check of the model as well as a tool to help group variables. Why group variables? As we shall see below, often two models that look very different can be shown to be quite similar when put in the same form. Recognizing such similarities can help avoid much unneeded effort.

Models are useless unless one can do something with them, which generally requires implementation into a computer code of some sort. Often this step has a major influence on the form of the model – if a model cannot be implemented or would require too much computer time to use, then it is not useful. There is thus often a balance between the desired accuracy of a model and the ability to actually use it in a calculation.

After implementation, the next step is the real point of CMSE, namely to calculate something with the model, i.e., to interrogate how the model works. This may be a *validation* step, in which predictions of the model are compared to available experimental data, theory, etc. to assess the quality of the model, the range of its validity, sensitivity to parameters, etc. One often uses this comparison to tune the model to make it more accurate and robust. It is not uncommon to go back to the construction of the model at this point to adjust its form so that it better meets the needs of the calculations. Of course, the reason for model development is to use the model to calculate some material property or function. What that "something" is depends on the problem. We note that a critical component of this step is to display the results in such a way so as to show its important features.

In this text, we shall use the process in Figure 1.2 often as we create models. While we shall rarely show the explicit links to these steps, they underlie all of what we do.

Before closing the discussion of model building, we would like to emphasize the importance of the Verification and Validation process (V&V). While validation is an attempt to assess the quality of the model to describe some behavior, *verification* is the process of ensuring that the computer code actually calculates what was planned. The goal is to model materials response. To ensure that we are doing so accurately requires both a good model and a proper implementation of that model into code. Too often, one or both of these processes are shortchanged, leading to poor simulations. If one is basing an engineering decision on simulations based on a model, the V&V process may be, quite literally, life-saving.

## 1.5  SUMMARY

Computational materials science and engineering is a field that is growing in capabilities and importance. The goal of this text is to introduce students to the basics of the most important methods used to simulate materials behavior. We are not intending to create experts, but rather to serve as the first introduction to what is a very exciting field.

# Part One

## Some basics

# 2 The random-walk model

Before introducing more complex methods, we start with a model of a fundamental materials process, the *random-walk* model of diffusion. The random-walk model is one of the simplest computational models in materials research and thus can help us introduce many of the basic ideas behind computer simulations. Moreover, despite its simplicity, the random-walk model is a good starting place for describing one of the most important processes in materials, the diffusion of atoms through a solid.[1]

## 2.1 RANDOM-WALK MODEL OF DIFFUSION

Diffusion involves atoms moving from site to site under the influence of the interactions with the other atoms in the system. An atom typically sits at a site for a time long compared with its vibrational period and then has a rapid transit to another site, which we will refer to as a "jump". To describe that process properly requires much more detail than we now have in hand (though we shall rectify that situation somewhat in the forthcoming chapters). Thus, we will take a very simple model that ignores all atomic-level details and that focuses just on the jumps.

Consider the simple example of a single atom moving along a surface, which we will assume consists of a square lattice of sites with a nearest-neighbor distance of $a$. Diffusion occurs by a series of random jumps from site to site in the lattice, as shown in Figure 2.1. We can understand the basic physics by considering the energy of the interaction between the diffusing atom and the underlying solid, which we show schematically in Figure 2.2a. The atom will vibrate around the bottom of one of the wells until there is sufficient energy in a direction that enables a jump to an adjacent site. A schematic view of the energy surface along the lowest-energy path between sites is shown in Figure 2.2b.

We cannot determine a priori when the atom will jump from its site nor can we predict to which adjacent site it will jump – it is a *stochastic* process.[2] From our knowledge of kinetics,[3] we know that the rate, $k_{jump}$ (which has the units of the number of jumps/time), that the atom

---

[1] A brief review of diffusion is given in Appendix B.7. More information is given in the recommended reading at the end of the chapter.

[2] A stochastic process is randomly determined, with a random probability distribution or pattern that may be analyzed statistically but cannot be predicted except in an average way. We will encounter many stochastic processes throughout this text.

[3] A brief introduction to kinetic rate theory is given in Appendix G.8.

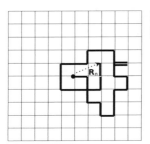

**Figure 2.1** First 27 jumps in a random walk on a square lattice. The arrow indicates $\mathbf{R}_n$, where $n = 27$, as defined in Eq. (2.1).

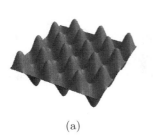

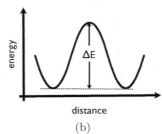

(a)

(b)

**Figure 2.2** Interaction of an atom with a surface. (a) Interaction potential of atoms to a surface showing barriers between low-energy wells. (b) Schematic diagram of barrier between two wells.

jumps from one site to another depends on the barrier height through an Arrhenius-type relation, a process that is usually referred to as an *activated process*. Examining Figure 2.2a, we see that the energy barriers between a site and its next-nearest neighbors are much higher than between the atom and its neighboring sites. Thus the rate of jumps to next-nearest-neighbor sites would be extremely low relative to that of the rate of a jump to one of its nearest neighbors and only jumps to nearest neighbors need to be considered.

Suppose we start an atom at a point in the square lattice, which we shall take for convenience to be the site $(0, 0)$. It can make a random jump, $\mathbf{r}$, to one of the four nearest neighbors, which are located at $(0, 1)a$ (up), $(0, -1)a$ (down), $(1, 0)a$ (right), or $(-1, 0)a$ (left).[4] From that site, it jumps randomly left, right, up or down, repeating the process until it has made some number (say $n$) jumps. If we traced the jumps on the lattice, it would look something like what is shown in Figure 2.1. That path through space is often called a *trajectory*. At one point, the path in Figure 2.1 reverses itself and, at another, it crosses the previous trajectory – there are no restrictions on where the trajectory goes in the plane. Since all the jumps are in random directions, this sequence of jumps is typically referred to as a *random walk*. If we were to place the atom back at the origin and start the walk again, the sequence would likely look very different than that in Figure 2.1, since every jump is chosen randomly from the four directions.

The idea that atoms jump randomly from site to adjacent site on a lattice is the basis for a very simple model for diffusion, where all we will need to know is the lattice (and its lattice lengths) and the jump rate $k_{jump}$, which we will assume has been supplied to us in some way.

---

[4] We use vector notation with the vector $\mathbf{r}$ describing the jump being given by $(x, y)a$, indicating a movement of a distance $a\,x$ along the $\hat{x}$ direction and $a\,y$ along the $\hat{y}$ direction on the square lattice. We will denote all vectors with **bold** type. For a review of vectors, see Appendix C.1.

These quantities are the input to the model (see Section 1.4). To define the output will require a bit more effort.

Our objective is to calculate the movement of the atom during its diffusion, so we need to be able to track its position. Since these are stochastic events that we assume are uncorrelated[5] (an assumption that we will revisit below), we can consider each jump as a random event with an average jump rate $k_{jump}$, i.e., there are, on average, $k_{jump}$ jumps/time.

Since the origin of the jump is arbitrary, it is easiest if we assume that the atom starts at the the point $(0, 0)$ at time $t = 0$. The position after one jump is given by $\mathbf{r}_1$, which would be one of the four nearest-neighbor sites chosen at random. The position after the second jump is at one of the nearest neighbors of site 1. Again, we do not know which of the four directions it moves from site 1. We can repeat this process to create a sequence of jumps. The position after the $n$th jump, which we denote by $\mathbf{R}_n$, is the vector sum of each jump in the sequence,

$$\mathbf{R}_n = \mathbf{r}_1 + \mathbf{r}_2 + \cdots + \mathbf{r}_n = \sum_{k=1}^{n} \mathbf{r}_k , \qquad (2.1)$$

i.e., a set of $n$ random selections of the four possibilities in $\mathbf{r}$.

We now have in hand the basis of the random-walk model of diffusion. With this approach we could track, for example, how an atom would diffuse across a surface. Examining more than one random walk, we see that diffusion is a stochastic process, with each sequence of jumps creating a distinct path along the surface. We need, however, a way to test whether the random-walk model is a reasonable description of actual diffusion. To do that, we need to connect results from the model to something we can measure. For diffusion, the obvious quantity would be the diffusion coefficient.[6]

## 2.2 CONNECTION TO THE DIFFUSION COEFFICIENT

A classic result from statistical physics is that the diffusion coefficient is related to a quantity called the *mean square displacement* by the relation:[7]

$$D = \frac{1}{6\,t} \langle R^2 \rangle , \qquad (2.2)$$

where $t$ is time and $\langle R^2 \rangle$ is the mean square displacement.[8] This expression for $D$ is applicable at macroscopic time scales, i.e., for very long times $t$ on an atomic scale. Since we know that the diffusion coefficient $D$ is not, in general, time dependent, Eq. (2.2) tells us something very important about $\langle R^2 \rangle$ – it must be linear in time, i.e., $\langle R^2 \rangle \propto t$, which we shall see below is true for a random walk.[9]

---

[5] Two events are uncorrelated if neither event affects or depends on the other.

[6] For a discussion of diffusion and the diffusion coefficient, please see Appendix B.7.

[7] See discussion leading up to Eq. (B.51).

[8] Hereinafter, the notation $\langle \rangle$ will denote an average.

[9] This is an example of usefulness of *dimensional analysis*. We learned the time dependence of the mean square displacement by just seeing how time enters into each term in Eq. (2.2).

Figure 2.1 shows one sequence of jumps in a two-dimensional random walk on a square lattice. After $n$ steps, the distance the atom has traveled from its starting position at $(0, 0)$ is the length of the vector $\mathbf{R}_n$, which is defined in Eq. (2.1). This distance is called the displacement and is the square root of the dot product of $\mathbf{R}_n$ with itself, i.e.,

$$R_n = \sqrt{\mathbf{R}_n \cdot \mathbf{R}_n} \,. \tag{2.3}$$

The square of the distance traveled (i.e., the square of the displacement) is just the square of $R_n$, or

$$R_n^2 = \mathbf{R}_n \cdot \mathbf{R}_n \,. \tag{2.4}$$

Suppose that another sequence of the same number of random jumps is created. This new sequence would be similar to that in Figure 2.1, but different. At the $n$th step, $R_n^2$ would also be different than in the first sequence. If we generated $N$ sequences of jumps (with the same number of jumps in each), then we would have $N$ values of $R_n^2$ for each step $(n)$ in the sequence. The mean square displacement $\langle R_n^2 \rangle$ is the average of $R_n^2$ over all $N$ sequences.

The beauty of the random-walk model is that we can calculate the mean square displacement analytically by some simple algebraic manipulations coupled with some thinking about averages. Combining Eq. (2.1) and Eq. (2.4), the displacement after $n$ jumps is

$$R_n^2 = \mathbf{R}_n \cdot \mathbf{R}_n = (\mathbf{r}_1 + \mathbf{r}_2 + \cdots + \mathbf{r}_n) \cdot (\mathbf{r}_1 + \mathbf{r}_2 + \cdots + \mathbf{r}_n) \,. \tag{2.5}$$

When taking the dot product, each term in $\mathbf{r}_1 + \mathbf{r}_2 + \cdots + \mathbf{r}_n$ will multiply itself once, so we will have the terms $r_1^2 + r_2^2 + \cdots + r_n^2$. To simply the equation, we will use the notation for a summation[10] to write that sum as $\sum_{k=1}^{n} r_k^2$. There will be two instances of each dot product between unlike terms, i.e., between $\mathbf{r}_j$ and $\mathbf{r}_k$ where $j \neq k$. So we will also have the terms $2\,\mathbf{r}_1 \cdot \mathbf{r}_2 + 2\,\mathbf{r}_1 \cdot \mathbf{r}_3 + \cdots + 2\,\mathbf{r}_2 \cdot \mathbf{r}_3 + \cdots + 2\,\mathbf{r}_{n-1} \cdot \mathbf{r}_n$. This latter set of terms is rather cumbersome, but we can simplify it by writing out the sum and collecting terms. Without belaboring the algebra, we can write those terms in a short form as

$$\sum_{k=1}^{n-1} \sum_{j=i+1}^{n} \mathbf{r}_k \cdot \mathbf{r}_j = \left(\mathbf{r}_1 \cdot \mathbf{r}_2 + \mathbf{r}_1 \cdot \mathbf{r}_3 + \cdots + \mathbf{r}_1 \cdot \mathbf{r}_n\right)$$
$$+ \left(\mathbf{r}_2 \cdot \mathbf{r}_3 + \mathbf{r}_2 \cdot \mathbf{r}_4 + \cdots + \mathbf{r}_2 \cdot \mathbf{r}_n\right)$$
$$+ \left(\mathbf{r}_3 \cdot \mathbf{r}_4 + \mathbf{r}_3 \cdot \mathbf{r}_5 + \cdots + \mathbf{r}_3 \cdot \mathbf{r}_n\right)$$
$$+ \cdots + \mathbf{r}_{n-1} \cdot \mathbf{r}_n\Big)$$

and the square of the displacement vector becomes

$$R_n^2 = \sum_{k=1}^{n} r_k^2 + 2\sum_{k=1}^{n-1} \sum_{j=i+1}^{n} \mathbf{r}_k \cdot \mathbf{r}_j \,. \tag{2.6}$$

For a square lattice, all jumps are of the same length ($r_k = a$), so $r_k^2 = a^2$. Based on the definition of the dot product, $\mathbf{r}_i \cdot \mathbf{r}_j = r_i r_k \cos\theta_{ij} = a^2 \cos\theta_{ij}$, where $\theta_{ij}$ is the angle between

---

[10] Eq. (C.9).

the two vectors, $\mathbf{r}_i$ and $\mathbf{r}_j$. Since we have $n$ terms in the first sum, each with the value $a^2$, the value of that sum is $na^2$. Putting it together (and taking the $n$ out of the sums), we have that the square of the displacement vector is

$$R_n^2 = na^2 \left( 1 + \frac{2}{n} \sum_{k=1}^{n-1} \sum_{j=i+1}^{n} \cos \theta_{kj} \right). \tag{2.7}$$

To calculate the diffusion coefficient, we need to evaluate the average of $R_n^2$ over many random-walk sequences of jumps, remembering that each jump $\mathbf{r}_k$ in each sequence is a random move to one of the nearest-neighbor sites. Since the average of a constant value is just that constant value, averaging the expression for $R_n^2$ in Eq. (2.7) gives the expression

$$\langle R_n^2 \rangle = na^2 \left( 1 + \frac{2}{n} \left\langle \sum_{k=1}^{n-1} \sum_{j=i+1}^{n} \cos \theta_{kj} \right\rangle \right). \tag{2.8}$$

The average is taken step by step over many independent sequences of jumps. For each step, there are just as many chances to move left as right and to move up as down. Thus, the average over $\cos \theta_{kj}$ must equal 0 and the second term vanishes.[11] We *predict* the very simple result that

$$\langle R_n^2 \rangle = na^2 \,, \tag{2.9}$$

i.e., there is a linear relation between the mean square displacement and the number of jumps. This relation does not depend upon the type of lattice used nor whether the random walk is in one, two or three (or any number of) dimensions.

Equation (2.9) is a remarkably simple relation, showing how the mean square displacement is related to the number of jumps in a random walk. The time dependence can be found by remembering that jumps occur with an average rate $k_{jump}$ (number of jumps/time). The *average* time for $n$ jumps is then $t = n/k_{jump}$, so we have

$$\langle R_n^2 \rangle = k_{jump} \, a^2 \, t \,, \tag{2.10}$$

with the linear dependence on time that we expected from dimensional analysis.

From the relation in Eq. (2.2), the diffusion coefficient is

$$D = \frac{k_{jump} a^2}{6} \,. \tag{2.11}$$

Note that we cannot actually evaluate $D$, because we do not have a value for the jump rate $k_{jump}$. If we could determine that in some other way, then we have a simple prediction for the diffusion coefficient.

Equation (2.11) is strictly true only for a random walk in which there are no correlations between successive jump directions. In a real system, when an atom jumps to a new site, the atoms around the new site and around the previous site will relax to somewhat altered positions. Thus, there is likely to be a slight tendency for atoms to jump back to their previous positions.

---

[11] As long as the system is symmetric, this term will vanish.

We do not include such correlated motions in the simple random-walk model. We can approximately correct for such correlated motions by introducing a scaling factor $f$ such that

$$D = \frac{k_{jump}a^2}{6}f,\tag{2.12}$$

where $f = 1$ for a random walk and $f < 1$ for most real systems.

What we have developed in this section is a model for a very simple problem – atoms diffusing along a surface. We are more interested in the important problem of atoms diffusing in a solid. Before doing that, however, there is another quantity that we can calculate that provides additional, and illustrative, information about diffusion.

## 2.2.1 End-to-end probability distribution

The random-walk model is simple enough that we can write analytic expressions for important descriptions of the random-walk sequence that go beyond the mean square displacement. For example, an important quantity that says much about the average behavior of a random walk is the probability distribution[12] of the vector defining the end position of an atom after $n$ jumps relative to the initial position of the atom. We will call that quantity $\mathcal{P}(\mathbf{R}_n)$, where we explicitly indicate that it is a function of the vector $\mathbf{R}_n$. We will not show how to derive $\mathcal{P}(\mathbf{R}_n)$, but will just discuss some of its properties.

Before writing down the expression for $\mathcal{P}(\mathbf{R}_n)$, consider what it means. For convenience, assume that we have an atom that is diffusing in one dimension along the $x$ axis, starting at $x = 0$. There is an equal probability that in any jump it can move 1 site either to the right or to the left along the axis, with the distance between sites being $a$. After $n$ jumps, it will reach a position $x_n$. Now suppose we generate many equivalent random trajectories (sequences of jumps). After $n$ jumps, the position of the atom in one of the trajectories will be at somewhere along the $x$ axis with an equal probability of being left or right of its initial position. If we average over enough trajectories,[13] we can determine the likelihood (probability) of finding $x_n$ on the lattice.

Since $x_n$ could be positive or negative, averaging over many trajectories will tend to cancel out $x_n$ and we find that the highest probability is at $x_n = 0$. A complete analysis shows that the probability of $x_n$ for a random walk along the $x$ axis is given by

$$I(x_n) = \left(\frac{3}{2\pi na^2}\right)^{1/2}\exp\left\{-\frac{3x_n^2}{2na^2}\right\},\tag{2.13}$$

which we show in Figure 2.3a. The functional form of $I(x_n)$ is called a *Gaussian* distribution, which we will encounter often in this text. The peak is at $x_n = 0$ and the probability that $x_n$ is non-zero goes rapidly to zero in both the positive and negative directions. This function gives us the probability of where *on average* the atom will end its random walk, but does not provide direct information about any individual trajectory. We can see, for example, that $I(x_n)$

[12] Please see Appendix C.4 for a brief review of probabilities.
[13] We shall discuss what we mean by "enough" a bit later in this chapter.

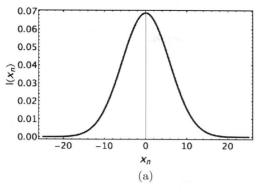
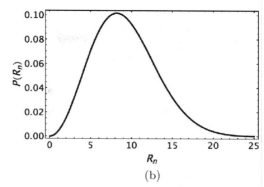

(a)                    (b)

**Figure 2.3** End-to-end distribution functions. Both plots were made assuming that $n = 100$ and $a = 1$. (a) Probability that the random jump is $x_n$ away from the start after $n$ steps ($I(x_n)$). (b) Probability distribution $\mathcal{P}(R_n)$ for the end-to-end distance, $R_n$.

is vanishingly small for a straight trajectory, which would have $x_n = n\,a$. (For the case shown in Figure 2.3a, with $n = 100$ and $a = 1$, $I(x_n = n\,a) \sim e^{-150}$.) That does not mean that it is impossible for a random walk in one dimension to yield a straight trajectory – it is only very very unlikely.

The expression given for $I(x_n)$ is, it turns out, approximate. The Gaussian function in Eq. (2.14) goes to zero quickly, but does not actually reach 0 until $x \to \infty$. Thus, the expression in Eq. (2.13) has a finite (though very small) probability that the end-to-end distance is actually greater than the total length (i.e., $x_n = n\,a$), which is, of course, impossible. However, in any practical sense, the Gaussian distribution is "exact".

In three dimensions, the probability distribution takes the form

$$\mathcal{P}(\mathbf{R}_n) = I(x_n)I(y_n)I(z_n), \tag{2.14}$$

where $\mathbf{R}_n = (x_n, y_n, z_n)$ and with expressions similar to Eq. (2.13) for $I(y_n)$ and $I(z_n)$.

$\mathcal{P}(\mathbf{R}_n)$ gives the probability that the vector $\mathbf{R}_n$ is at a position $(x_n, y_n, z_n)$. A more useful quantity would be the probability distribution for the end-to-end distance of the diffusion path, i.e., a measure of how far the atom has diffused in $n$ steps. Since $\mathbf{R}_n$ is a vector, $\mathcal{P}(\mathbf{R}_n)$ includes angular information that is not needed to determine an end-to-end distance. We need to average out the angles, which we can do by changing $(x, y, z)$ in $\mathcal{P}(\mathbf{R}_n)$ to spherical polar coordinates and integrating over the angles. Once we do that, we find that the end-to-end probability distribution is

$$\mathcal{P}(R_n) = \left(\frac{3}{2\pi n a^2}\right)^{3/2} 4\pi R_n^2 \exp\left\{-\frac{3R_n^2}{2na^2}\right\}. \tag{2.15}$$

We show $\mathcal{P}(R_n)$ in Figure 2.3b.

Note the difference between $I(x_n)$ and $\mathcal{P}(R_n)$ shown in Figure 2.3. $R_n$ is a distance and therefore always greater than or equal to zero. $\mathcal{P}(R_n)$ tells us that the greatest probability of the end-to-end distance in a sequence of $n$ jumps is at the peak value of $\mathcal{P}(R_n)$, which is at $R_n > 0$. It is convenient to have an analytical expression, because we can then determine the

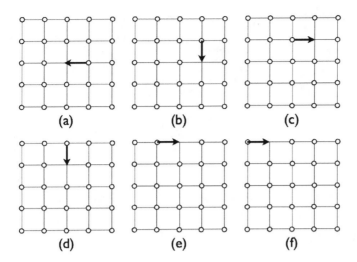

**Figure 2.4** Sequence of random atomic jumps to fill a vacancy.

value of the peak probability by finding the maximum value of the function, which we do by taking the derivative of $\mathcal{P}(R_n)$ with respect to $R_n$, setting it equal to zero, and solving for $R_m^{max}$, the distance of the peak probability. In this case, $R_n^{max} = \sqrt{2n/3}a$.

Average properties can be calculated directly from the probability distribution, as discussed in Appendix C.4. The mean end-to-end distance $\langle R_n \rangle$, for example, is given by

$$\langle R_n \rangle = \int_0^\infty R_n \mathcal{P}(R_n) dR_n = \sqrt{\frac{8n}{3\pi}} a \qquad (2.16)$$

and the mean square end-to-end distance (the mean square displacement) is

$$\langle R_n^2 \rangle = \int_0^\infty R_n^2 \mathcal{P}(R_n) dR_n = na^2, \qquad (2.17)$$

which is exactly what we determined in Eq. (2.9).

## 2.3 BULK DIFFUSION

We have developed a simple model for diffusion on an empty lattice, which is not really a problem that is very interesting from a materials perspective. Consider, however, an atom moving through a solid. Ignoring lattice defects (for example, grain boundaries or dislocations) and assuming that no interstitial sites can be occupied, then the only way an atom can jump to another lattice site is if that site is unoccupied, i.e., if the adjacent site is a vacancy. Of course, when an atom jumps into a vacancy, it fills that vacancy, leaving behind a vacancy in its former spot, as shown in Figure 2.4.

Suppose there is only one vacancy in a solid. The only atoms that can move are ones that are adjacent to that vacancy. Each jump, however, "moves" the vacancy, such that, over time, the vacancy moves through the system. Thus, while diffusion could be described by monitoring the moving atoms, in practice, it is far more efficient to just track the moving vacancy. Not surprisingly, this type of diffusion is referred to as *vacancy diffusion*.

Vacancy diffusion can be modeled as a random walk of the vacancies. In two dimensions, the model looks exactly like the case of the surface diffusion discussed above, with a vacancy taking the place of the atom. The analysis of a random walk in three dimensions is the same as in two dimensions, with the mean-square displacement following the relation in Eq. (2.10). As we shall see below, the only changes from one crystal lattice to another are the directions of the random jumps to the nearest-neighbor sites, with the average properties remaining unchanged.

Diffusion in real systems in the absence of a concentration gradient is usually referred to as *tracer diffusion*, which is the spontaneous mixing of atoms. This type of diffusion can be characterized experimentally using isotopic tracers, hence the name. Assuming no significant isotopic effects on the motion of atoms, tracer diffusion is generally assumed to be identical to self-diffusion. The relation between the tracer diffusion coefficient and the vacancy diffusion coefficient is $D_t = X_v D_v$, where $X_v$ is the mole fraction of vacancies from Eq. (B.23). Note that both $X_v$ and $D_v$ (through the jump rate) are highly temperature dependent.

With these results, we now have the next steps in the modeling sequence shown in Figure 1.2 completed. While we have not specified each step, we have the input (lattice and jump rate) and output (diffusion coefficient), we have identified the mechanisms (random jumps), we have targeted the precision (since we cannot predict $k_{jump}$ we cannot predict a value for $D$ so these will be qualitative, not quantitative, results), we have constructed the model (the sequences of random jumps), and we have done a dimensional analysis ($D$ must be independent of time). It is now time to develop the computer code.

## 2.4   A RANDOM-WALK SIMULATION

In this section we discuss how to implement the random-walk model of vacancy (or surface) diffusion onto a computer. It is difficult to be very specific, as each programming language has different syntax, names for commands, etc. Thus, we will just sketch the basic ideas and leave specific implementation for the computer-based exercises.[14]

Before going into detail, we should point out an obvious deficiency in the random-walk model of diffusion. While we can relate the model to the diffusion coefficient through Eq. (2.11), we cannot actually calculate $D$. The jump rate $k_{jump}$ is a process that is atomistic in nature and cannot be calculated with the random-walk model. Indeed, it is an input parameter, which we shall take as $k_{jump} = 1$ for the rest of this discussion. Any other choice for $k_{jump}$ would just scale the time of the results and not provide any additional information.

The first step is to define the lattice on which the vacancy will move, which could be any lattice type in two or three dimensions. Once that has been done, one possibility would be to set up a list of all possible lattice points that the vacancy could visit during a simulation. The question would be how big to make that list. The maximum distance the vacancy can move in

---

[14] Sample computer codes for these exercises are available online at http://www.cambridge.org/lesar.

any direction is $n$, the total number of jumps. Of course, having a vacancy move in a straight line for a large number of steps would be unlikely. However, to ensure that the vacancy cannot move "off" the lattice during the simulation, the size of the lattice will need to be some fraction of $n$ in each direction.

There are a couple of problems with creating such a lattice. First, we do not know how big to make it. We can overcome that problem by making the lattice large enough for all possible trajectories. Consider, for example, a two-dimensional square lattice with a random walk of $n$ steps that starts at the origin. The maximum movement along the $x$ axis would be for a straight trajectory of $n$ jumps in either the plus or minus directions, which would require a lattice of $2n$ sites, with the same size needed along the $y$ axis. The total number of sites in the system would thus be $4n^2$. Granted, the probability that a trajectory would be straight is very small, but this assumption ensures that a random walker would not reach the edge of the lattice. The fraction of sites visited in an $n$-jump walk would then be $n/(2n)^2 = 1/(4n)$. If $n$ is reasonably large, then the fraction of sites visited would be quite small. This approach is computationally very inefficient and thus not the best way to proceed. We will see in later chapters, however, that for many, if not most, simulation methods, creating a lattice on which to do calculations will be the best approach.

We can avoid creating a lattice in this case by remembering how we began the calculation of the mean square displacement in Eq. (2.1). We determined the positions of the atom (or vacancy) over time by simply adding the vectors for the random jumps. We can follow the same procedure on the computer, generating a list of the positions after each jump. For a random walk on a square lattice, each jump will occur to one of the four nearest neighbors. The position vector of the vacancy at step $(k)$ will be written as the vector $\mathbf{R}_k$, while the individual random jump for the $k$th step is $\mathbf{r}_k$.

Assume that at the start, $\mathbf{R}_0 = (0, 0)$. Now define the four possible lattice jumps as

$$\Delta \mathbf{r}_1 = (1, 0)\, a \qquad (2.18)$$
$$\Delta \mathbf{r}_2 = (0, 1)\, a$$
$$\Delta \mathbf{r}_3 = (-1, 0)\, a$$
$$\Delta \mathbf{r}_4 = (0, -1)\, a\,,$$

where $a$ is the lattice constant (which we take as 1 for now). The definition of the jumps defines the symmetry of the lattice, since each jump will be along one of these vectors. If we wanted to model diffusion on a triangular lattice, for example, there would be possible jumps to the six nearest-neighbor positions.

For the first jump, we would randomly select one of the four possibilities in Eq. (2.18). To do this, we would employ a *random number generator*, which is a computer code that creates long lists of pseudo-random numbers.[15] Random numbers will be used often in the simulations in this book and the creation of a "good" set of random numbers is of prime importance. How they are created and a discussion of what qualifies as "good" and "bad" are in Appendix I.2.

---

[15] They are called "pseudo" random because they are not really random, but they approximate a random distribution.

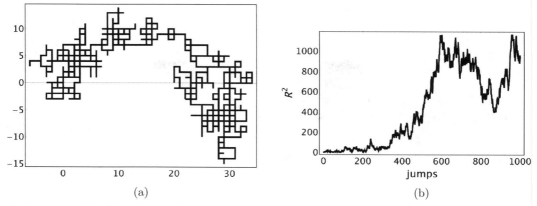

**Figure 2.5** Results of a random walk on a square lattice in two dimensions with lattice parameter $a = 1$. There were $n = 1000$ jumps. (a) The trajectory of the random walk. (b) The square displacement at each point along the trajectory.

For the first jump, a random number in the range 1–4 is generated, which chooses one of the jumps in Eq. (2.18). Throughout this text, we will denote such random numbers by the symbol $\mathcal{R}$. The new position would then be

$$\mathbf{R}_1 = \mathbf{R}_0 + \Delta\mathbf{r}_\mathcal{R} . \tag{2.19}$$

For example, if $\mathcal{R} = 2$, then $\Delta\mathbf{r}_\mathcal{R} = \Delta\mathbf{r}_2 = a\,\hat{y}$. To calculate the position at the next step, $\mathbf{r}_2$, we would pick another random number $\mathcal{R}$ and would add $\Delta\mathbf{r}_\mathcal{R}$ to $\mathbf{R}_1$. This procedure is exactly what we did in Eq. (2.1) in the derivation of the mean square displacement. By repeating the procedure over and over (a task for which computers are very well suited), a list of positions as a function of number of jumps would be created, as in Eq. (2.1).

In Figure 2.5a we show the position after each jump for a single trajectory generated with the procedure just described. Examine the figure carefully. This walk happened to restrict itself mostly to the right of the $y$ axis and is somewhat linear. Another walk would look very different, as you can see in the computer exercises. Since this is a random process, we cannot predict what a trajectory will look like.

Since we have the positions at the end of each jump, we can compute the square of the displacement along the trajectory, which, since the trajectory was started at $(0, 0)$ is $R_k^2 = x_k^2 + y_k^2$, where $\mathbf{R}_k = (x_k, y_k)$. For the walk in Figure 2.5a, the square displacement is shown in Figure 2.5b. Note that it is not linear, as expected from the expression in Eq. (2.9). Why? Because it is the result from a single trajectory and not an average. Eq. (2.9) only relates the mean (average) square displacement to the number of jumps.

Examining $R_k^2$ in Figure 2.5b tells us quite a bit about the random walk in Figure 2.5a. At first, the vacancy did not move far from the origin. Indeed, after the first 200 steps, it is still within about five lattice lengths from the origin. The vacancy begins to move away rapidly, then reverses its course somewhat, returning a bit closer to where it started before moving away again.

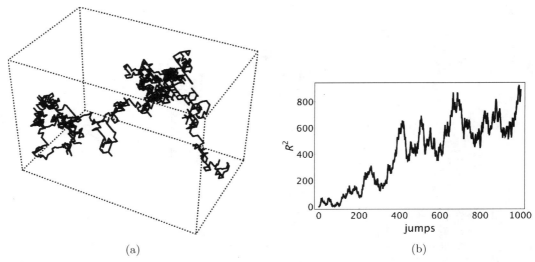

(a)                                        (b)

**Figure 2.6** Results of a random walk on a three-dimensional face-centered cubic lattice with lattice parameter $a = 1$. There were $n = 1000$ jumps. (a) The trajectory of the random walk. (b) The square displacement at each point along the trajectory.

We could repeat the simulation for any lattice in one, two or three dimensions (or higher, though the physical meaning might be unclear). All that would change is the definition of the jumps in Eq. (2.18). Suppose we wanted to model a random walk in a three-dimensional face-centered cubic ($fcc$) lattice.[16] We would start a trajectory at $\mathbf{R}_0 = (0, 0, 0)$ and then perform a series of random jumps. We would need to define the 12 vectors to the nearest neighbors and choose a random integer from 1 through 12. All else would be the same. In Figure 2.6a we show results of a random walk on a face centered cubic lattice with $a = 1$. Looking carefully, you can see that the symmetry of the jumps reflects the $fcc$ structure. The square of the displacement of the trajectory is shown in Figure 2.6b, where again we see that it is not at all linear, as now should be expected since it is not an average, but reflects the properties of a single trajectory.

To compare the expected values for the mean square displacement, we need to average over many trajectories, which could be done in a few ways. We could imagine a simulation in which we have many vacancies diffusing simultaneously. We could track the displacement of each of them relative to their starting points and then average. The problem with this approach is that, unless the system is enormous (such as found in a real crystal), we might find two vacancies adjacent to one another, a situation not accounted for in the analysis that led to Eq. (2.9). Another approach would be to run many simulations, each containing just one vacancy. Each trajectory would be different, because it would be generated with a different sequence of random numbers. The mean square displacement would then be the average over all the trajectories of the square displacements as a function of the number of jumps, i.e., we would average $R_k^2$ for all $k$ over all the trajectories, yielding $\langle R_k^2 \rangle$. We employed the latter approach in generating these examples. To be specific, we average $R_1^2$ over all trajectories, then $R_2^2$, and so on, until we have mapped

---

[16] For a brief review of crystal structures, please see Appendix B.2.

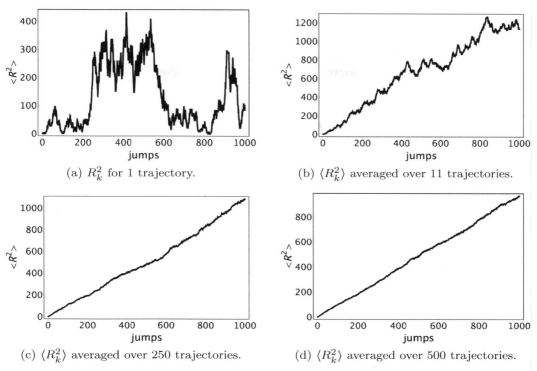

(a) $R_k^2$ for 1 trajectory.

(b) $\langle R_k^2 \rangle$ averaged over 11 trajectories.

(c) $\langle R_k^2 \rangle$ averaged over 250 trajectories.

(d) $\langle R_k^2 \rangle$ averaged over 500 trajectories.

**Figure 2.7** The evolution of the mean square displacement for a $n = 1000$ random walk with $a = 1$ as a function of the number of trajectories in the averaging.

out the mean square displacement for all $n$ jumps. Suppose we have $m$ trajectories, then the mean square displacement is

$$\langle R_k^2 \rangle = \frac{1}{m} \sum_{i=1}^{m} R_k^2(i) \qquad k = 1, 2, 3, \ldots, n, \tag{2.20}$$

where $R_k^2(i)$ is the value for $R_k^2$ in the $i$th trajectory.

Figure 2.7 shows the convergence of the mean square displacement to the expected straight line as a function of the number of jumps seen in Eq. (2.9). This calculation happened to have been based on a random walk on a square lattice, but the results would be quite similar for other two- or three-dimensional lattices. In Figure 2.7a, we show the result for the first trajectory in the averaging process. It shows a trajectory that moves rapidly away from where it started and then rapidly back. In Figure 2.7b, we have averaged over 11 trajectories, including that in Figure 2.7a. While not completely smooth, it is easy to see that the mean square displacement is approaching a linear function of the number of jumps. Figure 2.7c and Figure 2.7d correspond to 250 and 500 trajectories in the averaging process, respectively, and show a steady approach to a straight line. In Figure 2.8 we show the result for $\langle R_k^2 \rangle$ after averaging over 2000 trajectories. Within a very small statistical error, the relation between $R_k^2$ and $k$ is exactly the straight line that we expected from Eq. (2.9). Since $a = 1$, the slope is also as expected.

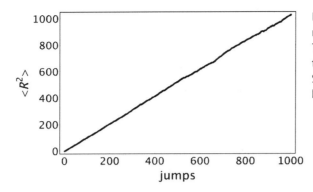

**Figure 2.8** Mean square displacement for a random walk on a square lattice with $a = 1$. The values were averaged over $m = 2000$ trajectories, each consisting of $n = 1000$ jumps. Simulations on any other lattice with the same lattice parameter would be identical.

The difference between the results in one run shown in Figure 2.7a and those averaged over many runs in Figure 2.8 is typical of what will be seen throughout this text. In many of the methods that we employ in studying materials, running one simulation cannot be expected to give "typical" results and averaging over many simulations will be required.

We have calculated an average quantity, the mean square displacement, that is related to a measurable quantity, the diffusion coefficient. Calculating average quantities to compare with experiment is a common feature of many of the methods we will discuss in this text. With computational simulations, however, we can often do more than just determine the average quantities. For example, we can often determine the probability of a structure or, for some methods, events. We can do that for the random walk model for the end-to-end distance, as described in Section 2.2.1. Because we have an analytical solution in Figure 2.3b, we can verify our simulations by comparing to that solution.

The probability $\mathcal{P}(R_n)$ in Eq. (2.16) tells us the likelihood of finding that the end-to-end distance (in a jump sequence of length $n$) is $R_n$. From the calculation of $\langle R_k \rangle^2$ we have a list of $R_n$ values for all the trajectories. To turn those numbers into a probability distribution is not difficult and is based on the basic ideas from Appendix C.4.

Consider Figure 2.3b, which shows a continuous function for $\mathcal{P}(R_n)$ over some range of values of $R_n$. From our $m$ simulations, however, we have a discrete, and finite, set of values for $R_n$. The best we can do is to create a discrete representation of $\mathcal{P}(R_n)$, which we do by dividing the $m$ values into *bins* that each represent a finite range of values of $R_n$. The quality of the calculated $\mathcal{P}(R_n)$ will depend on the number of bins (and hence their range in $R_n$) and the size of $m$. Details of the binning procedure are discussed in Appendix I.3.

A calculated probability distribution for the end-to-end distance is shown in Figure 2.9 for $m = 2000$ trajectories, each with $n = 1000$ jumps on a lattice with $a = 1$. We divided the data into $n_{bin} = 20$ equally spaced bins. In the figure we compare the simulation results to Eq. (2.16). Overall, the agreement is excellent, though with some scatter, as expected, in the calculated values.

As shown in Figure I.4 in Appendix I.3, by increasing the number of bins, we shrink the bin size, which would potentially improve the agreement between the discrete version of $\mathcal{P}(R_n)$ and the continuous equation. Indeed, if we shrunk the bin size to an infinitesimal size, then we could,

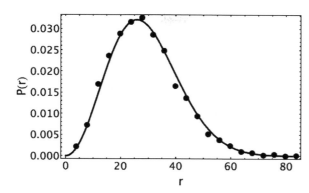

**Figure 2.9** Points are the probability distribution $\mathcal{P}(R_n)$ for the end-to-end distance, $R_n$, calculated from the simulation with $m = 2000$, where $m$ is the number of trajectories in the average. We divided the data into $n_{bin} = 20$ equally spaced bins. The solid curve is from Eq. (2.16). Both plots were made assuming that $n = 1000$ and $a = 1$.

in principle, recover the continuous function. There is one issue, however. We divided the results from 2000 simulations into 20 bins to generate the points in Figure 2.9, an average of about 100 points per bin. The scatter in the figure is statistical in nature and arises from the relatively small number of points per bin. Going to more bins requires more data to reduce the statistical fluctuations. Thus there is a tradeoff whenever one calculates probability distributions in this way: more bins mean a finer, more continuous, result, yet will require many more simulations to lower the statistical uncertainties. We will run into this situation many more times in this text.

## 2.5 RANDOM-WALK MODELS FOR MATERIALS

Since we can analyze a priori the random-walk model, it may seem somewhat a waste of time to actually do random-walk simulations. As shown here, they do, however, provide a good introduction to some of the features common to all simulations, i.e., issues of implementation, the importance of averaging over trials, etc.

Putting aside the relevance of actually doing simulations, we can discuss whether the random-walk *model* is useful. Of course, the most glaring problem is that while we can relate the mean square displacement to the diffusion coefficient, since we do not know the jump rate $k_{jump}$, we cannot actually calculate a diffusion coefficient. Moreover, we have already discussed how short-time atomic relaxations might inhibit an atom from revisiting a site after vacating it. This inhibition leads to the correlation term $f$ in Eq. (2.12), which we also cannot describe with a random-walk model. In short, the random-walk model does not really tell us much about diffusion in a normal bulk lattice.

There are cases, however, in which one might want to restrict the probability of visiting certain sites. For example, suppose one were trying to model a system that had some fast diffusion channels. An example might be a thin film crossed by threading dislocations in which diffusion along the dislocation core (which typically has a smaller density than the bulk lattice) might be much faster than in the bulk. Diffusion in this system would not be well described by a truly random walk. However, by varying the rate $k_{jump}$ between sites (e.g., sites along the channel could have faster diffusion rates than outside the channel) we could examine, in a general sense, how the presence of such diffusion channels affects the overall diffusion rate in the system. In this case, we would need the values of $k_{jump}$ as input parameters. Calculations that used this basic idea are described in [344]. While we could do that heuristically now, in Chapter 9 we will discuss a method, the Kinetic Monte Carlo method, that is designed to model such systems.

## 2.6  SUMMARY

In this chapter, we reviewed the basics of the random-walk model for diffusion and showed how to construct a simulation method to evaluate it. We used this simple model to introduce a number of important features of simulations, for example the need to average over many trajectories, how to calculate probability distributions, etc. In upcoming chapters, while we will expand on these simple ideas to create much more sophisticated simulation methods, many of the basic features will be similar to what we have discussed here.

### Suggested reading

Some books that you might find useful:

- A good treatment of the basics of diffusion is given in *Diffusion in Solids* by Shewmon [287].
- Another excellent book on diffusion is Glicksman's *Diffusion in Solids: Field Theory, Solid-State Principles, and Applications* [123].

# 3 Simulation of finite systems

In almost all methods used to model materials, the system will be described by a set of discrete objects of some sort. Those objects might be atoms and the goal may be to calculate the cohesive energy by summing interatomic interaction potentials. The objects do not have to be atoms, however. We may want to sum the interactions between spins or dislocations or order parameters or whatever. Learning how to calculate these sums is thus fundamental to essentially all materials modeling and simulation.

In modeling a material we typically face a rather major complication – we are trying to model a macroscopic system that contains large numbers of objects. For example, a bulk sample of a material may include many moles of atoms. Modeling the behavior of all those atoms would be computationally impossible. To approximate the (effectively) infinite systems, we use various boundary conditions, mostly based on introducing a repeating lattice. How one sums the interactions between the objects within the framework of these boundary conditions is the focus of this chapter.

## 3.1  SUMS OF INTERACTING PAIRS OF OBJECTS

We will often encounter systems that consist of objects that interact with each other in some way. The classic example is the cohesive energy of a solid, which is determined from the sum of the interactions between the constituent atoms and molecules.[1] The simplest case is when the interactions occur only between pairs of objects and depend only on the distance between the pairs. For convenience, we denote the interactions between a pair of objects by $\phi(r)$, where $r$ is the (scalar) distance between them.

Consider a pair of objects, which we will denote by $i$ and $j$. If object $i$ is located at a position defined by the vector $\mathbf{r}_i$ and $j$ is located at $\mathbf{r}_j$, then the vector between the objects, $\mathbf{r}_{ij} = \mathbf{r}_j - \mathbf{r}_i$, where the subscript $ij$ indicates the vector from $i$ to $j$ (note the order).[2] The distance between the objects is determined in the usual way, $r_{ij} = (\mathbf{r}_{ij} \cdot \mathbf{r}_{ij})^{1/2}$.

Now consider the sum of the interactions between a set of objects. To foreshadow Chapter 5, we will call this sum $U$.[3] This sum has some important attributes. First, since

---

[1] Interatomic potentials are covered in detail in Chapter 5.
[2] Please see Appendix C.1 for a discussion of vectors and their manipulations.
[3] If $\phi(r)$ is an interaction potential between atoms, then $U$ is just the potential energy, as discussed in Chapter 5.

interactions are energies, which are scalar quantities, $\phi_{ij} = \phi_{ji}$; the interaction of $i$ with $j$ is the same as $j$ with $i$. Also, each object only interacts with each other object once and an object does not interact with itself. For example, if there were four identical objects, then

$$U = \phi(r_{12}) + \phi(r_{13}) + \phi(r_{14}) + \phi(r_{23}) + \phi(r_{24}) + \phi(r_{34}). \tag{3.1}$$

Now suppose we have a system of $N$ objects, where $N$ is large. We certainly cannot write out all the interactions as we did in Eq. (3.1), so we need a shorthand notation. There are a number of equivalent ways to write these sums. A common way, though not efficient in terms of actual computation, is to write

$$U = \frac{1}{2} \sum_{i=1}^{N} \sum_{j=1}^{N} {}' \phi_{ij}(r_{ij}), \tag{3.2}$$

where the $'$ indicates that the $i = j$ terms are not included in the sums and the factor of $\frac{1}{2}$ compensates for having two copies of each interaction in the sums.[4] Eq. (3.2) is sometimes written as

$$U = \frac{1}{2} \sum_{i=1} \sum_{j \neq i} \phi(r_{ij}). \tag{3.3}$$

It is easy to calculate how many *unique* interaction terms there are in Eq. (3.1). In an $N$ atom system, each atom can interact with $N - 1$ others, so there are $N(N - 1)$ terms. However, that overstates the total number of interactions by a factor of 2, since each interaction is counted twice but should be counted only once. Thus, the total number of unique pairs is $N(N - 1)/2$. In a calculation in which all $N$ atoms are included in the interactions sums, the sums become computationally challenging as $N$ becomes large.

Another common way of writing the sum of interactions comes directly from the way Eq. (3.1) is written,

$$U = \sum_{i=1}^{N-1} \sum_{j=i+1}^{N} \phi(r_{ij}). \tag{3.4}$$

Note the range of the indices and compare with Eq. (3.1). Eq. (3.4) can be written in an even more compact form as

$$U = \sum_{i} \sum_{j>i} \phi(r_{ij}). \tag{3.5}$$

[4] If $N = 4$, then the terms in Eq. (3.2) are

$$U = \frac{1}{2} \Big\{ \{\phi(r_{12}) + \phi(r_{13}) + \phi(r_{14})\} + \{\phi(r_{21}) + \phi(r_{23}) + \phi(r_{24})\}$$

$$+ \{\phi(r_{31}) + \phi(r_{32}) + \phi(r_{34})\} + \{\phi(r_{41}) + \phi(r_{42}) + \phi(r_{43})\} \Big\}$$

$$= \phi(r_{12}) + \phi(r_{13}) + \phi(r_{14}) + \phi(r_{23}) + \phi(r_{24}) + \phi(r_{34}), \tag{3.6}$$

in agreement with Eq. (3.1).

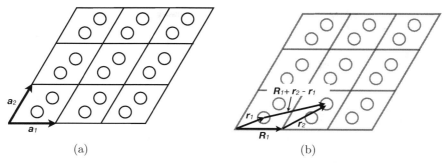

(b)

**Figure 3.1** Two-dimensional lattice with a basis. (a) There are two atoms per unit cell. (b) Vectors defining atomic positions and interatomic distances. Position vectors within the unit cell are denoted by $\mathbf{r}_i$. Shown is the vector connecting atom 1 in one cell to atom 2 in a cell separated from the first by the lattice vector $\mathbf{R}$.

The advantage to the latter two expressions is that they avoid including terms more than once, which is how we usually implement the sums on a computer.[5]

## 3.2  PERFECT CRYSTALS

Suppose that we want to calculate the interaction energies (or forces) between a set of objects that are arrayed in a periodic structure. The most obvious example would be atoms on a perfect lattice.[6] We show in Figure 3.1 an example of a two-dimensional lattice. There are two atoms associated with each cell in that lattice, one at the origin of the cell and one in the middle. The distribution of atoms within each cell is referred to as its *basis* – this structure has a basis of 2. In Figure 3.1b we show a lattice vector, denoted by $\mathbf{R}_1$, connecting one cell to another, as well as a vector, $\mathbf{R}_1 + \mathbf{r}_2 - \mathbf{r}_1$, connecting an atom in one cell with one in another cell.

If all the objects are identical, for a perfect lattice Eq. (3.2) takes the form

$$U_{cell} = \frac{1}{2} \sum_{\mathbf{R}} \sum_{i=1}^{n} \sum_{j=1}^{n} {}' \phi_{ij}(|\mathbf{R} + \mathbf{r}_j - \mathbf{r}_i|), \qquad (3.7)$$

where $\sum_{\mathbf{R}}$ indicates a sum over all lattice vectors. $|\mathbf{R} + \mathbf{r}_j - \mathbf{r}_i|$ is the distance from atom $i$ in the central cell to atom $j$ in the cell located at $\mathbf{R}$, as shown in Figure 3.1. The $'$ indicates that $i = j$ terms for $\mathbf{R} = (0, 0, 0)$ (the central unit cell) are not included. For example, object 1 does not interact with itself in the central cell, but does interact with all other "object 1"s in other cells.

$U_{cell}$ in Eq. (3.7) is the energy *per unit cell*. It is the sum of the interactions between all the objects in the central unit cell with each other and with the objects in the other cells in the

---

[5] The reader should verify that the expressions in Eq. (3.2)–Eq. (3.5) are all equivalent, as they are all used at some point in the text.

[6] In Appendix B.2 we provide a quick review of lattice vectors, reciprocal lattice vectors and the basic properties of crystal structures.

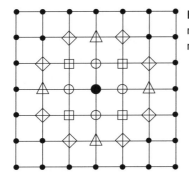

**Figure 3.2** First four neighbor shells in a square lattice: nearest neighbors are denoted by a ◯, next-nearest neighbors by ▢, the third neighbor shell by a △, and the fourth shell by a ◇.

lattice. It does not include the interactions between objects in other unit cells with objects in any other cell but the central cell. $U_{cell}$ is thus the energy per the $n$ objects in the unit cell. Since different crystal structures may have different numbers of objects in their basis, it is easier to compare calculated values by determining the lattice sums on a per object basis, i.e.,

$$u = U_a = \frac{1}{n} U_{cell} \, , \tag{3.8}$$

where we have shown two of the various notations people use to distinguish between the energy per object (atom) and the energy per cell.

Since we will see expressions like Eq. (3.7) many times in this text, they deserve some discussion. To make the discussion a bit clearer, consider a two-dimensional square lattice with a unit cell parameter $a$ and only one atom per cell, as shown in Figure 3.2.[7] The lattice vectors on a square lattice have the general form $\mathbf{R} = a(n_1\hat{x} + n_2\hat{y})$, where $n_1$ and $n_2$ are integers in the range $-\infty \rightarrow \infty$. The sum over $\mathbf{R}$ is then

$$\sum_{\mathbf{R}} = \sum_{n_1=-\infty}^{\infty} \sum_{n_2=-\infty}^{\infty} \, . \tag{3.9}$$

In three dimensions, for a cubic lattice $\mathbf{R} = a(n_1\hat{x} + n_2\hat{y} + n_3\hat{z})$ and a sum over $n_3$ would be required.[8]

The basic form of Eq. (3.9) is perhaps more understandable if we sum over the *shells* of neighbors, i.e., sets of neighbors that are the same distance from the central atom. In Figure 3.2, we show 4 nearest neighbors at a distance of $a$, 4 next-nearest neighbors at $\sqrt{2}a$, 4 next-next-nearest neighbors at $2a$, 8 in the fourth neighbor shell at $\sqrt{5}a$, etc. Thus, the first few terms of $U_{cell}$ from Eq. (3.7) are

$$U_{cell} = 2\phi(a) + 2\phi(\sqrt{2}a) + 2\phi(2a) + 4\phi(\sqrt{5}a) + \cdots \, , \tag{3.10}$$

remembering the factor of 1/2 in Eq. (3.7).

---

[7] Extension to non-cubic systems is not difficult. Details are given in Appendix B.2.4.

[8] The reader should write out a few terms of Eq. (3.7) for a two-dimensional square lattice. Assume only one object per unit cell ($n$=1) and then add a basis.

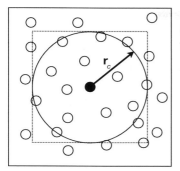

**Figure 3.3** System of interacting atoms showing a potential cutoff. The dashed box shows an optimal region for testing whether an atom is within the cutoff.

## 3.3    CUTOFFS

Equation (3.10) is written as a sum over objects in neighbor shells that are increasingly farther away from the central atom. The interaction terms, $\phi(r)$, decrease with distance, so $\phi$ in each term in Eq. (3.10) adds a smaller value to the sum. In most systems encountered in this text, the range of the functions $\phi(r)$ (or equivalent) is finite, such that if $r$ is large, then the value of $\phi$ is sufficiently small that we can truncate the sum at some cutoff distance $r_c$, with only small to moderate error.[9]

We show a cutoff distance in Figure 3.3, in which only those atoms that lie within the circle of radius $r_c$ with respect to the atom at the center of the circle are assumed to interact with the center atom. When a cutoff distance, $r_c$, is included, the expression in Eq. (3.7) must be modified to exclude all interactions between objects whose separations are greater than $r_c$.

While the interactions may be small at distances greater than $r_c$, they are not zero, leading to an error in the calculated energy. We can estimate that error as follows:

(a) A rough estimate of the number of neighbors in a shell is obtained by ignoring the structure of the lattice and assuming that the number of atoms in a region is approximately the number density $\rho$ times the volume of that region.

(b) For a shell of neighbors at a distance $r$, the volume can be taken as the surface area of a sphere of radius $r$ times the width of a shell, $\delta r$, so that the number of atoms in a shell a distance $r$ from a center atom is $4\pi r^2 \rho\, \delta r$.

(c) The net contribution to the interaction energy of the objects in that shell is

$$\delta U \approx 4\pi r^2 \rho\, \phi(r)\, \delta r\,, \tag{3.11}$$

where $\phi(r)$ is the interaction potential. As we sum over the lattice, each shell of neighbors is increasingly farther away, but on average contains more atoms. Thus, there is a balance between the value of the potential at each neighbor shell and the total number of neighbors in that shell. Equation (3.11) is an approximate result because we have neglected the actual distribution of the atoms.

---

[9] For functions $\phi(r)$ that go to zero with increasing $r$ too slowly, cutting off the potential is incorrect and other approaches must be used. Details as well as a definition of what is meant by "too slowly" are discussed in Section 3.6.

(d) A crude estimate of the error caused by the introduction of a cutoff distance can be obtained by converting $\delta r$ in Eq. (3.11) to a differential and integrating from $r_c$ to infinity, i.e., by integrating over all the interactions now excluded from the sum through the introduction of a cutoff distance:

$$\Delta U \approx 4\pi\rho \int_{r_c}^{\infty} r^2\phi(r)dr \,. \tag{3.12}$$

As we shall see in Chapter 5, many interaction potentials can be approximated as a function that depends on the inverse of the distance raised to some power, i.e., $\phi(r) \propto 1/r^n$. Inserting this form into Eq. (3.12), we have

$$\Delta U \approx 4\pi\rho \int_{r_c}^{\infty} r^{2-n}dr = \frac{4\pi\rho}{3-n}r^{3-n}\Big|_{r_c}^{\infty} = \frac{4\pi\rho}{n-3}r_c^{3-n} \tag{3.13}$$

as long as $n \geq 4$. As we shall see in Chapter 5, an important contribution to the energy of many molecular and atomic systems is a long-ranged interaction of the form $1/r^6$ (the van der Waals energy). For those interactions, the error introduced by a cutoff of that part of the potential goes as $1/r_c^3$.

Note that in Eq. (3.13), when $n = 3$ the integrand goes as $1/r$, whose integral is $\log(r)$, which diverges at infinity. The integral also diverges for $n < 3$. Thus, it is only possible to evaluate the total energy as a simple lattice sum for interactions with $n \geq 4$ [138]. For interactions with $n \leq 3$, other numerical methods must be used, as we shall discuss in Section 3.6.

When evaluating lattice sums, there is a balance between accuracy and computational speed. The smaller the value of $r_c$, the fewer are the neighbors in the sum in Eq. (3.7), yielding a faster calculation. By the same token, a smaller cutoff introduces increased errors because of neglected terms in the sum, as estimated in Eq. (3.12). In practice, $r_c$ is generally chosen to balance computational speed and accuracy.

## 3.4 PERIODIC BOUNDARY CONDITIONS

Materials models and simulations can be based on many types of objects, from atoms to "spins" to order parameters. The goal in most applications of these models is to describe the properties of a macroscopic system, which can be considered as being infinitely large relative to the system sizes that we can actually model in a reasonable amount of time on a computer. We thus need a way to *mimic* the real (essentially infinite) system, starting with a system of manageable size and then using *boundary conditions* to approximate the effects of the rest of the material.

A common approach is to employ *periodic boundary conditions*, in which a macroscopic system is described as an infinite array of equivalent finite systems. The objects of interest are placed in a finite-sized volume, referred to as the *simulation cell*, which is then *replicated* throughout space, as shown in Figure 3.4 for a two-dimensional system of point particles. A key feature in periodic boundary conditions is that whatever properties the objects in the simulation cell have, all the replicas of that object also have. For example, if the particle marked (1) in

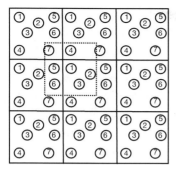

**Figure 3.4** Periodic boundary conditions. The central cell is replicated to fill space. As a particle in the central cell moves, its images move in the same way. The dashed box indicates the effective cutoff for the nearest-image convention for particle 1.

Figure 3.4 has a certain value associated with it, then all other (1)s in the system have the same value. With periodic boundary conditions, if a particle moves within the central cell, then all replicas of that particle move in the same way in their cells. When a particle leaves the simulation cell, its replica comes in from the opposite side, so the number of particles in the simulation cell is constant. When summing the interactions between the particles, the periodic replicas serve to represent the particles in the system outside the simulation cell.

Systems with periodic boundary conditions are very much like the perfect lattices we discussed above. Objects within the simulation cell interact with each other, as well as with all their replicas throughout the system. The relative positions of the replicated cells are described with lattice vectors, just as in the perfect crystal, as are the positions of the objects within the replicated cells. Simulation cells can be any shape, as long as that shape can be replicated to fill space. Thus, *any crystal lattice system can serve as a simulation cell*, with the choice of lattice type depending on the problem being studied. Sums over the lattice then take the same form as the expression for the potential energy given in Eq. (3.7), where now **R** is a lattice vector for the central simulation cell (and much larger than for a perfect crystal).

The use of periodic boundary conditions is a very powerful tool in simulating the bulk behavior of materials. However, there are issues that must be considered. The use of periodic boundary conditions introduces correlations between cells. If an object moves in the simulation cell, all its replicas move as well. If an object has a value in one cell, its replicas in all other cells do as well. The length scale of these correlations is the size of the cell. These correlations inhibit long-wavelength fluctuations, which can make calculations of quantities that depend on larger scales inaccurate.

Errors introduced by periodic boundary conditions can be checked by increasing the size of the central cell and then repeating the calculation – a bigger simulation cell should be more accurate because the correlations are less. Indeed, checking for system size effects is a common step used to validate a simulation.

A more subtle issue is that the fundamental simulation cell must be *commensurate* with the system of study, i.e., the symmetries of the structure must match at the periodic boundaries. For example, suppose one were modeling a system of up and down spins in which the lowest energy has the nearest neighbor spins aligned in an antiparallel way (e.g., ↑↓). Modeling the lowest-energy structure using periodic boundaries requires that it be maintained across the

(a)                                    (b)

**Figure 3.5** A central cell and the adjacent cells across the periodic boundaries, denoted by the dotted boxes. (a) There are an equal number of spins in the central cell and the spins maintain their antiparallel alignment at the cell boundaries. (b) With an odd number of spins, it is impossible to have an antiparallel alignment at the boundaries and thus impossible to model the low-energy structure.

boundaries. In Figure 3.5a, with an even number of spins in the central cell, the lowest-energy structure is maintained at the cell edge. However, when an odd number of spins is put in the central cell and then replicated to fill space, as shown in Figure 3.5b, at each boundary we have parallel spins, which are not in the minimum energy configuration. Thus, a system modeled with simulations based on Figure 3.5b would not be the same as that in Figure 3.5a.

## 3.5   IMPLEMENTATION

Every evaluation of the mathematical operators needed in the calculation of a distance or an energy takes computer time, so we only want to calculate the interactions in Eq. (3.7) that we actually need in the calculation. Introducing a cutoff distance helps reduce the computational load, but care must be taken in how that cutoff is implemented.

Suppose we are considering the interactions with one of the atoms in the central cell ($i$) and all other atoms with a cutoff distance set at $r_c$. The procedure is to pick an atom, calculate its distance from the atom of interest, check to see whether that distance is less than $r_c$, and, if so, calculate the potential and add to the sum. The challenge is that calculating distances to atoms outside $r_c$ only adds computational time, since those interactions are ignored. A simple approach is to limit the search for atoms that *might* be within the cutoff distance, which minimizes those that are outside $r_c$. A simple way to handle this problem is to consider an imaginary box centered on atom $i$, as shown in Figure 3.3. If the box is set up so that its side has a length $2r_c$, then the atoms that can interact with the atom at the center are within the box and the sum over $j$ in Eq. (3.7) only includes atoms within the box and no others. It is still necessary to calculate all the distances and to check the cutoff criteria, since the atoms at the box corners will be outside the cutoff range. However, the search is restricted, saving some computer time.

A more efficient approach is to introduce a list of possible neighbors and to only consider them in the lattice sums. This list, called, appropriately, a *neighbor list*, is determined at the beginning of a calculation and then updates throughout the calculation. In Figure 3.6 we show a simple version of such a list. At the beginning of a calculation, a list of atoms within a distance $r_u$ ($> r_c$) is created. In the first summation of the interactions, only those atoms on that list need to be considered. So far, this has not saved computer time. However, in subsequent evaluations of the sums, as long as no atom has moved more than $r_u - r_c$, we can use the same neighbor list, saving considerable computer time. Once an atom has moved a distance greater than $r_u - r_c$, then the neighbor list must be recalculated. For solids, for example, in which the

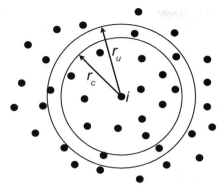

**Figure 3.6** Construction of a list of neighbors for atom $i$. The potential cutoff is $r_c$, while the upper limit for distances within the list is $r_u$, as discussed in the text.

distances moved by atoms are typically small, neighbor lists may not need to be recalculated in a simulation. When simulating systems with large atomic motions, like fluids, a balance will be needed when choosing $r_u$ – too large a value will include many atoms in the sums that are beyond $r_c$, while too small a value will lead to more numerous updates of the list. There are a number of other ways to create lists, especially for periodic boundary conditions. These are discussed in more specialized texts such as [109].

For very small systems, such as those in the sample computer exercises that accompany this text, we can use a particularly simple method to do the calculations. Consider Eq. (3.7). For all particles in the simulation cell ($i$) we sum over all other particles in the system (the sums over $j$ and $\mathbf{R}$). To simplify the calculation, we imagine a box centered on each site with a side that has the length of the simulation cell, as shown in Figure 3.3. Each atom in the cell interacts only with those atoms that lie within that box. This method is called the *minimum image convention*, because each particle interacts only with the nearest image of all other particles in the simulation cell, as indicated in Figure 3.4. The maximum cutoff is less than or equal to half the size of the simulation cell, $r_c \leq 1/2\, a_{cell}$. For large systems, the effective cutoff distance is far too large in the minimum image convention and methods such as discussed above must be used to ensure an efficient search for atoms within the cutoff.

## 3.6 LONG-RANGED POTENTIALS

As discussed in Section 3.3, for systems made up of objects that interact with potentials with the form $1/r^n$ with $n \leq 3$, we cannot calculate the total interaction by simply summing the pairwise interactions. Such interactions are generally referred to as being *long ranged*. The sums in Eq. (3.7) do not converge to a finite value or, at best, converge conditionally.[10] A classic

---

[10] A conditionally convergent series is one that converges, but not absolutely, i.e., the series does not converge if the absolute value of what is being summed is used. A good example is the sum

$$1 - \frac{1}{2} + \frac{1}{3} - \frac{1}{4} + \cdots = \sum_{n=1}^{\infty} \frac{(-1)^{n+1}}{n}, \tag{3.14}$$

which does converge to a finite value ($\ln(2)$), while the series $\sum_{n=1}^{\infty}(1/n)$ does not.

example is an ionic solid, whose properties are dominated by electrostatic interactions of the form $q_i q_j / r_{ij}$, where the charges on each ion are $q_i$.[11] The energy, which includes a sum over the electrostatic interactions, is finite because the positive ions are located next to the negative ions and the sums over the interactions are conditionally convergent. However, the interaction terms are so long ranged that we cannot sum them directly – we need special techniques.

It is important to remember that for long-ranged potentials, the interactions cannot be truncated at any distance without invalidating the calculation [138]. Thus, all possible interaction terms must be included in the evaluation of the energy. Since there are an infinite number of such terms, we cannot do those sums directly. Much work has gone into developing alternative approaches to direct summations, especially for ionic systems. In Appendix 3.8 of this chapter we discuss approaches commonly used in the literature, specifically the Ewald method and a particular variant of a class of methods called hierarchical-tree methods. We also discuss an approximate method that is easy to implement and efficient but seems to yield accurate results.

## 3.7 SUMMARY

In this chapter, we introduced the basic methods of calculating interactions between objects. We introduced the perfect lattice and lattice vectors. We showed how to extend the idea of a lattice to the use of periodic boundary conditions as a way to mimic infinite systems. To cut down on computer time, we discussed how to cut off the interactions at a prescribed distance. Finally, we mentioned the issues involved when using long-ranged potentials and, in an appendix, presented computational approaches to evaluate them.

### Suggested reading

A number of texts provide excellent discussions of this important topic. A few that I like are:

- D. Frenkel and B. Smit, *Understanding Molecular Simulations* [109]. This is an excellent book describing the basics of atomistic simulations.
- While focused on simulations of liquids, the book by M. P. Allen and D. J. Tildesley, *Computer Simulation of Liquids* [5], is a good basic guide to how to perform simulations.

## 3.8 APPENDIX

In this section we introduce three methods to calculate (or approximate in one case) the energy of long-ranged interactions. We do not include all methods developed to deal with these interactions. For example, the particle mesh approach, described in [109], has advantages for some types of systems. However, the methods described here are commonly used in materials research. As an example, we shall consider the evaluation of the electrostatic energy.[12] For more

---

[11] These systems will be discussed in more detail in Chapter 5.
[12] Please see Appendix E for a description of electrostatic interactions.

details and discussion, as well as modifications for calculating the force, please see other, more specialized, texts, for example [109].

Suppose we have a system of $N$ point charges in a periodically repeated cell. The total electrostatic potential energy is[13]

$$U_e = \frac{1}{2} \sum_{\mathbf{R}} \sum_{i=1}^{n} \sum_{j=1}^{n} {}' \frac{q_i q_j}{|\mathbf{R} + \mathbf{r}_j - \mathbf{r}_i|} \,. \tag{3.15}$$

The prime indicates no $i = j$ term when $\mathbf{R} = 0$, which is as seen in Eq. (3.7). The energy terms go as $1/r$ at long range, so the sums in Eq. (3.15) are not absolutely convergent. They are, however, conditionally convergent – they will converge to the correct answer if they are summed correctly and will either not converge or converge to the wrong answer if summed incorrectly. The next sections show how to evaluate these sums.

## 3.8.1  The Ewald double sum method

The strategy in a commonly used method, called the Ewald method, is to convert the conditionally convergent sums to convergent sums. The Ewald method is probably the most-used approach for ionic systems. As described, it is restricted to systems with periodic boundary conditions (e.g., a crystal lattice). We will not show the derivation here as it is covered in many texts. A particularly lucid derivation is given in the book by Frenkel and Smit [109].

The challenge of evaluating the electrostatic energy is that the terms are very long ranged. The Ewald method works by adding and subtracting an artificial Gaussian charge distribution to each lattice site. By both adding the charge and then subtracting it, the system is unchanged, but, as we shall see, we can use the new terms to rearrange the equations to create convergent, rather than divergent, sums.

The form of the Gaussian charge distribution is

$$\rho_i^G(\mathbf{r}) = \sum_{i=1}^{n} q_i \left(\frac{\alpha}{\pi}\right)^{3/2} e^{-\alpha(\mathbf{r}-\mathbf{r}_i)^2} \,, \tag{3.16}$$

where the charge distributions are centered at the atom sites ($\mathbf{r}_i$) and have the same charge as the atom at those sites. The parameter $\alpha$ governs the width of the distribution. A graph of a Gaussian distribution is shown in Figure C.5b(a).

The Ewald method starts by adding/subtracting $\rho_i^G(\mathbf{r})$ to/from the charges at each lattice site, as shown schematically in Figure 3.7. The system of point charges is shown on the left with a line to indicate a $+$ or $-$ charge. Mathematically, the point charges are described by a series of Dirac delta functions[14] located at each lattice site

$$\rho_i(\mathbf{r}) = \sum_{i=1}^{n} q_i \delta(\mathbf{r} - \mathbf{r}_i) \,. \tag{3.17}$$

---

[13]  Please note that, for convenience, we leave out the factor $k$ in these expressions. As discussed in Appendix E, the value of $k$ depends on the units used in the calculation.

[14]  See Appendix C.5.1.

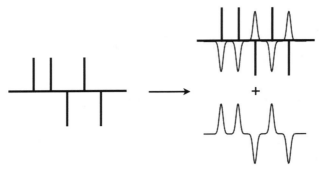

**Figure 3.7** Schematic view of the procedure used in Ewald method. Adapted from [109].

On the right we show at the top the *subtraction* of the Gaussian charges, while at the bottom we show the addition of the charges. The net charge for the top right figure in Figure 3.7 is

$$\rho_i^{total}(\mathbf{r}) = \sum_{i=1}^{n} q_i \left\{ \delta(\mathbf{r} - \mathbf{r}_i) - \left(\frac{\alpha}{\pi}\right)^{3/2} e^{-\alpha(\mathbf{r}-\mathbf{r}_i)^2} \right\}. \tag{3.18}$$

The electrostatic potential energy can be derived from the charge distribution. The $\frac{q_i q_j}{|\mathbf{R}+\mathbf{r}_j-\mathbf{r}_i|}$ terms from Eq. (3.15) are replaced by

$$\frac{q_i q_j \mathrm{erfc}(\sqrt{\alpha}|\mathbf{R} + \mathbf{r}_j - \mathbf{r}_i|)}{|\mathbf{R} + \mathbf{r}_j - \mathbf{r}_i|}. \tag{3.19}$$

The complementary error function $\mathrm{erfc}(x)$ is shown in Figure C.5b, in which we see that it goes to zero very quickly. Thus the conditionally convergent sums in the actual energy expression are converted to absolutely convergent sums when the Gaussian distributions are subtracted.

We now must evaluate the terms that arise from the interactions between the Gaussian distributions on the bottom right of Figure 3.7. Without going into detail, this term can be evaluated by transforming the charge distributions from the direct to the reciprocal lattice,[15] which becomes

$$|\rho(\mathbf{k})|^2 = \sum_{i=1}^{n} \sum_{j=1}^{n} q_i q_j e^{-i\mathbf{k}\cdot(\mathbf{r}_j-\mathbf{r}_i)}$$
$$= \sum_{i=1}^{n} \sum_{j=1}^{n} q_i q_j \cos\left(\mathbf{k} \cdot (\mathbf{r}_j - \mathbf{r}_i)\right), \tag{3.20}$$

where we use the real part of the exponential. The energy from this term is

$$\frac{2\pi}{V} \sum_{\mathbf{k}\neq 0} \frac{1}{k^2} |\rho(\mathbf{k})|^2 e^{-k^2/4\alpha}. \tag{3.21}$$

The sum is over the *reciprocal lattice* vectors, which for a cubic simulation cell go as $\mathbf{k} = 2\pi/L(k_x, k_y, k_z)$ where $L$ is the size of the simulation cell and $(k_x, k_y, k_z)$ are integers. This expressions has terms that go to zero as $\exp(-k^2/4\alpha^2)$ for large $k$, so it converges quickly.

---

[15] For a discussion of the reciprocal lattice, see Appendix B.2.3.

The total energy in Eq. (3.15) is the sum of those terms plus a correction term that corrects for a self energy, as described in [109],

$$U_e = \frac{1}{2} \sum_{i=1}^{n} \sum_{j=1}^{n} {\sum_{\mathbf{R}}}' \frac{q_i q_j \operatorname{erfc}(\sqrt{\alpha}|\mathbf{R} + \mathbf{r}_j - \mathbf{r}_i|)}{|\mathbf{R} + \mathbf{r}_j - \mathbf{r}_i|}$$
$$+ \frac{2\pi}{V} \sum_{\mathbf{k} \neq 0} \frac{1}{k^2} |\rho(\mathbf{k})|^2 e^{-k^2/4\alpha^2}$$
$$- \left(\frac{\alpha}{\pi}\right)^{1/2} \sum_{i=1}^{n} q_i^2 . \tag{3.22}$$

We note that we assume the system is embedded in a conductor, i.e., in a medium with infinite dielectric constant. Otherwise, an additional term is needed, as described in [109]. Similar expressions are available for the interaction between a lattice of dipoles [109].

The sum over $\operatorname{erfc}(\sqrt{\alpha}r)$ converges more quickly for large values of $\alpha$ while the sum in the reciprocal lattice converges faster for small $\alpha$. Thus, there is a balance between efficiency of the sums and $\alpha$ must be chosen to optimize the overall calculation. Overall, the Ewald method takes of the order of $O(N^{3/2})$ computer operations, where $N$ is the number of atoms in the central cell [77]. However, there are some restrictions on its use. For large system sizes, the lattice constant for the simulation cell is large and thus the reciprocal lattice vectors are short. Thus, the sum in the reciprocal lattice becomes cumbersome. The particle-mesh Ewald method, which speeds up the calculation of the reciprocal lattice terms, has a computational burden of the order of $N \log N$, and is thus more useful for large systems [77]. All in all, however, the Ewald method is a good approach when modeling relatively small, periodic, solids with long-ranged potentials.

## 3.8.2 Fast Multipole Method

Because of the limitations of the Ewald sum for large systems, much work has gone into finding other, less computationally costly, methods to sum long-range interactions. One of the more interesting of these approaches is the Fast Multipole Method (FMM) [130, 131]. The FMM is a hierarchical method that takes $O(N)$ operations per calculation of the interaction. Here we describe briefly the general approach for using the FMM to study Coulomb systems in three dimensions. This discussion is based on that in [338].

When two sets of particles are "well-separated", as shown in Figure 3.8, it is possible to determine the number of terms needed in a multipole expansion[16] of the field of one set of particles to calculate the forces on the particles in the other within a given accuracy. In this definition, two regions are well-separated if dislocations in each region are closer to each other than to dislocations in the other region. This is not to say that being well-separated is needed for the convergence of a multipole expansion. Convergence only requires that the sets be separated. Note that a slightly different definition is used for well-separated cells in the multipole methods described next.

---

[16] Please see the discussion of multipole expansions in Appendix E.3.

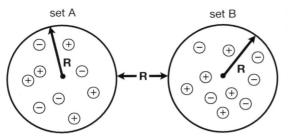

**Figure 3.8** Well-separated sets of charged particles.

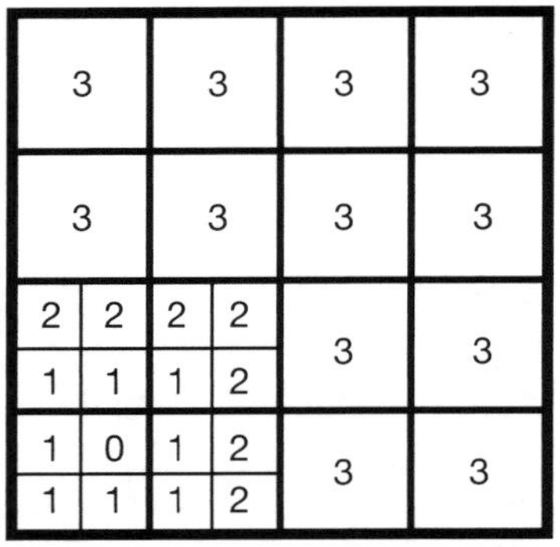

| | | | |
|---|---|---|---|
| 3 | 3 | 3 | 3 |
| 3 | 3 | 3 | 3 |

**Figure 3.9** Schematic diagram for interactions in a three-level hierarchy.

The multipole methods create a hierarchy of cells and can be used in any dimension. For clarity, we discuss a two-dimensional system, with a typical three-level two-dimensional hierarchy shown in Figure 3.9. The highest level, 0, is the simulation cell itself. That cell is then divided into 4ths to create level-1 cells. The level-1 cells are divided again into 4ths to create level-2 cells, which in turn are divided to create level-3 cells. We say that a level-$(n + 1)$ cell derived by dividing a level-$n$ cell into 4ths is the *child* of the level-$n$ cell, which in turn is the *parent* of the four level-$(n + 1)$ cells. For an $M$-level system, there are a total of $\sum_{n=0}^{M} 4^n$ cells. In three dimensions, eight children are created from each parent and there would be a total of $\sum_{n=0}^{M} 8^n$ cells for an $M$-level system. In this type of hierarchy, the next-nearest neighbor cells are the first that are considered "well-separated" – this holds at all levels.

The FMM achieves its $O(N)$ computational speed by an efficient calculation of the multipole moments and the interactions. As a concrete example, consider a system of charged particles in a level-3 system (Figure 3.9). The interaction of particles in cell 0 with other particles in 0 and with those in the nearest-neighbor level-3 cells (marked "1") are evaluated as a direct summation of the electrostatic interactions. The next-nearest neighbor level-3 cells (marked "2") are well separated from cell 0 and thus their interactions with particles in cell 0 are approximated with a multipole expansion. The level-2 cells marked "3" are well separated from the level-2 parent of cell 0, thus the particles in cell 0 interact with the multipoles of the other level-2 cells marked "3".

The multipole moments of the various cells are calculated as shown in Figure 3.10. The moments of the lowest level cells (level-3 cells in this example) are computed directly by

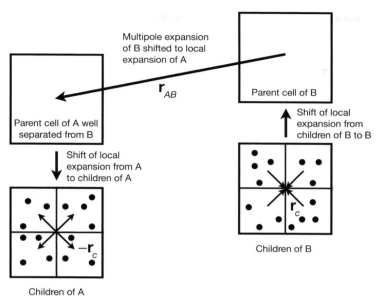

**Figure 3.10** Example of the steps needed in calculating the interactions of particles in cell 0 in Figure 3.9 with the multipoles of the next-nearest neighbors (3) of the parent of 0.

summing over the ions, as discussed in Appendix E.3. The moments of the level-2 cells (marked "3" in Figure 3.9) are computed by shifting the multipole expansion from the center of the level-3 cells to that of their parent level-2 cells, yielding simple expressions for the level-2 multipoles that do not require summing over the particles.[17] The multipole moments of the level-1 cells (i.e., parents of the level-2 cells) would be calculated in the same way. The interactions of the level-3 particles (i.e., in 0) with the multipoles of a level-2 cell (B) are calculated by shifting the center of the multipole expansion of the potential of the level-2 cell to the well-separated parent (level-2) of the cell in which the particle sits, creating what is usually referred to as a local expansion centered at the parent cell (A). This local expansion is then shifted to be centered at the level-3 cell (0) and the net interaction determined by adding the terms for each particle.

Details for the application of this method, with proper equations for evaluating the various multipole moments and their shifts, are given elsewhere [131, 339]. The FMM, at its best, includes $O(N)$ computations, so is faster than the Ewald methods, though there is a computational cost of creating the multiples. The FMM has been used for many applications [131], including galactic simulations, charged particles, and dislocations [338]. It can be used either with or without periodic boundary conditions, so does have that additional advantage over the Ewald method.

---

[17] Consider the moments of a distribution calculated with respect to the origin. The dipole moment from Eq. (E.13b) is $\boldsymbol{\mu} = \sum_{i=1}^{N} q_i \, \mathbf{r}_i$. The dipole moment relative to a shifted position $\mathbf{R}$ is just

$$\boldsymbol{\mu} = \sum_{i=1}^{N} q_i \, (\mathbf{R} + \mathbf{r}_i) = \mathbf{R} \sum_{i=1}^{N} q_i + \sum_{i=1}^{N} q_i \, \mathbf{r}_i = Q\mathbf{R} + \boldsymbol{\mu} \, . \tag{3.23}$$

The shifted dipole depends on the charge in the region as well as the dipole. The shifted quadrupole moment would depend on the quadrupole moment, the dipole, and the charge, and so on.

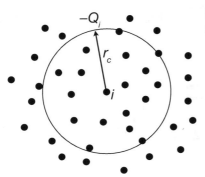

**Figure 3.11** Schematic diagram of the spherically truncated Coulomb potential.

### 3.8.3 Spherically truncated Coulombic potential

Wolf and colleagues [356] presented an approximate method for performing Coulombic inter-actions that is quite attractive. The basic idea is to introduce a cutoff distance in the calculation with a correction term to approximate the terms in the sum that are neglected in the truncation. It is not a formally correct method, as were the two methods just outlined, but has been shown to yield reasonably accurate values for the Coulombic contribution to the cohesive energy and avoids the pitfalls of the use of a cutoff with no corrections.

The basic idea is shown schematically in Figure 3.11. Each atom $i$ interacts with all ions within the cutoff distance $r_c$. As we have discussed, using that interaction energy is incorrect for long-ranged potentials. The basis for developing a correction term is to remember that the overall system is charge neutral. Thus, if the net charge within $r_c$ around atom $i$ is $Q_i$, then the net charge of the remaining ions in the system must be $-Q_i$, where

$$Q_i = \sum_{j}^{r_{ij} < r_c} q_j ,$$

(3.24)

and the sum is over all atoms within $r_c$ of $i$. $Q_i$ needs to be determined for each atom. To simplify matters, the effect of the external ions is approximated by a sphere of radius $r_c$ and charge $-Q_i$. A fundamental result from electrostatics is that the electrostatic potential inside a charged sphere with charge $-Q_i$ and radius $r_c$ is $-Q_i/r_c$, so that the net electrostatic energy between ion $i$ and the external charges is $-q_i Q_i/r_c$. The net Coulomb energy is thus a sum of the interactions of each atom with those internal to the cutoff plus the correction arising from ions outside the cutoff, i.e.,

$$E_{coul} = \sum_{i=1}^{N} \left( \sum_{j \neq i}^{r_{ij} < r_c} \frac{q_i q_j}{r_{ij}} - \frac{q_i Q_i}{r_c} \right).$$

(3.25)

# Part Two

# Atoms and molecules

# 4 Electronic structure methods

The behavior of a material can be related to the types of bonding between the atoms, whether it be metallic, covalent, ionic, etc. That bonding represents the distribution of electrons around the nuclei. Covalent bonds have a localized electronic distribution between atoms and are generally strong and directional. Materials with strongly covalent bonds include important semiconductors, such as silicon, gallium, and diamond. Metallic systems, in contrast, may have a degree of directionality to their bonding, but the dominant feature is a delocalized sea of electrons. Ionic bonds are dominated by the strong electrostatic interactions between the ions. Fundamentally, the properties of each material start with its bonding.

A fundamental description of bonding requires a calculation of the electronic distributions. The class of methods that yield such information are called *electronic structure methods*. In this chapter, we shall briefly review the basics of these methods, pointing out their inherent approximations. There are numerous books devoted to the fundamental theories behind these methods – embodied in quantum mechanics – as well as many texts devoted to electronic structure methods themselves [167, 219, 251, 254]. We can at best give a brief guide to this topic needed for discussions later in the text and as well as for a basis for understanding and evaluating this fascinating field.

Not so many years ago, practitioners of electronic structure calculations typically used home-grown computer codes, which often required heroic efforts on the parts of the programmers. These days, numerous codes are available, ranging from being free to being quite expensive and thus almost no one writes their own codes any more. The good news is that electronic structure calculations are thus now widely available and accessible to most researchers. The negative side is that there are many chances to do poor-quality calculations. Please be warned that the discussion in this text will not adequately prepare you to avoid all the pitfalls associated with these calculations.

This section may be a bit challenging for those without much experience in quantum mechanics. We have provided an all-too-brief introduction in Appendix F. The interested reader is urged to go beyond that simple description and read some of the texts listed at the end of this chapter. A nice tutorial on electronic structure calculations, with greater sophistication than this chapter, is available from [240].

## 4.1 QUANTUM MECHANICS OF MULTIELECTRON SYSTEMS

The basis for all quantum mechanical calculations of the electronic structure of a material is the Schrödinger equation,

$$\mathcal{H}\Psi = E\Psi, \tag{4.1}$$

where $\mathcal{H}$ is the Hamiltonian operator, $E$ is the energy, and $\Psi$ the wave function.[1] A key point of quantum mechanics is that $\mathcal{H}$ is an operator and thus behaves quite differently than an equivalent function in classical systems. Some details about operators are given in Appendix F.3.

The Schrödinger equation describes the energy of the electrons and nuclei in a material. Electrons interact with the positively charged atomic nuclei through an electrostatic potential. From the point of view of the electrons, the nuclei are fixed,[2] thus the electron interaction with a nucleus ($\alpha$) can be considered as an external potential that takes the form $v = -Z_\alpha/r_{i\alpha}$, where $Z_\alpha$ is the nuclear charge and $r_{i\alpha}$ is the distance from the nucleus $\alpha$ to the $i$th electron. If a system has $N$ electrons and $M$ nuclei, the Hamiltonian is

$$\mathcal{H} = \sum_{i=1}^{N} \left( -\frac{1}{2}\nabla_i^2 + v_{ext}(\mathbf{r}_i) \right) + \sum_i \sum_{j>i} \frac{1}{r_{ij}} \tag{4.2}$$

where the electrostatic (Coulomb) potential acting on electron $i$ from the $M$ nuclei is a sum over all $M$ nuclei

$$v(\mathbf{r}_i) = -\sum_{\alpha=1}^{M} \frac{Z_\alpha}{r_{i\alpha}}. \tag{4.3}$$

These equations are in *atomic units*, in which the fundamental constants $m$, $e$, $\hbar$, etc. are all equal to 1. The unit of energy is the hartree, the unit of length is the bohr, etc., as discussed in more detail in Appendix F.8.

The wave function $\Psi$ depends on the position of the electrons, $\mathbf{r}$. We will be more specific later, but for now let us just write the wave function as a function of the position of the $N$ electrons as $\Psi = \Psi(\mathbf{r}^N)$, where $\mathbf{r}^N$ is shorthand notation for the quantity $\{\mathbf{r}_1, \mathbf{r}_2, \ldots, \mathbf{r}_N\}$. The electron *density* of the system, $\rho$, is the number of electrons per unit volume,

$$\rho(\mathbf{r}_1) = N \int \ldots \int |\Psi(\mathbf{r}^N)|^2 d\mathbf{r}_2 \ldots d\mathbf{r}_N, \tag{4.4}$$

where we integrate over all electron coordinates but one. The normalization of the wave functions is $\int \rho(\mathbf{r}_1)d\mathbf{r}_1 = N$.

The Schrödinger equation in Eq. (4.2) is the fundamental equation describing the electronic structure of materials. If it could be solved exactly, the solution would be the wave function and

---

[1] More details are given in Appendix F and Eq. (F.6).

[2] Essentially all calculations are done based on the Born-Oppenheimer approximation, in which it is assumed that since electrons are much lighter and smaller than the atomic nucleus, they move fast enough that they can respond "instantaneously" to changes in the atomic positions. Thus, the nuclear coordinates can be taken as fixed when determining the electron distributions and energies.

energy, giving a complete description of the electronic properties. However, we cannot solve these equations exactly except for some simple problems.[3] The main objective of this chapter is to introduce the approximations used in modern electronic structure methods.

### 4.1.1 Classes of methods

There are classes of methods that assume an approximate wave function and find variational solutions to Eq. (4.2). These methods go by many names, depending on the approximations used for the wave functions. For example, in the *Hartree* approximation, the total wave function consists of single particle functions for each electron, which leads to an inaccurate, but easy to solve, set of equations. The electronic wave functions suffer from not being antisymmetric with respect to the exchange of electrons, as discussed in Appendix F.6.2.

In *Hartree-Fock* calculations, the wave functions are constructed so that they are antisymmetric with respect to the exchange of electrons and the results from these calculations are much superior to that with the simpler Fock method. Hartree-Fock theory neglects, however, the *correlation* energy, which accounts for the energy associated with the motions of the electrons being correlated so that they stay apart from each other (Appendix F.6).

Hartree-Fock and other, similar, methods were the mainstay of electronic structure calculations (especially in quantum chemistry) for decades, but have largely been supplanted by an approach based on a *density functional theory* (DFT) formulation of quantum mechanics. In DFT, the energy is written as a function of the electronic density, which is a function of the position, thus the energy is a *functional* of the density.[4] Given its ubiquitous use, DFT will be the focus of the rest of this chapter. Those interested in more traditional methods are urged to peruse the book by Kaxiras listed in the Suggested readings at the end of the chapter.

## 4.2 EARLY DENSITY FUNCTIONAL THEORIES

The earliest attempts to use the electronic density as a fundamental parameter in an electronic-structure calculation were published in 1927 independently by Thomas [309] and Fermi [99] (note that Schrödinger's equation was only published in 1926). They were each searching for a simple way to solve for the electronic structure of atoms. By analogy with classical mechanics, they assumed that there were three components to the energy: the attraction between the electrons and the nucleus of the atom, the kinetic energy of the electrons, and the repulsion between the electrons. They then assumed they could write the energy not in terms of wave functions, but rather in terms of functions of the electronic density of the system, $\rho(\mathbf{r})$.

Electrons interact with the positively charged atomic nuclei through an electrostatic potential, which for an atom takes the form $v_{ext} = -Z/r$, where $Z$ is the nuclear charge and $r$ is the distance from the nucleus to the electron. The interaction energy of all the electrons with the

[3] A few examples are discussed in Appendix F.5.
[4] See Appendix C.6 for a description of the properties of functionals.

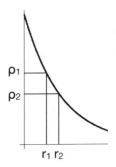

**Figure 4.1** Local-density approximation. Although the electronic density is rapidly varying, the energy expression at each density is based on a *uniform* electron gas with that electronic density.

nucleus is found by evaluating the integral of the electronic density as a function of $r$ times the interaction of an electron a distance $r$ away from the nuclei, i.e.,

$$\int \rho(\mathbf{r}) v_{ext}(\mathbf{r}) d\mathbf{r} . \tag{4.5}$$

In quantum mechanics, the kinetic energy is calculated with the $-(1/2)\nabla^2$ term in the Hamiltonian in Eq. (4.2). In the Thomas-Fermi model, the kinetic energy density is approximated by the expression for the energy of a uniform electron gas with density $\rho$, as described in Appendix F.5.1. The total kinetic energy of the electron gas, $T_{eg}$, is the integral over space of the kinetic energy density and in atomic units takes the form (from Eq. (F.24))

$$T_{eg}[\rho] = C_F \int \rho^{5/3}(\mathbf{r}) d\mathbf{r} , \tag{4.6}$$

where $C_F = (3/10)(3\pi^2)^{2/3}$.

There is an aspect to Eq. (4.6) that deserves some discussion. In Figure 4.1 we show a schematic view of the electronic density near the nucleus of an atom. At a distance $r_1$ in that figure, the density is $\rho_1$, and at $r_2$, it is $\rho_2$. In Eq. (4.6), the kinetic energy density at the two distances will be proportional to $\rho_1^{2/3}$ and $\rho_2^{2/3}$, respectively. However, the expression for the kinetic energy density was derived assuming a gas of electrons with uniform density everywhere (see Appendix F.5.1). Applying the electron-gas expression at each point regardless of how the density is actually varying is an approximation. However, it is an approximation that is essential to many electronic structure calculations. Since the density functionals are applied at each point and depend only on the values of the electronic density at that point, this approach is referred to as a *local-density approximation* (LDA).

The final term in the Thomas-Fermi (TF) model is the classical electrostatic (Coulomb) interaction between the electrons, a term that occurs so regularly it has its own symbol, $J$.[5] $J$ is given by an integral over the electronic distribution

$$J[\rho] = \frac{1}{2} \iint \frac{\rho(\mathbf{r}_1)\rho(\mathbf{r}_2)}{|\mathbf{r}_2 - \mathbf{r}_1|} d\mathbf{r}_1 d\mathbf{r}_2 . \tag{4.7}$$

The total energy in the TF model is

$$E_{TF}[\rho] = T_{eg}[\rho] + J[\rho] - Z \int \frac{\rho(\mathbf{r})}{r} . \tag{4.8}$$

---

[5] Described in Appendix E.2.

$E_{TF}$ is a *functional* of the electron density $\rho$,[6] that is, it is a function of the function $\rho$.[7] The Thomas-Fermi model is thus a simple example of a *density functional theory* (DFT).

In quantum mechanics, there are corrections to the Coulomb term ($J$ in Eq. (4.8)) that arise from the quantum nature of electrons. One of those corrections comes from a quantum effect called *exchange*, which arises from the fact that two electrons of the same spin can never be found at the same place. Thus, electrons of the same spin avoid each other and the electron-electron repulsion energy is less for electrons with the same spin than for electrons with the opposite spin. This difference in energy between parallel-spin and anti-parallel-spin electrons is the exchange energy. The origins of the exchange energy are discussed in Appendix F.6.3, in which we show that the exchange energy for a uniform electron gas is given by an integration of the exchange energy density from Eq. (F.45) over space,

$$E_x[\rho] = -C_x \int \rho^{4/3}(\mathbf{r}) d\mathbf{r}, \qquad (4.9)$$

where $C_x = (3/4)(3/\pi)^{1/3}$.

In 1930, Dirac modified the Thomas-Fermi model Eq. (4.8) to include the exchange energy [92]. Called the Thomas-Fermi-Dirac (TFD) model, the total energy functional is

$$E_{TFD}[\rho] = T_{eg}[\rho] + E_x[\rho] + J[\rho] - Z \int \frac{\rho(\mathbf{r})}{r} d\mathbf{r}. \qquad (4.10)$$

The goal in both the Thomas-Fermi and Thomas-Fermi-Dirac models was to determine the ground state (lowest) energy and the ground state electronic density, which was accomplished by minimizing the energy with respect to the density. The idea that the ground state could be determined through such a minimization was an assumption, not proved until the Hohenberg-Kohn theorem described in the next section. The minimum can be found using the method of Lagrange multipliers with the constraints of fixed total number of electrons (the integral of the density), the requirement that $\rho(\mathbf{r}) \to 0$ as $\mathbf{r} \to 0$ (to avoid a singularity in the Coulomb energy), and that $\rho(\mathbf{r}) \to 0$ as $\mathbf{r} \to \infty$ (finite-sized atoms).[8]

---

[6] Some properties of functionals are described in Appendix C.6.

[7] Functionals have some interesting properties, which are reviewed in Appendix C.6. We will see functionals again in the discussion of the phase-field model in a later chapter.

[8] Consider the solution to the TFD model. We can include the constraint that $\int \rho(\mathbf{r}_1) d\mathbf{r}_1 = N$ through the use of Lagrange multipliers [10], which leads to

$$\delta \left\{ E_{TFD}[\rho] - \mu_{TFD} \left( \int \rho(\mathbf{r}_1) d\mathbf{r}_1 - N \right) \right\} = 0, \qquad (4.11)$$

where $\delta$ is a functional differential (see Appendix C.6 for a discussion of the calculus of functionals). Taking the functional derivatives (e.g., $\delta E/\delta \rho$), we have

$$\mu_{TFD} = \frac{\delta E[\rho]}{\delta \rho} = \frac{5}{3} C_F \rho^{2/3}(\mathbf{r}) - \frac{4}{3} C_x \rho^{1/3}(\mathbf{r}) - \phi(\mathbf{r}) = 0, \qquad (4.12)$$

where the electrostatic potential is

$$\phi(\mathbf{r}) = \frac{Z}{r} - \int \frac{\rho(\mathbf{r}_1)}{|\mathbf{r} - \mathbf{r}_1|} d\mathbf{r}_1. \qquad (4.13)$$

Eq. (4.12) can be solved in a number of ways. For example, a functional form could be introduced that includes the constraints that $\rho(\mathbf{r}) \to 0$ as $\mathbf{r} \to 0$ and $\rho(\mathbf{r}) \to 0$ as $\mathbf{r} \to \infty$ or strictly numerical solutions could be used.

The TF and TFD models are appealingly simple, being based on functionals from a uniform electron gas and used with a local density approximation. They are, however, based on a crude description of the kinetic and exchange energies, they neglect the correlation energy described in Appendix F.6, their electronic densities are not based on realistic wave functions, and so on. Given these approximations, the question then is: how accurate are these methods?

There are actually two questions we could consider: (1) how well do the results from the TF and TFD calculations match similar, but more accurate, calculations? and (2) how well do the results match the real energies of atoms and molecules? This type of question will face us throughout this text. Should the results calculated from models be compared to reality or other calculations? Often, it will be the latter. For example, the TF and TFD models are exactly that, models. They neglect important physics. Thus, they cannot be expected to match experiments or accurate calculations on real systems. So question (2) is not very relevant in this case. A more informative approach would to compare these model calculations with accurate calculations using other methods that most closely match these, which will give us a sense of how well these models work. In the case of the TF and TFD models, they can best be compared to accurate calculations using methods such as the Hartree-Fock approach mentioned in Section 4.1.1. The Hartree-Fock method solves the Schrödinger equation in Eq. (4.2) very accurately, using antisymmetric wave functions. The Hartree-Fock method does not, however, include the correlation energy. Thus, comparison of TF and TFD to Hartree-Fock results is a direct measure of how well these simple models capture the physics of multielectron systems without correlation.

It can be shown analytically that the Thomas-Fermi energy for neutral, closed-shell, atoms is [251]

$$E_{TF} = -0.7687 \, Z^{7/3} \,, \tag{4.14}$$

where $Z$ is the charge on the nucleus (and equals the number of electrons). In Table 4.1 we compare the Thomas-Fermi energy to accurate calculations with the Hartree-Fock method, where we plot $-E_{HF}/Z^{7/3}$. If the Thomas-Fermi result were correct, then the Hartree-Fock result should take the value 0.7687 from Eq. (4.14). From the table, it is clear that the Thomas-Fermi model yields energies with considerable error, overestimating the binding energy of the electrons, with relative errors ranging from about 35% for He to 13% for Rn.

While adding the exchange correction was supposed to improve the TF results, they actually make predictions of TFD worse than TF, which we can see directly from Eq. (4.10). The Thomas-Fermi model overestimates the binding energies of the atoms. Adding an exchange term that is negative will lower the energy even more, increasing the error.

The TF and TFD models do poorly when compared to accurate results without correlation energy. A number of suggested improvements have been proposed to these models, including the use of gradient terms in the kinetic energy [251]. While the results are somewhat better, they are still not adequate. Thus, there is something fundamentally incorrect about the models. The major error arises from the treatment of the kinetic energy, which, in the TF and TFD models, ignores the shell structure of the electrons [219].

Despite their failings, we discuss the TF and TFD models because they represent the first of a class of methods that are based on what is called *density functional theory* (DFT) for calculating electronic structure. There has been a long history of these and similar methods, all based on an

Table 4.1 **Comparison of the energy of the Thomas-Fermi model with calculated energies $E_{HF}$ for rare-gas atoms [70]. The calculated energies were determined with the Hartree-Fock method, so do not contain the correlation energy. From [251]**

| Atom | Z | $(-E_{HF}/Z^{7/3})$ |
|------|---|---------------------|
| He | 2 | 0.5678 |
| Ne | 10 | 0.5967 |
| Ar | 18 | 0.6204 |
| Kr | 36 | 0.6431 |
| Xe | 54 | 0.6562 |
| Rn | 86 | 0.6698 |

*assumption* that one could find a true functional of the density that would describe the energy of a system of electrons and nuclei. This assumption was a leap of faith. Indeed, these methods were often mocked as not being based on good theory. It was not until the Hohenberg-Kohn theorem in the early 1960s that one knew for certain that such a density functional actually existed.

## 4.3 THE HOHENBERG-KOHN THEOREM

In a remarkable theorem, Hohenberg and Kohn [149] (HK) showed that the total energy, $E$, of a system of electrons in an external potential (in this case the Coulomb potential from the nuclei in a solid) is given *exactly* as a functional of the electronic density $\rho$. They showed further that the density that minimizes $E[\rho]$ is the ground-state electronic density and that other ground-state properties are also functionals of the ground-state density. Through this theorem, they proved that TF, TFD, and other such approaches had a theoretical justification – the Hohenberg-Kohn (HK) theorem made density functional theory well-founded.

The HK theorem, for all its importance, begs an important question. While the HK theorem states that a functional $E[\rho]$ exists, it fails to tell us what that functional is or how to find it. What the HK theorem does do, however, is to say that it is worth looking for such a functional.

Models such as TF, TFD, etc. were attempts at finding the correct functional. They were not very good, perhaps, but they were a step in the right direction. In the next section, we will discuss a more accurate approach that is the basis for most calculations today.

## 4.4 KOHN-SHAM METHOD

In this section we introduce the Kohn-Sham (KS) approach to solving the quantum mechanics of multielectron systems [173]. The use of this model, coupled with advances in treating the correlation energy, solves many of the inadequacies of the TF/TFD models and is the basis of most DFT calculations done today. Much of this discussion is based on that in [251]. A more recent review [240] is also quite useful. For more advanced readers, other recent books may be of help [102, 219].

The functional $E[\rho]$ represents the total electronic energy as a functional of the electron density $\rho$, and is a sum of the contributions from the external potential and the electronic energies,

$$E[\rho] = F[\rho] + \int v_{ext}(\mathbf{r})\rho(\mathbf{r})d\mathbf{r}, \qquad (4.15)$$

where

$$F[\rho] = T[\rho] + V_{ee}[\rho]. \qquad (4.16)$$

$F[\rho]$ is a sum of the kinetic energy $T[\rho]$ and the interaction energy between the electrons $V_{ee}[\rho]$. We write $V_{ee}[\rho]$ as the classical Coulomb integral (Eq. (E.10)) plus a correction term

$$V_{ee}[\rho] = J[\rho] + (V_{ee}[\rho] - J[\rho]). \qquad (4.17)$$

The focus is now on finding approximations for the kinetic energy $T$ and the part of the electron-electron interaction that is not described by the classical Coulomb energy $J$, i.e., $V_{ee}[\rho] - J[\rho]$.

As noted earlier, a major source of error in the TF and TFD models arises from the kinetic energy not reflecting the discontinuities inherent in the shell structure of the electrons. To avoid these difficulties, KS assumed that electron density of a system with $N$ electrons could be written as a sum of *one-electron orbitals* $\psi_i$ as

$$\rho(\mathbf{r}) = \sum_{i=1}^{N} |\psi_i(\mathbf{r})|^2. \qquad (4.18)$$

The significance of writing the density in one-electron orbitals is that (1) finding solutions to the Schrödinger equation is greatly simplified and that (2) the discontinuities in the shells are natural outcomes of the solutions. As shown below, the multielectron problem is reduced to finding a set of solutions to one-electron problems.

Kohn and Sham defined the kinetic energy functional to be

$$T_{KS}[\rho] = \sum_{i=1}^{N} \langle \psi_i | - \frac{1}{2}\nabla_i^2 | \psi_i \rangle, \qquad (4.19)$$

where we use the notation from Eq. (4.26). The point is that KS assumes a simple orbital wave function picture for the density and then approximates the kinetic energy with the exact kinetic-energy function operating on those approximate wave functions. The electronic part of the density functional in Eq. (4.16) is then given by

$$F[\rho] = T_{KS}[\rho] + J[\rho] + E_{xc}[\rho], \qquad (4.20)$$

where

$$E_{xc}[\rho] = T[\rho] - T_{KS}[\rho] + V_{ee}[\rho] - J[\rho]. \qquad (4.21)$$

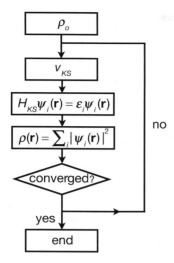

**Figure 4.2** Steps in the solution to the Kohn-Sham method. Adapted from [240].

The term $E_{xc}$ in Eq. (4.21) is called the exchange-correlation energy. It includes all the corrections between the sum of the kinetic and Coulomb energies and the correct answer. The hope is that $E_{xc}$, as a correction term, is small relative to the rest of the terms in Eq. (4.20). The task is then to develop a good form for $E_{xc}$, as discussed in the next section.

The KS method uses an iterative solution, shown schematically in Figure 4.2. One initially guesses at a set of wave functions $\psi_i^0$, from which is constructed an initial electron density $\rho_o$ from Eq. (4.18). The effective KS potential based on this density is found by taking the functional derivative of $E[\rho]$ yielding[9]

$$v_{KS}(\mathbf{r}) = v_{ext}(\mathbf{r}) + \int \frac{\rho(\mathbf{r}_1)}{|\mathbf{r} - \mathbf{r}_1|} d\mathbf{r}_1 + v_{xc}(\mathbf{r}), \qquad (4.22)$$

where $v_{xc} = \delta E_{xc}/\delta\rho(\mathbf{r})$ and $E_{xc}$ is the exchange-correlation function of choice.

Since $v_{KS}$ is defined by the electron density from the previous step in the iteration cycle, the Hamiltonian includes no direct interactions between electrons – it describes each electron as moving in an external field based on a fixed electron distribution, $\rho(\mathbf{r})$. The Hamiltonian is

$$\mathcal{H}_{KS} = -\frac{1}{2}\nabla^2 + v_{KS}(\mathbf{r}), \qquad (4.23)$$

from which we solve

$$\mathcal{H}_{KS}\,\psi_i(\mathbf{r}) = \epsilon_i\,\psi_i(\mathbf{r}). \qquad (4.24)$$

Solving Eq. (4.24) yields a new set of $N$ orbital wave functions $\psi$ (and the 1-electron orbital energies $\epsilon$), from which a new $\rho$ is found, and then a new $v_{KS}$. Equation (4.24) is solved again, and the process is repeated until a self-consistent $\rho$ is found, by which we mean a $\rho$ that does not vary more than a prescribed amount from one iteration to the next.

---

[9] Functional derivatives are described in Appendix C.6.

Typically, one expands the wave function $\psi$ in a set of basis functions,

$$\psi = \sum_j c_j \phi_j \,. \tag{4.25}$$

To solve Eq. (4.24), we form the Hamiltonian matrix and then diagonalize it to find the eigenvectors and eigenvalues.[10] The Hamiltonian matrix elements are

$$H_{ij} = \int \phi_i^*(\mathbf{r}) \left\{ -\frac{1}{2}\nabla^2 + v_{KS}(\mathbf{r}) \right\} \phi_j(\mathbf{r}) d\mathbf{r} \,. \tag{4.26}$$

The energy of the KS method is not the sum of the "orbital" energies $\epsilon_i$ from Eq. (4.24) – there are additional terms that must be added to arrive at the correct total energy. The total energy is [240][11]

$$E = \sum_i^{n_{occ}} \epsilon_i - \int \left[ \frac{1}{2} \int \frac{\rho(\mathbf{r}_1)}{|\mathbf{r} - \mathbf{r}_1|} d\mathbf{r}_1 + v_{xc}(\mathbf{r}) \right] \rho(\mathbf{r}) d\mathbf{r} + E_{xc} \,. \tag{4.27}$$

It is important to emphasize that the iterative method in Figure 4.2 converges to the minimum-energy state *for choice of the basis set employed in the calculation.*[12] A poor basis set will lead to poor calculations of the ground-state of the system. Indeed, one test of a basis set is to compare the ground-state energy with that calculated with other basis sets, with the descriptions that yield the lowest energy usually being preferred.

## 4.5   THE EXCHANGE-CORRELATION FUNCTIONAL

In Eq. (4.21) is defined a term called the *exchange-correlation* functional, $E_{xc}[\rho]$. This term includes all the corrections to the approximate Kohn-Sham Hamiltonian. There are two main types of exchange-correlation functionals in use, those based on the local-density approximation (LDA) and methods that include corrections to LDA, usually in the form of gradients of the density.

The TFD model is an example of a local-density approximation for the exchange function $E_x$, given in Eq. (4.9). Common approximations for $E_{xc}$ start with $E_x$ and add a correlation term,

$$E_{xc} = E_x + E_c \,. \tag{4.28}$$

There have been a number of models used for $E_c$. A common one is a simple functional of $\rho$, much as is $E_x$. Exacting quantum Monte Carlo calculations[13] were performed for a uniform

---

[10] Working with basis sets is described in Appendix F.7.

[11] Note that $\int E_{xc} d\mathbf{r} \neq \int v_{xc} \rho(\mathbf{r}) d\mathbf{r}$.

[12] That the minimum energy corresponds to the ground state is a consequence of the Hohenberg-Kohn theorem and the structure of the Kohn-Sham equations.

[13] The quantum Monte Carlo method is an approach to solving the Schrödinger equation very accurately as described in [60].

electron gas and the correlation energy was calculated as the total energy minus the kinetic, coulomb, and exchange energies [59]. Accurate expressions for the energy were obtained at high and low electron densities and a variety of schemes were proposed to parameterize these results [256, 257]. The advantage to these methods is that $E_{xc}$ remains a simple functional of $\rho$ and so the same types of manipulations can be followed to derive $v_{xc} = \delta E_{xc}[\rho(\mathbf{r})]/\delta\rho(\mathbf{r})$. A disadvantage is that in these schemes, $E_{xc}$ is still calculated in the local-density approximation.

The most common corrections to $E_{xc}$ are still local theories, but are based on not just the local value of the electron density but also on the local value of the gradient of the electron density. These methods are typically designated as *generalized gradient approximations* (GGA). In these methods, the exchange-correlation functional is assumed to have the form (for example in [191, 258])

$$E_{xc} = \int \rho(\mathbf{r})\epsilon_{xc}(\rho, \nabla\rho)d\mathbf{r},\qquad(4.29)$$

where $\nabla\rho(\mathbf{r})$ is the gradient of $\rho(\mathbf{r})$. The derivation of the expression for $v_{xc}$ is not difficult, as the functional derivatives are straightforward.

Various other approximations for $E_{xc}$ have been proposed,[14] but LDA and GGA are the most commonly used today for solids. As discussed below, both LDA and GGA functionals yield band gaps that are too small, so there is a focus on finding new functionals. Hybrid approaches, in which there is an interpolation between the Hartree-Fock approximation and the GGA [191], are increasingly being used. While initially developed for chemical applications, they have also been implemented for solids, in which they yield greatly improved descriptions of the electronic structure.

## 4.6 WAVE FUNCTIONS

The first step in a calculation is to choose the wave function. As discussed in Eq. (4.25), one typically writes the wave function as a sum over a set of functions, referred to as the basis set. What goes into the choice of basis set depends on the answers to a couple of questions:

(a) Can results with sufficient accuracy be found with the choice of basis set?
(b) What is the computational cost for achieving converged results with the desired accuracy?

Not surprisingly, the choice of wave function depends largely on the system of interest.

For solids, the wave functions should reflect the periodic symmetry of the system – they must satisfy Bloch's theorem [15] as described in Appendix F.9[15]

$$\phi_{\mathbf{k}}(\mathbf{r} + \mathbf{R}) = e^{i\mathbf{k}\cdot\mathbf{R}}\phi_{\mathbf{k}}(\mathbf{r}),\qquad(4.30)$$

---

[14] Including meta-GGA methods which include the Laplacian ($\nabla^2\rho$) in the functional [238].
[15] For more discussion of periodic systems, please see Section 3.4, in which periodic boundary conditions are introduced.

where $\mathbf{R}$ is a direct lattice vector. A general set of functions that satisfy this condition are *plane waves*

$$\phi_\mathbf{k}(\mathbf{r}) = e^{i\mathbf{k}\cdot\mathbf{r}} \sum_\mathbf{G} c_\mathbf{G} e^{i\mathbf{G}\cdot\mathbf{r}}, \qquad (4.31)$$

where $\mathbf{G}$ is a reciprocal lattice vector. The energy of a specific plane wave is proportional to $k^2$, as seen in the solution to the particle in a box in Appendix F.5.1. For application in electronic structure calculations, a set of wave functions with a choice of $k$ values is chosen and the coefficients in the wave function are determined as described in Figure 4.2.

Plane wave basis sets are common solutions for free electrons with boundary conditions that restrict the periodicity of the wave function[16] and offer many advantages in density-functional calculations for solids. The integrals involving plane waves are easier to evaluate and code than more complicated basis functions. They form a complete set and, by including sufficient numbers of them, yield convergence to an arbitrary accuracy. In practice, a finite number of plane-wave functions is used, with the convergence of the properties governed largely by a single parameter, the maximum energy plane wave used in the basis set.[17] Calculations based on a plane-wave basis generally converge smoothly to the minimum-energy solution. Since all functions in a plane-wave basis are mutually orthogonal, their use avoids a problem called "basis-set superposition error" that arises when non-plane-wave basis sets are used, as discussed below. Many more details are given in [240].

Plane waves can also be used for finite systems, such as atoms or molecules, by means of a "supercell" approach, in which the finite system is placed in a unit cell of a fictitious crystal, which is then made large enough such that the atoms and molecules in the central unit cell do not interact with their replicas in other cells. Supercells are used to model not just finite systems, but any system for which three-dimensional periodicity is broken. Examples include surfaces or interfaces (with two-dimensional periodicity), nanowires (one-dimensional periodicity) and quantum dots (with no periodicity).

While plane wave basis sets are common, they are not necessarily the best choice for all problems. For example, when describing solids consisting of molecules or systems with strongly directional bonding, it is common to use a basis of a finite number of atomic orbitals rather than a set of plane waves. These orbitals are centered at the atomic nuclei, making this choice of basis set similar to (but not actually the same as) the Linear Combination of Atomic Orbitals (LCAO) approximation that has been in use for many years to describe molecular electronic structure. While hydrogenic-like orbitals conform more closely to actual electronic distributions, approximating these with a linear combination of Gaussian functions offers many advantages owing to the ease of calculating the various integrals needed to evaluate the Hamiltonian matrix elements. In Figure 4.3 we compare a hydrogenic-orbital (called a Slater function) to an equivalent Gaussian function. Note the significant differences both at short and long range. The hope is that even with the Gaussian functions, many fewer basis set functions

---

[16] See, for example, the particle in a box in Appendix F.5.1.

[17] To avoid problems associated with the rapidly varying electronic density near the nuclei of the atoms, approximate functions, called pseudopotentials, are used to approximate the core regions, as discussed in the next section.

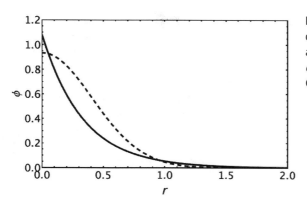

**Figure 4.3** Comparison of Slater and Gaussian oribtals. Solid curve is a normalized atomic-orbital-type (Slater) orbital of the form $e^{-\alpha r}$ and dashed curve is a normalized Gaussian orbital of the form $e^{-\alpha r^2}$, with $\alpha = 2$.

will be needed than in plane wave calculations because each function corresponds more closely to the actual form of the electronic distribution. Thus, non-plane-wave basis approaches tend to be computationally fast. Another advantage of the use of non-plane-wave basis functions is that they do not need to be periodic, which enables the calculation of properties of aperiodic or semiperiodic supercells.

The actual choice of functions that make up a basis set depends on a number of factors, with the final choice usually being governed by a balance between accuracy and computational time. The danger is that since one does not use a complete set of functions, the solutions found in the minimization of the energy may be incorrect owing to having excluded the actual ground state from the solution by the choice of basis set. It is also sometimes required to add "ghost" atoms, which are atomic-like orbitals that are not centered on an atom but instead are added in an ad hoc fashion to describe a region of space near atoms. For example you might add extra orbitals near point defects, surfaces, or in the covalent bond-charge. Details of these methods are beyond the scope of this text.

Calculations based on a finite basis set often suffer from "basis set superposition error", which arises from the overlap of their basis functions as two atoms come close to each other. When this occurs, the wave function of an atom begins to include contributions from the wave function of different atoms, leading to an overall error in the calculation. There are a couple of ways to eliminate or reduce this error, the details of which we leave to more advanced texts.

Another method that does not rely on plane waves is based on real space grids. At each point on the grid the wave function is just a numerical value and the Kohn-Sham Hamiltonian is solved numerically. This approach can be very fast and is easily extendable to non-periodic systems.

## 4.7 PSEUDOPOTENTIALS

It is common to think about the electronic structure of atoms, whether in a solid or not, by dividing the electrons into two groups, the outer, valence, electrons, and the electrons at the

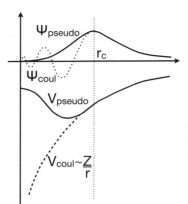

**Figure 4.4** Schematic view of a pseudopotential and a pseudo wave function, as described in the text. The bottom part of the plot shows the potential, with the term $V_{coul}$ representing the bare Coulomb potential from the core electrons and $V_{pseudo}$ the pseudopotential. In the top part of the figure we show the wave functions, with $\Psi_{coul}$ showing the oscillations in the wave function that arise because the wave functions of the valence electrons must be orthogonal to those of the core elections. By taking the core electrons out of the problem, the oscillations can be eliminated, leading to $\Psi_{pseudo}$, the smooth wave function that arises from the pseudopotential. $r_c$ is the core radius.

inner core. The inner shell electrons are very tightly bound to the nucleus and play little to no role in the chemical binding between atoms. In many ways, the inner shell electrons plus the nucleus can be considered as an essentially inert core, with a charge that is the normal charge on the nucleus minus the charge associated with the inner core electrons. The advantage to this point of view is that the inner core electrons can then be considered as frozen and only the outer, valence, electrons need be included in the calculation of the electronic structure. Of course, since the inner electrons are not point charges, they cannot be ignored completely.

The effects of the inner electrons are modeled by the use of *pseudopotentials*, in which each nucleus plus its core electrons is treated as a frozen core that does not change in response to changes in its environment. The interactions of the valence electrons with this core are described by a potential function, called the pseudopotential, that may be constructed from first-principles calculations to reproduce the behavior of the nucleus and core electrons exactly. The accuracy of pseudopotential calculations has been justified by comparison with calculations that include all of the electrons. Pseudopotential methods are much faster than all-electron calculations because the tightly-bound core electrons do not enter directly into the problem [290].

The simplest are the so-called empirical pseudopotentials, in which a functional form is assumed and parameters are chosen to match experiment. *Ab initio* pseudopotentials are determined from accurate calculations of the free atom Schrödinger equation. The empirical pseudopotentials are not commonly used today, with first-principles pseudopotentials, based on local solutions to the wave functions near the atomic nuclei, being the preferred approach. A detailed description of pseudopotentials can be found in [240].

In Figure 4.4 we show a schematic view of how a pseudopotential acts in the system. The bottom part of the plot shows the potential, with the term $V_{coul}$ representing the bare Coulomb potential from the core electrons and $V_{pseudo}$ the pseudopotential, in which the Coulomb potential is offset by the core electrons. In the top part of the figure we show the wave functions, with $\Psi_{coul}$ showing the oscillations in the wave function owing to the steep Coulomb potential and $\Psi_{pseudo}$ showing the smooth wave function that arises from the pseudopotential. $r_c$ is the core radius.

Table 4.2 **Comparison of some bulk properties of bulk Si obtained with LDA and GGA [240].** *a* is the lattice constant, *B* the bulk modulus, and $E_c$ the cohesive energy. Experimental results are from [192]

|  | LDA | GGA | Experiment |
|---|---|---|---|
| *a* (Å) | 5.378 | 5.463 | 5.429 |
| *B* (Mbar) | 0.965 | 0.882 | 0.978 |
| $E_c$ (eV/atom) | 6.00 | 5.42 | 4.63 |

Table 4.3 **Comparison of some bulk properties of bulk Cu obtained with LDA and GGA [240].** *a* is the lattice constant, *B* the bulk modulus, and $E_c$ the cohesive energy. Experimental results are from [277]

|  | LDA | GGA | Experiment |
|---|---|---|---|
| *a* (Å) | 3.571 | 3.682 | 3.61 |
| *B* (Mbar) | 0.902 | 0.672 | 1.420 |
| $E_c$ (eV/atom) | 4.54 | 3.58 | 3.50 |

## 4.8 USE OF DENSITY FUNCTIONAL THEORY

The density functional methods described in this chapter can reproduce the structural properties of materials with reasonable accuracy. Typically, LDA calculations underestimate equilibrium lattice parameters by about 1-2%, while use of the GGA leads to larger bond lengths than LDA, with predicted values that can be somewhat larger than experiment. Similarly, LDA predicts a cohesive energy larger than the experimental value, with the GGA correcting the predicted value.

For example, in Table 4.2 we compare calculated values (LDA and GGA) with experimental values of structural properties for bulk Si, a covalently bonded material. The LDA results are too bound, with a lattice parameter about 1% low, and the GGA results are a bit too large, though closer to experiment. The LDA and GGA results for the cohesive energy are both too large. In Table 4.3 we show similar results for a metallic system, bulk Cu. Again, the LDA results for the lattice parameter are too small (about 1%), while the GGA results are too large. Both methods give cohesive energies that are too large, though GGA yields results that are closer to experiment. The bulk modulus is not well calculated with either method for either system, being too small, though LDA does somewhat better than GGA.

While the trends in the differences in calculations based on LDA and GGA in Table 4.3 seem to generally hold, they are not always followed. Indeed, there is no theory that explains the difference between the results of the two methods. It is somewhat surprising that GGA-based calculations seem no more accurate that LDA-based ones. One might expect that including any additional information, such as the charge gradient, would result in a more accurate calculation.

The Hohenberg-Kohn theorem, and the Kohn-Sham method based on it, is only valid for the ground state energy. Thus it is not surprising that DFT does not do well in calculating the energy of excited states. For example, band gaps are predicted to be too small by approximately a factor of 2 in most cases. In Si, for example, the experimental gap is 1.17 eV and that calculated with LDA is 0.45 eV and the value with GGA is 0.61 eV [240].

One thing DFT does very well is to determine the electron density, which can lead to new insights into how materials are structured and behave. As just one example, in a study using DFT to study the potential of certain Bi compounds ($BiAlO_3$ and $BiGaO_3$) as piezoelectric materials [18], it was found that the piezoelectric response is a result of the stereochemical activity of the Bi lone pairs, which causes large displacements of the Bi ions from their positions in a centrosymmetric phase. This insight was gained from an analysis of plots of the electron localization.

Other limitations of DFT center on its computational complexity, which limits the number of atoms that can be included in a calculation. While that number changes with time as computers are made faster and algorithms are improved, suffice it to say that there are many problems for which these calculations are not well suited. A particularly important problem that causes challenges for these methods is to model systems at finite temperatures ($T > 0$). Some methods have been proposed that extend the methods to that regime and have been shown to work well [56]. However, the size limitations of the calculations limit their applicability to many problems. In later chapters we will introduce methods that will enable us to model very large systems of atoms at finite temperature. We will, however, have to give up the direct calculation of the electronic states and replace them with approximate descriptions of the bonding in materials.

Despite the limitations, density functional calculations have become ubiquitous in materials research, with applications that range from biological systems to alloy development in metals. Here we list just a few examples, showing the range of applicability of these methods.[18]

- Structure and thermodynamics: application of *ab initio* methods to calculate structural properties has become routine, as witnessed by a study in which they used DFT to create tables of the ground state energy and structures of 80 binary alloys [75]. A common use of these methods is to calculate thermodynamic properties of systems, one example being a study of intermetallic compounds and solution phases in Sn/Zn [325], and another being a study of the equation of state and elastic properties of a series of metals [31].
- Extreme conditions: another common use of these methods is to calculate properties of materials in regimes for which experimental data might be limited, for example a study of the high-pressure properties of ammonia borane ($NH_3BH_3$) [202].

---

[18] The reader is urged to look in the literature for application of electronic structure methods to systems that interest them.

- Defect structure and properties: defect structures are also commonly studied with these methods. One example from a myriad of studies is an examination of the fracture properties of Al/TiN interface [369]. DFT calculations have also been used to shed light on dislocation core structures, for example in aluminum [360].
- Biomaterial applications: biological applications are also now common, an example being a study of how oxygen enters a protein structure [69].
- New materials: electronic structure calculations have also been extremely influential in defining the basic physics of new types of materials. One example is the work of Nicola Spaldin, which ignited interest in the multiferroics [145].

These are just a few examples that indicate the range of applications of electronic structure calculations.

## 4.9 SUMMARY

In this chapter we introduce the basics of the density functional theory of the electronic structure of solids. After reviewing some of the historical approaches, we describe in some detail the basis of most electronic structure calculations done today, the Kohn-Sham method. We discuss the use of the local-density approximation (LDA) as well as improvements on it through the use of generalized gradient approximations (GGA).

These methods, while not perfect, provide a direct way to calculate the properties of materials with the fewest possible assumptions. They enable us to see where electrons are and let us avoid some of the pitfalls that arise from the use of approximate descriptions of the bonding between atoms, namely the interatomic potential functions described in the next chapter. However, electronic structure methods are computationally costly and thus can only be used for relatively few atoms at a time. Thus, despite their utility, we often must use the approximate methods discussed in later chapters.

### Suggested reading

There are many books and articles available that discuss this topic, with a wide range of readability, for example:

- An excellent and comprehensive text by E. Kaxiras, *Atomic and Electronic Structure of Solids* [167].
- A basic description of how to do electronic structure calculations is given in *A Primer in Density Functional Theory* [102].
- A recent text by Martin, *Electronic Structure: Basic Theory and Practical Methods*, is an excellent guide to both basics and applications [219].
- Some older, but very useful, texts include: March, *Self-Consistent Fields in Atoms* [217]; and Parr and Yang, *Density-Functional Theory of Atoms and Molecules* [251].

# 5 Interatomic potentials

Calculating the properties of a solid based on atoms requires a description of the energetics and forces between those atoms. This energy could be calculated by solving the quantum mechanics of all the nuclei and electrons in a system, as discussed in Chapter 4. In this chapter, however, we discuss a less computationally intensive approach, in which we develop and use models for the interactions between atoms, which are generally based on simple functional forms that reflect the various types of bonding seen in solids. We shall see that these functions are, by their nature, approximate, and thus the calculations based on them are also approximations of the materials they are designed to describe. However, the use of these potentials, though lower fidelity, will enable us to model much larger systems over much larger times than possible with the more accurate quantum mechanical methods.

## 5.1  THE COHESIVE ENERGY

The goal of most atomistic-level simulations is, in part, to calculate quantities that define the energetics and thermodynamics of materials. The most fundamental of those quantities is the potential energy, which is the sum of the energetic interactions between the atoms. At 0 K, that energy is the cohesive energy, which is defined as the energy required to assemble a solid from its constituent atoms and molecules.

Consider a system of $N$ atoms. The cohesive energy, $U$, is the negative of the energy needed to take all the atoms and move them infinitely far apart, i.e.,

$$U = E(\text{all atoms}) - \sum_{i=1}^{N} E_i , \tag{5.1}$$

where $E(\text{all atoms})$ is the total energy of the system and $E_i$ is the energy of an individual isolated atom.

Our goal is to develop simple analytical potentials that *approximate* the interaction energies between atoms.[1] The fundamental entities are the atoms and molecules that make up the solid, with the details of the electrons and nuclear charges being approximated in the analytical potentials. Thus, the simple potential functions are, in some sense, an average over the electrons and represent a great simplification over having to deal with the electrons individually. The

---

[1]  Intermolecular interactions are covered in Chapter 8.

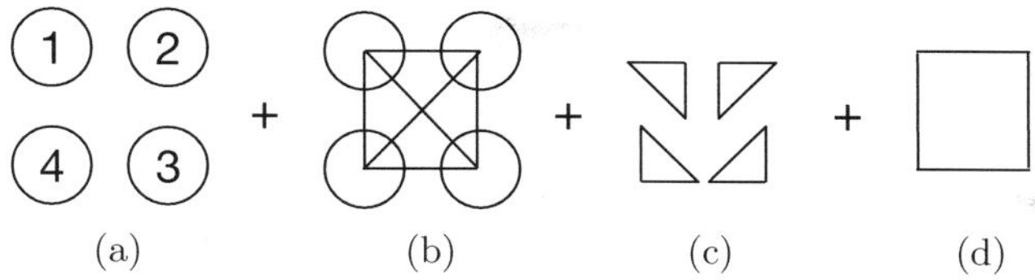

**Figure 5.1** Hierarchy of interactions between atoms: pair interactions.

procedure of developing models at one scale by averaging over properties at a lower scale is a theme in materials modeling, and will be revisited in later chapters.

The cohesive energy $U$ in Eq. (5.1) can be formally expanded in a series of terms that depend on the individual atoms, pairs of atoms, triplets of atoms, etc., as in

$$U = \sum_{i=1}^{N} v_1(\mathbf{r}_i) + \frac{1}{2} \sum_{i=1}^{N} \sum_{j=1}^{N}{}' \phi_{ij}(\mathbf{r}_i, \mathbf{r}_j) + \frac{1}{6} \sum_{i=1}^{N} \sum_{j=1}^{N} \sum_{k=1}^{N}{}' v_3(\mathbf{r}_i, \mathbf{r}_j, \mathbf{r}_k) + \cdots, \qquad (5.2)$$

where the ′ indicates that the $i = j$ terms and the $i = j = k$ terms are not included in the second and third sets of sums, respectively. This expression is shown schematically for a system of $N = 4$ atoms in Figure 5.1:

(a) Figure 5.1a shows the four atoms of the system. The first term in Eq. (5.2), $v_1$, represents the effect of an external field (for example, an electric, magnetic, or gravitational field) on the atoms. We will generally not need to be concerned with this term in this text.

(b) Figure 5.1b represents a sum over the interaction of all pairs of atoms, as indicated in Figure 5.1b. $\phi_{ij}(\mathbf{r}_i, \mathbf{r}_j)$ is a function of the atomic positions that represents the interaction between the pair of atoms $(i, j)$, located at $\mathbf{r}_i$ and $\mathbf{r}_j$, respectively. Most of our discussion will center on these types of interaction. $\phi_{ij}(\mathbf{r}_i, \mathbf{r}_j)$ is called a *pair potential* and, since it is an energy, $\phi_{ij} = \phi_{ji}$.

(c) Figure 5.1c arises from the interaction of triplets of atoms, $v_3(\mathbf{r}_i, \mathbf{r}_j, \mathbf{r}_k)$. These types of interactions are called *three-body* interactions. Any interactions that involve more than two atoms at a time are called *many-body* interactions. While pair potentials are adequate to describe many materials, there are important classes of materials whose interaction potentials involve many-body terms, e.g., metals, covalent solids, etc., as shall be discussed later in this chapter.

(d) Figure 5.1d is representative of a four-body interaction, i.e., $v_4(\mathbf{r}_1, \mathbf{r}_2, \mathbf{r}_3, \mathbf{r}_4)$. If there were more than four atoms in this example, evaluating the contributions of these terms to the energy would consist of sums over four quartets of atoms.

## 5.2 INTERATOMIC POTENTIALS

Interatomic potentials are the potentials between atoms – we are leaving molecular systems to a later chapter. The potential forms we will discuss should reflect the *bonding* between the atoms. In Figure 5.2 we show a schematic view of the bonding in simple types of solids. The simplest bonding occurs in rare-gas solids (He, Ne, Ar, Kr, Xe) as shown in Figure 5.2a,

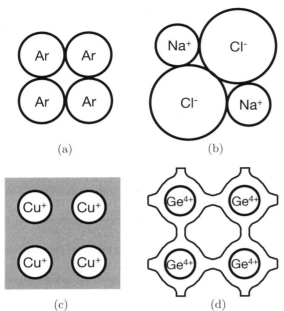

**Figure 5.2** Schematic view of bonding in atomic solids. The sizes of the atoms and ions are roughly equal to the appropriate atomic or ionic radii. Adapted from [15]. (a) Rare-gas solids. The inner core electrons and nucleus are surrounded by a closed shell of valence electrons. (b) Ionic solids. The inner core electrons and nucleus are surrounded by a closed shell of valence electrons. Note the difference in size between the cation (positive charge) and anion (negative charge). (c) Metals. The inner core electrons and nucleus form a positive ion, which is surrounded by a gas of electrons. (d) Covalent solids. The atoms have an ionic core with strong directional bonds that connect the atoms.

which are all closed-shell atoms. There is essentially no directionality to the bonds and, thus, the interatomic potentials depend entirely on the distance between the atoms. In Figure 5.2b we show a prototypical ionic material, $Na^+Cl^-$. Again, the atoms are closed shell and thus the bonding is essentially non-directional. Metals (Figure 5.2c) have very different electronic structures. The atoms are ionized (e.g., $Cu^+$ in the figure), with the valence electrons distributed throughout the system. In simple metals there is again no (or very little) directionality to the bonds, but the interaction between the delocalized electrons must be included in the description of the bonding. Finally, in Figure 5.2d we have a covalent crystal, with strong directional bonds connecting the atoms. In all cases, accurate descriptions of the interatomic potentials must reflect the nature of the electronic structure and, thus, the interactions. Of course, in real systems it is not so simple. Most systems are not strictly one type or another – they may be metals with some covalency, or partially ionic covalent crystals. In these cases, the simple models discussed in this text may need to be modified.

### 5.2.1 Basic forms of interatomic interactions

Before discussing models for interatomic interactions, a few words about the origins of the interactions are in order. For now, we will restrict the discussion to neutral atoms with non-bonding interactions. Extensions to ionic and covalently bonded systems will be discussed later in the chapter.

Suppose we have two atoms, $i$ and $j$, located at $\mathbf{r}_i$ and $\mathbf{r}_j$, respectively. The *interaction energy* between the two atoms is defined as the difference in energy between the energy of the pair of atoms, $E(i + j)$, and the energies of the individual atoms, $E(i)$ and $E(j)$, separated at infinity:

$\phi_{ij}(\mathbf{r}_i, \mathbf{r}_j) = E(i + j) - E(i) - E(j)$. We know, based on simple reasoning, something about the form of $\phi_{ij}$. If the separation between the atoms, $r_{ij} = |\mathbf{r}_j - \mathbf{r}_i|$, is sufficiently small, then the atoms must repel each other or matter would collapse. For larger separations, however, there must be a net attractive interaction between atoms or matter would not form solids and fluids at normal pressures.

## Interactions at short range

There are a number of ways that we can understand the repulsive nature of short-range interactions. Suppose we have two neutral atoms being brought together. Each atom consists of a positively charged nucleus and a distribution of elections. When the atoms are far enough apart that there is no overlap of the charge distributions, the net electrostatic interaction is zero, since the atoms are neutral. Once the atoms are close enough so that their charge distributions overlap, increasing the Coulomb repulsion between the ions.

The electrostatic interaction is not, however, the only origin of the repulsive interactions between atoms. There is a quantum mechanical effect that arises when the charge distributions overlap and the electrons are forced to occupy a smaller volume. From the discussion of the particle in a box in Appendix F.5.1, the energy scales as $L^{-2}$, where $L$ is the linear region in which the electrons are confined. As the volume in which the electrons are constrained decreases, the energy increases, leading to a repulsive interaction between the atoms. Since this effect arises from the need to maintain orthogonal wave functions (as in the particle in a box) and since only two electrons can be in an energy state, it is an example of the Pauli exclusion principle.

The density of electrons around an atom decreases exponentially with distance. The short-range interactions can be modeled with the general form

$$\phi_{SR}(r) = Ae^{-\alpha r}. \tag{5.3}$$

This energy represents a *repulsive* contribution to the energy, i.e., as the atoms are brought together the energy increases – there is a repulsive force between them.

## Nonbonding, nonionic interactions at long range

The analysis just given describes the interactions at short range, when the electronic distributions of the interacting atoms overlap. There are important interactions between atoms separated by distances long compared to the size of their electronic distributions, with the result that there is an additional attractive term that must be added to the short-range repulsive term described above. It arises from the fluctuations of the electronic clouds and is called the *dispersion energy*. It is also often referred to as the *van der Waals* energy, after the great Dutch physicist.

The *Drude* model for dispersion (van der Waals) forces is based on a model by London and described in detail in [146]. Other, more formal, methods for deriving these forces are discussed in many other places, including [159].

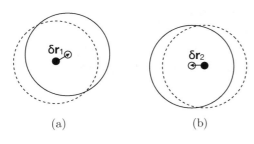

**Figure 5.3** A schematic view of dispersion interactions. The nuclei are shown as small filled circles surrounded by dashed circles that indicate the equilibrium distribution of valence electrons. The large solid circles represent the valence electron cloud centered (small open circles) a distance $\delta r$ away from the nucleus.

Electrons in an atom are not static – they fluctuate around their nucleus, destroying the spherical symmetry. These fluctuations can create a situation in which there may be more electrons on one side of the nucleus than the other, i.e., the fluctuations create an instantaneous dipole moment on the atom.[2] In Figure 5.3 we show a highly schematic representation of such fluctuations in which the nuclei are shown as small filled circles surrounded by dashed circles that indicate the equilibrium distribution of the valence electrons, where for simplicity we assume that all $N$ valence electrons are located in a spherical shell (with charge $-N$) centered on the nucleus (with charge $N$).[3] If the valence electrons move, either by a fluctuation or induced by an external field, their center of charge moves to the small open circles, located $\delta r$ away from the nucleus. The sphere of charge of the moved electron shell is shown as the solid large circles.

There must be, of course, a restoring force that binds the electrons to their nucleus. Since the fluctuations are small, a harmonic binding force can be used, with energy $1/2\,k\delta r^2$. The force constant is $k = N^2/\alpha$, where $N$ is the number of valence electrons and $\alpha$ is the polarizability of the atom. $\alpha$ is a measure of the tendency of a charge distribution to be distorted from its normal shape by an external electric field.[4] There is a quantum zero-point energy associated with the harmonic motion, with a value for an isolated atom of $3\hbar\omega/2$, where $\omega = \sqrt{k/m}$ and $m$ is the effective mass of the electrons.[5]

With two interacting atoms, the energy includes the zero-point motions of the shells as well as the electrostatic interactions. With no fluctuations, the electrostatic interactions would be zero – the net charge on each atom is 0.[6] When fluctuating, the instantaneous dipole on one atom interacts with the instantaneous dipole on the other. These dipole-dipole interactions change the motion of the fluctuating charges, leading to a change in the frequency of the fluctuations. This change in frequency leads to a change in the zero-point energy of the fluctuations, thus

---

[2] For a discussion of the electrostatic moments, please see Appendix E.3.

[3] The core electrons are assumed to be so tightly bound to the nucleus that their fluctuations are negligible.

[4] Suppose the atom is put in an external electric field $\mathbf{E}$. The outer shell containing $N$ electrons will be shifted from the nucleus by some displacement $\delta r$. It the shell is bound by a harmonic force, then the restoring force would be $-k\delta\mathbf{r}$, which must equal the total electrostatic force of $-N\mathbf{E}$ (force equals charge times field). Thus, $k\delta\mathbf{r} = -N\mathbf{E}$ or $\delta\mathbf{r} = -(N/k)\mathbf{E}$. This displacement produces a dipole moment (Eq. (E.13)) of $\mu = -N\delta\mathbf{r} = (N^2/k)\mathbf{E}$, which, from the definition of the polarizability $\alpha$, allows us to associate $\alpha = N^2/k$.

[5] From Eq. (F.26) the energy for a one-dimensional harmonic oscillator in the ground state ($n = 0$) is $\hbar\omega/2$. In three dimensions, there would be harmonic degrees of freedom in each direction, each with a contribution of $\hbar\omega/2$ to the zero-point energy.

[6] Assuming neutral atoms. Additional energetic terms from induced moments arise when ions are involved.

making a contribution to the interatomic interaction potential. Without presenting the formal derivation,[7] we find that for interacting, non-ionic, closed-shell systems, the leading term of the long-range attraction goes as

$$\phi_{vdw}(r) = -\frac{A}{r^6} \, , \tag{5.4}$$

where $A$ is some material-dependent constant.

$\phi_{vdw}$ in Eq. (5.4) goes by a number of names, as mentioned above. For this text, we will generally refer to it as the van der Waals energy. These interactions are very weak relative to other types of bonding. They are, however, extremely important for many systems, especially molecular systems, for which they often provide much of the binding energy.

### 5.2.2 A note on units

In Appendix A.2 we discuss various energy units and how one can convert between them. Units used to describe interatomic potentials depend in large part on the types of bonding the potentials are describing. For closed-shell rare-gas atoms, in which the energies are small, it is not uncommon to use Kelvin (K) to describe the energies, though you will also see erg, joules, and electron volt (eV) used. For metals, electron volts are the most common unit. Chemical units, such as kcal/mol or kJ/mol, are also sometimes used. The important thing is to be able to convert from whatever unit in which a potential is reported to one that is consistent with how you want to use it.

## 5.3 PAIR POTENTIALS

The potential functions that we describe next are models – approximations to the real interactions. They will generally be empirical, with parameters that must be determined in some way. In the end, the models may not even be particularly accurate. Thus, when one examines any calculations based on approximate potentials, it is good to treat them with at least some skepticism. So why study these functions? We will see that even when these potential functions are not particularly good representations of the real interactions between atoms, calculations based on them can shed light on important materials processes, providing understanding of materials structure and behavior not possible otherwise.

### 5.3.1 The Lennard-Jones potential

The Lennard-Jones (LJ) potential is commonly used because it is simple and yet provides a good description of central-force interatomic interactions [161]. While it was developed to describe the general interaction between closed-shell atoms and molecules, the Lennard-Jones potential has been used to model almost everything.

---

[7] For a more detailed derivation, see [146, 170].

## Constructing the Lennard-Jones model

The Lennard-Jones potential provides a good example of the procedure for model building described in Section 1.4:

(a) Following Figure 1.2, we first need to define the input and output. Our goal is to create a potential that will describe the interaction energy $\phi(r)$ (the output) between two spherical atoms and thus will depend (except for materials-specific parameters) only on the distance $r$ between them (the input).

(b) To create the model, the physical mechanisms that govern how the molecules interact with each other must be identified. We have discussed that at short range there must be a repulsive interaction between the atoms. At longer range, since matter is bound, there must be some sort of attractive interaction.

(c) Our goal is to find a generic form for a potential that can be applied at least *semi-quantitatively* to many materials. This requirement defines in an approximate way both how accurate the model must be and how its quality as a model can be assessed by comparing predictions based on the model to experimental data.

(d) We have already seen in Eq. (5.4) that for interacting, non-ionic, closed-shell systems, the leading term of the long-range attraction goes as $-1/r^6$. The long-range form of the potential is thus assumed to take the form $-A/r^6$, where $A$ is a constant that depends on the types of interacting atoms. When developed by Lennard-Jones, the form of the short-range repulsion was not known, though he assumed that it should should dominate the attractive part at short range and decay with $r$ faster than $r^{-6}$ at long range. Strictly for convenience, the short-range interaction was assumed to take the form $B/r^{12}$, where $B$ is again some constant specific to the types of interacting atoms. The short-range term mimics the exponential form discussed in Eq. (5.3). We will compare the two forms in Figure 5.6b. The choice for an exponent of 12 is rather arbitrary, but it leads to a very simple expression that is computationally convenient. The net potential has the form

$$\phi(r) = \frac{B}{r^{12}} - \frac{A}{r^6},\tag{5.5}$$

the input is $r$ and the output is $\phi$.

(e) Since the goal is to have a potential that is generic and easy to use, the Lennard-Jones potential is usually rewritten into a dimensionally more useful form that depends on two different parameters: $\sigma$, which is the distance at which the potential is zero, $\phi(\sigma) = 0$, and $\epsilon$, the absolute value of the minimum of the potential. With these parameters, the Lennard-Jones potential takes the form

$$\phi(r) = 4\epsilon\left(\left(\frac{\sigma}{r}\right)^{12} - \left(\frac{\sigma}{r}\right)^6\right).\tag{5.6}$$

The parameter sets are related by $\sigma = (B/A)^{1/6}$ and $\epsilon = A^2/4B$.

(f) In this and later chapters we will implement the Lennard-Jones potential into codes and use it to simulate various properties of materials. We will also examine how well the potential actually models real materials and will suggest improvements.

The Lennard-Jones potential is easy to characterize. For example, the point where the potential is zero is $r = \sigma$, i.e., $\phi(\sigma) = 0$. The position of the potential minimum $r_m$ is found by solving for the value of $r$ such that $d\phi/dr = 0$. The minimum is characterized by $r_m = 2^{1/6}\sigma$ and $\phi(r_m) = -\epsilon$. $\epsilon$ is usually referred to as the well depth. Thus, two parameters completely define the potential, with $\sigma$ having units of distance and $\epsilon$ having units of energy.

The Lennard-Jones potential is sometimes written in an equivalent form in terms of $r_m$ instead of $\sigma$

$$\phi(r) = \epsilon\left(\left(\frac{r_m}{r}\right)^{12} - 2\left(\frac{r_m}{r}\right)^6\right), \tag{5.7}$$

where, not surprisingly, $\sigma = 2^{-1/6}r_m$.

In this and later chapters it will be useful to express the Lennard-Jones potential in *scaled (or reduced) units*, in which the energy is expressed in units of $\epsilon$ and distances in units of $\sigma$. The reduced variables are written as $\phi^* = \phi/\epsilon$ and $r^* = r/\sigma$. With these definitions the Lennard-Jones potential becomes

$$\phi^*(r^*) = 4\left(\left(\frac{1}{r^*}\right)^{12} - \left(\frac{1}{r^*}\right)^6\right). \tag{5.8}$$

As we will discuss in detail in a discussion about molecular dynamics in Chapter 6, any calculation performed using the reduced potential $\phi^*$ is valid, with the appropriate substitution of parameters, for any interaction described by the Lennard-Jones potential, i.e., the calculation can be performed in reduced units and the final energies are found by scaling by the appropriate $\epsilon$ and the distances by the appropriate $\sigma$.

## Evaluating the Lennard-Jones model for gas-phase atoms

The prototypical systems described by the Lennard-Jones potential are the rare-gas atoms (He, Ar, ...). They are closed-shell, so no bonding takes place, and their long-range interactions are dominated by van der Waals forces. Thus, if the Lennard-Jones model should work well for any material, it should be for these.

In Figure 5.4 we compare the experimental interatomic potential between two Ar atoms with the best fit of that potential to a Lennard-Jones potential. The experimental curve was determined in a molecular beam scattering experiment [16]. Comparison of the experimental curve with the Lennard-Jones potential shows that the Lennard-Jones potential is somewhat too stiff (i.e., the repulsive wall is too steep) and the well is a little too shallow. While the shape of the Lennard-Jones potential is reasonable, it is not overly accurate, which is not surprising given its simplicity and reliance on just two parameters. Also, as discussed in Section 5.2.1, the actual form of the short-range potential is best modeled with an exponential. The $1/r^{12}$ form of the Lennard-Jones potential is not a particularly good representation of that form.

Table 5.1 **Values of the Lennard-Jones parameters for the rare gases [32]**

|  | Ne | Ar | Kr | Xe |
|---|---|---|---|---|
| $\epsilon$ (eV) | 0.0031 | 0.0104 | 0.0140 | 0.0200 |
| $\sigma$ (Å) | 2.74 | 3.40 | 3.65 | 3.98 |

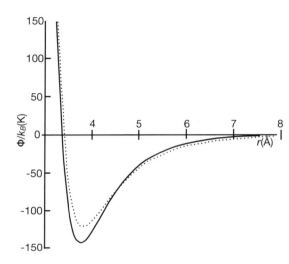

**Figure 5.4** Solid curve is the experimental Ar-Ar potential while the dashed curve is the best Lennard-Jones fit to that potential (adapted from [16]).

Based on scattering data (as well as other thermodynamic results), Lennard-Jones parameters have been determined for the interaction between like rare-gas atoms (i.e., He-He, Ar-Ar, . . .). These parameters are given in Table 5.1. The magnitude of the interactions is quite small, so we should not be surprised that solids consisting of these atoms are very weakly bound, showing, for example, very low melting points.

## Evaluating the Lennard-Jones model for solids

We can use the methods from Chapter 3 to calculate the zero K cohesive energy $U$ for solids with interactions described by the Lennard-Jones potential. Using Eq. (3.7), for example, we could sum the interactions, truncating at some cutoff distance, as shown in Figure 3.3. The equilibrium structure at 0 K is that which gives the lowest cohesive energy. For example, in a material that has a cubic lattice structure, there is only one lattice constant, $a$. Its equilibrium value, $a_o$, corresponds to the minimum in the potential energy, i.e., $a_o$ is the value of $a$ such that $\partial U/\partial a = 0$.

There is a class of systems, however, for which we do not need to do any summations at all; the sums have been evaluated by others and tabulated. Using special numerical methods, the lattice summation of potentials of the form $1/r^n$ for $n \geq 4$ has been performed for the basic perfect

Table 5.2 **Calculated and experimental properties of the rare gas solids.
Parameters taken from [15]**

|  |  | Ne | Ar | Kr | Xe |
|---|---|---|---|---|---|
| $r_o$ (Å) | Experiment | 3.13 | 3.75 | 3.99 | 4.33 |
|  | Theory | 2.99 | 3.71 | 3.98 | 4.34 |
| $u_o$ (eV/atom) | Experiment | −0.02 | −0.08 | −0.11 | −0.17 |
|  | Theory | −0.027 | −0.089 | −0.120 | −0.172 |
| $B_o$ (GPa) | Experiment | 1.1 | 2.7 | 3.5 | 3.6 |
|  | Theory | 1.8 | 3.2 | 3.5 | 3.8 |

cubic structures: face-centered cubic ($fcc$), body-centered cubic ($bcc$), and simple cubic. No cutoffs were assumed, so these sums include all possible interactions.

In Appendix 5.11 we show how we can use the analytical sums to develop an *exact* form for the 0 K structure, cohesive energy, and bulk modulus, $B = V(\partial^2 U/\partial V^2)$, for systems described by any potential that is based only on terms of the form $1/r^n$, for example the Lennard-Jones potential. We emphasize that these sums are available only for the perfect cubic lattices and potentials of the form $1/r^n$ for $n \geq 4$.

Using Eq. (5.40) and the potential parameters in Table 5.1, the zero-temperature cohesive properties of the rare-gas solids can be calculated based on the use of a Lennard-Jones potential. All of these materials form $fcc$ solids at equilibrium, as do all systems described by a Lennard-Jones potential. In Table 5.2, we show the calculated equilibrium nearest-neighbor distance, $r_o$, the equilibrium cohesive energy per atom, $u_o$, and equilibrium bulk modulus, $B_o$, for a series of rare-gas solids. We compare those values with low-temperature experimental results in Table 5.2.[8] All in all, the simple Lennard-Jones potential (with parameters determined for gas-phase interactions) does a remarkably good job in describing the basic properties of rare-gas solids, with the largest errors for Ne, which, being a very light atom, has a large zero-point motion, which tends to lead to larger lattices.[9]

In Figure 5.5a we show the dependence of the cohesive energy per atom on the volume per atom in scaled units, in which we also identify the position and value of the energy minimum, corresponding to the equilibrium states of the system given in Eq. (5.40). Note that these two parameters only identify a single point on the cohesive energy curve. The bulk modulus, on the other hand, is a measure of the *shape* of the cohesive energy at the equilibrium point and, as such, is a much more informative measure of how well a potential describes the interactions.

The cohesive energy curve $U(V)$ in Figure 5.5a was calculated at $T = 0$ K, which makes the thermodynamics simple. For example, at constant temperature and volume, the free energy

[8] Please see Appendix A for a discussion of units, including GPa (=$10^9$ Pascals).
[9] Zero-point motion is described in Appendix F.5.2.

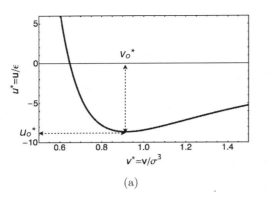
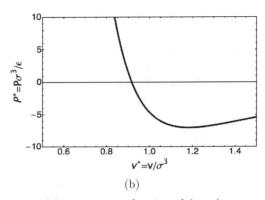

(a)                                    (b)

**Figure 5.5** (a) Cohesive energy per atom of the Lennard-Jones solid $u^* = u/\epsilon$ as a function of the volume per atom $v^* = v/\sigma^3$, from Eq. (5.39). $v_o^*$ and $u_o^*$ are the equilibrium values from Eq. (5.40). The bulk modulus is proprtional to the curvature of this curve at equilibrium. (b) The pressure for the Lennard-Jones solid at 0 K.

is the Helmholtz free energy $A$ and the pressure is $P = -(\partial A/\partial V)_{NT}$. At 0 K, $A = U$, and so the pressure can be calculated from Figure 5.5a as $P = -(\partial U/\partial V)_{NT}$. Based on Eq. (5.39), in which the analytic form for the cohesive energy as a function of the volume is given, the pressure was calculated and plotted in Figure 5.5b. Note that $P = 0$ at $v^* = v_o^*$ as expected. Once we have the pressure, the Gibbs free energy is $G = U + PV$ at 0 K.

From the agreement between experiment and calculations in Table 5.2, we see that the Lennard-Jones potential describes the shape of interactions between rare-gas atoms reasonably well, at least near equilibrium. The Lennard-Jones potential does much less well for systems under pressure, in which the nearest-neighbor atoms are forced up the repulsive wall of the potential. Moreover, the rare gas solids are not representative of most materials. The Lennard-Jones potential, while useful, does not describe the interactions between atoms in most materials very well at all, so other forms are needed if we want accurate calculations of the properties of materials.

### 5.3.2 The Mie potential

The Lennard-Jones potential has two parameters, $\epsilon$ and $\sigma$, that can be adjusted to improve the comparison between calculated quantities and experiment. Table 5.2 shows comparison with three quantities. We could adjust the two parameters in the Lennard-Jones potential to match two of those quantities, for example $r_o$, which depends only on $\sigma$, and $u_o$, which depends only on $\epsilon$. With only two parameters, we cannot necessarily fit $B_o$ as well.

Improvements can be made to the Lennard-Jones potential by adding additional parameters.[10] One approach would be to have the exponents in the $1/r^{12}$ and $1/r^6$ parts of the potential be adjustable parameters, i.e., instead of $r^{-12}$, perhaps $r^{-11}$ would be a better description of the

---

[10] Adding parameters corresponds to the "iterate and improve" phase of Figure 1.2.

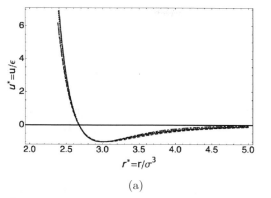
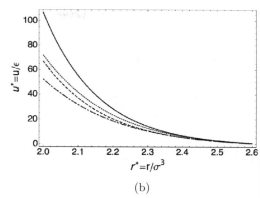

(a)                                     (b)

**Figure 5.6** Comparisons between pair potentials, all calculated with the potential minimum at $r_m = 3$ and the Lennard-Jones $\sigma = 2^{1/6}r_m$. The well depth was $\epsilon = 1$ for all cases. Shown are the Lennard-Jones potential (solid curve), the Mie potential with $m = 10$ and $n = 6$ (the long-dashed curve), the Morse potential with $\alpha = 2.118$ (dot-dashed curve), and the Born-Mayer potential with $\alpha = 4.88$ (dotted curve). (a) The well region. The potentials are all forced to have the same minimum and all go to zero at $r = \sigma$. (b) The short-range part of the potentials.

short-range interactions in some systems. A variation of the Lennard-Jones form, called the Mie potential (or the $mn$ Lennard-Jones potential), takes this approach, replacing $1/r^{12}$ with $1/r^m$ and $1/r^6$ with $1/r^n$. Writing the Mie potential as

$$\phi_{mn}(r) = \frac{\epsilon}{m-n}\left(\frac{m^m}{n^n}\right)^{\frac{1}{m-n}}\left(\left(\frac{\sigma}{r}\right)^m - \left(\frac{\sigma}{r}\right)^n\right) \tag{5.9}$$

yields an equivalent form to the Lennard-Jones potential. The Mie potential thus has four parameters: $\sigma$, $\epsilon$, $m$, and $n$. The parameters $\sigma$ and $\epsilon$ have the same meaning as in the standard Lennard-Jones potential: $\phi_{mn}(\sigma) = 0$ and $\phi_{mn}(r_m) = -\epsilon$, where the minimum is located at

$$r_m = (m/n)^{1/(m-n)}\sigma . \tag{5.10}$$

If $m = 12$ and $n = 6$, the standard Lennard-Jones potential is obtained.[11] The $mn$ potential yields somewhat better results for some systems, especially those with repulsive potentials that are less steep than the $1/r^{12}$ term in the standard Lennard-Jones potential. We compare the Mie potential with others in Figure 5.6.

The lattice sums in Appendix 5.11 of this chapter for cubic, body-centered cubic and face-centered cubic lattices are tabulated for any potentials of the form $1/r^n$. Thus, they are equally applicable to the Mie potentials as to the standard Lennard-Jones potential. For example, in Eq. (5.41) expressions are given for the equilibrium cohesive properties for systems described by the Mie potential.

---

[11] It is useful to rewrite the Mie potential in Eq. (5.9) in terms of $r_m$ instead of $\sigma$. What is the relation between $\sigma$ and $r_m$?

### 5.3.3 Other pair potentials

The basic form of any pair interaction will be similar to that of the Lennard-Jones potential, with a short-range repulsion and an attractive well at long range. Like the Lennard-Jones form, these potentials best describe systems made up of closed-shell atoms, ions, and molecules. Here we describe two of the more commonly used forms.

The *Born-Mayer* potential, which is sometimes referred to as the exponential-6 (or *exp-6*) potential, is

$$\phi(r) = Ae^{-\alpha r} - \frac{C}{r^6}, \tag{5.11}$$

where $A$, $C$, and $\alpha$ are positive constants that depend on the identities of the interacting atoms. $\alpha$, with units of 1/length, governs the steepness of the repulsive wall. An advantage over the Lennard-Jones potential is that the exponential function that describes the repulsive part of the potential matches what we expect from the electronic distributions of the interacting atoms (Eq. (5.3)) and is thus more physically reasonable than the very repulsive $1/r^{12}$ term in the Lennard-Jones potential. The long-range part of the potential is the van der Waals $-1/r^6$ attraction. The Born-Mayer potential is often used for ionic systems, where the electrostatic interactions between the ions are included, as discussed in Section 5.4.

At times, it is convenient to write the *exp-6* potential as

$$\phi(r) = \frac{\epsilon}{\alpha r_m - 6} \left\{ 6e^{-\alpha(r-r_m)} - \alpha r_m \left( \frac{r_m}{r} \right)^6 \right\}, \tag{5.12}$$

where the position of the minimum is $r_m$ and the well depth is $\epsilon$. The parameters in Eq. (5.11) are related to $r_m$, $\epsilon$ and $\alpha$ through the relations $A = 6\epsilon \exp(\alpha r_m)/(\alpha r_m - 6)$ and $C = \alpha r_m^7 \epsilon /(\alpha r_m - 6)$. Note that this form is restricted to values of $r_m$ and $\alpha$ such that $\alpha r_m - 6 > 0$.

Arising from a simple theory of diatomic bond potentials, the *Morse potential* takes the form

$$\phi(r) = \epsilon \left\{ e^{-2\alpha(r-r_m)} - 2e^{-\alpha(r-r_m)} \right\}, \tag{5.13}$$

where, as usual, $r_m$ is the position of the minimum, $\epsilon$ is the well depth, and $\alpha$ governs the shape of the potential. The Morse potential has a very soft repulsive wall and is sometimes used to model interatomic interactions in metals. Note that the interaction at long range falls off much more quickly than the van der Waals energy of Eq. (5.4).

In Figure 5.6 we compare the basic pair potentials: the Lennard-Jones potential (Eq. (5.6)), the Mie potential with $m = 10$ and $n = 6$ (Eq. (5.9)), the Born-Mayer potential (Eq. (5.11)), and the Morse potential (Eq. (5.13)). In all cases, the minimum was set at $r_m = 3$, with the well depth $\epsilon = 1$. To make the comparison more clear, the $\alpha$ parameters in both the Born-Mayer and Morse potentials were adjusted until the value of $r$ for which $\phi(r) = 0$ matched the $\sigma$ value in the Lennard-Jones potential, $\sigma = r_m/2^{1/6}$.[12] Given the constraints placed on the potentials such that their minimum and $\sigma$ values all agree, it is not surprising that in Figure 5.6a, which

---

[12] The relation of the $\sigma$ value to $r_m$ is slightly different for the Mie potential, being given by Eq. (5.10), where we find that $\sigma = 0.88r_m$ for the 10-6 Mie potential, as opposed to $0.89r_m$ for the Lennard-Jones potential.

focuses on the well region, we do not see significant differences between the potentials. The potentials are very different, however, at shorter ranges, as shown in Figure 5.6b. In order of "steepness", from most steep to least, the potentials are: Lennard-Jones, the Born-Mayer potential, the Mie 10-6 potential, and the Morse potential. The shape of this part of the potential affects the curvature at the well (and, thus, properties such as the bulk modulus) as well as the high-pressure properties, in which the atoms are squeezed close together. In general, the Lennard-Jones potential is too stiff, so one of the others is likely to be more appropriate for most applications. How one chooses the form and the parameters is discussed in Section 5.9.[13]

### 5.3.4 Central-force potentials and the properties of solids

Interatomic potentials that are functions only on the distance between the pairs of atoms are referred to as *central-force potentials*. All the potentials in this section are of that form. There are limitations, however, to the kind of materials that can accurately be described by central-force potentials. For example, perfect crystals with a single component whose interactions are described by central-force potentials will always take on simple structures like face-centered cubic, body-centered cubic, etc. Thus, central-force potentials cannot be used to model pure systems with more complicated structures.

There is also experimental evidence that using a central-force potential cannot provide an adequate description of the elastic behavior of solids. For example, as discussed in Appendix H.4, any deviation from the Cauchy relation $c_{12} = c_{44}$ for systems with symmetric lattice positions is a measure of the deviation from central-force interactions in the material.[14] One way to express that deviation is to consider the ratio of the elastic constants $c_{12}/c_{44}$, which should equal 1 for systems that interact only through central forces. In Table 5.3 we give experimental values for $c_{12}/c_{44}$ for a number of materials.

From Table 5.3, we see that a material whose interactions are described with the Lennard-Jones potential satisfies the Cauchy relation, as expected. A simple rare-gas solid, argon, also shows very small deviations from $c_{12}/c_{44} = 1$, indicating that a central-force pair potential description should also be effective in describing the elastic properties. Of the other materials in Table 5.3, only for NaCl does the Cauchy relation hold, with large deviations for metals (with $c_{12}/c_{44} > 1$) and for covalently bonded materials (with $c_{12}/c_{44} < 1$).[15] Thus, to accurately model most materials will generally require more complex potentials than those described in this section.

---

[13] It may be useful for the reader to compare the basic forms of pair potentials, for example: (a) plot the Mie potential for a series of values of $m$ with $n = 6$. Compare the repulsive part of the potential. Repeat for a series of values for $n$ with $m = 12$. (b) Compare the Lennard-Jones potential (LJ) with the Born-Mayer potential in Eq. (5.12) for a series of values of $\alpha$ (keeping $r_m$ the same for both potentials). (c) Repeat using the Morse potential of Eq. (5.13).

[14] $c_{12} = \lambda$ and $c_{44} = \mu$ are elastic constants, as discussed in Appendix H.

[15] The relation $c_{12}/c_{44} < 1$ for covalently bonded materials can be readily explained by a consideration of the bonding, as described in [334].

Table 5.3 **Comparison of values for $c_{12}/c_{44}$ for a number of materials. The value from calculations on a Lennard-Jones (LJ) potential are from [267], the data for Ar is from [168], while all other data are from Appendix 1 of [147]**

| Material | $c_{12}/c_{44}$ |
|----------|-----------------|
| "LJ"     | 1.00            |
| Ar       | 1.12            |
| Mo       | 1.54            |
| Cu       | 1.94            |
| Au       | 4.71            |
| NaCl     | 0.99            |
| Si       | 0.77            |
| MgO      | 0.53            |
| diamond  | 0.16            |

## 5.4   IONIC MATERIALS

Ionic solids typically consist of closed-shell ions with little charge in the interstitial regions (Figure 5.2). The simplest approach to describe the interactions between two ions, $i$ and $j$, in an ionic material is to start with a simple pair potential, $\phi_{ij}$, which could be any of the forms presented in this chapter. We add to that potential the electrostatic, or Coulomb, interaction between the two ions

$$k\frac{q_i q_j}{r_{ij}}, \tag{5.14}$$

where $r_{ij}$ is the distance between the two ions and $q_i$ and $q_j$ are the charges on ion $i$ and $j$, respectively. $k$ is a parameter whose value depends on the units, as discussed in Appendix E.1.[16] The net cohesive energy is

$$U = \frac{1}{2}\sum_{i=1}^{N}\sum_{j=1}^{N}{}' \left( \phi_{ij}(r_{ij}) + k\frac{q_i q_j}{r_{ij}} \right). \tag{5.15}$$

The sum of the Coulomb terms in Eq. (5.15) involves long-ranged interactions and requires special methods, which are discussed in Section 3.6.

---

[16] A discussion of the units and form of the Coulomb potential is given in Appendix E.

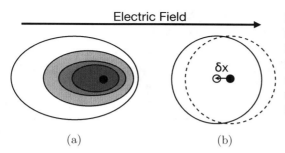

**Figure 5.7** (a) Schematic (exaggerated) view of distortion of the valence electron charge in an atom in an electric field. The small black circle represents the nucleus plus core electrons. (b) The shell model representation of (a). A spherical charge whose center is held by a harmonic potential (spring) to the nucleus/core electrons.

Since all ionic materials consist of at least two types of atoms, at least three distinct pair potentials, $\phi_{ij}$, will be needed in Eq. (5.15). For example, in NaCl, potentials are needed to describe the interactions between pairs of $Na^+$ ions, between pairs of $Cl^-$ ions, and between pairs consisting of $Na^+$ and $Cl^-$ ions. A common choice for $\phi_{ij}$ is the Born-Mayer potential in Eq. (5.11).

Equation (5.15) misses, however, an important bit of physics. Unlike solids made up of neutral atoms, ionic systems have large electric fields that can cause significant distortions of the electronic distributions around the ions (polarization) as shown schematically in Figure 5.7a. For perfect, highly symmetric crystal structures, these distortions may not matter, as the electric fields are essentially radial. However, as the ions in the lattice vibrate, or if they are in asymmetric positions (e.g., near defects), field gradients can be formed and distortions of the local electronic structure can be large. Since these distortions change with structure, a fixed potential cannot model changes in atom position very well, including lattice vibrations (phonons). The goal is then to find a way to account for these changes in electronic structure without sacrificing (too much) the simplicity of using the pair potential. The result is a commonly used approach called the *shell model* [90].

The shell model is based on the same simple harmonic model of the *polarizability* discussed in Section 5.2.1 to describe van der Waals interactions. In Figure 5.7a we show an exaggerated distortion of an electronic distribution around an atom in an electric field. Since the nucleus plus core electrons (the small black circle) has a positive charge and the electrons are no longer spherically distributed around them, a dipole moment will be created, where the dipole moment is $\mu = \alpha\mathbf{E}$. In the shell model shown in Figure 5.7b, an atom is represented as a massive core (nucleus plus core electrons) connected to a massless shell of charge $-N$ such that the total charge on the ion (core plus shell) is $q$, and the expected total charge is $q - N$ (e.g., $-1$ for $Cl^-$). The the center of the shell is connected to the core by a spring with force constant $k$, as shown in Figure 5.7. The force constant is related to the polarizability, $\alpha$, of the ion by $\alpha = N^2/k$.[17]

The total cohesive energy in the shell model is Eq. (5.15) plus a term that represents the energy to distort the shell by an amount $\delta x$, which we assume is the harmonic energy $(1/2)k\,\delta x^2$. It is important to note, however, that distortion of the shell changes the interaction between nearby atoms, as the core electrons and the shell electrons are no longer centered at the same place. The Coulomb energy must thus be partitioned into terms corresponding to core-core, core-shell,

---

[17] See footnote 4 in this chapter.

and shell-shell interactions and a repulsive pair potential is assumed to act only between the shells (i.e., they are the outermost electrons). The parameters in the pair potential are usually determined by fitting calculated properties to experimental data. Since anions are usually much more tightly bound than the larger, negatively charged, cations, they are usually modeled without including a shell. Recent comparison with accurate electronic structure calculations indicates that inclusion of the shell model provides a more accurate description than with a "rigid ion" model [313]. More details are given in [144].

## 5.5 METALS

Metals consist of cores (nuclei plus the core electrons) surrounded by electrons that are delocalized throughout the solid (as in Figure 5.2c). Simple pair interactions do not provide a particularly good model for the binding and properties of metals. Other forms of potentials that attempt to build in some aspects of the background electron distribution have been developed and some have proven remarkably adept at describing the properties of metallic systems.

### 5.5.1 Pair potentials for metals

Based on the picture of metallic bonding given in Figure 5.2c, it would be surprising if a pair potential description of the interatomic interactions could adequately describe the thermodynamic properties of both the ions (the nuclei) and the background gas of electrons. We would expect there to be a contribution from the interaction of the ions, mediated by the electron gas, as well as a contribution from the electron gas itself. If nothing else, we might expect there to be a volume dependence to the energy of the electron gas with the same general form as seen in the particle in a box discussed in Appendix F.5.1.

Consider the question of how well a pair potential can actually capture the bonding properties of a metal, starting with the Lennard-Jones potential. The $\sigma$ and $\epsilon$ values can be chosen such that the calculated equilibrium cohesive energy (at 0 K) and lattice length match experiment, as is discussed above. For the rare-gas solids, the calculated bulk modulus, which is a measure of the curvature of the potential at the minimum, was in reasonable agreement with experiment. For metals however, the calculated bulk moduli show considerable disparities relative to experimental results, indicating that the shape of the cohesive energy near its minimum is poorly captured with a Lennard-Jones potential. Thus, finite-temperature properties, in which atoms vibrate around their equilibrium positions and explore more of the potential surface, will not be well described. Calculated results for metal properties can be improved somewhat by using a Mie potential with the exponent of the short-range interaction term being smaller than $m = 12$. Using $m = 10$, for example, yields somewhat more accurate predictions of the thermodynamics, but the shape of the short-range part of the potential surface remains too repulsive and the bulk modulus and dependence on pressure are usually still too large. A softer potential, involving an exponential repulsive wall, is generally preferred. One model that works as well as, or better than, most is the Morse potential of Eq. (5.13). There does not seem to be any justification for this choice other than it yields somewhat better agreement with experiment for the elastic properties (e.g., the bulk modulus) of metals, though, as discussed in Section 5.3.4, no central-force potential can be expected to accurately capture all of the elastic properties. Application of simple pair potentials to the properties of defects in metals is also problematic. Even if a pair potential could work well for a perfect solid, they are less likely to be able to describe the properties of defects, at which the electronic density may be modified from that in the uniform solid, as well.

Since pair potentials typically do not provide a good description of the true nature of the metallic bonding, they cannot generally be used to provide quantitative predictions of metal properties.

## 5.5.2 Volume-dependent potentials

A pair potential description of the potential energy of metals neglects the energy associated with the delocalized electrons. The energy of this "electron gas" depends in some way on the average electron density. Thus, adding a (as yet unspecified) function of the electron density to a sum of pair terms seems to be a reasonable approach to improve the description of a metal. Since determining the electron density requires an expensive electronic-structure calculation, an approximate form is often sought. Remembering the difficulties in fitting the bulk modulus along with the equilibrium structure and energy, it would seem that adding a volume-dependent term to a sum of pair potentials might be a reasonable solution. Given that electron density is clearly a function of the volume of the system $V$, a form such as

$$U = \frac{1}{2} \sum_{i=1}^{N} \sum_{j=1}^{N}{}' \phi(r_{ij}) + U_{eg}(V),$$

(5.16)

where $U_{eq}(V)$ is a volume-dependent potential, is often used.

There could be a number of choices for $U_{eg}$. A simple one is to use

$$U_{eg}(V) = U_{known}(V) - \frac{1}{2} \sum_{i=1}^{N} \sum_{j=1}^{N}{}' \phi(r_{ij})$$

(5.17)

where $U_{known}(V)$ is a known volume-dependent cohesive energy from some source.[18] $U_{eg}(V)$ is calculated for a perfect system (thus ensuring that volume dependence of the cohesive energy is correct in that limit), with the expectation that the potential would be applied to other material properties, such as defects.

Use of a volume-dependent potential yields good agreement with bulk, volume-dependent, properties but offers no advantages over simple pair potentials for describing many other important materials properties, especially the energy of defects, such as grain boundaries, surfaces, dislocations, etc. For one, if there are no changes in volume, $U_{eg}$ does not come into play at all. More importantly, a term that depends only on bulk volume represents the energy of a uniform electron gas. In a solid near a defect, however, the electronic density is considerably

---

[18] For example, the Universal Binding Curve of the next section.

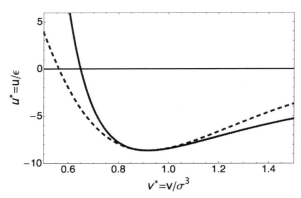

**Figure 5.8** Comparison of the cohesive energy from the Universal Binding Curve (dashed) to that of the Lennard-Jones solid (solid curve) from Figure 5.5a. The UBC curve was determined by fixing the equilibrium values for the volume, energy, and bulk modulus to that of the Lennard-Jones potential.

modified from that in the uniform solid. To describe the properties of defects properly requires a model that incorporates, in some way, a change in electron density near the defect.

### 5.5.3 Universal Binding Curve

The basic cohesive properties of many metals can be reasonably well described with an empirical relation, the so-called *Universal Binding Curve* (UBC) of Rose *et al.* [277]. Note that this is not a potential – it is an empirical fit to data from which one can approximate the total energy of a metallic solid. Foiles fit the energy-volume curves for $fcc$ metallic systems with the form [105]

$$U_{UBC} = -E_{sub}(1 + a^*)e^{-a^*}, \tag{5.18}$$

where $E_{sub}$ is the absolute value of the sublimation energy at zero pressure and temperature (i.e., the cohesive energy), and $a^*$ is a measure of the deviation from the equilibrium lattice constant $a$,

$$a^* = \left(\frac{a}{a_o} - 1\right)\Bigg/\left(\frac{E_{sub}}{9Bv}\right)^{\frac{1}{2}}, \tag{5.19}$$

where $B$ is the equilibrium bulk modulus ($B = V(\partial^2 U/\partial V^2)_{NT}$) and $v$ is the equilibrium volume per atom. $a$ is the $fcc$ lattice constant and $a_o$ the equilibrium (zero pressure and temperature) lattice constant. While $U_{UBC}$ is not exact, it does represent the binding curves for many materials very well [277], in part because it includes information about the shape of the potential through the direct inclusion of the bulk modulus. Note that since the pressure can be found from standard thermodynamics through $P = -(\partial U/\partial V)_{NT}$, Eq. (5.18) can also be used to determine the pressure-volume relation at 0 K. The UBC can be used very effectively as the known volume-dependent term in Eq. (5.17).

In Figure 5.8 we compare the Universal Binding Curve from Eq. (5.18) and Eq. (5.19) with the cohesive energy curve calculated for an $fcc$ solid described by the Lennard-Jones potential. For ease of comparison, we fixed the equilibrium values for the volume, energy, and bulk modulus in the UBC to the Lennard-Jones values. The UBC curve is much more shallow than that for a Lennard-Jones solid, even when the equilibrium bulk modulus was used. For metals, with a much smaller bulk modulus than that found with a Lennard-Jones potential, the UBC curve would be very shallow.[19]

### 5.5.4 Embedded-atom model potentials (EAM)

As we have seen, pair potentials cannot adequately describe the energetics of the distributed electrons in metals. Adding a volume-dependent term, while enabling more accurate equation of state calculations, cannot reflect any changes in local bonding for asymmetric sites, for example near grain boundaries or dislocations. A number of models have been developed that can, in principle, treat defect properties of metals more correctly. The idea is to add to a sum of pair potentials an energy functional of the local electron density.[20] Since the electronic density reflects the atomic positions, it varies throughout the system and reflects the local atom distributions. The hope is that these approaches will model the properties of metals better than methods that depend only on the average electronic density. While there are many variants of this type of potential, they are usually referred to as *embedded-atom model* (EAM) potentials after the first of their kind [78, 79].

Embedded-atom model potentials are motivated by density functional theory and have the general form

$$U = \sum_i F_i \left[ \sum_{j \neq i} f_{ij}(r_{ij}) \right] + \frac{1}{2} \sum_{i=1}^{N} \sum_{j=1}^{N}{}' \phi_{ij}(r_{ij}), \qquad (5.20)$$

where $f$ is some function of the interatomic distance representing an approximation of the electron density, and $\phi$ is a pair potential. The exact form of these functions depends on which model one uses. Each model arises from different assumptions and derivations and has different functional forms [106].

$F$ is a functional of a sum over functions that depend on the local positions of the atoms. As we shall see, $F$ is a non-linear function. Therefore, $F$ cannot be written as a sum of pair potentials and thus represents a true many-body interaction. We note that while the forms of the various embedded-atom model potentials are based on theoretical aspects of the binding of metals, most of them are used in an empirical fashion, with parameters fit by comparing calculated to experimental properties of the materials of interest.

To make the discussion a bit more specific, we will describe in some detail the original embedded-atom model (EAM) [80]. Our goal is to show the basic physics assumed in the model and we will not attempt to describe the many variations of the model that exist in the literature.

---

[19] Copper has the following material parameters: $E_{sub} = 4.566 \times 10^{-12}$ ergs, $a_o = 2.892$ Å, and $B = 9.99 \times 10^{11}$ ergs/cm$^3$. (a) Determine values for $\epsilon$ and $\sigma$ in a Lennard-Jones potential that matches as best you can these properties. (b) Plot the Universal Binding Curve Eq. (5.18) for copper, using the parameters: $E_{sub} = 4.566 \times 10^{-12}$ ergs, $a_o = 2.892$ Å, and $B = 9.99 \times 10^{11}$ ergs/cm$^3$. (c) Compare the UBC to a cohesive energy curve based on the Lennard-Jones potential for copper.

[20] A functional is a function of a function; cf. Appendix C.6.

In the EAM potential, the $F$ functional of Eq. (5.20) is chosen to represent the energy to embed an atom $i$ in a uniform electron gas of density $\bar{\rho}_i$. $\bar{\rho}_i$ is the local electronic density evaluated at the site of atom $i$ and is approximated as a sum of contributions of the electron densities $\rho_j(r_{ij})$ from atoms that neighbor $i$, i.e.,

$$\bar{\rho}_i = \sum_{j \neq i} \rho_j(r_{ij}), \tag{5.21}$$

where the notation $\sum_{j \neq i}$ indicates that we sum over neighbors of atom $i$ but not atom $i$ itself. Thus the form of the EAM potential is

$$U_{EAM} = \sum_i F_i(\bar{\rho}_i) + \frac{1}{2} \sum_{i=1}^{N} \sum_{j=1}^{N} {}' \phi_{ij}(r_{ij}). \tag{5.22}$$

In practice, the various versions of the EAM potential typically have different choices for the functions $F$, $\rho_j$ and $\phi$. Generally, these functions depend on parameters, which are usually determined by fitting[21] predictions of the model to experimental or calculated values (e.g., for cohesive energies, defect energies, etc.) [80]. For example, consider the pair potential $\phi$. In the original formulation of the EAM, a purely repulsive potential based on the electrostatic interactions between the ions in the solid was used. Later workers employed a Morse potential [328]. For the atomic electron densities $\rho_j$, the original EAM assumed an electron density for isolated (i.e., gas phase) atoms as determined from electronic structure calculations. Others assumed a set of parameterized hydrogenic orbitals [328]. Other choices have also been used for both of these functions. One way to choose the embedding function $F$ is to require that the overall cohesive energy curve for a perfect crystal agree with the Universal Binding Curve of Eq. (5.18). Other variations of EAM use a parameterized function for $F$.[22]

To give a concrete example, we show in Figure 5.9 a flow chart of how to construct a typical EAM potential for Ni based on the use of the Universal Binding Curve to define $F$. The first step is to choose functions for the pair potential $\phi$ and the atomic electron density function $\rho$. Assume, for example, that $\phi$ is a Morse potential and that $\rho$ represents the electronic density of a 4s hydrogen orbital, $\rho(r) = r^6 e^{-\beta r}$.[23] There are three parameters in $\phi$ ($\epsilon$, $\alpha$ and $r_m$) and one parameter in $\rho$ ($\beta$). Initial values will be chosen for these parameters based on known electronic densities (from atomic calculations) and the lattice parameter and cohesive energy for Ni.

The pair potential contribution to the energy is

$$U_p = \frac{1}{2} \sum_{i=1}^{N} \sum_{j=1}^{N} {}' \phi_{ij}(r_{ij}), \tag{5.23}$$

[21] See Section 5.9 for a discussion of fitting parameters.
[22] Baskes introduced a simple, completely analytical, form for the embedded-atom model in Eq. (5.22) [24]. The pair potential was a Lennard-Jones potential, the atomic electronic densities were written as $\rho(r) = \exp[-\beta(r - 1)]$, and the embedding function was $A\bar{\rho}[\ln(\bar{\rho}) - 1]$. Only nearest neighbors were included.
[23] This form of the density was used in [332] in a study of NiAl, though they added an additional term to ensure that $\rho(r)$ decreases monotonically throughout the range of interaction distances.

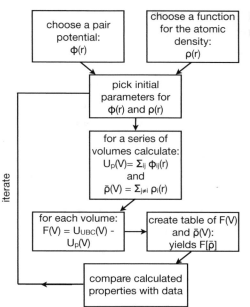

**Figure 5.9** How to construct a typical embedded-atom model potential.

which can be evaluated using the methods described in Chapter 3 for a range of values for the lattice parameter, thus creating a numerical tabulation of $U_p(V)$, where $V$ is the volume. During the same calculation, the embedding density at each site $\bar{\rho}$ will be calculated using Eq. (5.21), thus yielding a numerical tabulation of $\bar{\rho}(V)$.

For each volume entry in the table, we thus have $U_p$ and $\bar{\rho}$. We define $F$ as the energy needed to ensure that the total cohesive energy as a function of volume matches the Universal Binding Curve of Section 5.5.3, which is accomplished by defining $F$ *numerically* as

$$F(V) = U_{UBC}(V) - U_p(V). \qquad (5.24)$$

Using Eq. (5.24), $F$ can then be calculated for each volume. Thus, we will have created a table that includes $F$ and $\bar{\rho}$ at each volume, establishing the numerical functional $F[\bar{\rho}]$.

Based on this potential, we can then calculate a variety of properties. Since we have tied the overall potential to the Universal Binding Curve, the basic cohesive properties as a function of volume are set. Typically, EAM potentials are used to calculate defect properties, in which the atomic positions are not symmetric. Thus, one could also evaluate defect energies for comparison with experiment, such as the surface structure and energy, the vacancy formation energy, etc. Elastic constants are a measure of the shape of the potential surface, so also provide important tests of any model.[24] The parameters in the model can be determined by adjusting them until calculated properties fit the experimental values as closely as possible, which is the iteration step in Figure 5.9.

Development of EAM potentials for alloy systems is somewhat more complicated. The parameters for the pure solids are typically used to describe interactions between the same

---

[24] Elastic constants are defined in Appendix H.2.

atoms in an alloy, i.e. in NiAl the parameters for pure Ni define the Ni–Ni interactions while those for pure Al define the Al–Al interactions. To obtain parameters for the cross potentials requires some additional information and approximations [328].

Because the potentials in these cases are spherically symmetric, the embedded-atom model potentials work reasonably well for metals with completely full or completely empty $d$ bands. The pair functional methods as described do not generally give an accurate description of the bonding for materials with angular components in their interactions, since they do not build in any directionality in the electron densities.

A major limitation to EAM-type potentials is that they cannot reflect dynamic changes in the bonding that might arise from a changing local environment. For example, there may be charge transfer between unlike atoms in alloys or between atoms near defects that could change dynamically as the atoms move. Embedded-atom model potentials cannot handle such effects directly. Finnis and Sinclair introduced a type of potential similar in many ways to the embedded-atom model described in this section. Finnis-Sinclair potentials are based on the bond-order potentials described in Section 5.6.2 [101].

As described here, the embedded-atom model includes no angular terms and thus cannot adequately model covalently bonded materials. In the modified embedded-atom-model (MEAM) potential [23], the same basic form of the EAM potential is employed, but the electron densities $\rho_j$ are written in terms of functions with an angular dependence. The parameters in the density are then included in the fits to various material parameters. MEAM potentials have been used to model many types of covalent, or partial covalent, materials [169]. This topic is covered further in Section 5.7.

# 5.6 COVALENT SOLIDS

In covalent materials, such as silicon and carbon, the bonding is strongly localized in specific directions and interatomic potentials must reflect this strong directionality, as shown schematically in Figure 5.2d. A number of approaches have been suggested to describe these interactions, though only a few representative examples will be discussed here.

We will use silicon as our principal example. Not only is silicon of great technological interest, but it also has many structures, each with its own types of bonding. For example, for systems with only three Si atoms, the most stable form is a molecule with the angle between the atoms being 180°. For four Si atoms the angle is 120°, while for five Si atoms, the most stable form is a tetrahedron with an angle of 109°. The low-pressure form of solid Si is in the diamond structure, consisting of tetrahedral bonding, but other structures exist at higher pressures. Amorphous silicon is of increasing importance and consists of a continuous random network with not all atoms being 4-fold coordinated. While we could solve directly for the electronic structure using the methods of Chapter 4, the sizes of systems that can be studied with these methods are somewhat limited.

Developing a model that can capture a wide range of bonding types seen in silicon, and similar materials such as carbon, is challenging, especially if the goal is to describe the dynamic changes

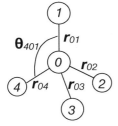

**Figure 5.10** Schematic view of the local tetrahedral bonding in solid Si.

in bonding that occur as structures transform. Some of the methods and models discussed next are sufficiently flexible to describe such phase transformations, while other models are limited to applications to specific structures.

## 5.6.1 Angular-dependent potentials

Perhaps the most obvious way to include angular-dependent potentials is to build them directly into the functional forms. As an example, consider a well-known potential for silicon, the Stillinger-Weber potential, that was designed specifically to model the tetrahedral form of the solid [296]. The authors directly built into the potential the basic structure of the solid: each Si atom is surrounded by four other Si atoms in the shape of a tetrahedron, with the equilibrium angle between the bonds set to the standard $109.47°$ for a tetrahedron, as shown in Figure 5.10.

The interaction between an atom and its four nearest neighbors was described based on the four vectors $\mathbf{r}_1$, $\mathbf{r}_2$, $\mathbf{r}_3$, and $\mathbf{r}_4$ connecting the atom to its neighbors, as indicated in Figure 5.10. The authors included a pair potential between the atoms that depended only on the distance to the neighbors. They also included terms that depend on the angles between the bonds ($\theta_{jik}$), which can be easily determined from $\mathbf{r}_{ij} \cdot \mathbf{r}_{ik} = 2r_{ij}r_{ik} \cos \theta_{jim}$. The challenge, however, is that to determine the angles requires consideration of three atoms, the atom at the center and the vector to two of its neighbors. This type of interaction is called a three-body potential. A computational challenge is that there are many more three-body terms than pair terms. All of the terms are taken as simple functions, whose parameters were found by fitting to experiment.[25]

While these types of potentials can provide a very accurate description of the properties of the material and structure for which it is created, they are typically limited to those materials. The Stillinger-Weber potential works well for the perfect tetrahedral solid but not at all well for the liquid, nor for any other solid structures. Thus, the ability to model such important materials properties as phase transformations, defects, etc. is limited. Other, more flexible, potentials have been developed that make such calculations possible.

---

[25] For example, the energy associated with an angle between atoms 0, 1, and 2 in Figure 5.10 was written as $h(r_{01}, r_{02}, \theta_{102}) = Af(r_{01}, r_{02})(\cos \theta_{102} + 1/3)^2$, where $A$ is a constant and $f(r_{01}, r_{02})$ is some function of the distances between the atom pairs. The angular term is a simple harmonic function centered on $\theta_{102} = 109.47°$ ($\cos 109.47 = -1/3$).

## 5.6.2 Bond-order potentials

*Bond-order potentials* provide a flexible, highly adaptable, method for describing dynamic, complex bonding by incorporating in an approximate way the local coordination around an atom [259]. The general expression for the energy of each atom in a periodic system as a function of the nearest-neighbor separation $r$ is [1]

$$E_i = \frac{1}{2} \sum_{j=1}^{Z_i} \left[ q V_R(r) + b V_A(r) \right], \tag{5.25}$$

where $q$ is a parameter that depends on the local electronic density, $V_R$ and $V_A$ are repulsive and attractive interactions, respectively, and $Z$ is the number of nearest neighbors. The total potential energy is a sum of the atomic energies

$$U = \sum_i E_i. \tag{5.26}$$

The $b$ parameter in Eq. (5.25) is the *bond order*, which controls the strength of the chemical bond. For example, if a carbon-carbon bond has a bond order of three, as in a triple bond, the bond is stronger than a bond of order 2. To represent the bonding in a solid, it seems reasonable to assume that the strength of the bonding between nearest-neighbor atoms should decrease with the number of nearest neighbors, $Z$, as there are only so many electrons that can be distributed in the bonds. Abell assumed that $b = cZ^{-1/2}$, where $c$ is a parameter. With this choice of $b$, the bond-order potential takes the form [1]

$$E_i = \frac{1}{2} \sum_{j=1}^{Z_i} \left[ ae^{-\alpha r} - \frac{c}{Z^{1/2}} e^{-\gamma r} \right]. \tag{5.27}$$

At first glance Eq. (5.27) seems to be nothing more than a sum of pair potentials. The difference arises from the inclusion of the $Z^{-1/2}$ term. $Z$ is a measure of the number of neighbors within some prescribed distance. The more neighbors, the smaller is $b$, and the "amount" of bonding between atoms is less. Thus, the attractive part of the potential is smaller as well. If that number changes, for example near a defect or in a specific structure of a covalent solid, then the interaction changes. Thus, the bond-order form reflects, at least at some level, the local environment and the local bonding. Variations of this model have been used to describe solids, liquids, condensed-phase reacting species [42], etc.

### Tersoff potential

Tersoff used a variation on a bond-order potential to approximate the bonding in the many different states of silicon, from liquid to amorphous silicon to the perfect diamond structure [308]. To achieve this flexibility, the model gives up some accuracy. However, that one can model quite complex changes in bonding with such a simple potential is, indeed, encouraging. A few years after Tersoff introduced his potential for Si, Brenner proposed an extended version of the Tersoff potential that can be used to model diamond and other forms of carbon [41, 42]. A mark of the success of these types of potential is their wide use in modeling covalent materials, including Si, Ge, C and their alloys.

The total binding energy of the system was written as a sum over individual bonds, in which the energy of each bond included a repulsive pairwise contribution and an attractive contribution given by the product of the bond order and a pairwise bond energy. The bond order incorporated the local environment through both the local coordination (number of bonds) and the bond angles, enabling the potential to distinguish between linear, trigonal and tetrahedral geometries. In general terms, the form of the bond term is

$$\phi_{ij}(r_{ij}) = \left[ \phi_R(r_{ij}) - B_{ij}\phi_A(r_{ij}) \right], \tag{5.28}$$

where $\phi_R$ and $\phi_A$ are repulsive and attractive potentials, respectively. The function $B_{ij}$ reflects the bond order (i.e., strength of bond) joining $i$ and $j$ and is a *decreasing* function of the bond coordination

$$B_{ij} = B(\psi_{ij}), \tag{5.29}$$

where $\psi_{ij}$ is a function that reflects the number of neighboring atoms with some prescribed distance and the angles between those atoms, i.e., it reflects the nature of the binding of the atoms.[26]

To model the chemical vapor deposition (CVD) of diamond growth, Brenner added terms to the Tersoff potential to improve the description of hydrocarbon bonding [41]. These potentials are often referred to as "reactive empirical bond order" (REBO) potentials. The Brenner potential was successful and led to a prediction of a specific chemical reaction at the (100) surface of diamond during CVD growth that served as the first step of growth. How that reaction fits into the growth chemistry was well described by a kinetic Monte Carlo[27] study of the CVD process [27].

---

[26] To give a sense of the complexity of these potentials, consider the Tersoff potential of [308], which has the form

$$\phi_{ij}(r_{ij}) = f_c(r_{ij})[Ae^{-\lambda_1 r_{ij}} - b_{ij}Be^{-\lambda_2 r_{ij}}],$$

where $A$, $B$, $\lambda_1$, and $\lambda_2$ are constants, $f_c(r)$ is a cutoff function that limits the range of the potential to the nearest neighbors, and $b_{ij}$ is the bond order, which reflects local bonding and determines the angular dependence of the potential. The cutoff function was taken as $f_c(r) = 1$ if $r < R - D$ and $f_c(r) = 0$ if $r > R + D$, thus providing a very sharp cutoff at $R$ with a width of $2D$ in which an interpolating function is used to ensure continuous derivatives, $f_c(r) = 1/2 - (1/2)\sin[(\pi/2)(r - R)/D]$. The bond order is

$$b_{ij} = (1 + \beta^n \zeta_{ij}^n)^{-1/2n},$$

where

$$\zeta_{ij} = \sum_{k \neq i,j} f_c(r_{ik})g(\theta_{ijk})\exp[\lambda_3^3(r_{ij} - r_{ik})^3]$$

and

$$g(\theta) = 1 + c^2/d^2 - c^2/[d^2 + (h - \cos\theta)^2].$$

$\theta_{ijk}$ is the angle between the $ij$ and the $jk$ bonds and $\beta$, $n$, $\lambda_3$, $c$, $d$, and $h$ are constants. Parameters for Si are given in [308].

[27] See Chapter 9.

The Brenner and Tersoff potentials have seen very wide use in the modeling of covalent materials. They are not, however, without flaws, failing, for example, to model the correct angular dependence of $\pi$ bonds, an obviously important shortcoming when modeling hydrocarbons [4]. Much work has been done to improve these potentials, for example by adding terms with different angular dependence [259]. All in all, however, the Tersoff and Brenner potentials have been very successful and show how a simple model, based on a good knowledge of physical mechanisms, can provide sufficient fidelity to enhance our understanding of complex phenomena.

## 5.7 SYSTEMS WITH MIXED BONDING

One of the challenges in all the previous methods is that they cannot model systems which are undergoing bonding changes accurately. While the bond-order potentials do reflect dynamic changes in bonding, they are still a relatively simplistic view of bonding. We could, of course, use the methods of Chapter 4 and perform the full quantum mechanical solution, but those methods are computationally more intensive than the use of analytical forms to describe the interactions and the system size that can be studied is limited. In this section, we introduce two approaches that can be used to model such systems: reactive force potentials and the tight-binding method.

### 5.7.1 Reactive force potentials

Two independently developed approaches, the Charge-Optimized Many Body (COMB) [364, 286] and Reactive Force Field (ReaxFF) [319] methods, employ variations on bond-order potentials that enable atoms to respond to their local environment to autonomously determine their charge. These methods are very useful for modeling a wide range of phenomena in which the atomic charges must adjust to changing conditions. Both COMB and ReaxFF, for example, allow the study of bond formation and bond breaking during chemical reactions, as well as the complex bonding at interfaces between different types of materials.

To be more specific, consider the basic form of these potentials. The overall energy of a system consists of a number of terms similar to what is found in most potentials, for example, terms describing Coulomb and van der Waals interactions. What is different about the reactive methods is in how they treat the bond. The COMB potential, for example, describes bonding with a term similar in form to Eq. (5.25),

$$E^{bond} = \frac{1}{2} \sum_i \sum_{j=1}^{Z_i} \left[ V_R(r_{ij}, q_i, q_j) + b^{effective} V_A(r_{ij}, q_i, q_j) \right], \tag{5.30}$$

where the attractive and repulsive energies depend on the charges, $q$, on each atom and $b^{effective}$ is a sum of bond-order terms reflecting the different aspects of a bond, including torsion, angle, conjugation, etc. The critical feature of both methods is the charge equilibration. They employ different approaches, but both seek a self-consistent set of charges. There is a self energy term, which depends on the electronegativity of the atoms, that reflects the change in atomic energy with change in the atomic charge. The end result is a set of potentials with great flexibility and thus a wider applicability than potentials with fixed charges.

As always, there is a computational penalty for flexibility. Both COMB and ReaxFF potentials require considerably more computation to evaluate than ordinary interatomic potentials. That said, using these potentials is considerably faster than the full DFT approaches in Chapter 4.

### 5.7.2 The tight-binding method

The tight-binding method starts with an expansion of the wave function in terms of atomic orbitals centered at each atom site. Because of the use of atomic orbitals, the tight-binding (TB) method is well suited for describing the electronic structure of non-metallic systems, especially covalently bonded materials. The tight-binding model uses a set of approximations to the quantum mechanics that enable very fast approximation of the electronic structure. Empirical tight-binding methods typically use data to help set the values of constants, while *ab initio* methods are based on theoretical calculations. Much of the work in developing a tight-binding method is the development of the correct basis set and the determination of the constants. Despite their approximate nature, however, tight-binding methods can be quite useful in describing systems with bonding that changes dynamically, for example, in liquid silicon. We refer the reader to the many papers and books on the subject, for example in [127, 298, 299].

## 5.8 WHAT WE CAN SIMULATE

The potentials described in this chapter are the basis for most atomistic simulations of materials. For calculations at 0 K, we can evaluate the lattice sums from Chapter 3 for whichever of the potentials we choose. For example, one could calculate cohesive energies for different crystal structures to determine the 0 K thermodynamics and phases. If one plots the Gibbs free energy at $0\,\text{K}\,(G = U + PV)$ for different structures, the intersection of the curves is the transition point. Alternatively, one could plot $U(V)$ versus $V$ for both structures and then the transition is found by finding the common tangent of the curves, as shown in Figure 5.11.[28] For more complicated structures, e.g., a general triclinic lattice, all six lattice parameters plus the atom positions with the cell could be varied to determine the lowest-energy structure. Energy minimization methods for materials are well described in much more detail in [57].

In addition to bulk material properties, a common use of these potentials at 0 K is to calculate defect properties, including free surfaces, grain boundaries, dislocations, and vacancies. Calculating surface energies does not require large changes in the lattice sum methods of Chapter 3.[29]

---

[28] At the transition point $G_1 = G_2$, or $U_1(V_1) + PV_1 = U_2(V_2) + PV_2$, in which case $P = (U_1(V_1) - U_2(V_2))/(V_2 - V_1)$ and the volume change is $\Delta V = |V_2 - V_1|$.

[29] There are some issues that must be dealt with to determine long-range energies in surfaces in ionic materials.

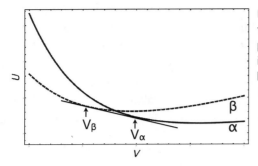

**Figure 5.11** Schematic view determining phase transitions. The energy $U$ versus volume $V$ curves are plotted for structures $\alpha$ and $\beta$. The common tangent indicates a transition from $\alpha$ to $\beta$ with the pressure given by the slope of the tangent and the volume change is $V_\beta - V_\alpha$.

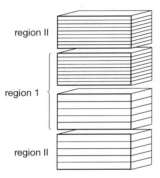

**Figure 5.12** Schematic view of simulation cell for modeling grain boundaries. The simulation region is marked I and the border regions are marked II. The grain boundary is a planar interface which is infinite in the $x$ and $y$ directions through the use of periodic boundary conditions in those directions. Atom planes are indicated by the thin lines. More details are in the text. Adapted from [363].

The biggest challenge with internal defects is the construction of the atomic positions within the defect as well as the creation of appropriate boundary conditions.

As an example of a defect calculation, consider the atomistic simulation of a grain boundary, which is the boundary between two crystals of the same composition but with different relative orientation as described in Appendix B. The structure and energy of grain boundaries are important in that they affect properties such as diffusion along boundaries, segregation to boundaries of impurities or alloyed components, and the driving force for boundary migration. Thus, determination of grain boundary energies through atomistic simulations has been the focus of numerous studies and papers.

The challenges of modeling grain boundaries are many, starting with the boundary conditions. The goal is to model a boundary between two infinite slabs of material. In Figure 5.12 we show a schematic view of one approach to modeling boundaries. In this example, we have the boundary between two crystallites with different orientations, indicated by the different width of the atom planes.[30] The simulation cell consists of the atoms that will be moved in a calculation, i.e., the degrees of freedom in the calculation. It is marked as region I. The two volumes marked region II are ideal crystals of the two crystal orientations. These regions are fixed during the simulation, thus mimicking an infinite system. An additional degree of freedom is the separation along $z$ between the two boundaries. In practice, the size of the regions must be checked to ensure that they do not have undue influence on the answer. The net grain boundary energy is

---

[30] These crystallites could, of course, be different materials. The same arguments would hold.

the energy of the combined system minus the energy of the two ideal systems. Details are given in [354, 355, 363] as well as in many other sources.

Another challenge to the simulation of the boundary shown in Figure 5.12 is the size of the system in the $x$ and $y$ directions. Boundaries have symmetry and some have periodic structures. It is essential to have a large enough system to include an appropriate number of repeat units to obtain converged results. Thus, boundary simulations tend to involve reasonably large numbers of atoms.

## 5.9 DETERMINING PARAMETERS IN POTENTIALS

All the approximate potentials described here have parameters that must be set in some way. Typically, these constants are determined such that properties calculated with the potential match a set of data from either experiment or more accurate calculations. The constants are then set and used to determine other materials properties or behavior. The set of properties used to determine the parameters are sometimes called a *training set*. The properties in the training set typically include some set of the cohesive energy and structure, energies of various surfaces, vacancy formation energies, elastic constants, stacking fault energies, etc.

A typical fitting procedure would be to assume some initial parameters, then calculate a set of material properties and compare them to the training set. The potential functions are non-linear and highly coupled, so robust methods must often be used to optimize the parameters. One procedure that is useful for models with many parameters is to fix some of the parameters, do an optimization of the other parameters, and then adjust the first set, and so on. The proof of the quality of a potential is not whether properties calculated with it match those in the training set, but rather the quality of calculations of properties not in that set.

It is important to remember that these potentials are all *models* – they are not the real material. Indeed, no calculations based on approximate potentials are really the "real" material. They are always models and may or may not be very good representations of the materials they purport to describe.

## 5.10 SUMMARY

This chapter summarizes some aspects of interatomic potentials, working from very simple pair interaction terms to more complex methods designed to model covalently bonded materials. We have just scratched the surface here, giving for each type of potential a single example chosen from a plethora of possibilities. These potentials will not actually work for all situations, so modifications are the norm. For example, systems may include more than one type of bonding at once. Creating good potentials requires both an understanding of bonding in the types of solids in which one is interested as well as a considerable amount of patience as one chooses the potential and, perhaps, seeks modifications to match the particular situation. However, the methods introduced here form the basis of how we model the interactions between atoms in solids.

## Suggested reading

A number of texts provide excellent discussions of this important topic. A few that I like are:

- An entire issue of the MRS Bulletin which was devoted to potentials and has good articles on almost all types of potentials discussed here. These articles have many references [329]. The May 2012 edition of the MRS Bulletin has a set of papers on reactive potentials.
- J. Israelachvili's delightful text, *Intermolecular and Surface Forces* [159], which is devoted to the interaction between atoms and molecules.

## 5.11 APPENDIX

The Lennard-Jones potential of Eq. (5.6) and the Mie potential (Eq. (5.9)) share a common form – they depend on functions of the interatomic separation, $r$, with terms only of the form $1/r^n$, where $n$ is an integer. A useful set of calculations from over 80 years ago makes it possible to calculate the properties of materials described by these potentials for the perfect lattices of the basic cubic structures with simple analytical expressions.

Consider a calculation of the 0 K cohesive properties of a solid consisting of atoms whose interactions are described by a Lennard-Jones or Mie potential, $\phi(r)$. To be a bit more specific, assume that the crystal structure is one of the basic cubic structures – simple cubic, body-centered cubic, or face-centered cubic. The cohesive energy per unit cell takes the form (from Eq. (3.7))

$$U_{cell} = \frac{1}{2} \sum_{\mathbf{R}} \sum_{i=1}^{n} \sum_{j=1}^{n} {}' \phi_{ij}(|\mathbf{R} + \mathbf{r}_j - \mathbf{r}_i|), \qquad (5.31)$$

where there are $n$ atoms in the unit cell and the prime indicates that there are no $i = j$ terms for a lattice vector of $\mathbf{R} = 0$. A simplification is possible if we rewrite Eq. (5.31) in the primitive lattice, in which there is just one atom per cell.[31] The cohesive energy per atom, $u = U/n$, is then just

$$u = \frac{1}{2} \sum_{\mathbf{R} \neq 0} \phi(R), \qquad (5.32)$$

where $R = |\mathbf{R}|$.

We can simplify Eq. (5.32) by rewriting the lattice vector $\mathbf{R}$ as a dimensionless number $\alpha(\mathbf{R})$ times the nearest-neighbor separation in the lattice $r$,

$$\mathbf{R} = \alpha(\mathbf{R})r . \qquad (5.33)$$

For example, in an $fcc$ structure, the distances to the nearest-neighbor shells of atoms are: $r$, $\sqrt{2}r$, $\sqrt{3}r$, ..., i.e., the $n$th neighbor shell is located a distance $\sqrt{n}r$ away. Thus, we can write

---

[31] The primitive lattices for the basic cubic structures are discussed in Appendix B.2.

the distance to the $n$th shell of neighbors in an $fcc$ structure as

$$R = \alpha(\mathbf{R}_n)r = \sqrt{n}r .$$ (5.34)

Similar forms for $\alpha$ are easily derivable for the other basic cubic systems.

For a Lennard-Jones or Mie potential, the interaction terms in $\phi$ depend only on the distance $R$ in the form $1/R^n$, which we can rewrite, as in Eq. (5.34)

$$\frac{1}{R^n} = \frac{1}{\alpha(\mathbf{R})^n} \frac{1}{r^n} .$$ (5.35)

The contribution of that term to the cohesive energy, $u$, is

$$\frac{1}{2} \sum_{\mathbf{R} \neq 0} \frac{1}{R^n} = \frac{1}{2} \frac{1}{r^n} \sum_{\mathbf{R} \neq 0} \frac{1}{\alpha(\mathbf{R})^n} = \frac{1}{2} \frac{1}{r^n} A_n ,$$ (5.36)

where we have pulled out of the sum the constant value $1/r^n$ and introduced the new sum

$$A_n = \sum_{\mathbf{R} \neq 0} \frac{1}{\alpha(\mathbf{R})^n} .$$ (5.37)

For $n \geq 4$, the sums $A_n$ are convergent and have been tabulated, taking the sums to $\infty$, for the $fcc$, body-centered cubic ($bcc$), and simple cubic ($sc$) lattices [161]. The values of $A_n$ are given in Table 5.4. The cohesive energy per atom of a material described, for example, with the Lennard-Jones potential can then be written as

$$u(r) = 2\epsilon \left[ A_{12} \left( \frac{\sigma}{r} \right)^{12} - A_6 \left( \frac{\sigma}{r} \right)^6 \right],$$ (5.38)

where $\epsilon$ and $\sigma$ are the standard Lennard-Jones parameters defined in Eq. (5.6), $r$ is the nearest-neighbor distance, and $A_6$ and $A_{12}$ are the lattice sums from Table 5.4.

It is sometimes convenient to rewrite Eq. (5.38) in terms of the volume per atom, $v$, instead of the nearest-neighbor distance. For each of the three crystal structures listed in Table 5.4, $v$ can be written in terms of the nearest-neighbor distance as $v = \beta r^3$, where $\beta = 1, 4/(3\sqrt{3})$, and $1/\sqrt{2}$, for the simple cubic, body-centered cubic and face-centered cubic lattices, respectively. We thus can write the volume dependence of the cohesive energy as

$$u(v) = 2\epsilon \left[ \frac{\sigma^{12} \beta^4}{v^4} A_{12} - \frac{\sigma^6 \beta^2}{v^2} A_6 \right].$$ (5.39)

This function is plotted in Figure 5.5a for an $fcc$ lattice.

Since $\epsilon$ and $\sigma$ are constants for a material and $A_{12}$ and $A_6$ are constants that depend only on crystal structure, analytic derivatives of Eq. (5.38) can be used to find the zero Kelvin equilibrium lattice constant, $r_o$, which is the value of $r$ for which $(\partial u/\partial r) = 0$. Given $r_o$, the equilibrium cohesive energy is just $u_o = u(r_o)$ from Eq. (5.38). We can also use the relation from thermodynamics for the bulk modulus, $B = V(\partial^2 U/\partial V^2)$, to calculate the equilibrium

Table 5.4 **Lattice sums ($A_n$) for cubic crystals. From [161].** $f_1 = 6 + 12(1/\sqrt{2})^n + 8(1/\sqrt{3})^n$, $f_2 = 8 + 6(\sqrt{(3/4)})^n + 12(\sqrt{3/8})^n$, **and** $f_3 = 12 + 6(1/\sqrt{2})^n + 24(1/\sqrt{3})^n$

| $n$ | Simple cubic | Body-centered cubic | Face-centered cubic |
|---|---|---|---|
| 4 | 16.5323 | 22.6387 | 25.3383 |
| 5 | 10.3775 | 14.7585 | 16.9675 |
| 6 | 8.4019 | 12.2533 | 14.4539 |
| 7 | 7.4670 | 11.0542 | 13.3593 |
| 8 | 6.9458 | 10.3552 | 12.8019 |
| 9 | 6.6288 | 9.8945 | 12.4925 |
| 10 | 6.2461 | 9.5645 | 12.3112 |
| 11 | 6.2923 | 9.3133 | 12.2009 |
| 12 | 6.2021 | 9.1142 | 12.1318 |
| 13 | 6.1406 | 8.9518 | 12.0877 |
| 14 | 6.0982 | 8.8167 | 12.0590 |
| 15 | 6.0688 | 8.7030 | 12.0400 |
| 16 | 6.0483 | 8.6063 | 12.0274 |
| 17 | 6.0339 | 8.5236 | 12.0198 |
| 18 | 6.0239 | 8.4525 | 12.0130 |
| 19 | 6.0168 | 8.3914 | 12.0094 |
| $n \geq 20$ | $f_1$ | $f_2$ | $f_3$ |

bulk modulus, $B_o$.[32] We find the simple results (with numerical values given for the $fcc$ lattice)

$$r_o = \left(\frac{2A_{12}}{A_6}\right)^{1/6} \sigma = 1.09\,\sigma$$

$$v_o = \left(\frac{2A_{12}}{A_6}\right)^{1/2} \beta\,\sigma^3 = 0.916\,\sigma^3$$

$$u_o = -\frac{\epsilon}{2}\left(\frac{A_6^2}{A_{12}}\right) = -8.6\,\epsilon \qquad (5.40)$$

$$B_o = -8\frac{u_o}{v_o} = 75\frac{\epsilon}{\sigma^3},$$

---

[32] Take the second derivative of $u(v)$ with respect to $v$ and then insert the expression for the equilibrium volume.

where $v_o = \beta r_o^3$ is the equilibrium volume. Comparison of the equilibrium values for the three structures, $fcc$, $bcc$, and $sc$, shows that systems described with Lennard-Jones potentials always have $fcc$ as their lowest-energy equilibrium structures.

There is no restriction of this approach to the Lennard-Jones potential – lattice sums for any potential with terms only of the form $1/r^n$ ($n \geq 4$) can be evaluated using Table 5.4. For example, the cohesive properties of systems described by the Mie potential in Eq. (5.9) can be written down as [196]

$$
\begin{aligned}
r_o &= \left(\frac{mA_m}{nA_n}\right)^{1/(m-n)} \sigma \\
u_o &= -\frac{\epsilon}{2}\left(\frac{A_n^m}{A_m^n}\right)^{1/(m-n)} \\
B_o &= -\frac{mnu_o}{9v_o} = \frac{mn\epsilon}{18v_o}\left(\frac{A_n^m}{A_m^n}\right)^{1/(m-n)},
\end{aligned}
\tag{5.41}
$$

where $v_o = \beta r_o^3$ and $m$ and $n$ are defined in Eq. (5.9). Note that the Lennard-Jones potential is recovered for $m = 12$ and $n = 6$.

These simple analytical results are very convenient, being able to describe the equilibrium structure for any material described with a Lennard-Jones potential by simply inserting the potential parameters for that system ($\epsilon$ and $\sigma$) in Eq. (5.40). Moreover, using this analysis coupled with simple models for the vibrational properties of solids, it can be shown that the finite temperature properties $T > 0$ for systems described by Mie potentials can be calculated analytically with high accuracy [196].

# 6 Molecular dynamics

The *molecular dynamics* method is one of the most used of all modeling and simulation techniques in materials research, yielding information about structure and dynamics on the atomic scale. The basic idea behind molecular dynamics is simple: calculate the forces on the atoms and solve Newton's equations to determine how they move. The molecular dynamics method was one of the first computer-based techniques employed in the studies of the properties of materials, dating back to pioneering work on the properties of liquids in the 1950s [2, 3].

The goal of this chapter is to provide a quick introduction of the basic methods, pitfalls, successes, etc. in molecular dynamics simulations, with an emphasis on applications to materials issues.

## 6.1 BASICS OF MOLECULAR DYNAMICS FOR ATOMIC SYSTEMS

Molecular dynamics (MD) is based on the ideas of classical mechanics as applied to atomic and molecular systems:[1] the total forces on all the atoms are calculated and Newton's equations are solved to determine how the atoms move in response to those forces. The *equilibrium* and *time-dependent* properties of the system are then calculated from the motions of the atoms.

Newton's second law states that the force on a particle equals its mass times its acceleration,

$$\mathbf{F}_i = m_i \mathbf{a}_i = m_i \frac{d^2 \mathbf{r}_i}{dt^2} . \tag{6.1}$$

From classical mechanics, we know that the force on a particle is equal to the negative of the *gradient* of the potential energy with respect to the position of that particle.[2] In Chapter 3 and Chapter 5, we describe how to calculate the potential energy $U$ of a system as a function of the position of the particles, $U(\mathbf{r}^N)$.[3] The force on atom $i$ is

$$\mathbf{F}_i = -\nabla_i U(\mathbf{r}_1, \mathbf{r}_2, \ldots, \mathbf{r}_N) = -\nabla_i U(\mathbf{r}^N) , \tag{6.2}$$

---

[1] A brief review of classical mechanics is given in Appendix D.

[2] This statement holds for a *conservative* force, i.e., one that depends only on the positions of the particles and not on their velocities.

[3] As a reminder, we use the notation $U(\mathbf{r}^N) = U(\mathbf{r}_1, \mathbf{r}_2, \ldots, \mathbf{r}_N)$, where $\mathbf{r}_i = (x_i, y_i, z_i)$.

where the gradient is taken with respect to the coordinates on atom $i$, i.e.,

$$\nabla_i U(\mathbf{r}^N) = \frac{\partial U(\mathbf{r}^N)}{\partial x_i}\hat{x} + \frac{\partial U(\mathbf{r}^N)}{\partial y_i}\hat{y} + \frac{\partial U(\mathbf{r}^N)}{\partial z_i}\hat{z}. \tag{6.3}$$

The simplest systems to describe are those whose atoms interact with central-force potentials, in which the interaction between the atoms depends only on the distance between them.[4] An example is the Lennard-Jones potential of Eq. (5.6). The potential energy for central-force potentials can be written from Eq. (3.2) as

$$U(\mathbf{r}^N) = \frac{1}{2}\sum_{i=1}^{N}\sum_{j=1}^{N}{}'\phi_{ij}(r_{ij}), \tag{6.4}$$

where the distance between atom $i$ and $j$ is $r_{ij} = |\mathbf{r}_j - \mathbf{r}_i|$ and the potential between them is $\phi_{ij}(r_{ij})$. The force on the $i$th atom is

$$\mathbf{F}_i = -\ggg \nabla_i U(\mathbf{r}^N) = \sum_{j\neq i}\left\{\frac{d\phi_{ij}(r_{ij})}{dr_{ij}}\frac{\mathbf{r}_{ij}}{r_{ij}}\right\} = \sum_{j\neq i}\mathbf{f}_{ij}(r_{ij}), \tag{6.5}$$

where

$$\mathbf{f}_{ij}(r_{ij}) = \frac{d\phi_{ij}(r_{ij})}{dr_{ij}}\frac{\mathbf{r}_{ij}}{r_{ij}} = \frac{d\phi_{ij}(r_{ij})}{dr_{ij}}\hat{r}_{ij} \tag{6.6}$$

is the force of atom $j$ on atom $i$ and $\hat{r}_{ij}$ is the unit vector $\mathbf{r}_{ij}/r_{ij}$, which is defined as $(\mathbf{r}_j - \mathbf{r}_i)/r_{ij}$.[5] Thus, the magnitude of the force between two atoms is proportional to the derivative of the potential with respect to the distance between the atoms and the force is directed along the vector connecting the atoms.

We can verify the sign of $\mathbf{f}_{ij}$ with a little thought. If the distance between two atoms, $r_{ij}$, is small, then the potential is repulsive and $d\phi/dr < 0$. Thus, atom $i$ will be forced away from atom $j$ and will move in the negative direction along $\hat{r}_{ij}$. The force of atom $j$ on atom $i$ is in the opposite direction of the force of atom $i$ on atom $j$, $\mathbf{f}_{ij} = -\mathbf{f}_{ji}$, as expected from Newton's third law.

The equation of motion for atom $i$ is

$$\frac{d^2\mathbf{r}_i}{dt^2} = \frac{1}{m_i}\mathbf{F}_i = \frac{1}{m_i}\sum_{j\neq i}\mathbf{f}_{ij}(r_{ij}). \tag{6.7}$$

The key to an effective simulation is to be able to solve Eq. (6.7) accurately and efficiently.

## 6.1.1 Numerical integration of Newton's equations

There are $3N$ coupled differential equations to be solved in Eq. (6.7), one for each coordinate of each atom. The basic approach for solving these equations is no different than finding the

---

[4] See Appendix D.4 for a discussion of central-force potentials and a derivation of these equations.
[5] We employ the notation: $\mathbf{r}_{ij} = \mathbf{r}_j - \mathbf{r}_i$, i.e., the vector is always defined as extending from point $i$ to point $j$.

solution for the harmonic oscillator in Appendix D.3 – after specifying the initial positions and velocities at time $t = 0$, a solution to the equations of motion in Eq. (6.7) needs to be determined, yielding $\mathbf{r}(t)$ for each atom. The equations in Eq. (6.7) are, however, highly coupled and no analytical solution exists. Thus, a numerical solution must be found.

The standard procedure for solving equations such as Eq. (6.7) on a computer is to break up the time $t$ into discrete intervals and then solve the equations of motion over those intervals. The intervals are referred to as *time steps* and denoted in this text by $\delta t$. One usually makes the further approximation that the properties of the system at $t + \delta t$ can be calculated from the properties at $t$. In solving Eq. (6.7) that means that the positions and velocities are found at $t + \delta t$ from the forces evaluated at $t$, which implicitly approximates the force $\mathbf{F}_i(t)$ on atom $i$ as a constant over the interval $t$ to $t + \delta t$. The acceleration $\mathbf{a}_i = d^2\mathbf{r}_i/dt^2 = \mathbf{F}_i(t)/m$ is thus also a constant over the interval $t$ to $t + \delta t$.

The assumption that $\mathbf{a}$ is constant over the time interval $t$ to $t + \delta t$ leads to a very simple solution to the equations of motion, which is found by integrating twice over time, yielding first

$$\mathbf{v}_i(t + \delta t) = \mathbf{v}_i(t) + \mathbf{a}_i(t)\delta t \tag{6.8}$$

and then

$$\mathbf{r}_i(t + \delta t) = \mathbf{r}_i(t) + \mathbf{v}_i(t)\delta t + \frac{1}{2}\mathbf{a}_i(t)\delta t^2 \tag{6.9}$$

where the position and velocity at the beginning of the time interval are $\mathbf{r}_i(t)$ and $\mathbf{v}_i(t)$, respectively.[6]

We could use Eq. (6.9) by starting from an initial set of positions and velocities at a time set to $t = 0$. From the initial positions, calculate $\mathbf{a}(0)$ from Eq. (6.5). Assuming that $\mathbf{a}$ is constant over $\delta t$, use Eq. (6.8) to find the velocities at $t + \delta t$ and Eq. (6.9) to find the positions at $t + \delta t$. Repeat for many time steps. While this approach is very straightforward, unfortunately, this simple procedure yields a very poor solution of the equations of motion. The problem with the solution in Eq. (6.8) and Eq. (6.9) is that the equations are not self-consistent.

The result of solving Eq. (6.8) and Eq. (6.9) is a discrete series of positions for each atom at each time step. We can, however, also estimate the velocities at each time step by finding the slope from the positions – the velocity of atom $i$ at time $t$ is approximately $\mathbf{v}_i(t) \approx (\mathbf{r}_i(t + \delta t) - \mathbf{r}_i(t - \delta t))/(2\delta t)$. We can then calculate the accelerations from the velocities in the same way, $\mathbf{a}_i(t) \approx (\mathbf{v}_i(t + \delta t) - \mathbf{v}_i(t - \delta t))/(2\delta t)$. However, the accelerations calculated in this way will not, in general, match those calculated directly from the forces through Newton's equations, i.e. $\mathbf{a} = \mathbf{F}/m$. There is numerical error that builds up over time. Thus, this approach does not yield self-consistent accelerations and forces.

In Figure 6.1a we show a calculation based on Eq. (6.8) and Eq. (6.9) for a one-dimensional harmonic oscillator with $k = m = \omega = 1$, $x(0) = 0$, and $v(0) = 1$. The exact result from

---

[6] If $\mathbf{a}(t)$ is constant over the interval $t$ to $t + \delta t$, then, since $\mathbf{a} = d\mathbf{v}/dt$, $v(t + \delta t) = \mathbf{v}(t) + \int_t^{t+\delta t} \mathbf{a}(t')dt' = \mathbf{a}(t)(t + \delta t - t)$, with a similar analysis to determine $\mathbf{r}$.

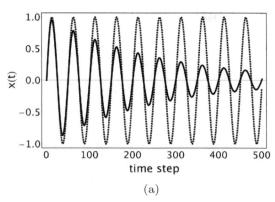
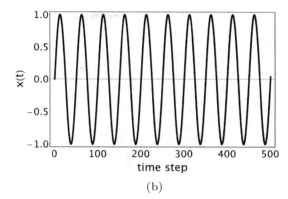

**Figure 6.1** Comparison of integration methods for the one-dimensional harmonic oscillator with $k = m = \omega = 1$, $x(0) = 0$, and $v(0) = 1$. Exact result from Appendix D.3 is given as the dotted curves. $\delta t = 2\pi/50$, i.e., there are 50 points per period of the oscillation. (a) The solid curve is from a calculation based on the simple solution in Eq. (6.8) and Eq. (6.9). (b) The solid curve is from a calculation based on the Verlet algorithm in Eq. (6.11). The solid curve overlays the exact solution.

Appendix D.3 for $x(t)$ is given as the dotted curves, while the solid curve is the integrated result with $\delta t = 2\pi/50$, i.e., there are 50 integration points per period of the oscillation. The inadequacies of this approach are clear – significant deviations of the calculated trajectory are apparent after 1 period. One can do better by moving to significantly smaller time steps, with 1000 points per period yielding reasonable, though not acceptable, results.

One could consider a better solution by introducing a further derivative of $\mathbf{r}(t)$, $\mathbf{b} = d\mathbf{a}/dt$. In this case, one would obtain an equation of motion for $\mathbf{a}$, and so on. The problem of not being self-consistent remains. Indeed, any approach based on Eq. (6.8) and Eq. (6.9) or variants of them does not yield an accurate solution to the equations of motion over time.

A number of methods that preserve self-consistency have been developed. A particularly simple approach that yields reasonably high-quality simulations is the *Verlet algorithm* [321, 322]. This algorithm is simple, yet robust, and, while not the best, is quite adequate for many purposes. The Verlet algorithm is that used in the sample MD code used in the examples shown later in this chapter.

The starting point of the Verlet algorithm is Eq. (6.9), which gives the position of atom $i$ *forward* in time from $t$, i.e. at $t + \delta t$. The position *backward* in time is found from the same equation, substituting $-\delta t$ for $\delta t$,

$$\mathbf{r}_i(t - \delta t) = \mathbf{r}_i(t) - \mathbf{v}_i(t)\delta t + \frac{1}{2}\mathbf{a}_i(t)\delta t^2 . \tag{6.10}$$

Adding Eq. (6.9) and Eq. (6.10) and rearranging yields

$$\mathbf{r}_i(t + \delta t) = 2\mathbf{r}_i(t) - \mathbf{r}_i(t - \delta t) + \mathbf{a}_i(t)\delta t^2 , \tag{6.11}$$

which is the equation employed in the Verlet algorithm for advancing the positions. The acceleration is determined at each time step from the force evaluation. The velocities can be *estimated* by a finite difference estimate of the derivative

$$\mathbf{v}_i(t) = \frac{\mathbf{r}_i(t + \delta t) - \mathbf{r}_i(t - \delta t)}{2\delta t} . \tag{6.12}$$

The velocities are not needed to calculate the positions, but are required to determine such quantities as the kinetic energy, $K = 1/2 \sum_i m_i v_i^2$, which is needed for establishing the conservation of the total energy, as discussed below.

The Verlet algorithm has many advantages, including its ease of programming and a reasonable accuracy. The positions are determined with an error proportional to $\delta t^4$, while the velocities are accurate to $\delta t^2$ [5]. Determining the appropriate size for $\delta t$ is discussed later in this section.

As in all MD calculations, calculations based on the Verlet algorithm start at $t = 0$ by establishing the initial positions and velocities, from which the initial forces (accelerations), potential energy, kinetic energy, etc. are determined. In the Verlet algorithm, the positions at time step $t + \delta t$ depend on both $t$ and $t - \delta t$ in Eq. (6.11). The first time step, from $t = 0$ to $t = \delta t$, thus requires the positions at $t = -\delta t$, which are found by use of Eq. (6.10) with $t = 0$.

In Figure 6.1b we show the solution to the harmonic oscillator using the Verlet algorithm with 50 integration points per vibrational period (same number of points used in Figure 6.1a). The Verlet results overlay the exact calculations, indicating a reasonably accurate solution to the differential equations. In the next section, we discuss a way to see if the solution is sufficiently accurate.

There are a few variations on the Verlet algorithm that yield more accurate velocities, and thus energies. For example, in the *velocity Verlet* algorithm [303], the velocities are calculated directly using an average of the accelerations. The form for the velocity Verlet alrgorithm is

$$\mathbf{r}_i(t + \delta t) = \mathbf{r}_i(t) + \mathbf{v}_i(t)\,\delta t + \frac{1}{2m}\mathbf{F}_i(t)\delta t^2 \tag{6.13}$$

$$\mathbf{v}_i(t + \delta t) = \mathbf{v}_i(t) + \frac{1}{2m}\big(\mathbf{F}_i(t) + \mathbf{F}_i(t + \delta t)\big)\delta t .$$

Note that the position at $t + \delta t$ depends only on quantities that are evaluated at $t$. From the position at $t + \delta t$, the force at $t + \delta t$ can be calculated and thus the velocity at $t + \delta t$. Once again, the calculation is started with a known set of positions and velocities and the algorithm applied sequentially.

A complicated, but accurate, set of algorithms are the so-called *predictor-corrector* methods, in which a solution to equations like Eq. (6.9) is found (the prediction) and then a term is added to bring them back into better self-consistency (the correction). The most well-known of these approaches is the Gear predictor-corrector algorithm. This method, and others related to it, can be quite accurate. However, they require some extra calculational steps and a great deal of memory, which once was an issue when trying to simulate large systems. The discussion of these approaches is beyond the scope of this text, but many descriptions are available in the books listed in the Suggested reading below.

The optimal size of the time step $\delta t$ is a balance between accuracy and computer time. Large time steps lead to larger numerical errors in the calculation, in part because the assumption that $\mathbf{F}$ is constant over the time interval becomes less reasonable. These inaccuracies can be reduced by employing a smaller time step. However, each time step requires a force evaluation, which consumes the largest amount of computer time in the calculation. There is thus a balance between efficiency (large $\delta t$), with fewer force calculations, and accuracy (small $\delta t$), with more force calculations. The hope is to have a robust enough method so that as-large-as-possible time steps, $\delta t$, can be used. Ultimately, the verification that the choice of time step is sufficient remains with the user. Criteria for evaluating the choice of time step are discussed in the next two sections.

### 6.1.2 Conservation laws

As discussed in Appendix D, the value of the Hamiltonian function

$$\mathcal{H}(\mathbf{r}^N, \mathbf{p}^N) = K(\mathbf{p}^N) + U(\mathbf{r}^N) \tag{6.14}$$

equals $E$, the internal energy of the system. For systems in which the potential energy does not depend on velocities (i.e., no frictional forces), $E$ is a constant – it does not vary with time. For standard molecular dynamics simulations, we thus expect $E = K + U$ to be conserved – to be constant over time within the numerical accuracy of the calculations. The degree to which a simulation follows this conservation law provides a stringent test of the quality of the calculation and guides the choice of both the method for integrating the equations of motion and the subsequent value for the time step.

It can be shown (using symmetry arguments) that for systems employing periodic boundary conditions, the total linear momentum [5]

$$\mathbf{P} = \sum_{i=1}^{N} \mathbf{p}_i \tag{6.15}$$

is also conserved, i.e., $d\mathbf{P}/dt = 0$. When choosing the initial velocities in a simulation, it is best to ensure that the total momentum $\mathbf{P} = 0$, which eliminates any overall drift of atoms during a simulation.[7]

### 6.1.3 Examining the reliability of a simulation

A key question in a molecular dynamics simulation is whether the algorithm used to integrate the equations of motion is of sufficient numerical accuracy. Trajectories will inevitably become inaccurate owing to the inherent numerical errors in any algorithm when integrating the equations of motion. These deviations are not of serious concern for most purposes unless they lead to errors in the average properties of the system. For standard molecular dynamics, we have

---

[7] If $\mathbf{P} \neq 0$, the center of mass of the atoms in the simulation will move in the direction of $\mathbf{P}$.

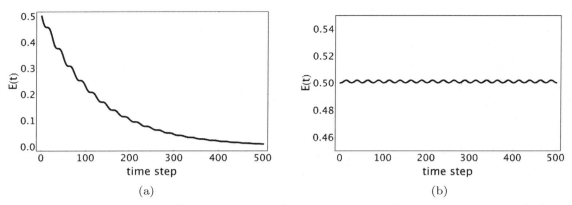

**Figure 6.2** Comparison of energy conservation between calculations with the two integration methods of Figure 6.1. The exact energy, $E_{exact}$, equals 1/2. (a) $E(t)$ versus $t$ for the calculation in Figure 6.1a. (b) $E(t)$ versus $t$ for the calculation in Figure 6.1b and is based on the Verlet algorithm.

an excellent way to monitor the simulations through the verification of the conservation laws discussed in the previous section, specifically the conservation of energy, $E$.

Because the equations of motion are not integrated exactly, $E$ will never actually be a constant. At their best, standard molecular dynamics simulations will have values for $E$ that will fluctuate around some average value and show no drift over time. The magnitude of the fluctuations in $E$ should be orders of magnitude smaller than fluctuations in the potential and kinetic energies. Experience states that having fluctuations in the energy of a few parts in $10^{-4}$ of its average value is generally sufficiently accurate, i.e., $(\max E - \min E)/\langle E \rangle \sim 10^{-4}$, where $\max E$ and $\min E$ are the maximum and minimum values of $E$ over the course of the simulation [5]. While this guideline is inexact, it seems to be a good guide for most simulations.

The assumption made in the last paragraph was that the energy fluctuated around a mean value. If things have gone awry in the simulation, either by an error in the program or too large a time step or the choice of a poor algorithm integrating the equations of motion, $E$ may show a drift in its value over time. If this occurs, the simulations are not valid and must be checked and modified until $E$ shows no drift in its value and small fluctuations.

In Figure 6.2 we show the calculated total energy $E(t)$ for the two solutions shown in Figure 6.1. In Figure 6.2a, which is from a simulation based on Eq. (6.8) and Eq. (6.9), the energy starts at the exact energy of $1/2$, but quickly drops to zero (a rather blatant example of a drift in $E$ over time). The energy is not conserved and thus this method is incorrect. In Figure 6.2b, which was calculated with the Verlet algorithm, the energy oscillates around the exact value, with a maximum relative deviation of the energy from the exact value of $\max(|E - E_{exact}|)/E_{exact} = 0.004$. While the value in Figure 6.2b may be somewhat larger than we might normally choose to ensure energy convergence, the solution is stable over time, oscillating around the exact solution and showing no discernible drift in its average value.

Once a method has been chosen, the accuracy is controlled by the time step $\delta t$. The fundamental time step in a molecular dynamics simulation must be chosen small enough so that the equations of motion are integrated accurately. The time scale is set by the fastest motion in the system, typically the shortest vibrational period. In a solid, the shortest period is typically of the order of a picosecond ($10^{-12}$ seconds) or less. How many time steps are needed to accurately integrate the equations of motion over a vibrational period depends in part on the method used. For the Verlet algorithm described above, of the order of 50 time steps per vibrational period are usually needed to obtain sufficient accuracy for energy conservation. Thus, the fundamental time step is of the order of $10^{-15} - 10^{-14}$ seconds or so.

The inherently small time step ultimately limits the utility of molecular dynamics. A nanosecond of real time would require a simulation with about a million time steps, a microsecond would need a billion, and so on. Standard molecular dynamics simulations are thus typically limited to short-time events.

### 6.1.4 Connection to thermodynamics

In a standard molecular dynamics simulation with a fixed number of particles, $N$, in a volume, $V$, the system is in the microcanonical ensemble, since the energy, $E$, is conserved.[8] The kinetic and potential energies are not fixed, but *at equilibrium* will fluctuate around their average values. Other thermodynamic quantities, such as pressure and temperature, can be calculated as well. Their average values, which in molecular dynamics are calculated by averaging over time, correspond to the equivalent thermodynamic variables of the system.

Thermodynamic values calculated with molecular dynamics are not, of course, exactly equal to the real thermodynamics of the material being studied. As noted in Chapter 5, any simulation is not of the real material, but rather calculates the properties of a model of that material, with the biggest error usually being the interatomic potential function. Beyond the errors in the potential, there are smaller errors caused by the simulation methods, probably the largest of which arising from the size of the system in the simulation.[9]

The simplest connection to thermodynamics is through the temperature, which is not a fixed quantity in a standard molecular dynamics simulation. There is no formal definition of an instantaneous temperature of a system. There is, however, a definition of the thermodynamic average temperature in terms of the average kinetic energy given in Eq. (G.19),

$$\langle T \rangle = \frac{2 \langle K \rangle}{3 N k_B} . \tag{6.16}$$

In a molecular dynamics simulation, instantaneous values of $K$ will be generated, from which $\langle K \rangle$, and thus $\langle T \rangle$, will be calculated. If we were to define an "instantaneous temperature function" as $\Theta(t) = 2K(t)/(3Nk_B)$, then, after the system has equilibrated, $\Theta$ will fluctuate around its average value. A limitation to standard molecular dynamics is the inability to directly set the temperature to a desired value. As will be seen in the example calculations in a later section, a molecular dynamics simulation is started from some set of initial positions and

---

[8] See Appendix G.5.5. Since the total momentum is conserved (Section 6.1.2), molecular dynamics simulations are not, strictly speaking, in the microcanonical ensemble. However, any differences are small and can be ignored [5].

[9] As discussed in Appendix G.6.1, there are differences that arise in average quantities as calculated in different ensembles that scale as $\mathcal{O}(1/N)$. We can expect uncertainties on that scale of the calculated thermodynamics.

velocities. The chosen velocities set the initial kinetic energy and, by Eq. (6.16), the initial temperature $\Theta(0)$. As the system evolves, however, the equilibrium value of the kinetic energy may well be very different than its initial value. If the goal of the simulation is to model a material under specific temperature conditions, then the inability to specify those conditions is limiting. Ways to either achieve a desired temperature or perform the simulation at a fixed temperature are discussed later in this chapter.

From Eq. (G.20), we can write the average pressure in a classic molecular dynamics calculation as

$$\langle P \rangle = \frac{N}{V} k_B \langle T \rangle - \frac{1}{3V} \left\langle \sum_{i=1}^{N} \mathbf{r}_i \cdot \nabla_i U \right\rangle, \tag{6.17}$$

where $\langle T \rangle$ is given in Eq. (6.16). From this equation, expressions for the pressure for systems whose interactions are described with central-force potentials are given in Eq. (G.21) and as[10]

$$\langle P \rangle = \frac{N}{V} k_B \langle T \rangle - \frac{1}{3V} \left\langle \sum_{i=1}^{N} \sum_{j>i}^{N} r_{ij} \frac{d\phi}{dr_{ij}} \right\rangle. \tag{6.18}$$

Comparing Eq. (6.18) with the equation for the force in Eq. (6.5) we see that they depend on similar terms and calculating the pressure requires little extra computational work.[11]

Any other thermodynamic quantities that depend explicitly on averages of functions of the atomic positions and momenta can thus also be determined in a molecular dynamics simulation. Below we will give some examples when we discuss an example calculation.

## 6.1.5 Initial conditions

Once the material to be studied has been chosen and the appropriate interatomic potentials developed and implemented within a molecular dynamics code, the problem to be studied must be identified and the appropriate boundary conditions and initial conditions must be chosen.

The problem sets the density (number of atoms and volume) as well as the shape of the simulation cell. If the system is a fluid, or in a cubic crystal structure, then typically a cubic simulation cell will be chosen. For a material in an $fcc$ structure, for example, a cube that is large enough to include an integer number, $n$, of $fcc$ unit cells would be appropriate. If the desired unit cell parameter is $a$, then the length of the side of the simulation cell would be $na$.

---

[10] Similar equations can be derived for other types of potentials as well. For example, for a pair-functional potential of Eq. (5.20) (e.g., the embedded-atom model), the pressure is

$$\langle P \rangle = \rho k_B T - \frac{1}{3V} \left\langle \sum_{i=1}^{N} \left[ \frac{\partial F}{\partial \bar{\rho}_i} \sum_{j\neq i} r_{ij} \frac{\partial \bar{\rho}_i}{\partial r_{ij}} + \sum_{j>i} r_{ij} \frac{\partial \phi_{ij}}{\partial r_{ij}} \right] \right\rangle. \tag{6.19}$$

[11] It is common to see plots of the instantaneous value of the pressure calculated at each time step, with its value fluctuating around its average value. This quantity is not, strictly speaking, the pressure, in the same way that $\Theta$ is not the actual temperature. It is the average values of these quantities that connect to their thermodynamic counterparts.

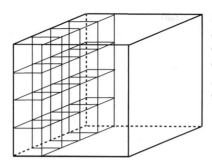

**Figure 6.3** Shown is a simulation cell containing one row of 4 unit cells in each direction (the other cells are not included so the figure will be less cluttered). The simulation cell is repeated through space with periodic boundary conditions. Atoms within the simulation cell are not restricted to their initial positions – the unit cells are just a way to start the simulation.

The net volume of the simulation cell would be $V = n^3a^3$. Since there are four atoms per $fcc$ cell, and since there are $n^3$ unit cells in the simulation cell, $N = 4n^3$, and the number density is $\rho = N/V = 4/a^3$, as expected.

It is important to emphasize that the unit cells are used only to place the atoms in their initial conditions. The unit cells then lose their identity and atoms within the simulation cell are free to move as needed in response to the interatomic forces. The simulation cell is repeated through space with periodic boundary conditions, so acts like a much larger "unit cell" with $N$ atoms and a volume $V$. We show a schematic constructing a simulation cell from a set of $4 \times 4 \times 4$ unit cells in Figure 6.3.

There is no restriction that says that the simulation cell must be cubic. It is perfectly fine to use any of the standard crystallographic cells such as orthorhombic, triclinic, etc. The choice depends on the crystal structure of the material being studied. If the crystal structure is orthorhombic, for example, then one must use an orthorhombic cell to represent the lattice. The distances and lattice sums will have to reflect the symmetry of the chosen structure using the methods of Chapter 3.

The system needs to be large enough to encompass the basic structures that are to be modeled. If one is interested in thermodynamics, then the number of atoms must be large enough to reduce size-dependent errors.[12] For modeling defects, then other criteria, such as reducing elastic effects at the boundary, must be considered. Other considerations of system size include ensuring that the structure is commensurate with the periodically repeated simulation cell, as displayed in Figure 3.5.

Once the number of atoms and the volume and shape of the simulation cell are chosen, the positions and the velocities of each particle must be given some initial values. The choice of the initial conditions is very important. Since the total energy is conserved, whatever energy the system starts with will be set for the entire simulation. If two atoms, for example, start very close together, then their interaction energy will be very high and the total potential energy of the system will be high as well, as will the total energy. Since the total energy is conserved, as the atoms move apart it will lower $U$ but the the kinetic energy will have to increase to keep $E$ constant. Since $T$ is proportional to $\langle K \rangle$, a poor choice of initial positions and velocities may preclude being able to model the desired thermodynamic conditions.

---

[12] See footnote 9.

Choosing initial positions of a simulation is usually straightforward. The key is to ensure that the initial structure is close to what is expected, which generally means that the system will not take too long to equilibrate to the final structure. For modeling the thermodynamics of solids, the initial positions are usually taken to be the normal atom sites. For modeling a fluid, since the structure is unknown and atoms should be started at low energy configurations, starting from atoms in a solid structure is best, letting the system melt spontaneously during the simulation.

Suppose, however, that the structure of interest is a defect, for example a grain boundary. Then one would want to use a simulation cell similar to that shown in Figure 5.12. The atoms should be placed in positions as close as possible to the minimum-energy structure of that defect. Of course, sometimes that structure is unknown and a trial-and-error procedure must be followed. A brief discussion of this topic is given in Appendix 6.9.1 for a simulation of a pair of dislocations.

The initial velocities can be chosen in a number of ways. They could, for example, be chosen from a Maxwell-Boltzmann distribution for a desired initial starting temperature $T_{init}$, as described in Appendix 6.9.2,

$$\rho(v_{ix}) = \frac{1}{\sigma\sqrt{2\pi}} e^{-v_{ix}^2/(2\sigma^2)}, \tag{6.20}$$

where $\sigma = \sqrt{k_B T / m_i}$. A simpler way would be to just pick the velocity components randomly over a small range – collisions will equilibrate these to a Maxwell-Boltzmann distribution in a few hundred time steps.

### 6.1.6 Steps in an MD simulation

The steps in an MD calculation are standard regardless of the system under study. There will be an initial time at the beginning of the run where the system will not have settled into equilibrium – there will be drift in the average values of calculated quantities. Once the system has reached equilibrium, then averages can be taken. Standard statistics can be used to measure the quality of the results.

Standard steps include:

(a) Initialize the positions $\{\mathbf{r}\}$ and momenta $\{\mathbf{p}\}$.
(b) Calculate the initial kinetic energy $K$, potential energy $U$, $E = K + U$, and other quantities of interest as well as the forces on each atom $\mathbf{F}_i$.
(c) For $n_{equil}$ time steps:
   (1) Solve the equations of motion for $\{\mathbf{r}_i(t + \delta t)\}$ and momenta $\{\mathbf{p}_i(t + \delta t)\}$ from values and forces at time $t$.
   (2) Calculate the kinetic energy $K$, potential energy $U$, $E = K + U$, and other quantities of interest as well as forces $\mathbf{F}_i$.
   (3) Check for drift of values that indicate that the system is not equilibrated.
   (4) When equilibrated, restart.

(d) For $n$ time steps:
  (1) Solve the equations of motion for $\{\mathbf{r}_i(t + \delta t)\}$ and momenta $\{\mathbf{p}_i(t + \delta t)\}$ from values and forces at time $t$.
  (2) Calculate the kinetic energy $K$, potential energy $U$, $E = K + U$, and other quantities of interest as well as forces $\mathbf{F}_i$.
  (3) Accumulate values of $K$, $U$, etc. for averaging.
(e) Analyze data: averages, correlations, etc.

## 6.2   AN EXAMPLE CALCULATION

To make the discussion a bit more specific, we take as an example a simple molecular dynamics simulation of the properties of a system made up of atoms interacting with the Lennard-Jones (LJ) potential described in Section 5.3.1. There have been thousands of published papers describing simulations of systems of atoms that interact with the Lennard-Jones potential, making "Lennard-Jonesium" probably the most simulated material.

The basic form of an LJ potential is

$$\phi(r_{ij}) = 4\epsilon \left[ \left( \frac{\sigma}{r_{ij}} \right)^{12} - \left( \frac{\sigma}{r_{ij}} \right)^{6} \right], \tag{6.21}$$

which can be written in a somewhat more compact form by introducing the *reduced units* $\phi^* = \phi/\epsilon$ and $r^* = r/\sigma$. Equation (6.21) becomes

$$\phi^*(r_{ij}^*) = 4 \left[ \left( \frac{1}{r_{ij}^*} \right)^{12} - \left( \frac{1}{r_{ij}^*} \right)^{6} \right]. \tag{6.22}$$

The force $\mathbf{f}_{ij}$ from Eq. (6.6) is

$$\mathbf{f}_{ij}^*(r_{ij}^*) = -\frac{24}{r_{ij}^2} \left[ \left( \frac{2}{r_{ij}^*} \right)^{12} - \left( \frac{1}{r_{ij}^*} \right)^{6} \right] \mathbf{r}_{ij}^*. \tag{6.23}$$

When employing the form in Eq. (6.22), the energy is determined in units of $\epsilon$ and the distances in units of $\sigma$. For use in a molecular dynamics simulation, new reduced units are required, which are listed in Table 6.1.[13]

### 6.2.1   Potential cutoffs

As discussed in Section 3.3, truncating the potential at some cutoff distance $r_c$ is essential for being able to do a calculation in a reasonable amount of computer time. For molecular dynamics

---

[13] It is useful to derive the reduced units for the pressure, density, temperature, and time for the Lennard-Jones potential. Check that all starred quantities are unitless. Hint: use the kinetic energy to define the time.

### Table 6.1  **Reduced units in Lennard-Jones systems**

| Value | In reduced units |
| --- | --- |
| Potential energy | $U^* = U/\epsilon$ |
| Temperature | $T^* = k_B T/\epsilon$ |
| Density | $\rho^* = \rho\sigma^3$ |
| Pressure | $P^* = P\sigma^3/\epsilon$ |
| Time | $t^* = t/t_o$, where $t_o = \sigma\sqrt{m/\epsilon}$ |

$\epsilon$ and $\sigma$ are defined in Eq. (6.21). All energies are in units of $\epsilon$, e.g., $E^* = E/\epsilon$, $K^* = K/\epsilon$.

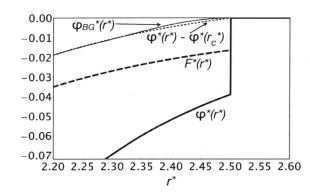

**Figure 6.4**  Force $F^*$, potential $\phi^*$, and shifted potential $\phi^*(r*) - \phi^*(r_c^*)$ for an LJ potential cutoff at $r_c = 2.5\sigma$. Note that there is a discontinuity in both the potential and force at the cutoff. $\phi^*_{BG}(r^*)$ is a potential that smoothly interpolates the potential and derivative to zero at $r_c^* = 2.5$, avoiding the discontinuities [43].

calculations, introducing a cutoff can cause discontinuities in the force, which can lead to a number of problems, including poor energy conservation.

In Figure 6.4, the potential and force for a Lennard-Jones potential is shown for a cutoff distance of $r_c = 2.5\sigma$, a short, but standard, value. There is a discontinuity at $r_c$ in both the potential $\phi$ and the force, remembering that the force is proportional to $-d\phi/dr$. The discontinuity in the potential can be removed by shifting the potential by a constant, $\phi(r_c)$, the value of the potential at the cutoff, creating a new potential, $\phi'(r) = \phi(r) - \phi(r_c)$, that goes to zero at $r_c$.[14] Shifting the potential does not, however, eliminate the discontinuity in the force, which, while small, can serve as a source/sink of energy as particles move back and forth across the cutoff radius. The presence of such discontinuities can adversely affect the energy conservation. One way around this problem would be to smoothly interpolate the Lennard-Jones potential so that the potential and its first few derivatives are zero at $r = r_c$. Such a potential is shown as $\phi^*_{BG}$ in Figure 6.4 [43]. These same issues arise with the use of other types of potentials, all of which include cutoffs, and any discontinuities must be taken into account.

---

[14] The total energy would need to be corrected at the end of the calculation to remove the shift.

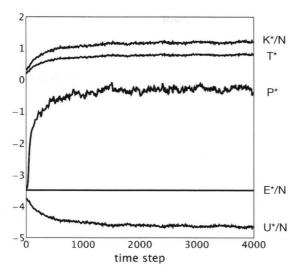

**Figure 6.5** Initial equilibration of the Lennard-Jones system described in the text. Shown are instantaneous values for $K^*/N$, $T^*$, $P^*$, $E^*$, and $U^*/N$. Note that while $K^*$ and $U^*$ vary greatly over the course of the first 1500 time steps or so, the energy $E^*$ is constant.

## 6.2.2 Analysis of molecular dynamics simulations

In many ways simulations are like experiments in that a great deal of "data" are generated that must be analyzed. Reliable answers depend on using high-quality statistical analysis. Only a sketch of such analysis is given in this text – much more detail is given elsewhere [5].

### Equilibration

Consider a model that consists of 864 atoms interacting with a Lennard-Jones potential in reduced units. A reduced density of $\rho^* = 0.55$ and a time step of $\delta^* = 0.005$ is used. The initial reduced temperature is $T^* = 0.2$. The cutoff was set to $r_c^* = 5$. The atoms were started in positions of a perfect $fcc$ lattice: 864 atoms represent a $6 \times 6 \times 6$ system of 4-atom $fcc$ unit cells and is commensurate with the periodic boundary conditions. Once the simulation starts, the atoms move away from the perfect lattice positions and, as we shall show below, the final thermodynamic state is a liquid.

The initial time steps are shown in Figure 6.5, where we show instantaneous values for the potential energy per atom $U^*/N$, the kinetic energy per atom $K^*/N$, and the total energy per atom $E^*/N$. We show variation of the temperature function and pressure function, which for convenience we denote by $T^*$ and $P^*$, respectively. Note that as we discussed above, only the average values of $T^*$ and $P^*$ have a thermodynamic meaning.

Over the course of the first 1500 time steps, there is a rapid variation of the thermodynamic variables, after which a steady convergence to what looks like an equilibrated state, i.e., a state in which the values settle to an average value with fluctuations about that average. Note that while $K^*$ and $U^*$ vary greatly over the simulation, the total energy $E^*$ is constant on the scale shown. If $E^*$ were not constant, then that would be a good indication that there is a problem with the integration of the equations of motion, perhaps owing to an over-large time step.

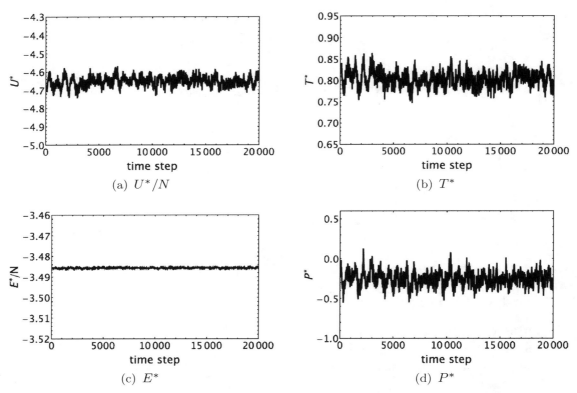

**Figure 6.6** Instantaneous values in an MD simulation of the Lennard-Jones system from Figure 6.5.

Figure 6.5 points out a drawback of standard molecular dynamics simulations. We started the simulation at a temperature $T^* = 0.2$. The final temperature is $T^* \approx 0.74$. Since standard MD has constant energy not temperature, we do not know ahead of time what the final temperature will be in a simulation. Similarly, since standard MD is at constant volume, we do not know what the final pressure will be. Since most actual experiments are at constant $T$ and $P$, this variation in an MD simulation can be a nuisance. Later in this chapter we will discuss ways to extend MD to other ensembles in which $T$ and/or $P$ are fixed.

## Averaging

In Figure 6.6 we show the values for $U^*/N$, $E^*/N$, $T^*$, and $P^*$ for 20 000 time steps at some time after the equilibration steps in Figure 6.5. Compare the scale on the plot of $E^*/N$ relative to that for $U^*$ or $T^*$ (which is proportional to $K^*$). The total variation of $E^*/N$ over the 20 000 time steps is $\Delta(E^*/N)_{max-min} = 0.0009$ (measured from its minimum value to its maximum), so $\Delta(E^*/N)_{max-min}/\langle E^*/N \rangle = 0.0003$. In contrast, the variation in $U^*/N$ (and, concurrently, $K^*/N$) over the same range is $\Delta(U^*/N)_{max-min} = 0.17$. Since $E = K + U$, Figure 6.6 shows how energy migrates from kinetic to potential in such a way as to keep the sum of the two a constant.

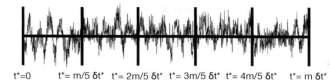

**Figure 6.7** Schematic of binning procedure. The total number of time steps ($m$) is broken into $n_b$ sections and averaged over each of those sections.

$t^*=0$   $t^*= m/5\,\delta t^*$   $t^*= 2m/5\,\delta t^*$   $t^*= 3m/5\,\delta t^*$   $t^*= 4m/5\,\delta t^*$   $t^*= m\,\delta t^*$

One can calculate the average values of the quantities in Figure 6.6 as

$$\langle U \rangle = (1/\tau_{max}) \sum_{\tau=1}^{\tau_{max}} U(\tau),\tag{6.24}$$

where $\tau_{max}$ is the total number of time steps. Since the set of values for each time step can be treated as data, standard statistical measures of those data can be determined. For example, the *variance* of $U$ is defined as

$$\sigma_U^2 = (1/\tau_{max}) \sum_{\tau=1}^{\tau_{max}} (U(\tau) - \langle U \rangle)^2 = \langle (U(\tau) - \langle U \rangle)^2 \rangle = \langle U^2 \rangle - \langle U \rangle^2 .\tag{6.25}$$

For Figure 6.6, $\langle E^* \rangle = -3.49$ and $\sigma^2(E^*) = 2.88 \times 10^{-8}$, $\langle U^* \rangle = -4.65$ and $\sigma_{U^*}^2 = 7.61 \times 10^{-4}$, $\langle T^* \rangle = 0.80$ and $\sigma_{T^*}^2 = 3.4 \times 10^{-4}$, and $\langle P^* \rangle = -0.253$ and $\sigma_{P^*}^2 = 8.58 \times 10^{-3}$. Note that the value for the pressure is negative, saying that this system is under a uniform tension. Decreasing the volume (increasing the density, since $N$ is fixed) would increase the pressure.

The variance of $U$ gives a measure of the size of the fluctuations in its values around its average value. It is related to a thermodynamic capacity, in much the same way that the fluctuations in $E$ are related to the heat capacity in the canonical ensemble, as discussed in Appendix G.6. From the discussion in Appendix G.6, we know that the relative error in $\langle U \rangle$, which is defined as the standard deviation of $U$, $\sigma_U$, divided by the average of $U$, $\langle U \rangle$, goes as $\sigma_U/\langle U \rangle \propto 1/\sqrt{N}$, where $N$ is the number of atoms in the system.

The variance of $U$ does not give a measure of how well its average quantities are evaluated in a simulation, however. A quantitative measure of the statistical error in a simulation is given by the variance of the *mean* of $U$, i.e., $\sigma_{\langle U \rangle}^2$. One way to evaluate the variance of the mean is to divide the data into smaller sets, then to average the data in each of those sets, creating a list of averages from which the variance can be determined. It is important to remember that the values of the data, for example $U$, are not independent – the positions and velocities at one time step are close to those at the previous time step and the properties of those states do not differ appreciably. The energies thus are *correlated* over some time span. To avoid these correlations in the data, the set of all time steps must be divided into subsets that include a sufficient number of time steps so that the properties at the end of the subset are not correlated with the properties at the beginning.

The usual parlance is to refer to the entire set of data as a "run" and to the subsets of data as "bins". The general procedure is shown schematically in Figure 6.7. Averages are found over each bin and the variance of the bin averages is calculated, with the requirement that there are enough steps in the bins that the data at the end of the bin are uncorrelated with those at the beginning. More details are given in [283].

If the data are divided into $n_b$ bins of equal length $\tau_b$, then average quantities are given by (using $U$ as an example)

$$\langle U \rangle = \frac{1}{n_b} \sum_{i=1}^{n_b} \langle U \rangle_i \,, \tag{6.26}$$

where the bin averages are

$$\langle U \rangle_i = \frac{1}{\tau_b} \sum_{\tau=\tau_o}^{\tau_o + \tau_b} U(\tau). \tag{6.27}$$

The variance of the bin averages $\langle U \rangle_i$, $\sigma^2(\langle U \rangle)$ is a statistical measure of the quality of the data. Other tests are available to test for drift in the bin averages (i.e., to examine whether the system has equilibrated) and other statistical measures of accuracy [5, 283]. This analysis is quite important to establish the quality and reliability of the results.

For example, binning the data in Figure 6.6 into 20 bins (1000 time steps/bin), and using the standard deviation $\sigma_{\langle U^* \rangle}$ as a measure of the error, we have for the potential energy $\langle U^* \rangle = -4.651 \pm 0.005$. Similarly, for the pressure the results are $\langle P^* \rangle = -0.25 \pm 0.02$, for the temperature $\langle T^* \rangle = 0.804 \pm 0.003$, and for the total energy $\langle E^* \rangle = -3.486 \pm 3 \times 10^{-5}$. Slightly different values would be obtained depending on the number of bins used. Note that there is a balance between ensuring that the time steps are uncorrelated and having enough bins so that the bin statistics are meaningful, as discussed in Appendix I.3.

## "Lennard-Jonesium" as a model for materials

While a Lennard-Jones potential may not be a terribly good description of the interatomic interactions in most materials, simulations based on the Lennard-Jones potential can provide very useful information that can be applied to a wide range of materials. The simulation whose results are shown in Figure 6.6 was performed in reduced units. There are no parameters in the potential in those simulations, which yielded results at a thermodynamic state of $\rho^* = 0.55$, $\langle T^* \rangle \approx 0.80$, and $\langle P^* \rangle \approx -0.25$. To scale these results for a specific system requires substituting the appropriate potential parameters in Table 6.1.

For example, from Table 5.1, the parameters for the interaction of Ne atoms are $\epsilon = 0.0031$ eV and $\sigma = 2.74$ Å, while for Xe the parameters are $\epsilon = 0.02$ eV and $\sigma = 3.98$ Å. Since $T^* = k_B T / \epsilon$, $T = T^* \epsilon / k_B$. For Ne, $T^* = 0.8$ corresponds to $T = 28.8$ K, while for Xe, $T = 165.7$ K. If we had the parameters for systems other than the rare-gas atoms, we would have a prediction for the point on the phase diagram for those materials as well. Note the advantage of employing reduced units – we did one simulation, yet have results for many systems. Of course, since the Lennard-Jones potential may not be a very good description of the interactions between atoms in the desired material, the predication may not be very accurate. However, in many cases this information can provide an acceptable starting point for a more accurate calculation.

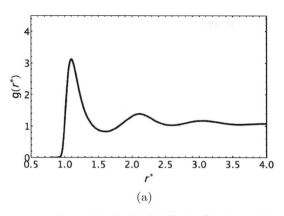
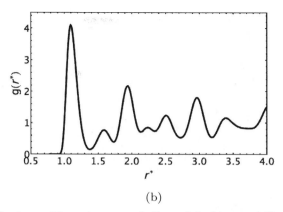

(a)                      (b)

**Figure 6.8** Radial distribution functions. (a) $g(r)$ for the equilibrated structure in Figure 6.6 with $\rho^* = 0.55$. (b) Results from a similar calculation as in $a$, but with a density of $\rho^* = 1.0$.

## Spatial correlation functions

In addition to thermodynamic quantities such as $U$, $T$, and $P$, atomic-level simulations can also provide information of the structure of a material, for example by considering *correlation functions*, as discussed in Appendix G.7. The pair distribution function $g(r)$, for example, gives the probability that for any atom there sits another atom a distance $r$ away from it. $g(r)$ is independent of angle and so it does not provide information on the direction that atom will lie, but only how far it is. However, $g(r)$ provides very useful information that can differentiate structures. Details on how to calculate $g(r)$ are given in Appendix 6.9.3.

In Figure 6.8a is shown the calculated $g(r)$ for the equilibrated structure in Figure 6.6. The form of $g(r)$ in Figure 6.8a is typical of a pair distribution function for a liquid. It shows a set of broad peaks that asymptotes to $g(r) = 1$. As discussed in Appendix G.7, integration of $g(r)$ over some range of $r$ gives the average number of neighbors in that range. Integrating over the first peak, which represents the "nearest neighbors" in the liquid, shows that there are on average about 13 neighbors.

Figure 6.8a provides some important information about the simulated system in Figure 6.6 – it is a liquid. Thus, we have determined that a system described with a Lennard-Jones potential, with thermodynamic conditions $\rho^* = 0.55$, $\langle T^* \rangle \approx 0.80$, and $\langle P^* \rangle \approx -0.25$, is a liquid. This kind of data provides a start on calculating the *phase diagram* of a system described by a Lennard-Jones potential. Having used reduced units, the Lennard-Jones phase diagram can then be applied to a range of materials that may or may not be well described by a Lennard-Jones potential.

The thermodynamic state in Figure 6.6 corresponds to a liquid. A solid could be formed by either lowering the temperature or increasing the density, which is equivalent to increasing the pressure. We took the latter approach, decreasing the volume so that the density increased to $\rho^* = 1.0$. To achieve a state with an average temperature somewhat close to that achieved in Figure 6.6, we did a series of simulations at different starting temperatures, settling on an initial temperature of $T_{init}^* = 1.4$. While we do not show the results, an equilibrated state was reached

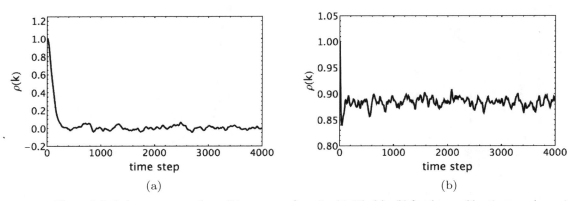

**Figure 6.9** Order parameters for solid structures from Eq. (G.48). (a) $\rho(\mathbf{k})$ for the equilibration run shown in Figure 6.5. with $\rho^* = 0.55$. (b) $\rho(\mathbf{k})$ for the solid structure shown in Figure 6.8b with a density of $\rho^* = 1.0$.

with $\langle T^* \rangle \approx 0.74$ and $\langle P^* \rangle \approx 1.8$. This would be equivalent to an experiment in which $T$ is kept reasonably constant and the pressure is raised from $-0.25$ to $1.8$ (in reduced units).

The hope is that we have created a solid. However, the thermodynamics alone cannot determine what phase it is in (unless the full phase diagram is already known for the Lennard-Jones potential). However, $g(r)$, as shown in Figure 6.8b, is that for a solid, with well-defined peaks indicating the shells of neighbors surrounding a central atom with a low probability of having atoms between the peaks.[15] If we calculated the average number of atoms in each of the peaks as well as the average distance of the peaks, we would see that there are 12 atoms in the first peak, 6 in the second, 18 in the third, etc. The peak positions are located at $(1, \sqrt{2}, \sqrt{3}, \ldots)a_o$, with $a_o \approx 1.12$. Based on the positions of the neighbors and the number of atoms in each cell, the system is clearly in an $fcc$ structure, as expected for systems interacting with a Lennard-Jones potential.

The radial distribution function indicates that a solid has been formed in a calculation with $\rho^* = 1.0$, $\langle T^* \rangle \approx 0.74$ and $\langle P^* \rangle \approx 1.8$. Another metric indicating whether the system is a solid or a liquid is the translational order parameter $\rho(\mathbf{k})$ of Eq. (G.48), where $\mathbf{k}$ is a reciprocal lattice vector of the solid structure. In Figure 6.9 we show the value of $\rho(\mathbf{k})$ using a reciprocal lattice vector of an $fcc$ lattice, $\mathbf{k} = (2\pi/a_o)(-1, 1, -1)$, where $a_o$ is the unit cell length, not the simulation cell size. Based on its definition, $\rho(\mathbf{k})$ will equal 1 for a perfect $fcc$ solid and should average to 0 for a liquid, in which there is no orientational order over time.

Figure 6.9a shows $\rho(\mathbf{k})$ calculated during the equilibration run shown in Figure 6.5. The system started as an $fcc$ solid (set by the choice of initial conditions) and quickly lost the $fcc$ ordering over the first 400–500 time steps. Comparison with Figure 6.5 shows that the system had not quite reached its equilibrium state by the time the $fcc$ ordering seems to have been lost. That $\rho(\mathbf{k})$ decayed to zero is suggestive that the system had melted, but it could have been

---

[15] The shape of $g(r)$ in Figure 6.8b is that for a solid at relatively high temperatures. At lower temperatures, the peaks would be sharper with essentially no probability of having atoms at distances between the neighbor shells, as shown in Figure G.5b. At zero temperature (neglecting zero-point motion) $g(r)$ would be a series of delta functions at the distances of the neighbor shells.

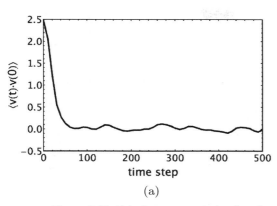
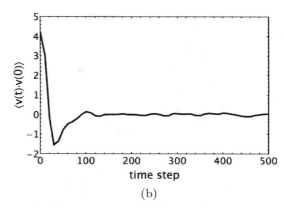

(a)                                                      (b)

**Figure 6.10** Velocity autocorrelation function from Eq. (6.28). Note that the statistics for these results are not particularly good. (a) $\langle \mathbf{v}(t) \cdot \mathbf{v}(0) \rangle$ for the simulation results shown in Figure 6.6 with $\rho^* = 0.55$. (b) $\langle \mathbf{v}(t) \cdot \mathbf{v}(0) \rangle$ for the solid structure shown in Figure 6.8b with a density of $\rho^* = 1.0$.

that another crystal structure had formed such that $\rho(\mathbf{k}) = 0$. $g(r)$ in Figure 6.8a shows that the system has formed a fluid.

The initial time steps of the simulation that led to the radial distribution function in Figure 6.8(b) are shown in Figure 6.9b. The simulation was started in perfect $fcc$, so the value is $\rho(\mathbf{k}) = 1$ at $t = 0$. As the atoms start to move, the perfect symmetry is lost and $\rho(\mathbf{k})$ rapidly decreased to about 0.9. The structure is clearly $fcc$ (or some other structure with that reciprocal lattice vector) with normal thermal vibrations leading to a slight loss of the average order.

## Time correlation functions

Correlation functions can tell us more than just structural information. They can also provide insight into the *dynamics* of the atomic motion. In Appendix G.7.1 we discuss the time correlation functions, which represent how values of a quantity at one time are correlated with its value at another time. Such functions provide a measure for many quantities, including the loss of order over time. We show in Appendix G.7.1 that the velocity autocorrelation function $c_{vv}(t)$ (the correlation of the velocity with itself) is proportional to

$$\langle \mathbf{v}(t) \cdot \mathbf{v}(0) \rangle \tag{6.28}$$

where the averages are over all the particles. We also show that the diffusion constant $D$ can be found by a time integral over $c_{vv}(t)$.

In Figure 6.10 we show the calculated values of $\langle \mathbf{v}(t) \cdot \mathbf{v}(0) \rangle$ for the liquid simulation from Figure 6.5 and the solid structure shown in Figure 6.8b. We see that $\langle \mathbf{v}(t) \cdot \mathbf{v}(0) \rangle$ rapidly decreases to zero over the first 300–400 time steps. These results show similar behavior to the simulation results for $c_{vv}(t)$ in Lennard-Jones liquids shown in Figure G.4. Note the dip below zero in Figure 6.10(b), which corresponds to the "backscattering" discussed in Appendix G.7.1.

The statistics in Figure 6.10 are poor. $\langle \mathbf{v}(t) \cdot \mathbf{v}(0) \rangle$ was calculated by averaging $\mathbf{v}(0) \cdot \mathbf{v}(t)$ over the atoms in an 864-atom system for one trajectory, i.e., we simply calculate $\mathbf{v}(0) \cdot \mathbf{v}(t)$ for

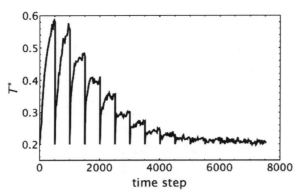

**Figure 6.11** Variation of $T^*$ in a velocity rescaling simulation.

every atom at every time step and averaged. Comparison to Figure G.4 shows that the statistics using this simple method are not particularly good. In Appendix 6.9.4 we discuss how to better evaluate $\langle \mathbf{v}(t) \cdot \mathbf{v}(0) \rangle$.

## 6.3 VELOCITY RESCALING

A practical weakness of simulations in the microcanonical ensemble is the inability to set a specific temperature. As shown in Figure 6.5, the system typically drifts away from the initial temperature as the atoms relax away from their initial positions. However, we often want to compare with experiments, which are performed at specific temperatures. In Section 6.4.1 we discuss a method by which molecular dynamics can be transformed into a canonical ensemble, in which $T$ is fixed. Here we want to discuss an approach in which the simulations remain in the microcanonical ensemble, but the system is driven to a desired temperature.

The easiest way to force the system to be a specific temperature is to rescale velocities. If the desired temperature is $T_s$ and the instantaneous temperature is $T(t) = 2K(t)/(3Nk_B)$, the system can be forced to take on the temperature $T_s$ by *rescaling* the velocities by[16]

$$v_{i\alpha}^{new} = \sqrt{\frac{T_s}{T}} v_{i\alpha} .$$  (6.29)

If, after some time, the average temperature has again drifted away from $T_s$, then the velocities can be rescaled again. The general approach is to rescale the velocities, allow for new equilibration, and then rescale again if needed. An example of velocity rescaling is shown in Figure 6.11. The desired temperature was $T^* = 0.2$. The system was started under the same conditions as the simulation in Figure 6.5, with the initial velocities derived from a Boltzmann distribution with $T^* = 0.2$. The temperature quickly rose, however, to about 0.6. After 500 time steps, the velocities were rescaled so the temperature was 0.2. It quickly rose again, only to be rescaled after another 500 time steps. After about 10 cycles of rescaling, the system was driven

---

[16] $T \propto K \propto \langle v^2 \rangle$, so $T_1/T_2 \propto K_1/K_2 \propto \langle v^2 \rangle_1/\langle v^2 \rangle_2$, or $\langle v^2 \rangle_2 \propto \langle v^2 \rangle_1 T_1/T_2$.

to have a stable temperature of $T^* \approx 0.2$, rather than the final temperature of 0.80 found in the unconstrained simulation.

After the system has achieved the desired temperature, the velocity rescaling should be turned off. The system should be allow to *equilibrate* after the last velocity rescaling before taking averages. Note that the averages are still in the *NVE* ensemble. Thus, velocity rescaling is not a means of creating molecular dynamics simulations at constant temperature, i.e., in the canonical ensemble. It is just a way to force the final temperature to a desired value.

## 6.4 MOLECULAR DYNAMICS IN OTHER ENSEMBLES

Systems studied with standard molecular dynamics methods discussed so far are in the micro-canonical (*NVE*) ensemble. The energy is fixed, so its conjugate variable the temperature is not. The volume is fixed, so its conjugate variable the pressure is not. Since the world generally is at constant temperature and constant pressure, the conditions in standard molecular dynamics can be very inconvenient for comparison with experimental data. There have been many efforts to extend molecular dynamics to systems at constant temperature and pressure. Velocity rescaling was one such approach, though the simulations remained in the microcanonical ensemble. More interesting approaches have been developed to convert the molecular dynamics method to different ensembles. To extend molecular dynamics to other ensembles requires a trick. The idea is rather intuitive, and goes back to basic ideas in statistical mechanics.

Suppose we want some quantity, such as the temperature, to be fixed in a molecular dynamics simulation. We could imagine that every time the temperature dropped below the desired value, we could add some kinetic energy (since we can associate the temperature with $K$) and every time the temperature exceeded the desired value, we could remove some kinetic energy, which is the procedure we followed (in a brute force way) in velocity rescaling. Now suppose we could do this automatically, by directly coupling the motion of the atoms to a heat bath such that energy flows into and out of the simulation. If this energy exchange is done properly, molecular dynamics simulations can be converted from being *NVE* to being *NVT* – from the microcanonical to the canonical ensemble.

### 6.4.1 Molecular dynamics in the canonical (*NVT*) ensemble

The Nośe thermostat [241], with a generalized version referred to as the Nosé-Hoover thermostat [154], takes molecular dynamics from the *NVE* to the *NVT* ensemble. The general idea is to introduce a new variable (a new degree of freedom) that represents (in some way) a connection of the system to some reservoir of energy – a heat bath – so that energy can flow into and out of that reservoir, constraining the system to stay at a fixed temperature $T$.

The Nosé method introduces an artificial force that couples the system to an external heat bath at a temperature $T_s$. This force is derived from an equally artificial potential energy, which is added to the overall Hamiltonian. To complete the new Hamiltonian requires yet another artificial term, the kinetic energy associated with the coupling force.[17] What Nosé showed is that for the correct choice of artificial potential and kinetic energy, the simulations are changed to the canonical ensemble [241].

In the Nosé method, the particle velocities are related to the time derivative of position by

$$\mathbf{v} = s\frac{d\mathbf{r}}{dt},\tag{6.30}$$

where the variable $s$ couples the system to the heat bath. Increasing (decreasing) $s$ increases (decreases) the particle velocities, thus changing the kinetic energy and temperature. Nosé introduced a fictitious potential $U_s$ and fictitious kinetic energy $K_s$ for the variable $s$,

$$U_s = \mathcal{L}\,k_B T_s \ln(s)$$

$$K_s = \frac{p_s^2}{2Q},\tag{6.31}$$

where $\mathcal{L} = 3N + 1$ is a constant chosen to ensure that the final results agree with the canonical ensemble [109]. The momentum associated with the variable $s$ is $p_s = Q\,ds/dt$, where $Q$ is a coupling parameter (a "mass"). The Hamiltonian associated with the bath coordinate is defined in the usual way as

$$\mathcal{H}_s = K_s + U_s.\tag{6.32}$$

The total Hamiltonian of the system of atoms plus the bath coordinates is

$$\mathcal{H}_T = \mathcal{H}_o + \mathcal{H}_s,\tag{6.33}$$

where $H_o$ is the usual Hamiltonian from Eq. (6.14).

From the total Hamiltonian $\mathcal{H}_T$, we can derive the equations of motion for the atom positions that include the dependence on the bath coupling parameter $s$. We can also derive an additional equation for $s$, which corresponds to the rate of energy flowing into and out of the reservoir. Using Eq. (6.7) and Eq. (6.33), the coupled equations of motion for the Nosé thermostat are [241]

$$\frac{d^2\mathbf{r}_i}{dt^2} = \frac{1}{ms^2}\mathbf{F_i} - \frac{2}{s}\frac{ds}{dt}\frac{d\mathbf{r}_i}{dt}\tag{6.34}$$

$$Q\frac{d^2s}{dt^2} = \sum_{i=1}^{N} m\left(\frac{d\mathbf{r}_i}{dt}\right)^2 s - \frac{\mathcal{L}}{s}k_B T.$$

A convenient way to implement these in a molecular dynamics calculation using the Verlet algorithm of Eq. (6.11) is given elsewhere [109].

$Q$ in Eq. (6.31) is a parameter that defines the strength of the coupling to the heat bath and is an input into the simulation. Too high a value for $Q$ results in slow energy flow and too low a value for $Q$ leads to poor equilibration. In practice, $Q$ is often varied until the system behaves as desired.

---

[17] There are a number of subtleties that we do not discuss here. Please see Section 6.1.2 in [109] for a more rigorous presentation.

The Nosé thermostat is widely used in modern simulations. Simulations done with this method yield a canonical ensemble distribution in positions and momenta, but do not necessarily give the correct short-term dynamics, owing to the factors of $s$ in the equation of motion for the particle positions.

### 6.4.2 Molecular dynamics in the isobaric (*NPH*) ensemble

In standard, microcanonical, MD, the volume is set and the pressure is determined only after the end of the simulation as an averaged quantity. If the goal is to compare with experiment, then a fixed pressure is usually desired. One procedure would be to vary the size of the simulation cell until the desired pressure is obtained, in much the same way as we scaled the velocity to obtain a desired temperature, although that approach is tedious. The simulation would remain in the microcanonical ensemble. Another limitation to normal molecular dynamics is that the *shape* of the simulation cell is fixed. For the relatively small systems usually studied with molecular dynamics, the fixed cell can inhibit the study of phase transformations in which the symmetry changes (e.g., most solid-solid phase transformations). A technique called the *Parrinello-Rahman method* was developed to enable simulations both at constant pressure (or stress) and with a simulation cell that can change with time [252]. For the constant pressure case, for example, the enthalpy ($H = E + PV = K + U + PV$) is constant for the simulation, not the energy, corresponding to the isobaric (*NPH*) ensemble.

In the same spirit as the Nosé thermostat, the Parrinello-Rahman method introduces terms that couple the simulation cell to an external field, in this case a field that represents a force acting on the simulation cell itself. We will not go into the details of this method here, but just sketch how the method works and the problems to which it can be applied. In Chapter 7 we will discuss a Monte Carlo version of this approach, which is easier to implement.

Consider first a molecular dynamics system under constant pressure $P_{app}$.[18] The basic idea is that pressure exerts a force per area on the simulation cell with a potential energy given by $P_{app}V_{cell}$, where $V_{cell}$ is the volume of the cell. The potential energy term is added to the Hamiltonian. As in the Nosé-Hoover thermostat, a kinetic energy term must be added as well, along with an effective "mass" that determines the overall dynamics of the changing volume. From this Hamiltonian the equations of motion can be derived [252].

In the simplest case, we apply an external fixed pressure and keep the shape of the system constant. This type of simulation would be appropriate for modeling the behavior of a fluid. The volume of the system as well as the calculated pressure (Eq. (6.18)) would generally show large variations until the system equilibrates, after which they would fluctuate around their average values. An important test is to ensure that the average calculated pressure, from Eq. (6.18), should equal the applied pressure.

If the goal is to model solid-solid phase transitions, then having a system that can undergo a shape change is essential. This can be accomplished by describing the simulation cell as a

---

[18] If the system is under a constant applied stress $\sigma_{app}$, then the potential energy is the strain energy of the system, as given in Eq. (H.17), with the stress given by $\sigma_{app}$.

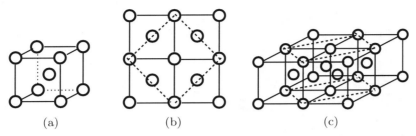

**Figure 6.12** The Bain transformation: $bcc \leftrightarrow fcc$ via a change in the face-centered tetragonal unit cell parameters. (a) A body-centered cubic unit cell. (b) Top view of four body-centered cubic unit cells showing a new face-centered tetragonal cell. (c) View of body-centered cubic as a face-centered tetragonal crystal.

triclinic lattice, such that all six lattice parameters (three lengths and three angles) are allowed to vary in the simulation, which is reflected in the equations of motion for the system. The geometry of the simulation cell and the positions of the atoms are described using the formalism introduced in Appendix B.2.4 for non-cubic lattices.

For example, suppose we want to model a system that can undergo a phase transition from $fcc$ to $bcc$ (for example, the Bain transformation in Fe). It is not possible to create a cubic system with a constant number of atoms in which such a transition can take place. However, if we allow for the length of the simulation cell sides to change, then such a transition is possible, as shown in Figure 6.12. An $fcc$ structure can become a face-centered tetragonal by changing the lengths of the sides of the simulation cell. If two of those lengths are $\sqrt{2}$ times the other, the system is equivalent to a $bcc$ structure, for example if $a = b = \sqrt{2}c$, as shown in Figure 6.12c. Thus, shape changes in the simulation cell enable the modeling of certain solid-solid transitions.

## 6.5 ACCELERATED DYNAMICS

As noted above, the fundamental time step in a molecular dynamics simulation must be of the order of $10^{-15}$–$10^{-14}$ seconds to achieve accurate integration of the equations of motion. The *computational* time of a conventional MD simulation scales linearly with the number of time steps since the number of computational operations needed to calculate the forces, update the energies, move the particles, etc. is the same for each time step. A calculation of a nanosecond will require of the order of $10^6$ time steps and a calculation for a microsecond will take 1000 times longer. Even with the fastest computers, calculations rarely extend beyond a nanosecond or so. Thus, conventional MD simulations are typically restricted to phenomena of a nanosecond or less. Indeed, since the length scale of a standard MD simulation is typically of the order of nanometers, processes that can be modeled well with MD have *rates* of the order of nanometers/nanoseconds or meters/second.[19]

[19] As an example, consider a molecular dynamics calculation of physical vapor deposition. A rate of deposition of a meter/second is many orders of magnitude faster than that achieved experimentally. Indeed, deposition at a rate of 1 meter/second could deposit Mt. Everest in about 2.5 hours, which is a trifle unrealistic.

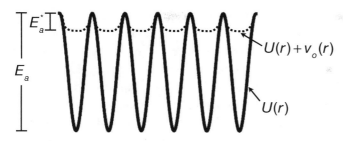

**Figure 6.13** Schematic view of the addition of a bias potential (dashed curves) to the actual potential (solid curves) in a one-dimensional system. The dynamics of the actual potential can be extracted from simulations on the biased potential, as described in [333].

There are, however, numerous issues in materials science that are atomistic in nature and that require much longer times. The classic example is diffusion, which in solids is a slow process, with activated jumps from site to site. The diffusion constant in a solid varies widely, but is generally small enough that in a nanosecond a particle on average has very little appreciable movement, even when located next to a vacancy.

A series of very clever methods have been proposed to extend the time scale in molecular dynamics. These methods can, for some problems, extend MD simulations even into the millisecond regime [333]. We will not go into much detail here, but will outline the ideas behind just one of the many methods.

Suppose the dynamics of a system are characterized by "infrequent events", in which the system makes an occasional transition from one potential basin to another.[20] In Figure G.6, we show a schematic view of a potential surface in a solid, whose contours represent the potential energy of an atom around its lattice site in a solid. The heavy lines represent lines of constant energy, with the energy decreasing towards the centers at each lattice site. In the middle is a basin in which an atom vibrates around its mean position, while at the right and left are barriers between the original basin and adjacent ones. The motion of the atom is shown as the light line. The atom spends most of its time moving around the first basin. Every once in a while, an atom develops sufficient energy in a mode that points toward an adjacent basin so that it can move through the dividing surface (the vertical lines) at the ridge top between the basins. The trajectory between basins is referred to as the *transition coordinate* or, because of rate theory's origin in chemistry, the *reaction coordinate*. The atom may cross the dividing surface to the other basin or it may recross back into its original basin.

The type of motion in Figure G.6 would be typical of, for example, diffusion of atoms from one site to another. As a particle moves from one basin to another, it traverses through the lowest-energy regions between the basins, which are the saddle points of the surface. If we knew the topology of the potential energy surface, especially at the saddle points, rate theory could be used to estimate the dynamics. However, the events that occur in real systems are complicated and the potential surfaces are too complex for that approach to be feasible.

Consider the one-dimensional model system shown in Figure 6.13. A particle can diffuse along the chain with its dynamics governed by the potential surface shown in the solid curves. From kinetics, the rate of jumping from one site to its neighbor is proportional to $\exp(-E_a/k_B T)$, where $E_a$ is the barrier height shown in Figure 6.13. If $E_a$ is large, then the jump rate becomes

---

[20] A brief review of kinetic rate theory is given in Appendix G.8.

extremely small and the particle would simply vibrate at the bottom of the potential well. A molecular dynamics calculation would show an oscillatory motion and, over the possible time scales in a simulation (e.g., nanoseconds), would show few, if any, jumps. If $E_a$ were relatively small and the temperature sufficiently high, some movement could take place.

A number of methods have been introduced to accelerate the dynamics of a molecular dynamics simulation [333]. One such method, called "hyperdynamics" [330, 331], is rather intuitive. A "bias" potential is created in such a way as to not change the shape of the potential surface near the saddle points. That bias potential is added to the real potential so that the net energy surface shown as the dashed lines in Figure 6.13 is created. The bias potential is designed so that the change in dynamics occurs mainly through a change in the barrier height. Since $E_a^* \ll E_a$ in Figure 6.13, the ratios of the rates of transitions on the biased surface relative to those on the actual surface go as $\exp(-(E_a^* - E_a)/k_B T)$ and are greatly enhanced. The system evolves on this revised potential surface and makes transitions from state to state with the correct relative probabilities, but at an accelerated rate. Time is no longer an independent variable, but is instead estimated statistically as the simulation evolves, ultimately converging on the correct result. In simulations on metal systems using embedded-atom interatomic potentials, for example, simulation times can be extended into the microsecond range with no advanced knowledge of the transitions the system will make. Particularly lucid explanations of this approach, as well as a number of other methods to accelerate the dynamics, are given in [260, 333].

## 6.6 LIMITATIONS OF MOLECULAR DYNAMICS

Before discussing applications of molecular dynamics in materials, it is worthwhile reminding ourselves about the limitations of the method. It is always important to do this sort of assessment before starting any modeling or simulation. Otherwise, it is impossible to know how seriously to take the results.

From Chapter 1, a simulation is a study of the response of a modeled system to external forces and constraints. For a molecular dynamics simulation, the model is the description of the interactions between the atoms (and molecules) in the system. Intermolecular potentials are approximations to the true interactions and, depending on system and complexity of the potential form, may or may not give a particularly accurate description of a specified material. Materials in which the bonding is distorted from spherical symmetry, for example in systems with covalent bonds, are generally less well described than those with simple bonding. If the bonding is dynamic, for example if there is a change in bonding depending on atomic positions, then simple descriptions are less likely to be applicable.[21]

In addition to inherent errors in the models for interatomic interactions, there are also limitations to the molecular dynamics method itself. The number of atoms that can be reasonably

---

[21] Molecular dynamics methods based on the electronic structure methods of Chapter 4 are becoming more common, though they are limited to small systems [56].

included in a molecular dynamics simulation is limited by the computational burden of calculating the forces and energy, which, when cutoffs are used, scales roughly as the number of atoms $N$.[22] The limitations are made worse by the simple geometry of the simulation cell. For $N$ atoms in a simulation cell with linear length $L$ and volume $V = L^3$, the density is $N/L^3$. Thus, to keep constant density, $L$ scales $N^{1/3}$ – to double the linear size of the cell requires $8N$ atoms to maintain constant density and a simulation would take roughly 8 times as long to run.

Until the development of parallel computers, simulations with a few hundred thousand atoms were considered very large. Today, multimillion atom simulations are relatively common [279] and, for those with access to very large parallel machines, simulations with multibillion atoms are possible.[23] A cubic micron of Cu contains about $10^{11}$ atoms, so simulations of roughly a micron in size are now feasible, if limited in time, as discussed next.

The time steps in molecular dynamics are very small, of the order of $10^{-15}$–$10^{-14}$ seconds. A long simulation will yield at best a few nanoseconds. These short time scales make the study of many physical processes quite challenging with standard molecular dynamics. There is hope, however, that some of these time constraints can be overcome for certain systems, as discussed in the section on accelerated dynamics.

With all these limitations – potentials, size, and time – there are many strengths to molecular dynamics. With the advent of new computers and the new accelerated dynamics approaches, the ability of MD to couple directly with experiment has greatly improved, yielding fundamental information not available experimentally. Indeed, there is now a direct link between experiment and modeling in some nanoscale materials. In the next section we list a few representative examples of applications of molecular dynamics.

## 6.7 MOLECULAR DYNAMICS IN MATERIALS RESEARCH

There are tens of thousands of papers in the literature employing molecular dynamics simulations to study the properties of materials. Indeed, there have been applications of molecular dynamics to all types of materials, and to essentially all phenomena for which molecular dynamics is at least somewhat applicable. Here we list just a few examples, based around a small set of topics.

- Structure and thermodynamics: the earliest use of molecular dynamics in 1957 was to study the structure and thermodynamics of liquids based on a system of hard spheres [2]. Since then, the application of molecular dynamics to determine the structure and thermodynamics of materials has been common for both crystalline and noncrystalline bulk materials, with so many papers that listing them is not practical. Given the size constraints of molecular

---

[22] If there are $M$ neighbors on average within the cutoff distance around each atom, then the calculation scales roughly as $NM$.

[23] A trillion-atom simulation has been reported for atoms interacting with a Lennard-Jones potential with a cutoff of $r_c = 2.5\sigma$. It was carried out on one of the fastest computers available at that time and modeled the system for a total time of 10 picoseconds ($\times 10^{-11}$ seconds) [118].

dynamics discussed in the previous section, applications to nanoscale materials have become common. A review of some of these results, along with detailed comparison with experiment, can be found in [20].

- Extreme conditions: simulations are often used to describe materials under extreme conditions, for example high pressures or high temperatures, for which the experiments might be difficult or impossible. Molecular dynamics simulations have been used to answer many questions about such regimes. Simulations can provide information on the equation-of-state of material,[24] the stability of various crystal phases under different conditions, and can be used to map out phase diagrams. Many of these simulations have been performed for geological applications to better understand the dynamics of the inner earth. For example, equation-of-state calculations for silica at high pressure [30], phase stability in $MgSiO_3$ and $CaSiO_3$ using *ab initio* methods from Chapter 4 [315], and phase diagrams for the high-pressure alkali halides [275], etc.

- Defect structure, properties, and dynamics: simulations can shed light on the structure, properties and dynamics of defects (e.g., vacancies, interstitials, dislocations, grain boundaries, surfaces) that cannot easily be resolved with experiment. Critical questions may include the atomic arrangements and energetics of the defect in both pure and alloyed systems. The dynamics of the defect are also of great importance, and at times quite amenable to calculation with molecular dynamics, whether it be the details of diffusion (which in solids essentially always involves a defect), microstructural evolution (e.g., grain growth), the interaction of defects (e.g., dislocations with grain boundaries), or many other phenomena. As noted above, nanoscale systems fit in well with molecular dynamics so many calculations have been done at that scale, some of which having been reviewed in [229].

The structure of bulk defects has been studied for decades with molecular dynamics, for example in [35]. Dislocation structures and interactions have also been well studied, as in work examining the short-range dynamical interactions between oppositely signed screw dislocations [301] (see Appendix 6.9.1). Other applications include the study of grain boundary migration in metals as is discussed in [317]. Emission of full and partial dislocations from a crack were studied in [307]. Conventional molecular dynamics was employed to determine the diffusivities in the *hcp* and *bcc* structures of zirconium, finding that diffusion is dominated by interstitial diffusion in the *hcp* structure and by both vacancy and interstitial diffusion in *bcc* [225]. Accelerated dynamics techniques have also been employed, as described in a study of defect kinetics in spinel structures [316]. Grain growth for nanoscale crystals has been studied by a number of groups, including [362]. These are just a few examples.

- Deposition: an important processing method in materials that is amenable to molecular dynamics study is deposition. Many simulations have been made of the deposition process using standard molecular dynamics. An example is a study of the stress and microstructure evolution during polycrystalline Ni film growth [249]. Care must be taken when looking at

---

[24] An equation of state is a thermodynamic equation describing the state of a system (e.g., $V$ or $U$) under given thermodynamic conditions (e.g., $P$, $T$).

simulations of this process, as in many published studies the simulated deposition rate far exceeds anything possible in the laboratory.

- . . . and many more.

## 6.8 SUMMARY

This chapter introduces the basic ideas and implementations of molecular dynamics, starting from Newton's equations. Methods for integrating the equations of motion are discussed, along with some examples of successful and unsuccessful methods. An example calculation based on the use of Lennard-Jones potentials was presented, along with discussions of the various stages of a molecular dynamics simulation. Analysis of simulation results was discussed, along with implementation methods to calculate correlation functions. Approaches to implement molecular dynamics in $NVT$ and $NPH$ ensembles were introduced, along with a brief discussion of the very important new methods to accelerate molecular dynamics time scales. Finally, an assessment of the limitations of molecular dynamics was discussed.

### Suggested reading

There are probably more books about molecular dynamics than on any other method used to model materials. Here are a few that I find useful:

- D. Frenkel and B. Smit, *Understanding Molecular Simulations* [109]. This is an excellent book describing the basics of atomistic simulations.
- While focused on simulations of liquids, the book by M. P. Allen and D. J. Tildesley, *Computer Simulation of Liquids* [5], is a good basic guide to how to perform simulations.
- Haile's book, *Molecular Dynamics Simulation: Elementary Methods* [135], is also quite useful.

## 6.9 APPENDIX

In this appendix we cover a number of basic methods used in molecular dynamics simulations, including how to set initial conditions for positions and velocities. We also present methods for calculating pair distribution and velocity autocorrelation functions.

### 6.9.1 Choosing initial positions

The initial positions and boundary conditions in a molecular dynamics calculation depend on what quantities one wishes to calculate. The most straightforward calculations are designed to calculate bulk thermodynamic or structural properties. In those cases, periodic boundary

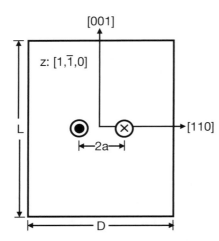

**Figure 6.14** Schematic representation of the geometry of the simulations of two screw dislocations in an *fcc* crystal. The two screw dislocations have Burgers vector $a/2$ [110] and $-a/2$ [110], and are separated by a distance $2a$. The periodic length in the $x$-direction is $D$ while that in the $y$-direction is $L$.

conditions are employed with a simulation cell with the symmetry of the problem of interest. For a liquid, cubic simulation cells are usually employed. Generally, cubic cells are chosen for solids with cubic symmetry as well. For solids with more complicated crystallographic symmetry, the simulation cell should be chosen with the same symmetry as the solid of interest. For example, a solid with triclinic symmetry is usually modeled with a simulation cell with the same symmetry. Since we are usually interested in systems with more than one unit cell, the simulation cell is generated as multiples of the fundamental cell, as shown in Figure 6.3. In the example in the text for a system of atoms described with a Lennard-Jones potential, we used 864 atoms, which was constructed from a $6\times6\times6$ array of 4-atom *fcc* unit cells, as discussed in Figure 6.3.

From a materials perspective, one of the main uses of atomistic simulations is the ability to study defects in materials, such as surfaces, grain boundaries, dislocations, and vacancies. For those cases, the initial positions and boundary conditions must be carefully chosen to match the problem of interest. For defects, these boundary conditions may be quite complicated.

In Figure 5.12, we show a simulation cell that could be used to model grain boundaries. As discussed in that figure, periodic boundary conditions are employed parallel to the boundary. Perpendicular to the boundary, however, the system is broken into simulation regions near the boundary and border regions that link the simulation region to the bulk structure of the crystal. Since the periodic cell must reflect the symmetry of the boundary (or lack thereof), these simulations generally require large simulation cells parallel to the boundary.

Atomistic simulations have also been used to examine the interaction of small numbers of dislocations.[25] For example, the simulation cell in Figure 6.14 was used to model the annihilation of two screw dislocations [301]. Typical simulations included about 50 000 atoms. Fully periodic boundary conditions were used.[26] The initial displacements for the dislocations were obtained

---

[25] For those not familiar with dislocations, please see Appendix B.5.

[26] The presence of a dislocation causes long-ranged displacement fields around the dislocation. With a single dislocation, those displacement fields would not match at a periodic boundary (i.e., the displacements of atoms from the dislocation in the central cell would not match those due to the replica dislocations at the boundary). With two dislocations of opposite sign, however, the displacement fields largely cancel at the edges of the cell.

from the known displacement fields from a screw dislocation, as derived from linear elasticity [147]. There are, however, issues in atomistic modeling of dislocations using periodic boundary conditions that arise from long-ranged stress fields. A method to avoid those issues has been developed [54].

### 6.9.2  Choosing initial velocities

The velocities for a molecular dynamics simulation are typically chosen either randomly over a small range or from the Maxwell-Boltzmann distribution given in Eq. (G.22). If the latter approach is taken, then each component of the velocity for each atom is chosen from a distribution

$$\rho(v_{ix}) = \frac{1}{\sigma\sqrt{2\pi}} e^{-v_{ix}^2/2\sigma^2}, \tag{6.35}$$

where $\sigma = \sqrt{k_B T_{int}/m}$ and $T_{int}$ is the desired initial temperature.

This type of probability distribution is referred to as a *normal* (or Gaussian) distribution, which has a general form

$$\rho(x) = \frac{1}{\sigma\sqrt{2\pi}} e^{-(x-\langle x\rangle)^2/2\sigma^2}, \tag{6.36}$$

where the mean is $\langle x\rangle$ and the standard deviation is $\sigma$. In Appendix I.2.1 we discuss a method by which we can generate random numbers in a normal distribution.

The velocity distribution is given as in Eq. (6.35) with $\sigma = \sqrt{k_B T_{inp}/m}$ and $\langle v_{i\alpha}\rangle = 0$ for each component $\alpha$ ($x$, $y$, or $z$) of the velocity of atom $i$. For a Lennard-Jones potential with reduced units, we have $\sigma = \sqrt{T^*}$. For each $v_{i\alpha}$, we use random numbers from a normal distribution calculated with the method described in Appendix I.2.1 to give our initial guess for $v_{i\alpha}$. Since we have a finite number of atoms, we normally do not have $\langle \mathbf{v}_i\rangle = 0$. However, we can ensure that the total velocity of the system is zero by finding $V_\alpha = \sum_{i=1}^{N} v_{i\alpha}$ and then shifting the velocities by $v_{i\alpha} = v_{i\alpha} - V_\alpha/N$.

In Figure 6.15 we show the calculated velocity distribution for an LJ system with $N = 864$ particles and the reduced temperature $T^* = 1$. The $v_{i\alpha}$ in Figure 6.15 do not quite match a normal distribution because we have a finite number of trials. We can force the initial temperature of the system to be exactly equal to $T_{inp}$ by rescaling the velocities using $v_{i\alpha} = \sqrt{T_{inp}/T}$, where $T_{inp}$ is the input temperature and $T$ is determined using Eq. (6.16).

### 6.9.3  Calculating the pair distribution function

The pair distribution function (pdf) of Appendix G.7.2 can be defined as

$$g(r) = \frac{V}{4\pi r^2 N^2} \left\langle \sum_i \sum_{j\neq i} \delta(r - r_{ij}) \right\rangle, \tag{6.37}$$

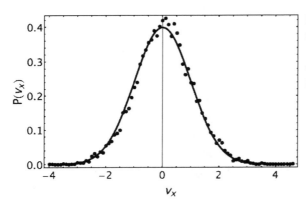

**Figure 6.15** Initial velocity distribution from 864 particles at a reduced temperature of $T^* = 1$. The symbols are the calculated velocity distribution and the solid line is calculated with Eq. (A-1-1).

where $\delta(r - r_{ij})$ is the delta function of Eq. (C.53). This is the conditional probability density of finding a particle at $r_{ij}$, given that particle $i$ is at the origin. Thus, $g(r)$ provides a measure of local spatial ordering in a fluid or solid. Note that since $g(r)$ depends only on distance, no information on the angular distribution is contained in it.

To determine $g(r)$, proceed as follows:

(a) The desired range of $r$-values in the $g(r)$ calculation ranges from 0 to $r_{max}$, where $r_{max}$ is the maximum distance for which $g(r)$ will be calculated. $r_{max}$ should not be greater than $L/2$, where $L$ is the side length of the basic cell.[27] Divide the range of $r$ values into $n_b$ intervals of length $\Delta r = r_{max}/n_b$. The choice of $n_b$ is a balance between statistics and resolution, as discussed in Appendix I.3. If $n_b$ is large, $\Delta r$ is small and there will be few points in that range and the statistics will be poor. If $n_b$ is small, $\Delta r$ is large and there will be poor resolution in the calculated $g(r)$. Typically $n_b = 50 - 100$ is sufficient.

(b) A given configuration $\{\mathbf{r}_1, \mathbf{r}_2, \ldots, \mathbf{r}_N\}$ is scanned to determine the distance $r_{ij}$ between each pair $(i, j)$ of atoms. A "distance channel" number $k = \text{int}(r_{ij}/\Delta r)$ is found, where int is a function that truncates the fractional portion of a number's magnitude with its sign preserved (e.g., int(1.6) = 1).

(c) In a histogram table $g(k)$, the corresponding value is then incremented by 1, $g(k) = g(k) + 1$. At the end of a calculation $g(k)$ will be a count of how many times the distance between atom pairs was between $k \Delta r$ and $(k + 1)\Delta r$. This procedure is repeated every $m$ MD steps, where $m$ is typically set to minimize computational cost and with the realization that the structures do not change very rapidly.

(d) At the end of the simulation, the histogram is normalized according to Eq. (6.37).

For more information on calculating continuous probability distributions, please see Appendix I.3.

---

[27] Because of the periodic boundary conditions, any distances beyond $L/2$ are related by symmetry to those already included.

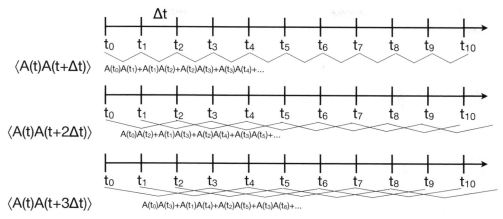

**Figure 6.16** Calculation of a time autocorrelation function showing how one set of time steps can be used to generate good averages using Eq. (6.38). At the top, averaging over each time pair $(t, t + \Delta t)$ yields $\langle A(t + \Delta t)A(t) \rangle$. In the middle, we accumulate all values to calculate $\langle A(t + 2\Delta t)A(t) \rangle$ and at the bottom $\langle A(t + 3\Delta t)A(t) \rangle$, and so on. Adapted from [172].

### 6.9.4 Calculating the time correlation function

There are a number of approaches for calculating time correlation functions, including fast Fourier transform methods, which can be quite efficient. In this section we will discuss another approach that does not require any special programming and that can be done quite easily within a molecular dynamics simulation. We will use the determination of an autocorrelation function as an example.

Consider some time-dependent quantity $A(t)$. The autocorrelation function is

$$\langle A(\tau)A(0) \rangle = \langle A(\tau + t_o)A(t_o) \rangle. \tag{6.38}$$

It does not matter what the time origin is when determining the autocorrelation function – everything is measured relative to an initial starting time.

The average is taken over all the atoms in the system. However, in an atomistic simulation, there are relatively few atoms (many fewer than Avogadro's number) so simply averaging over the atoms rarely gives sufficiently good statistics (cf. Figure 6.10). Thus we need to do something a bit different. We could do as we did in the random-walk model of Chapter 2 and perform a number of molecular dynamic runs from similar starting conditions and then average over the runs. That is not very efficient.

A better approach is to recognize from Eq. (6.38) that every molecular dynamics simulation can be used to generate many starting points to determine $\langle A(\tau + t_o)A(t_o) \rangle$ (discussion adapted from [172]). Such a procedure is shown in Figure 6.16, in which we generate accumulated values for $A(t + \Delta t)A(t)$, $A(t + 2\Delta t)A(t)$, $A(t + 3\Delta t)A(t)$, ... $A(t + n\Delta t)A(t)$ from which we can generate $\langle A(\tau)A(0) \rangle$ as a function of $\tau = n\Delta t$.

In practice, one first sets the time interval over which $\langle A(t)A(0) \rangle$ will be calculated, i.e., $(0, t_{max})$. Assume this range consists of $n$ time steps, i.e., $t_{max} = n\Delta t$. For convenience, drop the $t$ and replace it by time step $k$ ($t = k\Delta t$), identifying the value of $A$ at time step $k$ by $A_k$. We then need to evaluate $\langle A_0 A_k \rangle$ for $k = 1, 2, \ldots n$.

The procedure is relatively straightforward and is based on having stored, at each time step, the previous $n$ values of $A_k$. Then, at each time step $\ell$

(a) Evaluate $A_\ell$.
(b) Calculate $\sum_{i=1}^{n} A_\ell A_{\ell-i}$.
(c) Add $A_\ell$ to the list of $A$ values and drop $A_{\ell-n}$.

As in any method, there are tradeoffs. In this case, one balances the range of $\langle A(t)A(0) \rangle$ to be evaluated versus the storage requirements and the computational time to evaluate it, which goes as approximately $n^2$. To reduce that computational burden, a coarse-grained method has been developed, which is described in [109].

# 7 The Monte Carlo method

In this chapter we introduce the Monte Carlo method, a remarkably powerful approach that is the basis for three chapters in this text. For the purposes of this chapter, Monte Carlo provides an alternative to molecular dynamics for providing thermodynamic information about a material. It differs from molecular dynamics in that it is based on a direct evaluation of the ensemble average, as discussed in Appendix G, and thus cannot yield direct dynamical information, at least as described for the version of Monte Carlo in this chapter.

The Monte Carlo method was devised at Los Alamos in the 1940s to solve multidimensional integrals and other rather intractable numerical problems [227]. The method is based on statistical sampling and is called Monte Carlo in recognition of the very famous casinos there. It is not called Monte Carlo because of gambling – at least not entirely – it is named Monte Carlo at least in part because of its remarkable ability to solve intractable problems.[1]

## 7.1 INTRODUCTION

What is the Monte Carlo method? As first employed, it was a way to solve complicated integrals. As a simple example, Monte Carlo is used to evaluate the one-dimensional integral $\int_1^4 \ln(x)dx$ in Figure 7.1a. First, a region that includes the function to be integrated is defined. Random points in that region are chosen via a random-number generator.[2] The integrated function is just the *fraction* of the points that fall below the curve multiplied by the area of the sampled region.

In Figure 7.1b is shown the convergence of a Monte Carlo integration of $\int \ln(x)dx = -x + x\ln(x)$. The value of the integral over the interval $(1, 4)$ is approximately 2.5418. As is clear from Figure 7.1b, there is very poor convergence of the Monte Carlo integration of this function, with about 10 000 trials needed to obtain a reasonable estimate of the correct answer. We would never evaluate such an integral in this way in one dimension, since there are many other methods that would do better. However, Monte Carlo is often the method of choice for multidimensional integration [51].

---

[1] According to one of its originators, the late Nicholas Metropolis, the name "Monte Carlo" was named in part in honor of the uncle of another of its discovers, Stanislaw Ulam. It seems this uncle would borrow money from relatives because he "just had to go to Monte Carlo". He just had to go. Monte Carlo solved his problem as would no other solution. The Monte Carlo method is similar – it can often solve problems that cannot be solved any other way.

[2] Random numbers are discussed in Appendix I.2.

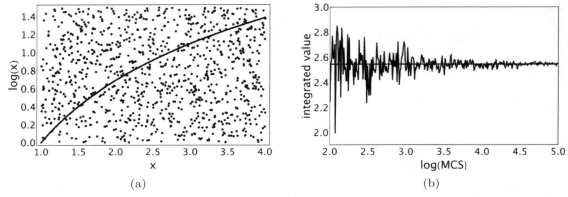

**Figure 7.1** An example of a Monte Carlo integration in one dimension. (a) Plotted is ln(x) versus x in the region $1 \le x \le 4$. In Monte Carlo integration, many random numbers are chosen in the integration region shown, which has area $A = 4 \times 1.5 = 6$. The value of the integral is the fraction of random numbers that are less than the curve times the area $A$. (b) The value of the computed integral $\int_1^4 \ln(x)dx$ as a function of the number of random numbers, called Monte Carlo steps (MCS). The exact answer is shown as the straight line and is approximately 2.5418.

The focus of this chapter is on another use of the Monte Carlo method. In Appendix G, we discuss two ways of evaluating thermodynamics from simulations: by time averaging or by ensemble averaging. The molecular dynamics method is based on time averaging. The Monte Carlo method enables us to determine thermodynamics by evaluating ensemble averages.

## 7.2 ENSEMBLE AVERAGES

One goal of Monte Carlo simulations is to calculate the thermodynamics of a system. Standard Monte Carlo describes systems with a constant temperature $T$, a constant number of particles $N$, and a constant volume $V$, and thus corresponds to the canonical ensemble.[3] As discussed in Appendix G, thermodynamic averages for systems in the canonical ensemble are given by expressions like

$$\langle B \rangle = \frac{\sum_\alpha e^{-E_\alpha/k_B T} B_\alpha}{\sum_\alpha e^{-E_\alpha/k_B T}} = \sum_\alpha B_\alpha \rho_\alpha \,. \tag{7.1}$$

where $B$ is some quantity (for example the potential energy), the subscript $\alpha$ indicates the possible configurations in the system, and the probability of being in a specific configuration is[4]

$$\rho_\alpha = \frac{e^{-E_\alpha/k_B T}}{\sum_\alpha e^{-E_\alpha/k_B T}} = \frac{e^{-E_\alpha/k_B T}}{Q} \,, \tag{7.2}$$

---

[3] The ensemble formation of statistical thermodynamics is discussed in Appendix G.5. Since the canonical ensemble refers to a system constrained to have a constant number of particles, $N$, constant volume $V$, and constant temperature $T$, it is often written as the $NVT$ ensemble.

[4] If Eq. (7.1) and Eq. (7.2) are unfamiliar to you, please see Appendix G.

where the partition function $Q$ is

$$Q = \sum_\alpha e^{-E_\alpha/k_B T} .$$ (7.3)

The relative probability between two states takes a very simple form. Suppose a system is in a state $\alpha$ with energy $E_\alpha$. The probability of the system being in a state $\beta$ with energy $E_\beta$ relative to the probability of being in $\alpha$ is

$$\frac{\rho_\beta}{\rho_\alpha} = \frac{e^{-E_\beta/k_B T}}{Q} \frac{Q}{e^{-E_\alpha/k_B T}} = e^{-(E_\beta - E_\alpha)/k_B T} .$$ (7.4)

Note that $Q$ cancels out of this expression. The relative probabilities of the two states are completely determined by the energy difference $\Delta E_{\alpha,\beta} = E_\beta - E_\alpha$

$$\text{if } E_\beta - E_\alpha \leq 0 \quad \text{then} \quad \frac{\rho_\beta}{\rho_\alpha} \geq 1$$

$$\text{if } E_\beta - E_\alpha > 0 \quad \text{then} \quad \frac{\rho_\beta}{\rho_\alpha} < 1 .$$ (7.5)

When applying Monte Carlo to atomic or molecular systems, the sum over discrete states of Eq. (7.1) is replaced by a set of integrals. For a system of $N$ atoms, from Eq. (G.16) and Eq. (G.17), the probability of an atomic configuration, $\{\mathbf{r}^N = \mathbf{r}_1, \mathbf{r}_2, \ldots, \mathbf{r}_N\}$, is

$$\rho(\mathbf{r}^N) = \frac{e^{-U(\mathbf{r}^N)/k_B T}}{Z_{NVT}} ,$$ (7.6)

where the *configurational integral* is

$$Z_{NVT} = \int e^{-U(\mathbf{r}^N)/k_B T} d\mathbf{r}^N .$$ (7.7)

The average value of a quantity that depends only on positions, for example the potential energy $U$, is

$$\langle U \rangle = \frac{1}{Z_{NVT}} \int e^{-U(\mathbf{r}^N)/k_B T} U(\mathbf{r}^N) d\mathbf{r}^N ,$$ (7.8)

The integrals are over the three coordinates of each atom, i.e., $d\mathbf{r}^N = dx_1 dy_1 dz_1 dx_2 dy_2 dz_2 \ldots dx_N dy_N dz_N$. Thus there are $3N$ coordinates that define all the possible configurations of the system.

In principle, evaluating Eq. (7.1) or Eq. (7.8) requires a list of all possible configurations. However, many (if not most) of the configurations of any system will be possible, but not probable. For example, in a system of atoms, if two atoms are very close to each other, the interaction energy $\phi$ between the two atoms would be very high and positive. Thus the potential $U$ and total $E$ energies would be large and the probability of being in that state, which is proportional to $\exp(-E/k_B T)$, would be vanishingly small. In essence, the system will essentially never take on that sort of configuration.

As discussed in Appendix G.6, the probability of finding a state with an energy $E$ is highly peaked around the mean of $E$. In Figure G.3 we show a schematic of the probability distribution of the energy in a thermodynamic system and show that the standard deviation of that distribution goes as $1/\sqrt{N}$, where $N$ is the number of atoms. In a thermodynamic system, $N$ is roughly Avogadro's number, so the distribution is so narrow that in that case we cannot differentiate the distribution of $E$ from a delta function.

Thus, $\rho_\alpha$ is very sharply peaked and, of all possible configurations, there are *relatively* few that are probable. The question is how, in evaluating the sums in Eq. (7.1), we can include only those that are probable and ignore those that are not. The key to the Monte Carlo method discussed in this chapter is a very clever way to sample only those configurations that are most likely. Since these are the important configurations, this type of approach is often called *importance sampling*. Part of the challenge in calculating Eq. (7.1) is that we cannot actually evaluate $Q$ and thus cannot determine $\rho_\alpha$. The method introduced in the next section avoids that problem as well.

## 7.3 THE METROPOLIS ALGORITHM

In Section 7.1 we discuss how the Monte Carlo method can be employed to evaluate integrals, especially multidimensional integrals. It does this by randomly sampling a volume that contains the region over which the integral is defined, keeping track of which points are in that volume and which are not. We have to follow a similar procedure to evaluate Eq. (7.8). The challenge is that the number of integrations needed is 3 times $N$, the number of atoms in the system (one for each $x$, $y$, and $z$ coordinate). A straightforward integration would thus be computationally impossible for $N$ of any significant size. Thus, numerically evaluating the partition function, $Q$, is not possible.

In 1953, Metropolis *et al.* [228] introduced a method that samples configuration space in such a way that a state, $\alpha$, occurs in the sampling with a probability $\rho_\alpha$. This approach, called the *Metropolis algorithm*, creates that sampling by focusing on the relative probabilities of the states. The outcome is a set of states with the correct probability, from which average quantities can be determined.[5]

The Metropolis algorithm is based on the idea that while we cannot know the actual probability of a state (since we cannot evaluate $Q$), Eq. (7.5) can be used to create a list of configurations through configuration space that has the correct probability distribution. This list is called a *trajectory* through configuration space. Their approach is to start a system at a configuration, then to make a *trial move* of the system to a new configuration, and to test, based on the probability of the new configuration relative to the starting configuration, whether the new configuration should be added to the trajectory or not. More specifically, suppose configuration $i$ has energy $E_i$. A trial is made to a new configuration $i + 1$ and then the energy in the new configuration, $E_{i+1}$, is calculated. The decision of whether to add $i + 1$ to the trajectory is based on the ratio of probabilities, $\rho_{i+1}/\rho_i$, from Eq. (7.4), $\rho_{i+1}/\rho_i = \exp(-\Delta E_{i,i+1}/k_B T)$, where $\Delta E_{i,i+1} = E_{i+1} - E_i$.

---

[5] For much more information about the Monte Carlo method and the Metropolis algorithm, see [188] or one of the other books listed at the end of the chapter.

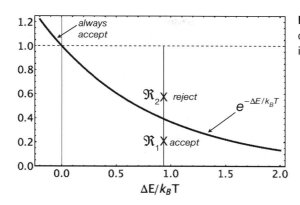

**Figure 7.2** Schematic of the acceptance/rejection criteria in the Metropolis algorithm, as described in the text. $\Re_1$ and $\Re_2$ are random numbers.

The basis of the Metropolis algorithm is that a move is accepted or rejected according to:

$$\Delta E_{i,i+1} \leq 0 \qquad \text{accept because the probability } e^{-\Delta E_{i,i+1}/k_B T} \geq 1$$

$$\text{or} \tag{7.9}$$

$$\Delta E_{i,i+1} > 0 \qquad \text{accept move with probability } e^{-\Delta E_{i,i+1}/k_B T}.$$

Repeating this procedure many times, we can generate a list of configurations $\{n\}$ with energy $E_n$ that have the correct overall probability.

The implementation of the Metropolis algorithm goes as follows:

(a) Suppose the system starts in configuration $i$. A trial move is made to configuration $i + 1$ and $\Delta E_{i,i+1} = E_{i+1} - E_i$ is calculated.

(b) Decide whether to accept or reject the trial move:

　1. If $\Delta E_{i,i+1} \leq 0$, then the relative probability $\rho_{i+1}/\rho_i \geq 1$, so the trial move is accepted and added to the trajectory.

　2. If $\Delta E_{i,i+1} > 0$, then the move will be accepted with probability $\exp(-\Delta E/k_B T)$. That decision is made by first generating a random number $\Re$ between $(0, 1)$. As shown in Figure 7.2, if $\Re \leq \exp(-\Delta E/k_B T)$ (e.g., $\Re_1$) then the move is accepted and if $\Re > \exp(-\Delta E/k_B T)$ (e.g., $\Re_2$), then the move is rejected.

(c) If the trial move is accepted, then the next configuration in the trajectory is the new state $(i + 1)$.

(d) If, however, the trial move is rejected, the next configuration is taken to be the same as configuration $i$. *Each trial generates a configuration for the trajectory, but it may be the same configuration as the one previous to it.* This point is critical, because otherwise an improper distribution will be obtained, as explained in more detail below.

(e) Repeat.

A few words are in order concerning item b in the list of steps in which a random number $\Re$ is compared to $\exp(-\Delta E/k_B T)$ to decide acceptance or rejection. Since both $\exp(-\Delta E/k_B T)$ and $\Re$ span between 0 and 1, the probability that a random number is less than $\exp(-\Delta E/k_B T)$

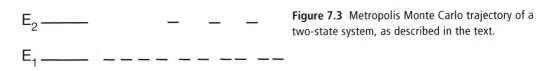

$E_2$ ——                    —  —  —

$E_1$ ——   — — — — —  — —  —— ——

**Figure 7.3** Metropolis Monte Carlo trajectory of a two-state system, as described in the text.

is just $\exp(-\Delta E/k_B T)$. While an individual trial may not seem to yield the correct probability, when considered over many trials the approach discussed in item b yields the correct result.

The average of a quantity $B$ is the average of its value over all the configurations in the series,

$$\langle B \rangle = \frac{1}{m} \sum_{\alpha=1}^{m} B_\alpha \,, \tag{7.10}$$

where $m$ is the total number of trial moves. It must be emphasized again that when a trial move is rejected, the next configuration in the trajectory is set to the configuration of the system *before* the trial move.

To make the discussion more concrete, consider the simple two-state system shown in Figure 7.3, in which there are two states with $\Delta E = E_2 - E_1$. Assume $\Delta E/k_B T = 1$ and $\exp(\Delta E/k_B T) = 0.367$, with the starting state as state 1. In the Metropolis algorithm, a series of random numbers are needed, which we take to be $\Re = \{0.431, 0.767, 0.566, 0.212, 0.715, 0.992, \ldots\}$. For the first trial move, $\Re = 0.431$, which is greater than $\exp(\Delta E/k_B T)$ so, by Figure 7.2, the move is rejected. We repeat state 1 in the list of configurations making the trajectory, so the current list is $\{1, 1\}$. The second trial has $\Re = 0.767$, which is again greater than $\exp(\Delta E/k_B T)$. The move is rejected and state 1 is repeated in the list of configurations, which now is $\{1, 1, 1\}$. The third trial has $\Re = 0.566$, so the trial move is again rejected, so the list becomes $\{1, 1, 1, 1\}$. The fourth trial move has $\Re = 0.212$ which is less than $\exp(\Delta E/k_B T)$ so the move is accepted and the system is now in state 2 and the list is $\{1, 1, 1, 1, 2\}$. Since $\Delta E < 0$ to go from state 2 to state 1, the fifth move is accepted, so the list is $\{1, 1, 1, 1, 2, 1\}$. The next random number is $\Re = 0.715$, the move is rejected, and the list is $\{1, 1, 1, 1, 2, 1, 1\}$. This procedure is repeated as many times as needed to obtain good statistics.

If many trials were made, we would expect to find that the fraction of occurrences of state 2 in the configuration list would be[6]

$$\rho_2 = \frac{e^{-E_2/k_B T}}{e^{-E_1/k_B T} + e^{-E_2/k_B T}} = \frac{e^{-\Delta E/k_B T}}{1 + e^{-\Delta E/k_B T}} \approx 0.269 \,, \tag{7.11}$$

which is easily verified with a trial calculation.[7] This example, while simple, shows why, when a move is rejected, the previous state is repeated in the trajectory. Otherwise, the lower energy states would not receive the correct weighting in the sampling.

---

[6] See Eq. (7.2).

[7] The reader is urged to try this, as it is a very good way to check your understanding of the Metropolis algorithm.

### 7.3.1 Sampling in the Metropolis algorithm

Consider a system in which there are many possible configurations. Over the course of a Monte Carlo calculation, the goal is to sample enough of the probable configuration space to obtain good averages. The Metropolis algorithm enables a decision of whether a move to a new configuration is accepted or rejected in a way that generates the correct probability within the canonical (*NVT*) ensemble for each configuration. However, it says nothing about how to choose those configurations.

There are two main issues to be considered. First, sampling of configurations must be done in such a way as to avoid biasing the system. Second, the sampling should be efficient, in that a move from a configuration to one that is very different is less likely to have a low energy change and thus is more likely to be rejected.

It is essential that sampling methods treat all configurations as equally probable. Any bias in the sampling of configurations will also bias the Metropolis algorithm, and thus the results of the simulation. For example, if a sampling method chooses a set of configurations more often than any others, then the final probability of being in those states would be higher than the canonical (*NVT*) ensemble would predict. The goal in Monte Carlo is to sample each state with uniform probability, letting the Metropolis algorithm sort out the correct probability in the canonical ensemble.

Consider a concrete example of a system of $Q$ levels, with energies $\{E_1, E_2, \ldots, E_Q\}$. For now, assume all levels have the same energy $E$, so that any attempts to change levels would be accepted in the Metropolis algorithm. In this way, we can test whether a sampling method treats all levels as equally probable.

In the Metropolis algorithm, the current state, $i$, is changed to some new state, $j$. A simple way to choose $j$ would be to add a random integer between 1 and $Q - 1$, $\Re_{1,Q-1}$, to $i$, i.e., $j = i + \Re_{1,Q-1}$. To ensure that $j$ is between 1 and $Q$, take $j = j - Q\,\text{Floor}[(j - 1)/Q]$, where $\text{Floor}[x]$ returns the greatest integer less than or equal to $x$.[8] We can test this method by doing a Metropolis Monte Carlo calculation in which, since the energies are all the same, all trials are accepted. We can then count the occurrences of each level in the list, which we show as a histogram in Figure 7.4a for a calculation with $Q = 5$. Note that all states have been visited the same number of times. Doing a similar Metropolis Monte Carlo calculation in which the $\{E_1, E_2, \ldots, E_Q\}$ are different yields the correct probability of being in each level when compared to what is expected from the canonical (*NVT*) ensemble (cf. Eq. (7.2)), as shown in Figure 7.4c.

Now consider a somewhat different choice of sampling method. It starts with the same initial choice for the new level, $j = i + \Re_{1,Q-1}$, but instead of using $j = j - Q\,\text{Floor}[(j - 1)/Q]$ to ensure that $1 \leq j \leq Q$, if the first choice for $j$ is greater than $Q$, it is simply set to $j = 1$. With all levels having energy $E$, we can then generate a list of visited levels and count how many times each level appears. The histogram of this calculation is shown in Figure 7.4b for

---

[8] This method treats the $Q$ levels like a clock that periodically repeats itself every $Q$ times. For example, if $Q = 10$ and $i = 3$, then a choice of $\Re = 6$ leads to $j = 9$, while a choice of $\Re = 8$ leads to $j = 11 - 10 * \text{Floor}[(11 - 1)/10] = 1$.

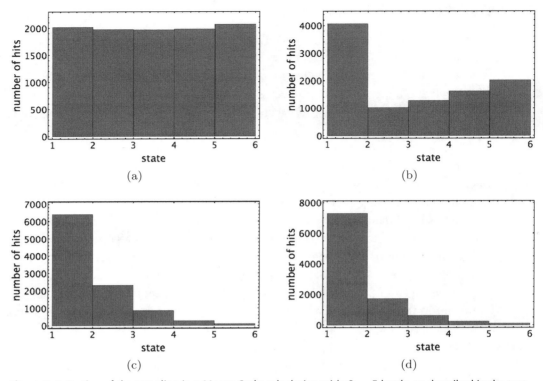

**Figure 7.4** Testing of the sampling in a Monte Carlo calculation with $Q = 5$ levels, as described in the text. (a) An unbiased sampling method in which all levels have the same energy and are sampled uniformly. (b) The biased sampling method discussed in the text showing a bias towards level 1. (c) A Metropolis Monte Carlo sampling based on the same sampling method with $Q = 5$ levels with energies {1, 2, 3, 4, 5}. The probabilities of the levels are in agreement with what is expected from the canonical ($NVT$) ensemble. (d) A Metropolis Monte Carlo sampling based on the same sampling method with $Q = 5$ levels with energies {1, 2, 3, 4, 5}. If the sampling were correct, it would agree with the (c).

$Q = 5$, which clearly shows that level 1 is visited far more often than any other level, when all should be visited an equal number of times. Doing a Metropolis Monte Carlo calculation with $Q = 5$ levels with energies {1, 2, 3, 4, 5} gives the probability in each state shown in Figure 7.4d. Comparison with the correct values in Figure 7.4c shows that using a biased sampling will lead to probabilities that do not agree with the canonical ($NVT$) ensemble. Thus, all properties calculated with biased methods are incorrect.

Of course, even if the random sampling approach used in Figure 7.4a is correct, it may not be the most efficient method for employing the Metropolis method. Imagine that the energies of the $Q$ states have equal spacing, $\Delta e_o$. If the current state was, for example, 1, with energy defined as $E_1 = e_o$, then the final state has an energy $E_Q = Q\Delta e_o$. In the Metropolis algorithm, the key quantity is the change in energy from the current state to the trial state, in this case $\Delta E = (Q - 1)\Delta e_o$. If $\Delta E \gg k_B T$, then the probability of accepting the trial move would be very small. Most of the trials with large changes in levels, and thus energy, would fail and would be, in some sense, wasted effort. A more efficient approach may be to only allow jumps between a state and states that are "adjacent" to it, i.e., trials could be restricted to $\Delta i = \pm 1$. As long as the sampling scheme remains unbiased, the method would be acceptable, but there would be more accepted trial moves and the system would be sampled more efficiently.

The key thing to remember is that the correctness of the Monte Carlo method depends on sampling all possible states uniformly and letting the Metropolis algorithm sort out which states are actually visited. All sampling algorithms should be thoroughly tested before a Monte Carlo simulation is run.

### 7.3.2 Updating the energy in the Metropolis algorithm

The Metropolis Monte Carlo method requires the calculation of the energy change $\Delta E$ for each trial move. The straightforward approach would be to calculate the total energy of the system after a trial move $E_j$ and then to subtract the total energy of the system from before the trial move $E_i$, $\Delta E = E_j - E_i$. For most problems this is hugely inefficient. Typically, interaction terms are short ranged, e.g., an interatomic potential with a cutoff. When a trial move is made, only very few interaction terms are altered, while calculating the total energy involves calculating all interaction terms. It is far more efficient to only calculate those terms that change when a trial move is made. Much effort in creating Monte Carlo calculations is spent in developing efficient methods to determine the change in energy $\Delta E$ and then to *update* the energy of the system if the trial move is accepted. This approach is used in the sample calculation discussed in this chapter and in the sample exercises online.

## 7.4 THE ISING MODEL

A classic problem in physics, the *Ising model*, provides an excellent introduction to the Monte Carlo (MC) method. The Ising model is simple, yet describes, in a phenomenological way, magnetism, phase transformations, etc. In many ways, the Ising model best fits into a later chapter in this book (Chapter 10) that focuses on application of Monte Carlo at the mesoscale, in which atoms are *not* the fundamental entities of the simulation. The Ising model, however, provides such a good introduction to Monte Carlo methods that we will use it as our first example. In Chapter 10 we will introduce a variant of the Ising model, the Q-state Potts model, that is often used in mesoscale simulations, with applications ranging from grain growth to the growth of fungi.

Suppose there is a lattice with $N$ sites and that each lattice site is occupied by a spin with two possible values, $s_i = 1$ or $s_i = -1$. In the Ising model, the energy of the system is given by the expression

$$E = -\frac{J}{2} \sum_{i=1}^{N} \sum_{j \in Z} s_i s_j + B \sum_{i=1}^{N} s_i , \qquad (7.12)$$

where the first term represents the interactions between spins and the second term is the interaction of the spins with an applied field $B$. $J$ is a constant that determines the strength of the interaction between the spins. The notation $\sum_{j \in Z}$ is a short-hand notation that indicates that only the $Z$ nearest neighbors of site $i$ are included in the second sum. The $1/2$ corrects for over-counting in the interactions, as discussed in Chapter 3. On a square lattice with nearest-neighbor interactions, for example, $Z = 4$.

The degree to which all spins align can be described by the *magnetization*

$$M = \frac{1}{N} \sum_{i=1}^{N} s_i, \tag{7.13}$$

i.e., the average value of the spins for a configuration.

If $J$ is positive and the applied field $B = 0$, the lowest-energy state has all spins at $+1$ or all at $-1$, i.e., there is a degenerate ground state with an energy of $E_{min} = -2NJ$. The absolute value of the magnetization equals 1 in the ground state. As the temperature is raised, the system disorders and at some temperature, called the Curie or critical temperature $T_c$, there is a loss of order and the magnetization, $M \to 0$.

The Ising model is difficult to solve analytically,[9] and so it has generally been studied with computer simulation, specifically with the Metropolis Monte Carlo method.

It is convenient to scale out the constants in the Ising model in the same way we did for the Lennard-Jones potential in Table 6.1. For the Ising model, the reduced units are found by dividing by the interaction parameter $J$, so that the reduced energy is $E^* = E/J$ and the reduced field is $B^* = B/J$. Similarly, the reduced temperature is $T^* = k_B T/J$. The reduced energy becomes

$$E^* = -\frac{1}{2} \sum_{i=1}^{N} \sum_{j \in Z} s_i s_j + B^* \sum_{i=1}^{N} s_i \tag{7.14}$$

and the canonical ($NVT$) probability goes as

$$e^{-E/k_B T} = e^{-E^*/T^*}. \tag{7.15}$$

All calculations for a given $T^*$ are the same. For example, a calculation with $T = 1\,K$ and $J = 2$ is exactly equivalent to one in which $T = 100\,K$ and $J = 200$. Recognizing the equivalence of simulations at the same reduced parameters can save a great deal of computation.

The thermodynamics of the Ising model are characterized by the average energy $\langle E \rangle$ and the average magnetization $\langle M \rangle$. In addition to standard thermodynamic quantities, various correlation functions can be defined.[10] The *spin-spin correlation function* says a great deal about the ordering of the spins,

$$c_{ij}(r_{ij}) = \langle s_i s_j \rangle = \left\langle \frac{1}{N_{ij}} \sum_{\ell m}^{r_{ij}} s_\ell s_m \right\rangle. \tag{7.16}$$

The notation $\sum_{\ell m}^{r_{ij}}$ indicates a sum over all pairs of spins in the lattice for which $\ell$ and $m$ are separated by the distance $r_{ij}$ and $N_{ij}$ is the total number of such pairs in the lattice. We discuss

---

[9] The one-dimensional solution is straightforward. The two-dimensional solution is quite difficult, but a solution was found by Lars Onsager in the 1940s.

[10] See, for example, Appendix G.7.

$c_{ij}$ for the Ising model a bit later in this section. We expect $c_{ij}$ to have a dependence on the separation between the spins of the form

$$c_{ij}(r_{ij}) = e^{-r_{ij}/\ell_p},$$
(7.17)

where $\ell_p$ is the *correlation length* of the spins.

### 7.4.1 Metropolis Monte Carlo simulations of the Ising model

It is straightforward to develop a finite-temperature Monte Carlo simulation of the Ising model to calculate such quantities as $\langle E \rangle$, $\langle M \rangle$, etc. To establish the system, a lattice must be chosen,[11] which defines the neighbors in Eq. (7.14), while the size of the system defines the number of spins $N$. Periodic boundaries are generally used. As in all simulations, an initial set of spin values must be assigned.

Once the system is created, there are only two parameters that are needed, the reduced field $B^*$ and the temperature $T^*$. A series of configurations is generated with the Metropolis algorithm:

(a) At step $n$ pick a lattice site $i$ at random. The current energy of the system is $E_{old}$.
(b) Make a trial change of spin. If $s_i = 1$ then change it to $-1$ and vice versa. Find the energy in the new configuration $E_{new}$.
(c) Find the change in energy $E_{new} - E_{old}$.
(d) Accept or reject according to the Metropolis algorithm in Eq. (7.9) and Figure 7.2.
    (1) If accept, the site $i$ has the new spin and the energy is $E_{new}$ for step $n + 1$. The values for any other calculated quantities (e.g., the magnetization) are set to their new values.
    (2) If reject, the site $i$ has the old spin and the energy is $E_{old}$ for step $n + 1$. The values for any other calculated quantities (e.g., the magnetization) are set to their old values.
(e) Accumulate averages, etc.
(f) Start again.

Note that just as in molecular dynamics in Section 6.2.2, the system must equilibrate before averages are determined.

A number of ways to sample the configurations have been suggested, not all of which are particularly good. Some have used a sequential sampling of variables, i.e., for an $N$-site system, an attempt is made to change each variable in a given order (e.g., site 1, then site 2, then site 3, . . .). This type of sampling is, in general, discouraged, as correlations may develop between the sites. The best practice is to choose a site for a trial move randomly, which is the method we employ in the examples below.

The standard convention is to define a Monte Carlo step (MCS) as a random sampling of $N$ sites, where $N$ is the total number of sites in the system. A simulation with a total of $M$ MCS thus ensures that, on average, each site has been sampled $M$ times.

---

[11] For example, a square or triangular lattice in two dimensions or one of the standard lattices (simple cubic, face-centerd cubic, . . .) in three.

Consider now an example calculation of an Ising model on a two-dimensional square lattice with nearest-neighbor interactions $Z = 4$. The energy can be written as

$$E^* = -\frac{1}{2}\sum_{i=1}^{N} s_i \sum_{j \in Z} s_j + B^* \sum_{i=1}^{N} s_i$$

$$= -\sum_{i=1}^{N} \left(\frac{1}{2}S_i - B^*\right)s_i, \tag{7.18}$$

where

$$S_i = \sum_{j \in Z} s_j \tag{7.19}$$

is the sum of the values of the spins at the nearest neighbors of site $i$. All dependence of the energy on other sites in the system is included in $S_i$.

The energy *change* that accompanies a spin flip $(s_k \to -s_k)$ at site $k$ can be written as[12]

$$\Delta E_k^* = 2s_k(-B^* + S_k). \tag{7.20}$$

For the square lattice, $S_k$ can take on five values: $+4, +2, 0, -2,$ and $-4$.[13] Since $\Delta s_i$ can be either $+1$ or $-1$, there are only ten possible values to the change in energy. Using the notation $\Delta E^*[\Delta s_i, S_i]$, we have

$$
\begin{array}{ll}
\Delta E^*[+1, +4] = 2(-B^* + 4) & \Delta E^*[-1, +4] = -2(-B^* + 4) \\
\Delta E^*[+1, +2] = 2(-B^* + 2) & \Delta E^*[-1, +2] = -2(-B^* + 2) \\
\Delta E^*[+1, 0] = -2B^* & \Delta E^*[-1, 0] = 2B^* \\
\Delta E^*[+1, -2] = 2(-B^* - 2) & \Delta E^*[-1, -2] = -2(-B^* - 2) \\
\Delta E^*[+1, -4] = 2(-B^* - 4) & \Delta E^*[-1, -4] = -2(-B^* - 4)
\end{array}
\tag{7.21}
$$

Rather than calculating the total energy of the system each time a spin is flipped and then subtracting the energy from the old configuration from the new energy to calculate the energy change, at each trial we just need to count the nearest-neighbor spins to determine $S_i$ and, knowing the value of $s_i$, use the table in Eq. (7.21) to find the energy change. This is an example of energy updating discussed in Section 7.3.2 and is a much faster approach than having to calculate the total energy of the system for each spin flip.

---

[12] Write out all the terms in the total energy that include $s_k$ from Eq. (7.14), ignoring the $B^*$ term. In the overall sum over $i$, the term $i = k$ yields the terms $(-1/2)s_k(s_1 + s_2 + s_3 + s_4) = (-1/2)s_k S_k$, where the four neighbors of $k$ are indicated with the subscripts 1, 2, 3, and 4. However, term $i = 1$ will include a term $(-1/2)s_1 s_k$, since $k$ is also a neighbor of 1. We find similar terms for the other neighbors of $k$, so the total energy associated with site $k$ is $-s_k S_k$. If the spin is flipped, $s_k \to -s_k$, then the new energy is $s_k S_k$ and the energy change is
$\Delta E = s_k S_k - (-s_k S_k) = 2s_k S_k$.

[13] If all neighboring spins are $+1$, then $S_k = 4$. If 3 are $+1$ and 1 is $-1$, then $S_k = 2$, and so on.

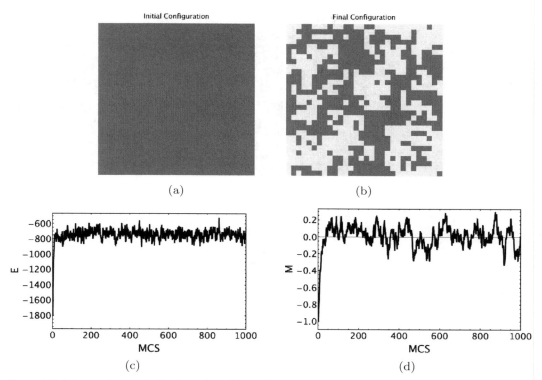

**Figure 7.5** Ising model results for $B^* = 0$, and $T^* = 3$. Dark gray squares: $s = -1$; light gray squares: $s = 1$. (a) The initial configuration, (b) the final configuration, (c) energy, (d) the magnetization.

## 7.4.2 Example simulations of the Ising model

In Figure 7.5 we show results of a short Monte Carlo calculation for the Ising model. It is a small system, consisting of a $30 \times 30$ site square lattice, at a reduced temperature of $T^* = 3$ and no applied field ($B^* = 0$). The simulation was started with all spins in the ground state with $s_i = -1$. The initial configuration, the final configuration after 1000 MCS, the energy, and the magnetization are shown in Figure 7.5a–d, respectively. One Monte Carlo step (MCS) is taken as the random flips of $N$ spins, where $N$ is the total number of sites.

Looking at the plot of magnetization versus MCS, we see that the system rapidly disorders ($M$ goes from $-1$ to approximately 0 in 25 MCS steps or so). The energy also quickly reaches its equilibrium value and fluctuates around that value. The final configuration shows regions with spins $+1$ and regions with $-1$. If we were to animate the results, we would see large fluctuations in the spin values.

In Figure 7.6, we show the same system, but with $T^* = 2$ (the applied field is still $B^* = 0$). Note that the magnetization settles down to a value of about $-0.9$ and that the final configuration consists almost entirely of spins with $s_i = -1$, with just a few spins with a value of $+1$. Clearly, something drastic has happened between $T^* = 2$ and $T^* = 3$. Accurate calculations show that there is a *critical point* for the two-dimensional Ising model (with no

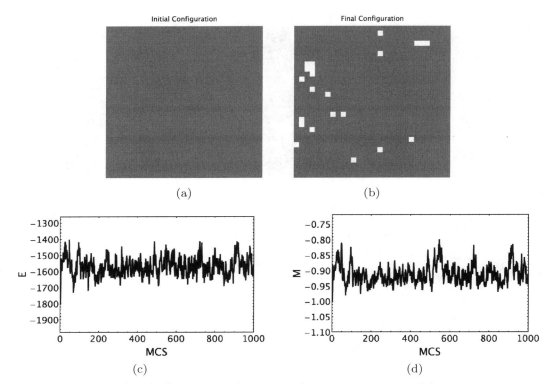

**Figure 7.6** Ising model results for $B^* = 0$, and $T^* = 2$. Dark gray squares: $s = -1$; light gray squares: $s = 1$. (a) The initial configuration, (b) the final configuration, (c) energy, (d) the magnetization.

applied field) at $T_c^* = 2.269$ [243]. In the online exercises, you will have the opportunity to explore this transition in greater detail.

Averages of a quantity are calculated by summing the value of the quantity over the $N$ configurations and dividing by $N$, as in Eq. (7.10). Analysis of the results follows the same procedure as for molecular dynamics in Section 6.2.2. For the Ising model, $\langle E \rangle$ and $\langle M \rangle$ are the most obvious quantities to calculate. We note that just as in molecular dynamics, the system must be equilibrated before averages are calculated.

In Figure 7.7 we show the calculated spin-spin correlation function $c_{ij}$ from Eq. (7.16) for the nearest-neighbor spins as a function of $T^*$. At $T^* = 0$, the system is completely ordered and $c_{ij} = 1$. As $T^*$ increases, there is a slow decrease in $c_{ij}$ until the system nears its critical point, at which point $c_{ij}$ drops quickly. Note, however, that it does not go to zero, as the magnetization does in Figure 7.5. $c_{ij}$ shows a slow decay towards zero as $T^*$ increases, indicating that the local ordering of spins remains despite the loss of magnetization.

### 7.4.3 Other sampling methods for the Ising model

There are other sampling methods that could be used. Instead of flipping spins on the same site during a trial, Kawasaki, for example, introduced a method in which a trial consists of an *exchange* of neighboring spin variables [165].

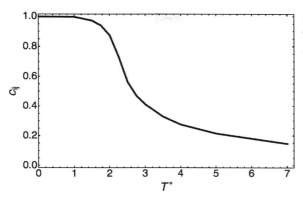

**Figure 7.7** Spin-spin correlation function $c_{ij}$ from Eq. (7.16) for the nearest-neighbor spins as a function of $T^*$.

In the region of the critical point, simulations employing the Metropolis and Kawasaki methods tend to become very slow owing to large regions of correlated spins. Flipping a single spin or exchanging two spins does not reflect the large-scale fluctuations in the configuration. One approach is to just speed up the calculation, which can be done by employing a method called the $N$-fold way, which we will discuss in Chapter 10. "Cluster" algorithms can also be useful in suppressing the slowing down as the critical point is neared. They do this by flipping not single spins, but instead, large regions of spins [302, 341, 357]. The cluster methods are extremely efficient when used to model systems near criticality.

## 7.5 MONTE CARLO FOR ATOMIC SYSTEMS

The Metropolis Monte Carlo method is not restricted to simulations of spins on a lattice. It can be applied to any system for which the energy can be expressed in terms of a set of variables.[14] Monte Carlo methods have been quite often employed in the study of atomic and molecular systems, being used to calculate thermodynamics, structure, phase transformations, ordering, etc. In this chapter, we discuss how to employ Monte Carlo to study atomic systems. Applications to molecular systems are briefly discussed in Chapter 8.

The basic approach to using Monte Carlo to study atomic systems is the same as for the Ising model – a trial move is made to a new configuration, the change in energy is calculated, and the Metropolis algorithm is employed to decide whether to accept that move or not. Of course, in an atomic fluid, or solid, the atoms are not fixed to a lattice site, as the spins in the Ising model are, so new methods for sampling configuration space are needed.

Application of the Monte Carlo method to an atomic system should yield the same thermodynamics as would a molecular dynamics simulation under the same thermodynamic conditions, $\rho_{MD} = \rho_{MC}$, $\langle T \rangle_{MD} = T_{MC}$, $\langle P \rangle_{MD} = \langle P \rangle_{MC}$, etc.[15] In molecular dynamics, a list of configurations as a function of time is generated. In Monte Carlo, a list of configurations based on

---

[14] Typically, the function that defines the energy in terms of those variables is called the *Hamiltonian* of the system.

[15] Except for some differences between the canonical ($NVT$) and microcanonical ($NVE$) ensemble discussed in Appendix G.6.1.

the canonical probability is generated. Based on the discussion of ergodicity in Appendix G, properties calculated from these lists should be essentially identical. The analysis of the data is the same and the same structural correlation functions can be calculated. One difference is that time-dependent properties, such as the velocity autocorrelation function, are restricted only to molecular dynamics.

It is important to note the physical significance of temperature in a Monte Carlo simulation. It is strictly a thermodynamic variable. Thus the temperature at which a phase transition occurs is meaningful and should agree with those calculated with molecular dynamics.[16] The evolution of the continuous position variables in Monte Carlo is not, however, physically meaningful. Only the average quantities are well-defined.

### 7.5.1 Simulations of atoms in the canonical (*NVT*) ensemble

All applications of the Metropolis Monte Carlo method start with the identification of the appropriate ensemble. For standard Monte Carlo, that is the canonical ensemble defined by constant $N$, $V$, and $T$ (though it is relatively simple to extend Monte Carlo to other ensembles, as will be discussed in a later section). Since atomic motions are continuous, the probability of being in a configuration is given by Eq. (7.6).

To employ the Metropolis algorithm to generate a series of atomic configurations that are consistent with the canonical ensemble follow the same basic procedure as in the Ising model:

(a) Pick an atom at random (atom $i$) from all $N$ atoms. Atom $i$ currently has position $\mathbf{r}_i(\text{old})$ and the system has energy $U(\text{old})$.

(b) Move atom $i$ to a new position $\mathbf{r}_i(\text{new})$ by a random displacement.

(c) Find the change in potential energy $\Delta U = U(\text{new}) - U(\text{old})$ owing to the displacement of atom $i$.

(d) Accept or reject the move with the Metropolis algorithm of Figure 7.2:
   (1) If accept the trial, the next entry in the Monte Carlo list of configurations has $\mathbf{r}_i = \mathbf{r}_i(\text{new})$ with energy $U(\text{new})$.
   (2) If reject the trial, the next entry in the Monte Carlo list of configurations has $\mathbf{r}_i = \mathbf{r}_i(\text{old})$ with energy $U(\text{old})$ – we repeat the configuration from before the rejected move.

(e) Accumulate averages based on Eq. (7.8) with values from the next entry in the Monte Carlo list of configurations.

While the basic structure of the Monte Carlo method for atoms is the same as for the Ising model, some of the details are, of course, very different. We will discuss in some detail steps (b) and (c).

---

[16] There are two caveats to the statement that phase transition temperatures calculated with Monte Carlo should agree with those in molecular dynamics. First, there are small corrections that arise from Monte Carlo being in the $NVT$ ensemble and molecular dynamics in the $NVE$ ensemble, as discussed in Appendix G.6.1. Second, phase transitions are dynamic events. The different ways that Monte Carlo and molecular dynamics evolve the degrees of freedom can lead to differences in the prediction of the thermodynamic states at which phase transitions occur.

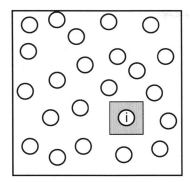

**Figure 7.8** Random move of an atom in a Monte Carlo simulation of an atomic system. A maximum distance $\Delta_{max}$ that a particle can move in any coordinate (the shaded area) is set. A new position for the atom is randomly chosen in that area.

## Sampling the atomic coordinates

In the Ising model, sampling the variables was straightforward – there were only two possibilities. The trial move for atoms in a solid or a fluid is more complicated, with a number of issues that must be considered.[17] First, the direction of the trial move must be random and unbiased. Second, the size of the move must be chosen in such a way that the system can be moved through configuration space as efficiently as possible.

Figure 7.8 shows a schematic view of a method for choosing a random atomic move. The idea is simple. After picking an atom at random (atom $i$), move it randomly to a position inside a small cube with a side of length $2\Delta_{max}$ centered on the position of $i$. The simplest way to do that move is to:

(a) Find three random numbers $\Re_1$, $\Re_2$, and $\Re_3$ in the range $(-1, 1)$
(b) Take

$$x_i(\text{new}) = x_i(\text{old}) + \Re_1 \Delta_{max}$$
$$y_i(\text{new}) = y_i(\text{old}) + \Re_2 \Delta_{max} \qquad (7.22)$$
$$z_i(\text{new}) = z_i(\text{old}) + \Re_3 \Delta_{max} .$$

There are other methods that could be used to move the atom, but in practice this approach is simple and works well.

$\Delta_{max}$ sets the maximum displacement in a Monte Carlo move and is a parameter that is input into the simulation. $\Delta_{max}$ is chosen to optimize the sampling of configuration space. Consider Figure 7.8. If the shaded box is made larger, then many trial moves could put the atom very near one of its neighbors. The energy of the new configuration would then be high and the change in energy large and positive. Such moves would not be likely to be accepted in the Metropolis scheme. Thus, while the moves could be large and configurations sampled quickly, many moves would not be accepted, limiting the overall effectiveness of the sampling. If the shaded box were made very small, then the energy changes would be small as well, and most moves would

---

[17] See Section 7.3.1.

probably be accepted. The problem with this approach is that very little movement would be made in each step, thus the sampling of the possible configurations would be limited.

There is no theoretical justification for a particular value, but $\Delta_{max}$ is usually set such that about 50% of the moves are accepted. This acceptance ratio seems to be a good balance between accepting enough moves to sample configuration space and making the moves big enough to sample a large enough region of configuration space.

## Calculating the change in energy

In step (c), the change in potential energy is required to determine whether a trial move is accepted. One approach would be to calculate the total potential energy of the system after the move $U(\text{new})$ and subtract from that the potential energy from before the move $U(\text{old})$, $\Delta U = U(\text{new}) - U(\text{old})$, where the total energy (for pair potentials) is

$$U(\mathbf{r}^N) = \sum_{i=1}^{N-1} \sum_{j=i+1}^{N} \phi_{ij}(r_{ij}).$$  (7.23)

If the interactions between the atoms are short-ranged, this approach would be very inefficient. A much better approach would be to include in the energy change only those terms that involve the coordinates of atom $i$. Terms in the potential energy that do not involve atom $i$ will not change and thus will have no effect on the value for $\Delta U$. For pair potentials, this energy change becomes

$$\Delta U = \sum_{j \neq i} \phi(|\mathbf{r}_j - \mathbf{r}_i(n)|) - \sum_{j \neq i} \phi(|\mathbf{r}_j - \mathbf{r}_i(o)|),$$  (7.24)

where $(n)$ indicates the new positions and $(o)$ the old positions. Implementation of Eq. (7.24) is relatively straightforward. Some subtleties arise when there is a cutoff distance in the potential – the list of atoms interacting with $i$ may be different in its new position than in its old position. In a computer program, one would typically write a routine that returns the energy change. It is always useful to test that routine by comparing to a direct calculation of $\Delta U$ based on Eq. (7.23).

## Example calculations for "Lennard-Jonesium"

Consider as an example simulations of atoms interacting with a Lennard-Jones potential of Eq. (5.6). Reduced units are used, as defined in Table 6.1.[18] It is important to emphasize again that all simulations under the same reduced conditions are identical. If someone talks about a simulation of Lennard-Jones "gold" and then another of Lennard-Jones "copper", be suspicious. The only difference between Lennard-Jones "gold" and Lennard-Jones "copper" is the choice of potential parameters $\epsilon$ and $\sigma$. For a simulation with reduced units, results for "gold" or

---

[18] The potential energy is $U^* = U/\epsilon$, the distance $r^* = r/\sigma$, the temperature $T^* = k_B T/\epsilon$, and the density becomes $\rho^* = \rho \sigma^3$.

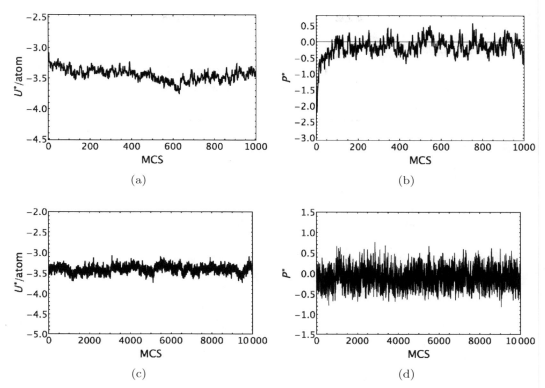

**Figure 7.9** Results from an MC simulation of a 108-particle LJ system at $T^* = 1$ and $\rho^* = 0.5$. The system started from a pure *fcc* lattice. (a) Potential energy per atom, $U^*$/atom for the first 1000 MCS. (b) Pressure, $P^*$ for the first 1000 MCS. (c) $U^*$/atom after equilibration. (d) $P^*$ after equilibration.

"copper" are obtained by simply rescaling the results with the appropriate choice of parameters. Of course, using different values for the cutoff distance will also change the results.

In Figure 7.9 we show results from simulations for a Lennard-Jones system with $T^* = 1$, $\rho^* = 0.5$, and $N = 108$. The system was started as a perfect *fcc* solid. Figure 7.9a shows how the energy changes over the first 1000 MCS, with a similar plot of the instantaneous pressure in Figure 7.9b. In Figure 7.9c and Figure 7.9d are shown the values for the energy and pressure, respectively, after equilibration.

As noted above, averages calculated with Monte Carlo should agree with those calculated with molecular dynamics under the same thermodynamic conditions (with the small ensemble corrections discussed in Appendix G.6.1). The results of a Monte Carlo calculation should be analyzed the same way as the molecular dynamics calculation of Section 6.2.2. We can calculate all the same quantities, such as the average potential energy, average pressure, etc. We can also calculate radial distribution functions and any other quantity that depends only on the positions of the atoms. All of these calculations are done exactly as in the molecular dynamics simulation, replacing the time sequence of configurations with the Metropolis list of configurations.

## 7.6  OTHER ENSEMBLES

It is relatively easy to extend the Monte Carlo method to new ensembles. The energy will change to reflect the new thermodynamic constraints and the sampling will need to reflect the new degrees of freedom. We will discuss two of the more important ensembles for studying materials problems.

### 7.6.1  The isobaric-isothermal ($NPT$) ensemble

The isobaric-isothermal ensemble is described in Appendix G.5.7 and refers to a thermodynamic system constrained to have constant number of particles $N$, constant pressure $P$, and constant temperature $T$. It is often referred to as the $NPT$ ensemble. The configurational part of the partition function in the $NPT$ ensemble is (from Eq. (G.26))

$$Z_{NPT} = \int dV e^{-P_{ext}V/k_B T} \int d\mathbf{r}^N e^{-U(\mathbf{r}^N)/k_B T} , \qquad (7.25)$$

where $P_{ext}$ is the externally applied pressure. The general procedure is the same as for constant volume simulations – $N$ atoms are added to a simulation cell of volume $V$ at constant $T$, the energy is calculated, and the Metropolis algorithm is used to move the atoms around. The simulation cell can either be cubic or a more complicated structure such as described in Appendix B.2.4.

The change to the $NPT$ ensemble comes in the form of dependence of the partition function on a new variable, the volume $V$. The energy depends on the volume in two ways, first through the $P_{ext}V$ term and second through the volume dependence of the potential energy $U(\mathbf{r}^N)$. The first term is obvious, the second deserves some comments.

Consider first the potential energy of a systems in a cubic simulation cell with periodic boundaries (from Section 3.2)

$$U(\mathbf{r}^N) = \frac{1}{2} \sum_{\mathbf{R}} \sum_{i=1}^{N} \sum_{j=1}^{N} {}'\phi_{ij}(\mathbf{R} + \mathbf{r}_j - \mathbf{r}_i) , \qquad (7.26)$$

where $\mathbf{R}$ is the lattice vector of the periodically repeated cubic simulation cell and $\mathbf{r}_i$ is the position of atom $i$ within that cell. Assume the simulation cell has a side of length $L$, so $V = L^3$. As discussed in Appendix B.2.2, we can rewrite the positions within the simulation cell as

$$\mathbf{r}_i = L\mathbf{s}_i , \qquad (7.27)$$

in which the $\mathbf{s}$ are the fractional coordinates of the atoms within the cell. Similarly, the simulation cell lattice vectors can be written as

$$\mathbf{R} = L(n_1\hat{x} + n_2\hat{y} + n_3\hat{z}) = L\mathbf{S} , \qquad (7.28)$$

where the $\{n\}$ are integers. Equation (7.26) can be rewritten as

$$U(\mathbf{s}^N, V) = \frac{1}{2} \sum_{\mathbf{S}} \sum_{i=1}^{N} \sum_{j=1}^{N} {}'\phi_{ij}\left[L(\mathbf{S} + \mathbf{s}_j - \mathbf{s}_i)\right] . \qquad (7.29)$$

As the volume changes, the *relative* positions of the atoms within the cell do not change, but the distances between the atoms scale linearly with $L$. Thus, there is an explicit dependence of $U$ on the volume.

Consider the integral over coordinates in Eq. (7.25). By rewriting $U$ as in Eq. (7.29), we can rewrite the integrals over $\mathbf{r}^N$ to integrals over $\mathbf{s}^N$,

$$\int d\mathbf{r}^N e^{-U(\mathbf{r}^N)/k_B T} = V^N \int d\mathbf{s}^N e^{-U(\mathbf{s}^N)/k_B T} \,, \tag{7.30}$$

where a factor of $V$ is picked up for each transformation of $\mathbf{r}$ to $\mathbf{s}$.[19] Thus, in scaled coordinates

$$Z_{NPT} = \int dV \int d\mathbf{s}^N e^{-\left(U(\mathbf{s}^N) + P_{ext}V - k_B T N \ln V\right)/k_B T} \,, \tag{7.31}$$

where we have combined all the energy terms in one exponential, using the relation $\exp(N \ln V) = V^N$ and noting that the minus sign in front of the $N \ln V$ term is cancelled by the other minus sign.[20]

To study systems in the *NPT* ensemble, the positions of each particle and the volume $V$ must be varied, with the external pressure $P_{ext}$ set as a parameter. The probability function used in the Metropolis algorithm is $\exp(-(U(\mathbf{s}^N) + P_{ext}V - k_B T N \ln V)/k_B T)$, thus the change in $U(\mathbf{s}^N) + P_{ext}V - k_B T N \ln V$ must be considered for each move. The positions $\mathbf{s}^N$ are sampled in the usual way, remembering that they are the fractional positions within the unit cell. The size of the cubic lattice is sampled by a trial move of the form $L(\text{new}) = L(\text{old}) + \Delta_L \Re$, where $\Delta_L$ is a parameter that sets a limit to the size of the changes in volume and $\Re$ is a random number on $(-1, 1)$. There is no theoretical justification for one specific method of sampling.

The trial moves of the atomic coordinates are generally made with a fixed volume, using a method such as in Figure 7.8, but in the scaled coordinate system ($\mathbf{s}$). A typical procedure would be to do a trial move for the lattice, followed by a number of Monte Carlo steps (MCS) for the coordinates, where 1 MCS corresponds to $N$ random atom moves. The frequency of the lattice moves relative to the coordinate moves is an input parameter, recognizing that a change in the lattice size requires a calculation of the total energy of the system, not an update energy. $\Delta_L$ is set so that roughly 50% of the lattice moves are accepted.

Cubic simulation cells are appropriate for simulations on liquids, which cannot support a shear stress. For solids, however, it is best to include fluctuations (or changes) in the shape of the simulation cell as well as the volume. The best way to include the shape is through the representation of non-cubic cells presented in Appendix B.2.4. Constant stress simulations can be done as well, substituting the elastic energy for the external pressure. We note that the advantage of the Monte Carlo method over molecular dynamics for simulations in this ensemble is that it is generally much easier to implement, avoiding the issues with the fictitious mass and energy terms of the Parrinello-Rahman method of Section 6.4.2.

---

[19] For each particle, we have $\int dx\,dy\,dz = \int L\,ds_x\,L\,ds_y\,L\,ds_z = L^3 \int ds_x\,ds_y\,ds_z$.
[20] A more formal derivation of this result is given in Section 5.4.1 of [109].

In Appendix B.2.4 we introduced the matrix $\mathbf{h}$ whose columns are the simulation cell lattice vectors $\mathbf{a}$, $\mathbf{b}$, and $\mathbf{c}$, i.e.,

$$\mathbf{h} = (\mathbf{a}, \mathbf{b}, \mathbf{c}). \tag{7.32}$$

The coordinates of the atoms were rewritten as

$$\mathbf{r}_i = \mathbf{h}\,\mathbf{s}_i, \tag{7.33}$$

which is analogous (and equivalent for cubic systems) to the scaling of variables for cubic systems ($\mathbf{r} = L\mathbf{s}$). Transforming the variables from $\mathbf{r}^N$ to $\mathbf{s}^N$ adds a factor of $V^N$ to the partition function, exactly as in Eq. (7.30).[21]

Distances between atoms are given by

$$r_{ij} = \mathbf{s}_{ij}^T \mathbf{G}\,\mathbf{s}_{ij}, \tag{7.34}$$

where $T$ indicates the transpose of the vector[22] and $\mathbf{G} = \mathbf{h}^T \mathbf{h}$ is the metric tensor of Eq. (B.22)

$$\mathbf{G} = \begin{pmatrix} a^2 & ab\cos\gamma & ac\cos\beta \\ ab\cos\gamma & b^2 & bc\cos\alpha \\ ac\cos\beta & bc\cos\alpha & c^2 \end{pmatrix}. \tag{7.35}$$

Sampling the variables is done the same way as for the cubic case. There is some question as to the best way to sample the volume, in that all possible shapes and sizes must be sampled uniformly. Some have suggested sampling the various components of the $\mathbf{h}$ matrix; however, we believe that sampling the components of the metric tensor $\mathbf{G}$ would be more appropriate. The goal is to choose the method that produces the least bias in the sampling.[23]

During simulations, whether with a cubic cell or a variable box, averages are calculated in the usual way. Of particular importance is to ensure that the average pressure equals the external pressure, $\langle P \rangle = P_{ext}$. If that does not hold, then there is an error somewhere in the program, perhaps in the calculation of the pressure. Any structural parameters can be calculated, including the volume, simulation cell parameters, etc. These types of simulations have been used to map out phase transitions in the same way as has the constant-stress molecular dynamics method of Section 6.4.2.

## 7.6.2 The grand canonical ($\mu VT$) ensemble

There are many situations in materials science in which one may not know the equilibrium distribution of species. The ensemble that allows for a change in the number of atoms is the *grand canonical ensemble*, which is described briefly in Appendix G.5.7.

---

[21] In this case, the transformation of coordinates is $d\mathbf{r}^N = (\det \mathbf{h})^N d\mathbf{s}^N = V^N d\mathbf{s}^N$, where the Jacobian of the transformation of $\mathbf{r}$ to $\mathbf{s}$ is the determinant of $\mathbf{h}$. From Eq. (B.15), $V = \det \mathbf{h}$.

[22] A transpose is defined in Eq. (C.21).

[23] Suppose one sampled the various lattice parameters ($a$, $b$, $c$, $\alpha$, $\beta$, and $\gamma$) directly. If the lattice lengths are large, then very small changes in the angles would lead to large changes in the absolute positions of atoms at the far side of the simulation cell. Sampling the metric tensor, on the other hand, avoids that situation.

The conjugate variable to the number of atoms is the chemical potential $\mu$ and the grand canonical ensemble is a natural function of $(\mu V T)$. From Eq. (G.28), the grand canonical partition function is

$$Q_{\mu VT} = \sum_{N=0}^{\infty} e^{(\mu N/k_B T)} Q_{NVT} , \tag{7.36}$$

where the full canonical partition function for a one-component system is (from Eq. (G.15))

$$Q_{NVT} = \frac{1}{N! \Lambda^{3N}} \int e^{-U(\mathbf{r}^N)/k_B T} d\mathbf{r}^N = \frac{V^N}{N! \Lambda^{3N}} \int e^{-U(\mathbf{s}^N)/k_B T} d\mathbf{s}^N \tag{7.37}$$

and $\Lambda = h/\sqrt{2\pi m k_B T}$. Note that on the right we have rewritten the coordinates as fractions of the simulation cell as in Eq. (7.30). We can rewrite Eq. (7.37) as

$$\begin{aligned} Q_{\mu VT} &= \sum_{N=0}^{\infty} e^{(\mu N/k_B T)} \int \frac{V^N}{N! \Lambda^{3N}} e^{-U(\mathbf{r}^N)/k_B T} d\mathbf{r}^N \\ &= \sum_{N=0}^{\infty} \int d\mathbf{r}^N e^{-\left(U(\mathbf{r}^N) - \mu N + k_B T\left[\ln(N!) + 3N \ln(\Lambda) - N \ln V\right]\right)/k_B T} . \end{aligned} \tag{7.38}$$

In a Monte Carlo simulation, there would be two types of trials, one to move the particles and the other to insert or remove a particle:

(a) For a particle move, the trial is accepted or rejected based on the value of $\exp(-\Delta U(\mathbf{s}^N)/k_B T)$, where $\Delta U(\mathbf{s}^N) = U(\mathbf{s}'^N) - U(\mathbf{s}^N)$, where $\mathbf{s}'^N$ indicates a set of coordinates in which one atom has been moved.

(b) For a particle insertion/deletion, the trial is accepted or rejected based on the value of the exponential $\exp(-\Delta \mathcal{U}/k_B T)$ of the change in $\mathcal{U} = U(\mathbf{r}^N) - \mu N + k_B T[\ln(N!) + 3N \ln(\Lambda) - N \ln V]$.

Frenkel and Smit provide the basic structure of a simulation in the grand canonical ensemble [109].

Use of the grand canonical ensemble for simulations in condensed systems is somewhat problematic. To allow the number of particles to vary requires moves in which particles are inserted into the system. While some clever methods have been proposed to do this, in general it is difficult to find enough space to put in an additional particle, especially if it is inserted at random. Acceptance rates are therefore quite low. A way around this problem is achieved by restricting the insertion of particles to a swap with a particle that is already there. For single-component systems, that obviously does not lead to anything interesting. For systems with two (or more) types of particles, this type of simulation can be very useful.

For a system with two components ($A$ and $B$) with the atoms restricted to $N$ lattice sites, such that $N = N_A + N_B$, Eq. (7.36) becomes

$$Q_{\Delta\mu VT} = e^{N\mu_B/k_B T} V^N \sum_{N_A=0}^{\infty} e^{(N_A\Delta\mu/k_B T)} \frac{1}{N_A! \Lambda_A^{3N_A}} \frac{1}{N_B! \Lambda_B^{3N_B}} \tag{7.39}$$

$$\times \int d\mathbf{s}_A^{N_A} d\mathbf{s}_B^{N_B} e^{-U(\mathbf{s}_A^{N_A}, \mathbf{s}_B^{N_B})/k_B T},$$

where $\Lambda$ is the de Broglie wavelength of Eq. (G.15). Note that $Q_{\Delta\mu VT}$ is a function of the difference in chemical potentials, $\Delta\mu = \mu_A - \mu_B$.[24]

There are two types of steps in the Metropolis algorithm, just as in a full grand canonical ensemble calculation, but in this case these are particle moves and particle swaps, in which an atom is inserted at a lattice site while the previous occupant is removed. The acceptance/ rejection of a move of a particle is based on the change in potential energy, while a particle swap will depend on the change in potential energy as well as on the number of each species.

Since we have introduced a restriction on $N$, this is not really the grand canonical ensemble but rather a restricted version of it.[25] The relative proportion of $A$ or $B$ type atoms is controlled by $\Delta\mu$. This ensemble has been used to study bulk properties and segregation to interfaces. Examples of these types of simulations are given in [284].

## 7.7 TIME IN A MONTE CARLO SIMULATION

In Monte Carlo simulations, there is, in general, no explicit time dependence. A set of configurations is generated with the correct probabilities and averaged quantities are determined by averages over those configurations. There are situations, however, in which time can be inferred from a Monte Carlo simulation, though linkages with "real" time are often not well defined. If the processes modeled by the Monte Carlo method can be associated with discrete events and a mobility or rate can be assigned to those events, then a Monte Carlo trial can be associated with a time change, with a proportionality constant that depends on the system, but may not be well determined. The Ising model, and the Potts model discussed in Chapter 10, are examples of systems in which the events are discrete (e.g., a flip from one spin state to another), and a time can be associated with the Monte Carlo sampling. Details are given in Section 10.2. The motions of atoms or molecules does not consist of discrete events, however, and no time can be associated with the Monte Carlo moves for simulations on those systems.

---

[24] The full term in $Q$ is $e^{(N_A\mu_A + N_B\mu_B)/k_B T}$. Since $N = N_A + N_B$ is a constant, we can rewrite $N_A\mu_A + N_B\mu_B = N_A\mu_A + (N - N_A)\mu_B = N\mu_B + N_a(\mu_A - \mu_B)$. $\mu_B$ is also a constant, so the thermodynamics depends only on the difference $\Delta\mu$.

[25] It is sometimes referred to as the "semi-grand-canonical ensemble".

## 7.8 ASSESSMENT OF THE MONTE CARLO METHOD

One could argue that Monte Carlo simulations offer no advantages over molecular dynamics yet have the disadvantage that there is a lack of dynamical information. For atomic systems with potentials from which the force is easily derived, that is largely a valid point of view. However, if one is interested in structure and thermodynamics, then Monte Carlo is a reasonable option. Monte Carlo simulations are easily extended to molecular or polymeric systems as well, some details of which are given in Chapter 8.

It is generally easier to extend Monte Carlo to other ensembles than it is with molecular dynamics. All one needs is an expression for the energy of the system in the new ensemble, where in molecular dynamics, artificial terms defining the energy and dynamics of external constraints must be added to the Hamiltonian, cf. Section 6.4.

There are a number of types of problems for which the Monte Carlo method is the only approach. Spin models, such as the Ising model, are not amenable to molecular dynamics simulations. Indeed, any system described by a Hamiltonian in which there are no easily defined forces must be modeled with the Monte Carlo method. Because only the energy is needed, the Monte Carlo method is particularly well suited for use with models based on collective variables, for example the spin models for grain growth discussed in Chapter 10. The basic principles of Monte Carlo are also not restricted to energy-based modeling. In Chapter 9 we introduce a method called kinetic Monte Carlo, in which the rates of processes take the place of energy, enabling long-time simulations of systems whose dynamics can be described as activated processes.

## 7.9 USES OF THE MONTE CARLO METHOD IN MATERIALS RESEARCH

There are many examples of the use of Monte Carlo to simulate important materials phenomena at an atomic scale and we cannot hope to review the wide range of applications in any substantive detail. Here we list just a few examples, based around a small set of topics.

- Structure and thermodynamics: in Section 6.4.2 we described the Parrinello-Rahman [252] method for doing molecular dynamics in a constant-pressure ensemble, pointing out the complexities in defining the constraints. As noted in Section 7.6.1, it is probably easier to apply variable-box boundary conditions in Monte Carlo than in molecular dynamics. To our knowledge, the first application of constant-stress Monte Carlo using the Parrinello-Rahman boundary came just two years after their paper, in a study of the Bain transformation in Fe (cf. Figure 6.12) [237]. Using simple potentials, they mapped out the temperature-dependence of the $bcc \leftrightarrow fcc$ transition. Given the size constraints of atomic-scale Monte Carlo calculations, applications to nanoscale materials have become common. A review of some of these results, along with detailed comparison with experiment, can be found in [20].

- Defect structure and properties: use of Monte Carlo simulations in the study of defects can produce important information on their structures and energetics. In many cases, 0 K simulations are done to find minimum structures followed by Monte Carlo simulations to find finite temperature properties, a procedure followed in a study of tilt boundaries in $fcc$ metals [274].
- Segregation: the semi-grand-canonical ensemble Monte Carlo method (cf. Eq. (7.39)) was used with an embedded-atom potential to study grain boundary premelting in Cu-rich Cu-Ag alloys [352]. They studied how the chemical composition within the boundaries approached that of the liquidus at the same temperature as the boundary disordered. Similar studies using molecular dynamics to study segregation to grain boundaries and surfaces are reviewed in [284].

## 7.10 SUMMARY

This chapter introduces the basic ideas and implementations of the Monte Carlo method in the canonical ($NVT$) ensemble, with an emphasis on the Metropolis algorithm for generating configurations. Details of the steps needed in a Monte Carlo calculation were discussed, with specific focus on the importance of unbiased sampling of configuration space. An example calculation of the properties of the Ising model was used to illustrate the basic methodology. Extension was made to atomic systems interacting with continuous interatomic potentials, using as an example systems of atoms whose interactions are described by a Lennard-Jones potential. Extension to other ensembles, for example the isobaric, isothermal ensemble and the grand canonical ensemble, were introduced.

### Suggested reading

Some general discussions of the Monte Carlo method that I find useful are:

- D. P. Landau and K. Binder, *A Guide to Monte Carlo Simulations in Statistical Physics* [188]. One of a series of books by Kurt Binder and coworkers on Monte Carlo methods.
- The book by D. Frenkel and B. Smit, *Understanding Molecular Simulation: From Algorithms to Applications* is an excellent source for Monte Carlo methods [109].

## 7.11 APPENDIX

From Appendix G.5.3, the *equilibrium* probability in the canonical ($NVT$) ensemble that a system is in the state $\alpha$ is $\rho_\alpha \propto \exp(-E_\alpha/k_B T)$. Here we show that the Metropolis algorithm yields systems within that ensemble.

Consider a system with a number of states, which we will indicate by a subscript. We start by remembering that equilibrium does not mean static – the system will always be changing, moving from state to state. Let $w_{ij}$ be the probability with time that the system goes from state $i$ to state $j$, i.e., $w_{ij}$ is a rate. The time rate of change of the probability of being in a state, $\rho_i$, is just

$$\frac{d\rho_i}{dt} = -\sum_j w_{ij}\rho_i + \sum_j w_{ji}\rho_j, \tag{7.40}$$

where the first term is the rate the system leaves state $i$ and the second term is the rate the system enters state $i$, summed over all other states in the system. (Note the change in the order of the indices of the $w_{ij}$ term.) If the system is at equilibrium, then the probability of being in state $i$ is constant, i.e., $d\rho_i/dt = 0$. Thus,

$$\frac{w_{ij}}{w_{ji}} = \frac{\rho_j}{\rho_i} = \frac{e^{-E_j/k_B T}}{e^{-E_i/k_B T}} = e^{-(E_j - E_i)/k_B T}, \tag{7.41}$$

Since sampling in the canonical ensemble depends on $\rho_j/\rho_i$, it requires that the ratio of the rates to and from a state, $w_{ij}/w_{ji}$, is given by $w_{ij}/w_{ji} = \exp(-\Delta(E_j - E_i)/k_B T)$. In the discussion of the kinetic Monte Carlo method in Section 9.2, we establish that the *rate* of going from state $i$ to state $j$ is proportional to the *probability* of going from $i$ to $j$, i.e., $w_{ij} = C$ probability$(i, j)$, where $C$ is some constant.

Consider now the probability of going between states $i$ and $j$ in the Metropolis algorithm. Based on Figure 7.2, for the two cases, $E_j - E_i \le 0$ and $E_j - E_i > 0$, we can write the probabilities as

$$\begin{aligned} \text{probability}(i, j) &= 1 && \text{if } E_j - E_i \le 0 \\ \text{probability}(i, j) &= e^{-(E_j - E_i)/k_B T} && \text{if } E_j - E_i > 0. \end{aligned} \tag{7.42}$$

Consider now the ratio of rates, $w_{ij}/w_{ji}$,

$$\frac{w_{ij}}{w_{ji}} = \frac{\text{probability}(i, j)}{\text{probability}(j, i)}, \tag{7.43}$$

where the constant $C$ cancels. For the two cases,

$$\begin{aligned} \frac{w_{ij}}{w_{ji}} &= \frac{1}{e^{-(E_i - E_j)/k_B T}} = e^{-(E_j - E_i)/k_B T} && \text{if } E_j - E_i \le 0 \\ \frac{w_{ij}}{w_{ji}} &= \frac{e^{-(E_j - E_i)/k_B T}}{1} = e^{-(E_j - E_i)/k_B T} && \text{if } E_j - E_i > 0. \end{aligned} \tag{7.44}$$

Thus, the procedure followed by the Metropolis algorithm is consistent with the canonical ensemble. Recognition that one can sample a system in a specific ensemble was a very important event in statistical mechanics.

# 8 Molecular and macromolecular systems

In this chapter, we discuss how to extend the methods introduced in the previous chapters from atomic to macromolecular systems. The basic ideas are the same, but there are additional complexities that arise from the molecular shapes. The simulation of molecular systems, especially polymeric and biological materials, is a very active field and we barely touch the surface here. For more information, please see the texts in the Suggested reading section.

After a review of the basic properties of macromolecules, the chapter continues with a discussion of some of the common approaches to model the interaction between the molecules, followed by descriptions of how molecular dynamics and Monte Carlo methods can be applied to molecular systems. When discussing systems of large molecules, such as polymers or proteins, however, it becomes challenging to include the full complexity of the molecules within a calculation. Thus, various models that approximate the physics have been developed. The chapter ends with a discussion of some of these approximate methods.

## 8.1 INTRODUCTION

Polymers (macromolecules) are large molecules made up of long chains of monomer units. In some biological molecules, the number of monomers ($N$) can be quite high, e.g., in DNA $N \sim 10^8$ in some cases. In other systems, $N$ can be of the order of a few hundred. The identity of the monomer units defines the overall properties of the polymer – DNA and RNA are made up of nucleotides, proteins are made up of amino acids, etc. In commercial polymers, the monomer units may be simple groups, such as $-CH_2-$ in polyethylene, or more complicated systems such as ethylene-vinyl acetate (EVA), which is a copolymer of ethylene and vinyl acetate. In block polymers, the monomers are ordered, i.e., there are regions where the monomers are of one type or another. Block copolymers, for example, are made up of two types of monomers, with one end of the polymer consisting of one type of monomer and the opposite end consisting of the other type of monomer. If one end of a block copolymer is hydrophobic (more attracted to itself or non-polar solvents than to water) and the other is hydrophylic (more attracted to polar solvents such as water than to non-polar molecules) then the polymer will naturally form many of the important structures found in biological systems, for example, micelles.

Polymer molecules are single macromolecules composed of repeating structural units, held together by covalent chemical bonds, such as shown in Figure 8.1a for polyethylene. Polymeric systems may consist of individual straight chains, as in Figure 8.2a, branched polymers as

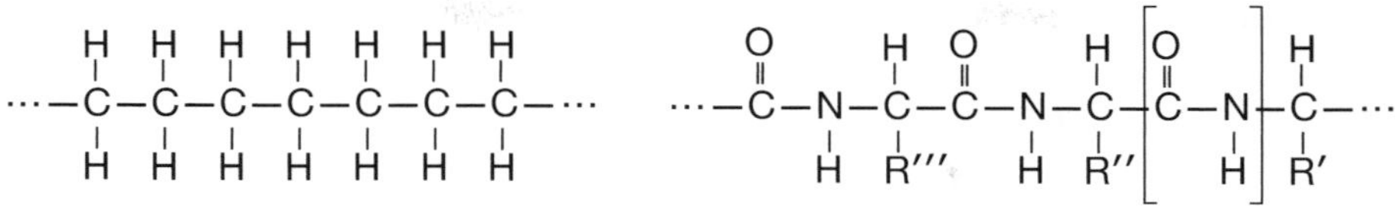

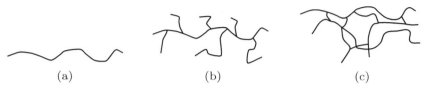

**Figure 8.1** Examples of macromolecules. (a) Polyethylene, a polymer molecule. (b) An example of a protein. R′, R″ and R‴ indicate side chains, the composition of which determines the type of amino acid. The CNOH group in brackets is a *peptide bond*, which links the amino acids along the chain.

**Figure 8.2** Schematic view of the chain structure of polymers. (a) Single chained. (b) Branched. (c) Cross-linked.

in Figure 8.2b, or can have cross links between strands, as in Figure 8.2c. The mechanical properties of the polymeric systems depend strongly on which type of polymeric structure is present, with systems of single-chained polymeric molecules being generally less rigid than those made up of branched polymers, which are in turn less stiff than cross-linked systems.

Polymeric systems have a range of properties that are the basis of many types of engineered systems. Plastics, for example, are usually made of organic polymeric systems, with their properties determined by both the repeating units that make up the polymer backbone as well as that of the side chains. By fine-tuning the composition, materials with a remarkable range of properties can be created, leading to the ubiquitous use of plastics in our world. The structure of the plastics can be either amorphous or partially crystalline and partially amorphous. In the latter case, they have a melting point and at least one glass transition, defined as the temperature at which the extent of localized molecular flexibility is substantially increased.

One of the most exciting developments in materials science and engineering over the past few decades is the confluence of materials and biology. New materials based on biological templates and the use of biology to make new materials have great promise. Macromolecules are the foundation of all biological systems. They may appear as a single molecule or as a collection of molecules held together by electrostatic or van der Waals forces. They express genetic information, they serve as chemical factories, and they provide the mechanical backbone for cellular structures. Cytoskeletal biopolymers (e.g., actin filaments and microtubules), for example, serve as the structural elements of the cell [156]. There are four basic types of biomolecular polymers: proteins, lipids, nucleic acids, and carbohydrates. There are also many non-polymeric biological molecules with large molecular mass.

As an example, consider the structure of proteins, as shown in Figure 8.1b. Proteins consist of polypeptides, which are single linear polymer chains of amino acids that are bonded together

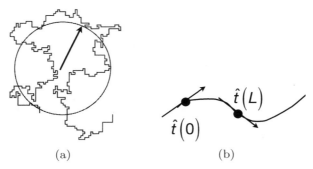

**Figure 8.3** Measures of polymer properties. (a) Radius of gyration, $R_g$, as a measure of the size of the molecule. This figure was generated using a self-avoiding random walk in two dimensions (described in Section 8.2), with the mean position and a circle of radius $R_g$ superimposed on the trajectory. (b) The persistence length is a measure of the loss in correlation of the tangents, indicated by arrows, along the polymer chain.

by peptide bonds between the carboxyl and amino groups of adjacent amino acid residues, as indicated by the brackets in Figure 8.1b. There are a number of primary amino acids, depending on the side chains, indicated by R′ in that figure. Most proteins fold into unique three-dimensional structures, called the *tertiary structure*, which are stabilized by a number of different types of interactions, including hydrogen bonds, bonds between two sulfur atoms (sulfide bonds), etc. The tertiary structure is what controls the basic function of the protein.

### 8.1.1 Properties of polymer solutions

For this text, we restrict the discussion to applications of modeling to polymer solutions. In this case, the interest is primarily on the properties of individual molecules, which includes their structures and conformations. One of the unique features of polymeric solutions is that the time scales of their relaxation processes can be very slow. For example, protein folding involves physical time scales from microseconds to seconds, which is much too long for standard dynamical simulation methods like molecular dynamics [183]. We will address methods by which we can model changes of these slow processes.

There are a number of ways to categorize the properties of a polymer chain. For example, one measure of the *shape* of a polymer is the *radius of gyration*, which is defined as the root mean square distance of the monomers in the polymer from the mean position of the monomers,

$$R_g^2 = \frac{1}{N}\left\langle \sum_{k=1}^{N} \left(\mathbf{r}_k - \langle\mathbf{r}\rangle\right)^2 \right\rangle, \qquad (8.1)$$

where the mean monomer position is

$$\langle\mathbf{r}\rangle = \frac{1}{N}\sum_{k=1}^{N} \mathbf{r}_k. \qquad (8.2)$$

Below we shall discuss $R_g$ in the context of random walk models. A view of $R_g$ is given in Figure 8.3a.

Another important measure is the *persistence length*, which is related to the mechanical properties of a polymer. The persistence length is defined as the length over which correlations in the direction of the tangent to the polymer chain are lost. Consider Figure 8.3b. The unit tangent at some point, called 0, along the chain, is denoted by $\hat{t}(0)$. Moving a distance $L$ along the chain, the unit tangent is $\hat{t}(L)$. The angle between the tangents, as a function of the distance $L$ along the polymer, can be determined from the dot product $\hat{t}(0) \cdot \hat{t}(L)$. The persistence length, $\ell_p$, is defined by averaging over many configurations of the polymer and is given by

$$\langle \hat{t}(0) \cdot \hat{t}(L) \rangle = e^{-L/\ell_p} . \tag{8.3}$$

From a mechanics perspective, semi-flexible polymer chains can be modeled by an approximation called the *worm-like chain model* [36, 156], which is also referred to as the Kratky-Porod model [180]. The worm-like chain model is based in a continuously flexible rod and works well for stiffer polymers in which nearby segments along the polymer chain tend to point in approximately the same direction. Within this simple mechanical picture, it is possible to relate the persistence length to the *stiffness parameter*, $\kappa_f$, of the polymer chain by[1]

$$\ell_p = \frac{\kappa_f}{k_B T} , \tag{8.4}$$

which in turn is related to the Young's modulus, $E$, of the bending motion of the chain by[2]

$$\kappa_f = \mathfrak{I} E , \tag{8.5}$$

where $\mathfrak{I}$ is the second moment of inertia of the cross section of the polymer. For example, for a circular rod, $\mathfrak{I} = \pi R^4 / 4$, where $R$ is the radius of the rod. The mechanical energy of the bending of the rod is

$$E_{bend} = \frac{\kappa_f L}{2R^2} . \tag{8.6}$$

Typical values for the Young's modulus are about 3 GPa for polystyrene and over 100 GPa for most metals. Microtubules, a macromolecular assembly that forms an important part of the cytoskeleton, has a Young's modulus of about 2 GPa [285].

## 8.2 RANDOM-WALK MODELS OF POLYMERS

Consider a very simple description of a polymer chain, namely as a connected set of $n$ linear segments with length $a$ that are sequentially placed on a lattice, connecting nearest-neighbor sites. Let us further suppose that these segments are put down randomly, with no restrictions on whether segments touch one another. This procedure is just a random walk on a lattice and we can use the analysis given in Chapter 2 to describe the properties of the polymer configurations created with this approach, which is generally referred to as a *random-chain model*.

An important quantity for a polymer is the average *end-to-end distance*. One measure of this quantity is the square root of the mean square displacement, $\langle R_{ee}^2 \rangle$, given in Eq. (2.9),

$$\langle R_{ee}^2 \rangle^{1/2} = \sqrt{n} a \quad \text{(random chain)} . \tag{8.7}$$

---

[1] An excellent description of the mechanical properties of biological molecular assemblies is given in [36].
[2] The Young's modulus is defined in Eq. (H.28).

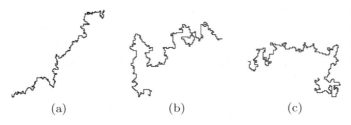

**Figure 8.4** First 500 jumps in three self-avoiding random walks on a square lattice. Compare with the normal random walk in Figure 2.5a.

(a)  (b)  (c)

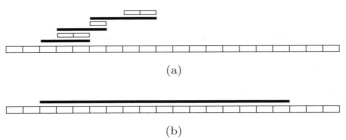

(a)

(b)

**Figure 8.5** Self-avoiding random walk in one dimension. (a) Random walk in one dimension. Black lines indicate movement to the right and thin white boxes represent movement to the left. (b) Self-avoiding random walk in one dimension.

Thus we see that in the random-chain model, the mean end-to-end distance is a linear function of the square root of the number of segments in the polymer chain, which is referred to as *ideal* scaling. Note that as we discussed in Chapter 2, this scaling relation holds in two dimensions or three and does not depend on the lattice.

While the random-chain model is very simple, the result in Eq. (8.7) is not a very good description of a real polymer chain. Why? Because a real polymer is made up of atoms and atoms have a volume associated with them, which excludes other atoms from occupying the same space. It is this excluded volume that is not considered in the random-chain model.

We can improve on the random-chain model by simply forbidding any chain to cross itself, an approach referred to as the *self-avoiding* chain or, more commonly, the self-avoiding walk (SAW). A self-avoiding walk is simply a random path from one point to another that never intersects itself, and thus incorporates, at least at some level, the idea of an excluded volume. We show an example of the first 500 steps of three simulations of self-avoiding walks on a square lattice in Figure 8.4. There are no crossings or retracing of jumps in the walks, in contrast with the standard random walk in Figure 2.5a.

A self-avoiding chain has very different scaling properties than a random chain, as can easily be seen for a chain in one dimension, an example of which is shown in Figure 8.5a. We expect from Eq. (8.7) that for any dimension, $\langle R_{ee}^2 \rangle^{1/2} = \sqrt{n}a$, which seems consistent with the shown trajectory in one dimension. The trajectory folds back on itself and so the average end-to-end distance at the end of the trajectory is far lower than the number of jumps in the trajectory.

Consider, however, the trajectory for a self-avoiding chain in one dimension, as shown in Figure 8.5b. That trajectory can never fold back on itself, so the end-to-end length is always $\langle R_{ee}^2 \rangle^{1/2} = na$. Thus, the scaling in one dimension goes as $n$ for a self-avoiding chain. In general, the scaling for a self-avoiding chain is $\langle R_{ee}^2 \rangle^{1/2} = n^v a$, where $v$ depends on dimension. Flory

[104] showed that the scaling exponent is approximately given by

$$\nu \sim \nu_{FL} = \frac{3}{2+d},$$  (8.8)

where $d$ is the dimensionality of the system [36]. $\nu_{FL} = 1$, 3/4, and 3/5 in one, two and three dimensions, respectively, results that simulations show are essentially exact. For polymers in four or higher dimensions (whatever that means), ideal scaling holds and $\nu = 1/2$.

At the simplest level, implementing a self-avoiding walk is straightforward. One could just pick random jumps, keeping a list of every site that has been visited. If a new jump takes the system to an already visited site, that jump is not allowed and another jump must be attempted. This approach might work well for short chains, though even in that case there could be situations in which no jump is possible. For example, if the beginning of the jump sequence in Figure 8.4, which is marked as a small filled circle, were at the end of a jump sequence, there would be no possible jumps that could be made and the trajectory would be terminated.

For large chains, the time it takes to sort through large lists of visited sites is prohibitive. As we saw for random chains, many trials are needed to obtain good statistics for determining properties such as the average end-to-end distance. Thus, much work has gone into finding efficient algorithms. One method, called the *pivot* algorithm, has proven quite successful. Details are given elsewhere [187, 212]. An implementation in Mathematica is also available [117].

## 8.3 ATOMISTIC SIMULATIONS OF MACROMOLECULES

While the random walk models show interesting behavior, they do not incorporate any of the fundamental chemistry and physics of the macromolecule. To go beyond those simple models, we can use many of the same procedures that we have already discussed for atomic systems, using intermolecular potentials (Chapter 5) as a basis for molecular dynamics (Chapter 6) or Monte Carlo (Chapter 7) simulations. Here we will just outline the main differences and challenges faced when extending those methods to molecular systems. For more details, please see the Suggested readings at the end of the chapter.

### 8.3.1 Interactions between molecules

Intermolecular potentials are based on the same basic physics as the interatomic potentials from Chapter 5. However, intermolecular potentials have the added complexity of having to incorporate the shape of the molecules, which, for larger molecules, may be quite complex. Molecules also have internal degrees of freedom (e.g., bond stretches and bends) that affect their structure and properties. We summarize the basic approaches to both of these in this section, starting with the most common way to model the interactions between molecules. In Appendix 8.8, we discuss an example of the interaction between small molecules, the $H_2O$–$H_2O$ interaction, that illustrates these ideas.

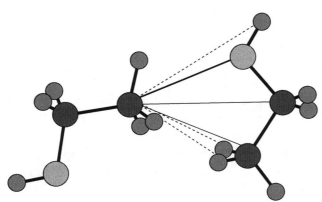

**Figure 8.6** Atom-atom interactions between two ethanol ($C_2H_5OH$) molecules: carbons in dark gray, oxygen atom in light gray, and hydrogen atoms in medium gray. The interactions between C and O are shown as dark solid lines, between C and C as light solid lines, and between C and H as dashed lines. For clarity, only some interactions are shown.

## Atom-atom potentials

Intermolecular interaction potentials must accurately reflect the position and chemical identity of the atoms within the molecules. The simplest, and most commonly used, form of potential consists of sums of pair potentials between the atoms within the molecules. This type of potential is appropriately called an *atom-atom* potential. An example is shown in Figure 8.6 for the interaction of two ethanol ($C_2H_5OH$) molecules.[3] For clarity, only some of the interactions are shown in the figure.

For two molecules, $a$ and $b$, consisting of $N_a$ and $N_b$ atoms, respectively, the atom-atom potential between the molecules is written as

$$V_{ab} = \sum_{i=1}^{N_a} \sum_{j=1}^{N_b} \phi_{ij}(\mathbf{r}_{bj} - \mathbf{r}_{ai}) + V_{ab}^{elect} , \qquad (8.9)$$

where $\phi_{ij}$ is a potential function appropriate for the types of atoms $i$ and $j$, $\mathbf{r}_{ai}$ is the position of the $i$th atom on molecule $a$, $\mathbf{r}_{bj}$ is the position of the $j$th atom on molecule $b$, and $V_{ab}^{elect}$ is the electrostatic energy between the molecules. Ignoring any cutoffs in the potential, there are $N_a N_b$ terms in the atom-atom potential, so this form becomes computationally challenging for very large molecules (e.g., polymers and proteins).

The bonding in a condensed-phase molecular system is most similar to the rare-gas molecules of Figure 5.2a – the electrons are tightly bound to the molecule with little to no electronic distribution between the molecules. The dominant long-range interactions between molecules are electrostatic interactions $V_{ab}^{elect}$ and van der Waals interactions. It should not be surprising then that the typical forms used for $\phi_{ij}$ are the Lennard-Jones (Eq. (5.6)) or exponential-6 (Eq. (5.11)) functions, which include the dominant van der Waals interactions at long range.

Parameters for $\phi_{ij}$ have been tabulated for many types of atom pairs. They usually reflect the local bonding of the atoms. For example, potentials between a carbon atom and an oxygen atom in Figure 8.6 would be different than between carbon and hydrogen or carbon and carbon. Indeed,

---

[3] The positions of the interacting sites can be moved off the atom positions, in which case it is usually called a *site-site* potential.

the potentials between two carbon atoms may also reflect the difference in local environments of the two atoms.[4]

Development of a set of self-consistent potential parameters is very challenging. The potential functions are simple, yet the bonding within molecules can be quite complex. Parameters in the potential are varied until properties calculated with the potentials match experimental data or quantum mechanical calculations. Different sets of potential parameters have been developed for different applications, for example for polymers, proteins, nucleic acids, etc. Two examples of potential parameters for proteins and polymers are the CHARMM22 [209] and Amber [262] packages. The CHARMM22 package, for example, uses the Lennard-Jones potential for $\phi$ in Eq. (8.9) coupled with a point charge model for the electrostatic energy, as discussed in the next section.

### Electrostatic energy terms

Charge distributions in molecular systems can be quite complicated. Even when the molecules have closed-shell electronic distributions, the distribution of chemical species within the molecule can lead to charge distortions, which in turn create local electrostatic moments.[5] For larger molecules, these continuous charge distributions are usually modeled as a sum of discrete charges whose electrostatic moments match those of the true molecule. In the latter case, the net interaction between molecules $a$ and $b$ takes the form

$$V_{ab}^{electro} = \sum_{i=1}^{M_a} \sum_{j=1}^{M_b} \frac{q_{ai}q_{bj}}{|\mathbf{r}_{bj} - \mathbf{r}_{ai}|} , \qquad (8.10)$$

where there are $M_{a(b)}$ charges on molecule $a(b)$, and $q_{ai}$ is the charge on the $i$th charge on molecule $a$. Note that the number and location of the charges does not need to match the sites in the atom-atom potential discussed above (though they generally are the same).

For small molecules such as water, both descriptions such as Eq. (8.10) as well as the multipole expansions of Appendix E.3 are used. Even for these small molecules, the best descriptions of their interactions typically use distributed charges as in Eq. (8.10). In practice, the charges need not be put on to the atoms, but can be distributed to best match the electrostatic moments of the charge distributions. For example, in Appendix 8.8, we discuss a model for $H_2O$ that includes charges both on and off the atom sites.

### 8.3.2  Intramolecular energy terms

Functions that described the interaction energy between atoms within the same molecule must also be incorporated into the overall energy expression.[6] These *intra*molecular energy terms can

---

[4] The chemical identity of neighboring atoms may alter the electronic distribution of the atom and thus change the interactions with other atoms.

[5] Please review Appendix E.3.

[6] Models in which the molecules are kept rigid are sometimes used, in which case the expressions in this section are omitted.

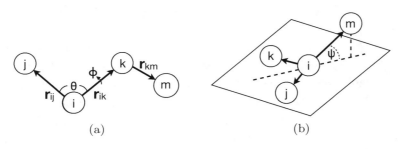

**Figure 8.7** Angles and distances used in intramolecular potentials.

(a)          (b)

be divided into four basic types, described with the angles and distances shown in Figure 8.7: bond stretching, valence angle bending, dihedral angle bending, and inversion angle bending. For larger, flexible, molecules, there may also be intramolecular contributions from the direct interactions of atoms on the same molecule that would take a form similar to that between molecules in Eq. (8.9). These terms would be included in the summations of the atom-atom potentials.

The bond stretching energy is a function of the deviation of the length of a bond from its equilibrium length. The basic form of the potential could, perhaps, best be described by a Morse potential described in Eq. (5.13), which is based on the dissociation potential of a diatomic molecule. Generally, however, approximations derived from a Taylor series of the correct potential around its equilibrium value are used. The usual approach is to stop at the harmonic term, which is the first non-vanishing term in the Taylor series. In this *harmonic approximation*, the bonds are considered as harmonic springs, as described in Appendix D.3. Expressions that include higher-order terms in the Taylor expansion (e.g., cubic, quartic, etc.) can also be used.

Consider the bond between atoms $i$ and $j$ in Figure 8.7a. The instantaneous length of that bond is $r_{ij} = |\mathbf{r}_{ij}|$, while its equilibrium length is denoted as $r_{ij}^o$.[7] The energy of a bond can be written as

$$U(r_{ij}) = \frac{k_{ij}}{2}(r_{ij} - r_{ij}^o)^2 + \frac{g_{ij}}{3}(r_{ij} - r_{ij}^o)^3 + \frac{h_{ij}}{4}(r_{ij} - r_{ij}^o)^4 + \cdots, \qquad (8.11)$$

where $k_{ij}$, $g_{ij}$, and $h_{ij}$ are the force constants of the bond. In the harmonic approximation, the energy of a bond is approximated with the first term in Eq. (8.11). The total contribution to the energy is a sum over all bonded pairs. The force constants tend to be large for covalent bonds, so the frequencies of vibrations are quite high – much higher than the fundamental intermolecular frequencies of the lattice vibrations, for example.

The bond angle, $\theta_{ijk}$, is the angle between two bonds $ij$ and $ik$ in Figure 8.7a, where

$$\theta = \cos^{-1}\left[\frac{\mathbf{r}_{ij} \cdot \mathbf{r}_{ik}}{r_{ij}r_{ik}}\right]. \qquad (8.12)$$

Typically the energy of the bond angle is approximated with a harmonic term of the form

$$U(\theta_{ijk}) = \frac{k_{ijk}}{2}(\theta_{ijk} - \theta_{ijk}^o)^2, \qquad (8.13)$$

---

[7] Note that we define, in the usual way, $\mathbf{r}_{ij} = \mathbf{r}_j - \mathbf{r}_i$.

where $\theta_{ijk}^o$ is the equilibrium angle and $k_{ijk}$ the force constant. The force constants of bond bending tend to be less than those for bond stretching.

The dihedral angle $\phi$ is the angle between the normals to the planes containing atoms $jik$ and $ikm$ in Figure 8.7a. Since we know the normal to the plane defined by three atoms is proportional to the cross product of the bond vectors, we can write the dihedral angle as

$$\phi = \cos^{-1}\left(\hat{n}_{ijk} \cdot \hat{n}_{ikm}\right), \tag{8.14}$$

where the unit normals are defined as

$$\hat{n}_{ijk} = \frac{\mathbf{r}_{ij} \times \mathbf{r}_{ik}}{|\mathbf{r}_{ij} \times \mathbf{r}_{ik}|} \tag{8.15a}$$

$$\hat{n}_{ikm} = \frac{\mathbf{r}_{ki} \times \mathbf{r}_{km}}{|\mathbf{r}_{ki} \times \mathbf{r}_{km}|}. \tag{8.15b}$$

The energy associated with a change in $\phi$ (a torsion) can be written in a number of forms. For example, to model potentially large-scale torsions, a planar potential of the form

$$U(\phi_{jikm}) = A\cos(\phi_{jikm} + \delta) \tag{8.16}$$

is appropriate, where $A$ and $\delta$ are constants. For smaller, harmonic motion,

$$U(\phi_{jikm}) = \frac{1}{2}k_{jikm}(\phi_{jikm} - \phi_{jikm}^o)^2 \tag{8.17}$$

would be an acceptable choice.

The *inversion* angle $\psi$ is shown in Figure 8.7b. It describes the energy associated with the arrangement of three atoms around a central one. An important example is the arrangement of three hydrogen atoms around a nitrogen. By inverting the nitrogen atom (think of an umbrella), a new, energetically equivalent, structure can be formed. $\psi$ is the angle between the bond vector $\mathbf{r}_{im}$ and the plane containing the other three atoms. Typical forms for the energy of this motion include a harmonic potential as well as planar potentials of the form

$$U(\psi_{ijkm}) = A\cos(\psi_{ijkm} + \delta). \tag{8.18}$$

Note that there would be a potential term for the inversion of each atom for which inversion is possible.

### 8.3.3 Exploring the energy surface

The total potential energy of a system of $N$ molecules is a sum over the atom-atom potentials of Eq. (8.9),

$$U = \frac{1}{2}\sum_{a=1}^{N}\sum_{b=1}^{N}V_{ab} + \sum_{a=1}^{N}V_{intra}(a), \tag{8.19}$$

where the sums over $a$ and $b$ are over all the molecules and $V_{intra}(a)$ is the intramolecular energy of molecule $a$ as discussed in Section 8.3.2. The force on a specific atom can be found from the gradient of the potential energy with respect to that atom's coordinates, as discussed in Section 6.1. The potentials and forces are usually truncated at a cutoff distance, as described in Section 6.2.1.

In Section 5.8 we discuss methods in which the zero-temperature properties of materials can be determined by minimizing the energy of a system with respect to structural parameters. The same types of simulations are commonly done for molecular systems. For solids consisting of small molecules, for example water, simulations to determine crystal structures and energetics would be done in essentially the same way as described in Section 5.8 for atomic systems, with the added complication owing to the degrees of freedom associated with the molecular orientations, which would need to be included as variables in an energy minimization. We present a simple example of orientational changes in the discussion of the Monte Carlo method applied to molecular systems in a later section.

Another important application is the mapping of the energy for configurational changes in a macromolecule, an important example being protein folding and binding. To understand these phenomena, we introduce a quantity referred to as an *energy landscape*, which is a mapping of all possible conformations of a molecule and their corresponding Gibbs free energies [335]. One way to think about protein folding is as a diffusion-like process, in which molecules reach their final configuration by moving down a funnel-shaped free-energy landscape [244]. The shape of the funnel guides the molecules to their final state. The dynamics are determined by the details of the underlying energy surface. These surfaces are not smooth and diffusion occurs through a series of peaks and valleys. Such surfaces are referred to as being *rugged*. Thus, to understand the folding of proteins requires knowledge of not only the energy of the initial and final states, but also the roughness of the energy surface that connects those two states [244].

Computer simulations can play an important role by mapping out the energy landscapes of phenomena like protein folding. Indeed, most of our current knowledge of the roughness of the energy surfaces comes entirely from simulations of small model proteins. Details of energy minimization and dynamics methods applied to this problem are given elsewhere [282]. Suffice it to say that the basics of the methods are based on the methods for describing energies described in this section.

## 8.3.4  Molecular dynamics

Applying molecular dynamics to macromolecular materials follows the same procedures outlined in Chapter 6. The force on an atom is calculated by taking the negative of the gradient of the energy in Eq. (8.19) with respect to the atom position in the usual way. Once one calculates the forces, the equations of motion can be solved based on the same algorithms as presented in that chapter.

Simulations of macromolecules can be performed in any of the ensembles introduced in Chapter 6, including the isobaric, isothermal ensemble (*NPT*). The steps of a simulation are the same as those discussed in Section 6.1.6, with the first step being the choice of an initial configuration of the molecules. Unlike the study of simple atomic systems, in which the creation of an initial configuration is not particularly difficult, choosing the configuration for a macromolecule can be quite complicated. Typically, the starting configuration may come from experiment, for example an X-ray diffraction study, or from a minimization procedure as outlined in Section 8.3.3. Analysis of the results also proceeds for any other molecular dynamics simulation, including determination of averages of thermodynamic quantities, measures of structure, spatial and temporal correlation functions, and so on.

There are a number of issues that arise in the study of macromolecular systems, which center, ultimately, on the very large number of atoms needed to model macromolecular systems as well as on the fundamental time step of the simulations. The systems sizes of simulations of macromolecules are usually quite large – a typical simulation may include tens to hundreds of thousands of atoms in the molecules as well as sufficient numbers of solvent molecules (often water) to fill in the simulation cell. In Section 8.4 we will discuss a method by which one can extend length scales in simulating macromolecules by introducing *coarse-grained* representations of the molecules, in which groups of atoms are treated as one center of force.

As in all molecular dynamics simulations, the size of the time step is set to obtain an accurate integration of the equations of motion. Regardless of the algorithm used, the time step must be considerably smaller than the period of the fastest motion in the system, which, for molecular systems, will be the intramolecular vibrations. The intramolecular vibrations are much faster than the motion of the centers of mass of the molecules, with periods that are typically in the range of about $10^{-14}$ seconds or smaller. These short time steps limit the ability of molecular dynamics to study large-scale molecular motion, whether it be the center of mass translation or the movement of large parts of a molecule relative to each other, for example the bending of a protein configuration. In the next section we discuss a method by which we can suppress the fast intramolecular motions. We will also see that the time scales for coarse-grained models are longer than for molecular dynamics of chemically realistic descriptions of the molecules, which arises, in part, from a suppression of the fast intramolecular vibrations because of the coarse-graining procedures.

## Constrained dynamics

We can achieve an increase in time scales in molecular dynamics of chemically realistic models by suppressing the fast vibrations. We start by recognizing that the fast internal vibrations are usually decoupled from the overall translational and rotational motion of the molecules. Thus, keeping the bonds and bond angles fixed should not greatly affect the overall properties of the system. We can keep those motions fixed by using an algorithm that introduces constraints into the system, which we will illustrate by a simple example.

Suppose that we have a system of diatomic molecules whose atoms are, for convenience, restricted to the $x$ axis, as shown in Figure 8.8a. The vibrations of the molecule are very fast relative to the translational motion of the molecule, so that the overall properties of the system can be accurately modeled even if the bond lengths are fixed.

Suppose that at time $t$ the atoms have positions $x_1(t)$ and $x_2(t)$, as shown in Figure 8.8a, where the separation between them is $d$, the known bond length. The forces on the two atoms, $F_1$ and $F_2$, will depend on the neighboring molecules, so there is no reason to think that they

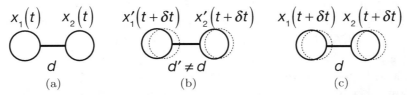

**Figure 8.8** Application of a constraint algorithm. (a) Atom positions at time $t$, with the interatomic separation equal to the bond length, $d$. (b) Solid circles indicate the positions at $t + \delta t$ after applying the Verlet algorithm without constraints, with the interatomic separation $d'$ not equal to the bond length. (c) Solid circles indicate the positions at $t + \delta t$ after applying the constraints, with the interatomic separation $d'$ now equal to the bond length.

would be the same. We can use the Verlet algorithm from Eq. (6.11) to determine the positions at time $t + \delta t$,

$$x_1'(t + \delta t) = 2x_1(t) - x_1(t - \delta t) + \frac{F_1}{m_1}\delta t^2$$

$$x_2'(t + \delta t) = 2x_2(t) - x_2(t - \delta t) + \frac{F_2}{m_2}\delta t^2 , \qquad (8.20)$$

where $m$ is the mass and the prime indicates unconstrained positions. These positions are shown in Figure 8.8b. The problem is that

$$d' = x_2'(t + \delta t) - x_1'(t + \delta t) \qquad (8.21)$$

does not necessarily equal the bond length, $d$. We can force the bond length to be fixed at $d$ by applying a constraining force [109, 280].

Define the energy

$$\sigma = d^2 - (x_2 - x_1)^2 , \qquad (8.22)$$

which is 0 when the instantaneous bond length, $x_2 - x_1$, equals the equilibrium value, $d$. We can derive a force from this energy by taking the negative of the gradient,

$$\mathbf{G} = -\frac{\lambda}{2}\nabla\sigma , \qquad (8.23)$$

where $\lambda$ is a parameter that we will determine below. For our one-dimensional problem, the constraining forces are

$$G_1 = -\frac{\lambda}{2}\frac{d\sigma}{dx_1} = -\lambda(x_2 - x_1) = -\lambda d$$

$$G_2 = -\frac{\lambda}{2}\frac{d\sigma}{dx_2} = \lambda d. \qquad (8.24)$$

Adding the constraining forces to the equations of motion in Eq. (8.20), we have

$$x_1(t + \delta t) = x_1'(t + \delta t) - \frac{\lambda}{m}d\,\delta t^2$$

$$x_2(t + \delta t) = x_2'(t + \delta t) + \frac{\lambda}{m}d\,\delta t^2 . \qquad (8.25)$$

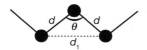

**Figure 8.9** Bond angle constraint as a fixed distance [280].

We now apply the constraint

$$\left(x_2(t + \delta t) - x_1(t + \delta t)\right)^2 = d^2,$$  (8.26)

which leads to the equation

$$\left(d' + \frac{2d\,\delta t^2}{m}\lambda\right)^2 = d^2,$$  (8.27)

or

$$\frac{4d^2\,\delta t^4}{m}\lambda^2 + \frac{4dd'\delta t^2}{m}\lambda + (d')^2 - d^2 = 0.$$  (8.28)

Solving the quadratic equation in Eq. (8.28) determines the value of $\lambda$ needed to constrain the molecular length. This is inserted in Eq. (8.25), yielding the constrained bond length, as shown schematically in Figure 8.8c. This procedure could be done at each time step for each molecule, keeping the bond length fixed.

Imagine applying this procedure to a polyatomic molecule. Each atom is bonded to two others, so there would be two constraint equations to keep the two bond lengths fixed. Bond angles can also be fixed. For $n$-alkanes, for example, the bond angle can be fixed by a distance criterion, as shown in Figure 8.9, introducing a constraint equation that couples each atom to its second-nearest neighbors along the chain by fixing the next-nearest-neighbor distance to $d_1 = 2d \sin(\theta/2)$, where $\theta = 109°28'$ for alkane chains. Thus, each atom will have two additional constraints. Solving for the coefficients in the constraint equations thus becomes a far more challenging computational procedure than in our simple example, involving solving a matrix equation at each time step. In the SHAKE algorithm [280], an approximate form of the equations is used and an iterative procedure followed to determine the appropriate coefficients. More details, and a description of the effect of using constrained dynamics on the ensemble averages, are presented in the book by Frenkel and Smit [109].

## 8.3.5 Monte Carlo

The Monte Carlo method of Chapter 7 is an important technique for the study of molecular and polymeric systems. It can be used with all-atom potentials, as described in Section 8.3.1, or for coarse-grained potentials as described in Section 8.4, or with lattice models from Section 8.5. The basic method is as described in Chapter 7: random movements of the degrees of freedom are attempted, with each trial move accepted or rejected using the Metropolis algorithm of Section 7.3.

Monte Carlo simulations of small molecules with rigid bond lengths and angles are a straightforward extension of those for atomic systems. Trial moves include centers of mass of the molecules as well as their orientations. For small linear molecules, such as $N_2$, sampling the orientations can be done by writing the positions of the atoms on each molecule in terms of its center of mass position $\mathbf{r}_i$ and the unit vector along the molecular bond, $\hat{n}_i = n_{xi}\hat{x} + n_{yi}\hat{y} + n_{zi}\hat{z}$. For example, for a diatomic molecule with bond length $r_b = 2\Delta$, the positions of the two atoms on molecule $i$ would be $\mathbf{r}_i \pm \Delta \hat{n}_i$. Sampling the position and orientation of each molecule can be broken into two parts: sampling the center of mass, which is done in the same way as the atomic position in Figure 7.8, and sampling the orientation, which can be done by adding a small random value to each of the three components of $\hat{n}_i$ and then renormalizing the vector. The maximum size of the orientation change is set in the same way as for the center of mass, by ensuring that the acceptance rate of the trials is about 50%.

Application of Monte Carlo to small, rigid, non-linear, molecules is a bit more complicated than the simple change in orientation of linear molecules. A number of alternative methods have been developed [5]. A common approach is to describe the orientational change in terms of a rotation through the Euler angles, described in Appendix C.1.4, in which a bond vector can be rotated about a fixed axis as described in Eq. (C.25). As discussed in [190], there are some subtleties in this method to ensure uniform sampling. Another approach is to use quaternions, which are four-dimensional vectors related to the Euler angles. They have the advantage that the rotation matrix of Eq. (C.26) is no longer dependent on trigonometric functions when written in quaternions. Details are given in [190]. Applying the Monte Carlo method to systems of non-rigid, small, molecules is not difficult, though there are some subtleties that are beyond the discussion in this text. The reader is referred to Section 4.7.2 in [5] for details.

There are considerable challenges when applying Monte Carlo to large, flexible, molecules, which we can see with a simple example. Suppose we are studying a long-chained molecule and attempt a trial move in which we change a bond angle from $\phi$ to $\phi + \delta\phi$. If the molecules are long, then even a small bond angle change at one end of the molecule can lead to a large change in the position of the atoms at the other end of the molecule. If the molecule is in a condensed phase, then such a large movement can lead to overlaps with other molecules within the system, which in turn will lead to a large energy change and a consequent low acceptance rate in the Metropolis algorithm. A variety of approaches have been suggested to overcome these difficulties, as described in the text by Leach in the Suggested reading section [190]. While such details are beyond what can be covered in this text, they enable application of Monte Carlo methods to critical problems in polymer conformation and structure.

## 8.4 COARSE-GRAINED METHODS

In Figure 1.1 we show time and length scales in the mechanical properties of materials. A similar figure could have been developed for macromolecular systems as well, again ranging from the subnanometer scale of electrons through the millimeter scale and higher, with times that range from femtoseconds to seconds [239]. As just one example, the mechanical folding and unfolding of the proteins in muscles has a gap of about six orders of magnitude between the time scales that can be modeled with full atomistic simulations and the time scales seen

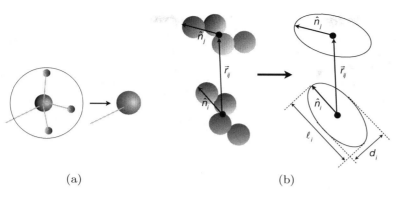

**Figure 8.10** Schematic view of coarse-grained potentials. (a) Representation of a methyl group (-CH$_3$) as a single point of interaction. (b) Representation of the atoms in a small chain as a united-atom potential.

(a)                                   (b)

in vivo [239]. In Part III of this book, we focus on a series of methods that enable us to go beyond the limitations of atomistic-level simulations to model phenomena in materials at a much larger scale. We base those methods on the idea that not all details from the atomic scale are important at the larger scales. In this chapter, we do something similar for macromolecular systems, first by employing a reduced model for the intermolecular interactions and then, in the following section, by restricting the molecules to a lattice. The approach in this section is called *coarse-graining* and represents a way to simplify the complexity of molecular interactions [74, 235].

The goal of a coarse-grained model is to capture the properties of a system to a desired accuracy, while reducing the complexity sufficiently to make computer simulations of that model feasible. The methods used will depend on the properties of interest, the amount of computation one wants to (or can) do, and the desired accuracy of the final predictions. There are many ways to go about creating coarse-grained models – we will discuss (briefly) just a few. More details are available elsewhere [48, 124, 181, 235, 239].

The basic idea behind coarse-graining for molecules is shown in Figure 8.10a. In the full atom-atom potential of Figure 8.6, each atom interacts with all other atoms. In a *united-atom model*, the hydrogens on a methyl group, for example, are combined with the carbon atom and the -CH$_3$ unit is represented by a single center of force, as indicated in the figure. This model will reduce the number of interacting particles in the methyl group from 4 to 1, with a proportional decrease in the computational burden. There is additional error introduced by this approximation, but the increase in speed owing to the fewer number of interactions will enable much larger systems to be studied. The simulator decides on the balance between accuracy and computational time.

Larger "united atom" groups can be created, as shown schematically in Figure 8.10b for a small-chain molecule. In the case shown there, replacing the group of atoms by a spherical "atom" would not necessarily be a particularly good representation of the interactions. Thus, an ellipsoidal "atom" could be employed [116]. The interactions between ellipsoids are described in terms of the parameters shown in Figure 8.10b. The Gay-Berne model, for example, describes the interactions between ellipsoids with a Lennard-Jones potential modified to reflect the shape of the ellipsoids. It has been used to model liquid crystal properties, using the length ($\ell_i$) and

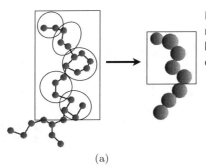

**Figure 8.11** Schematic view of coarse-grained potentials for large molecules. (a) Representation of a series of groups as connected beads. The beads are connected with spring chains so that the chain's behavior mimics the molecule.

(a)

width ($d_i$) as parameters to examine how the aspect ratio of the ellipsoids changes the behavior [81]. Details are given in [116].

In Figure 8.11a we show a higher-level reduced model of a large molecule. In that example, subunits of the macromolecule are described by united atoms, which in these types of models are often called "beads", or "pearls". In a *pearl-necklace* model, the distance between neighboring beads can vary, but only within a very tight limit. In *bead-spring* models, the beads are linked together with anharmonic springs. The beads could be spherical as in Figure 8.10a or ellipsoidal as in Figure 8.10b. Based on comparison between coarse-grained models and full-atom simulations, it seems that one can safely include about four to five carbon-carbon bonds within a single coarse-grained bond [124].

In bead-spring models the subunits are linked together with forces that enable the macromolecule to mimic the conformations of the actual molecule. Based on the number of C-C bonds in the coarse-grained model, these forces are usually bond vibrations (Eq. (8.11)) and bond bending (Eq. (8.12)), and not the torsional bonds (Eq. (8.16)).

## Determining the parameters

An excellent discussion of the steps needed in a coarse-graining procedure is given by Cranford and Buehler [74]. They identify the basic components as:

(a) the basic coarse-grained potentials
(b) a set of fully atomistic simulations that can be used to determine the parameters in the coarse-grained potentials
(c) a method to fit the results of fully atomistic simulations to determine the coarse-grained parameters
(d) a set of different fully-atomistic simulations or experiment for validation of the coarse-grained model.

When fitting the parameters, the most fundamental property that must be matched between simulations with the coarse-grained model and all-atom simulations is the energy, where energy equivalence is imposed between the coarse-grained model and the atomistic calculations. There are a number of ways that the energy could be defined when making that comparison, with the

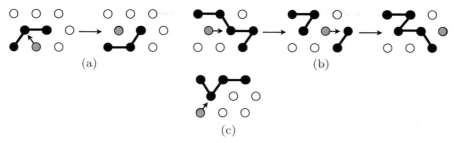

**Figure 8.12** Lattice model of polymers. Gray circles are vacancies. (a) An allowed move of a single bead. (b) A move involving multiple steps. (c) This move is not allowed, as it cannot occur without creating links that go beyond nearest neighbors. Adapted from [136].

choice affecting the final parameters in the coarse-grained potential. For example, one could compare the energy directly between the coarse-grained and all-atom calculations. Alternatively, an equivalence of the energy could be established by demanding that the mechanical properties, for example, the extension of a chain, the elastic properties, etc., be consistent between the coarse-grained and all-atom models [74]. Another approach is to focus not completely on the properties of the individual molecules, but to include in the fitting procedure a requirement that the structure of a system calculated with the coarse-grained potentials matches all-atom simulations. Such a structural match can be accomplished by demanding that the calculated pair correlation functions be equivalent between the two approaches. This latter procedure was used with great success by Ashbaugh and coworkers using a reverse Monte Carlo procedure[8] to optimize the parameters [11].

## 8.5 LATTICE MODELS FOR POLYMERS AND BIOMOLECULES

Lattice models of polymeric systems, such as the random-walk models presented in Section 8.2, offer another way to extend the length and scales of simulations of polymeric systems. The basic idea of these methods is shown in Figure 8.12, in which some portion of the polymer is represented as a bead that is placed on a lattice and connected to its neighboring beads along the polymer chain by a bond. The statistical mechanics of these beads approximates the thermodynamic properties of the full polymeric system. There are many variations of lattice models, many of which are described in [320].

To highlight some of the issues in lattice models, consider the two-dimensional model shown in Figure 8.12, which is based on work described in [136]. A polymer is broken into regions of a size less than the persistence length. These regions are represented as beads and placed

---

[8] See [291]. An initial guess of the coarse-grained parameters was chosen, creating a set of coarse-grained potentials, $\phi_{CG}(r)$. A simulation was run with those potentials in which the radial distribution functions were calculated, $g_{calc}(r)$. The potentials were updated according to $\phi_{CG}^{new}(r) = \phi_{CG}^{old}(r) + f\,k_B T\,\ln[g_{calc}(r)/g_{target}(r)]$, where $f$ is a constant. The simulations were run for a while with this potential, the $g(r)$s were calculated, and the potentials updated again, and so on until convergence was reached.

on a lattice and connected by bonds. Each bead reflects the local chemistry of the region of the polymer that it represents. Generally, the bonds are restricted to nearest neighbors. The lattice can be in any dimension, as long as it is space-filling. In Figure 8.12, a two-dimensional triangular lattice is used.

The beads can move to a different lattice site, subject to the constraint that the total length of the polymer is a constant. The total length does not refer to the end-to-end distance, which depends on the conformation of the molecule, but rather to the sum of all the bond lengths. That constraint changes the possible moves, as described in Figure 8.12. For example, in Figure 8.12b, each "move" involves multiple steps to keep the length constant. In Figure 8.12c, the moves are not allowed because they violate the restriction that the bonds must be nearest neighbors. The effects of being in a solution can be modeled by replacing some of the empty sites with solvent molecules.

The energy of the system is based on the interactions between the beads. For example, the bead-bead interactions can represent

(a) polymer-polymer interactions, including the interactions between different parts of the same polymer
(b) polymer-solvent interactions
(c) solvent-solvent interactions
(d) polymer-vacancy interactions, which models the surface tension of the polymer.

One can also include interactions between segments on the same polymer to mimic the normal configuration of the polymer, reflecting the actual bonding. For example, the angular dependence of these potentials could favor a $180°$ angle between the neighboring bonds, or whatever angle makes the most sense with the polymer being modeled and the symmetry of the lattice. Other interactions could include an external field, bead-surface interactions (to model the affinity of a polymer with a surface of some sort), etc.

Once the energy is found, standard Metropolis Monte Carlo methods based on moves such as shown in Figure 8.12 can be used. Applications can be made for such disparate problems as the basic polymer structure, calculating such things as the end-to-end distance, or the interdiffusion of polymers and solvent found in the "welding" of two polymers together [136]. With an appropriate choice of parameters, the properties of block copolymers have been studied, including such effects as the development of micelles, lamella structures, etc. [91]. Many other applications of such calculations have been made.

## 8.6 SIMULATIONS OF MOLECULAR AND MACROMOLECULAR MATERIALS

The literature describing simulations of molecular and macromolecular materials is vast. Some examples based on methods discussed in this text include:

• Polymer conformations: Monte Carlo simulations play a critical role in modeling the conformation of macromolecular systems, including such important processes as protein folding

[245]. A nice review of general procedures for optimization of complex structures is given in the book by Wales on *Energy Landscapes* in the Suggested readings [335].

- Solvent effects: interactions of polymers with solvents affect their properties, leading to, for example, the separation of oil and water, surfactants, etc. There are many challenges to simulating solvent effects, owing largely to the computational requirements of including the solvent molecules explicitly. Often, hybrid approaches are used, coupling molecular-level simulations with a continuum treatment of solvation [174]. Another approach has been to include the effects of the solvent molecules implicitly by modifying the intermolecular interactions [63].

- Structure and thermodynamics: one of the powerful aspects of the Monte Carlo method is that it can model any system for which one can define an energy in terms of a set of variables, regardless of whether a dynamical equation can be defined for those variables. This flexibility enables Monte Carlo calculations to be applied to systems for which molecular dynamics is inappropriate. One example involves a simple, and small, molecular system, $O_2$, that has 1/2-filled electron orbitals and an accompanying magnetic moment. Thus, the interactions between the molecules includes a pair potential such as described in Appendix 8.8 plus a magnetic interaction, which can be described with a simple model, called the Heisenberg model, in which the magnetic energy between a pair of molecules is $-J\mathbf{s} \cdot \mathbf{s}$, where $J$ defines the energy scale and $\mathbf{s}$ is the vector describing the magnetic moment. Monte Carlo calculations can be used to sample the molecular positions as described in Section 8.3.5 with a separate sampling of the spin variables. This approach was used to study a low-temperature, magnetically driven phase transformation in solid $O_2$ [195].

## 8.7 SUMMARY

We introduce the basics of how one models molecular systems. The chapter includes discussions of lattice models, starting with extensions of the random-walk model of Chapter 2. How one can describe the interactions between molecular systems is presented, followed by some comments on how one can apply molecular dynamics and Monte Carlo methods to systems of molecules. The chapter includes a discussion of how to extend calculations to very large molecules such as proteins, which requires a simplification of the molecular descriptions called coarse-graining.

### Suggested reading

- For a general discussion of the basic physics of polymers, the classic introductory text by Doi, *Introduction to Polymer Physics* [93], is very useful, as is the more in depth discussion in Doi and Edwards, *The Theory of Polymer Dynamics* [94].
- Leach's text, *Molecular Modelling: Principles and Applications* [190], gives an excellent introduction to molecular simulations, including much more detail than given here, to modeling molecular systems.

- For macromolecular systems, a useful concept is the energy landscape, discussed in Section 8.3.3. There is an excellent book on this subject by Wales, *Energy Landscapes* [335], which describes how many of the methodologies discussed in this text can be used to map out complex energy surfaces.
- Frenkel and Smit, *Understanding Molecular Simulation: From Algorithms to Applications* [109], provides a great deal of useful material.

## 8.8 APPENDIX

Probably the most important small molecule for modeling polymeric solutions is water, $H_2O$. Water seems simple – there are only three atoms, which form, of course, a plane. It would not seem to be that much more complicated than a linear molecule, for example $N_2$, except we know that water is a somewhat odd material. It has a much higher boiling point than would be expected for a system of light molecules – the boiling point of $N_2$ is $-195.8°C$, not $+100°C$. At $4°C$ water expands with either heating or cooling (which is rather important, or ice would sink).

The origin of water's rather odd behavior creates complications in modeling its properties. The cause of water's complexity arises from the structure of the water molecules. A simple way to understand that structure is to assume that the O atom in water is $sp^3$ hybridized, with the four orbitals in roughly a tetrahedron. Two of those orbitals bond with H atoms and the other orbitals have lone pairs of electrons. A consequence is that water has a large dipole moment; the oxygen end of the molecule is partially negative while the two hydrogen atoms are partially positive. The molecules of water are attracted to one another, with the slightly positive hydrogens attracted to the negative oxygens of other water molecules. This intermolecular attraction is termed "hydrogen bonding". These strong interactions between the water molecules make its behavior so unlike most materials.

There have been many models proposed for describing the interaction between water molecules and we will not discuss most of them here as they have been reviewed elsewhere [337]. An assessment of the accomplishments of these potentials has also been made [132]. The most successful potentials reflect the basic structure of the molecule. They include both the electrostatic interactions between the dipoles (and higher moments) as well as repulsive and van der Waals terms. While there are numerous potentials in the literature that describe interactions between water molecules, we discuss only three, closely related, potentials. These forms are popular, because they are simple and work well. They also illustrate some of the issues in contracting intermolecular interaction potentials.

In Figure 8.13, we show a simple model for a water molecule that is the basis of three examples of intermolecular potentials describing the interactions between model molecules [236]. There are six sites in this model; the O and H atoms (shown as black spheres) as well as three sites (M and L, shown as gray spheres) that contain extra charges that will be used to model the electron charge distribution. The M site is located along a line that bisects the ∠ HOH in the HOH plane. The L sites are located so that the molecule is an approximate tetrahedron.

Table 8.1 **Structural parameters for three water potentials**[a]

|  | TIP4P[b] | TIP5P[c] | Nada[a] |
|---|---|---|---|
| $r_{OH}$ (Å) | 0.9572 | 0.9572 | 0.980 |
| $r_{OM}$ (Å) | 0.15 |  | 0.230 |
| $r_{OL}$ (Å) |  | 0.70 | 0.8892 |
| $\angle$ HOH (°) | 104.52 | 104.52 | 108.0 |
| $\angle$ LOL (°) |  | 109.47 | 111.0 |
| $\epsilon_{OO}$ (K) | 78.0202 | 80.5147 | 85.9766 |
| $\epsilon_{HH}$ (K) |  |  | 13.8817 |
| $\sigma_{OO}$ (Å) | 3.154 | 3.12 | 3.115 |
| $\sigma_{HH}$ (Å) |  |  | 0.673 |
| $q_H$ (e) | 0.52 | 0.241 | 0.477 |
| $q_M$ (e) | −1.04 |  | −0.866 |
| $q_L$ (e) |  | −0.241 | −0.044 |

[a] Reference [236].
[b] Reference [162].
[c] Reference [214].

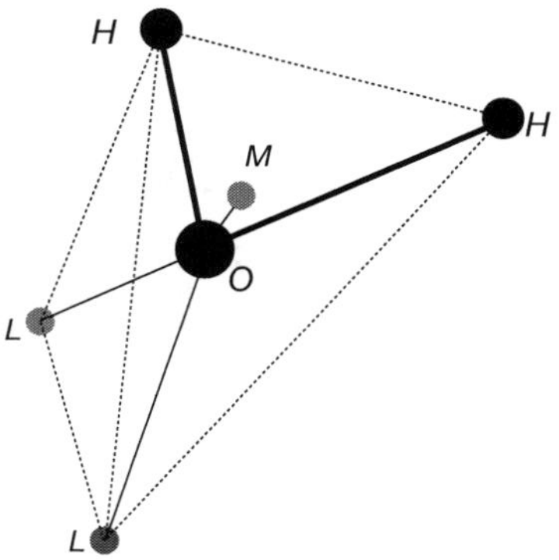

**Figure 8.13** Coordinates for water-water potential [236].

The magnitude of the charges on the sites as well as the lengths of the "bonds" to those charges and the angles between them serve as variational parameters that can be employed to optimize the potential.

In Table 8.1, we give parameters for three potentials, all based on the structure in Figure 8.13. The classic TIP4P potential is a four-site model, made up of the O and H atoms as well as the extra charge at M. There are charges at the H and M sites and a Lennard-Jones potential between the O atoms on different molecules. The TIP5P potential has, as the name suggests, five sites, excluding the M site but adding the two L sites at the corners of the (approximate) tetrahedron around the oxygen. Again, there is a Lennard-Jones potential between the O atoms on different molecules. The Nada potential [236] extends the model further, adding the M site to TIP5P and including additional Lennard-Jones interactions between H atoms on different molecules. Note that in none of these potentials is there a charge on the oxygen atom. We could easily imagine further enhancements of these models, adding charge to the oxygen, switching from Lennard-Jones (which we know is too repulsive) to a better site-site potential, adding a site-site potential between the H and O atoms, etc.

What governs our choice of potential? It depends on what the goal of the use of the potential will be, i.e., what is to be calculated. The potential by Nada in Table 8.1 was developed specifically to simulate ice and water near the melting point. They varied parameters until they could calculate the real melting point accurately and obtain the correct densities of both water and ice. They showed that the six-site model predicted that proton-disordered hexagonal ice is the stable structure at the melting point, as in real ice, while the TIP4P and TIP5P potentials failed to do so.

The point is that the authors applied earlier models (TIP4P and TIP5P) to a specific problem and found that those models were unable to reproduce experimental data under specific conditions. They then improved the model until it was accurate enough for their needs – they followed the general procedure for developing models discussed in Section 1.4.

Of course, just because the six-site model works near the melting point, it may not work well under other conditions, in the same way that TIP4P and TIP5P and all other water potentials have limitations of applicability. It is essential to remember that these are all *models*; they are not real water. Indeed, no calculations based on approximate potentials are really the "real" material. They are always models and may or may not be very good representations of real materials.

What about potentials for more complicated molecules? The general idea is the same; the potential is usually written as a sum of central force interactions between sites located on the molecule. For rigid molecules, i.e., molecules with atoms that are fixed relative to each other, there is no reason to restrict the interaction centers to the atoms. Indeed, to better represent an interaction, there may be more interacting sites than there are atoms in the molecule. In any case, the interacting sites may not be located at the atom sites, in which case the form is usually called a *site-site* potential. The first step is to choose the number and initial location of the interaction sites. The form of the site-site potential is then picked. Usually, a simple atom-atom form like those discussed above is used. The electrostatic interaction is modeled either as a sum over charges or in a multipole expansion.

It is important to remember that any intermolecular potential has been developed to be employed in simulations. Some of these are very high quality and some, not surprisingly, are not. The key is to find a form that works for the system of interest for the purpose of the calculations.

# Part Three

# Mesoscopic methods

# Kinetic Monte Carlo

In Chapter 7 we introduced the Monte Carlo method, with the focus being on calculating equilibrium properties based on sampling of the degrees of freedom in a Hamiltonian (energy) function. In this chapter, the focus is on the Monte Carlo method applied to *rates*. We shall see that for certain classes of problems, we can find an association between the Monte Carlo "time" and actual time, opening the door to a new class of simulation methods that can model time-dependent processes at time scales far beyond what is possible with standard molecular dynamics.

The fundamental input to the kinetic Monte Carlo method is a list of possible events such as a jump from one site to another in a diffusion problem, a chemical reaction, etc. Associated with each event will be a rate, which will be related to a probability that the event will occur. An understanding of rates is thus very important, so in Appendix G.8 we give a brief review of kinetic rate theory.

A number of researchers independently developed what has come to be known as the kinetic Monte Carlo method. The N-fold way as a methodology for accelerating the simulations of the Ising model is probably the first example [38], which will be discussed in Chapter 10. Voter introduced a similar approach as a way to study the dynamics of cluster diffusion on surfaces [327]. We will discuss his calculation later in this chapter as one of two examples of the complexities, and limitations, of the kinetic Monte Carlo approach.

## 9.1 THE KINETIC MONTE CARLO METHOD

Consider a system whose properties are dominated by thermally activated processes, such as diffusion. The evolution of such a system is discrete and stochastic – the system randomly moves from one configuration to another, with some average time between events. The rate the events occur is just the inverse of that average time between events. For such systems, if the rates of the possible events are known, we can apply the *kinetic Monte Carlo* method.

The basic idea behind the kinetic Monte Carlo method is an association of the probability of an event and its average rate, i.e., the number of times per second that it occurs. If there is more than one event that can take place, then at any time the probability of one event relative to all other events is proportional to its rate relative to the rates of other events. By stochastically sampling events based on their average probability (rate), then a chain of events can be followed

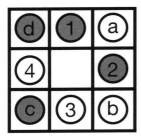

**Figure 9.1** The local environment around a vacancy on a two-dimensional square lattice showing a distribution of two species of atoms, indicated by white or gray circles. The numbered atoms indicate those that are nearest neighbors to the vacancy site and thus could jump into that site, while the atoms indicated by letters establish the local environment for the possible jumps.

over time. The time-dependent properties of the system can then, subject to some restrictions, be followed.

The overall procedure in a kinetic Monte Carlo calculation is straightforward, but often complicated. The first step is to identify all of the possible discrete events that can take place. For example, consider an alloy consisting of more than one type of atom, in which each species can diffuse through the system by vacancy diffusion,[1] jumping from site to site in the lattice. The energy barriers between sites will be different for each type of atom and thus the rate of jumping from site to site will not be the same for each species. It is even a bit more complicated than that, because the rate of jumps of a species will likely depend on the types of atoms in its neighborhood – a big atom may have a lower energy barrier for jumping when surrounded by small atoms than when it is located amid other big atoms. In the kinetic Monte Carlo method, the first step is to create a list of all the various types of jumps that could take place along with the rates of each type of jump.

As an example, consider the two-dimensional square lattice in Figure 9.1 in which there are two species of atoms surrounding a vacancy. Each of the numbered atoms could jump into the vacancy (since they are nearest neighbors), but would do so with a different rate, first because the species are different and second because the environment is different. For example, the potential surface that atom 1 would "feel" as it jumps into the vacancy depends on the identity of atoms near it during the jump. Its neighborhood is different from that of atom 2 and, even though 1 and 2 are the same species, the potential surface during a jump would be different for each atom and thus their jump rates would be different.

In the language of the kinetic Monte Carlo method, the jumps of atom 1 and atom 2 are considered to be distinct types of events, because they correspond to distinct local environments and thus distinct rates. In kinetic Monte Carlo, a list of all possible events, which in this case would include all possible local environments,[2] and their rates needs to be established before a calculation takes place.

Now consider the alloy at a specific time with a specific configuration of atoms. Vacancies will be distributed around the system, with each vacancy surrounded by a local configuration

---

[1] As discussed in Chapter 2.

[2] Listing all possible local environments is not particularly difficult in this case, since there are relatively few possible configurations. In other situations, however, the biggest barrier to applying the kinetic Monte Carlo method is the creating of the list of events.

of alloy species. Diffusion occurs by one of the atoms located next to a vacancy exchanging its site with the vacancy. The question is how to decide which atom of all the atoms in the system that could jump, does jump. In kinetic Monte Carlo, that decision is based on the relative rates of all possible events. For example, suppose the alloy is on the square lattice of Figure 9.1 and there are $N$ vacancies distributed throughout the system, none of which being located next to another vacancy. Since there are four nearest neighbors for each vacancy, there is a possibility of $4N$ jumps that could take place, one for each of the neighbors of each vacancy. Each of those jumps could occur at a certain rate, which depends on the distribution of the atoms surrounding the vacancies. The rates are established prior to the calculation. In kinetic Monte Carlo, the probability of a specific jump occurring is its rate divided by the total rate of all possible jumps that could take place with the atoms in their present configuration.

### 9.1.1 Choosing an event

Suppose a system of $N$ species can undergo a set of $M$ possible transitions, with each transition associated with a rate $r$. The probability of species $l$ undergoing transition $k$ is then proportional to $r_k$. In Figure 9.1, for example, there are two gray atoms, each with its own type of transition and its associated rate.

The total *activity* of the system is defined as the sum of the rates of all possible events that the system could undergo at any one time, i.e.,

$$A(t) = \sum_{k=1}^{M} r_k(t), \tag{9.1}$$

where $M$ is the total number of events that could take place in the system. In Figure 9.1, for example, there are four possible events, with their four rates. Each of these rates would be added to the activity, as would the rates of all possible jumps that could occur at every other vacancy in the system. It is important to remember that the activity at time $t$ depends on the configuration at time $t$. If atom 1 jumps in Figure 9.1, then the vacancy has moved and will be surrounded by a new set of atoms all with their own rates. The activity of that configuration will be different.

The probability of a specific event (for example, the jump of atom 1 in Figure 9.1) taking place at time $t$ is its rate divided by the sum of rates

$$P(l) = \frac{r_l}{A(t)}. \tag{9.2}$$

It is a normalized probability, since summing $r(l)$ over all possible events is just the activity $A(t)$ from Eq. (9.1).

To be more explicit, suppose we have a system with three species, each of which has two possible transitions. For simplicity, assume that all rates for each species are the same. Taking the rates as $r_1$ and $r_2$ for transitions 1 and 2, respectively, then the total activity is just $A = 3r_1 + 3r_2$ – there are three possible transitions of type 1 and three of type 2. Each of the

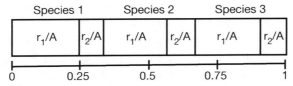

**Figure 9.2** Picking events in a kinetic Monte Carlo calculation. Each event is put sequentially on a line with its length equal to its probability. The total length of the line is then the sum of all the probabilities, which is just 1. A random number between 0 and 1 picks out a specific event with the correct probability. For example, if a random number between 0.25 and 0.333 were chosen, then reaction 2 for species 1 would be enacted.

six possible transitions is added to a list, weighted by their probability. In this simplified case, the list would have three entries with probability $r_1/A$ and three with $r_2/A$. A random choice is made from that list to determine which event takes place at that time step.

There are a number of ways to choose an event. One very simple approach is to put each event sequentially on a line with its length equal to its probability. The total length of the line is then the sum of all the probabilities, which is just 1, as discussed above. This procedure is shown on Figure 9.2 for the six events in the example discussed in the previous paragraph. A random number between 0 and 1 is chosen and where it falls on the line identifies the event that will be enacted in that time step.

To be more general, suppose that at any Monte Carlo step there are $M$ events that could take place. The probability of event $l$ is $P(l) = r_l/A$, where $A = \sum_{k=1}^{M} r_k$. The events are ordered in some way $\{1, 2, \ldots, M\}$. To chose an event, a random number $\Re$ is chosen on $(0,1)$ and the event $m$ such that

$$\frac{\sum_{k=1}^{m-1} r_k}{A} < \Re < \frac{\sum_{k=1}^{m} r_k}{A}, \tag{9.3}$$

is identified.

A kinetic Monte Carlo calculation would thus proceed in the following way:

(a) At a time $t$, determine which events could happen in a system and then add up the rates of each event, determining $A(t)$ from Eq. (9.1).
(b) Calculate the probability of each of the possible events by dividing the rate of that event by $A(t)$ as in Eq. (9.2).
(c) Create a list of the events, weighted by their probability, as in Figure 9.2.
(d) Pick a random number $\Re$ on $(0,1)$ and choose which event will occur, as in Figure 9.2.
(e) Enact the chosen event, changing the configuration of the system as required.
(f) Advance time by $\Delta t$: $t \to t + \Delta t$.
(g) Repeat steps.

There is one issue, however, that has yet to be addressed. What is the change in time, $\Delta t$, associated with an event? That question is answered in the next section.

## 9.2    TIME IN THE KINETIC MONTE CARLO METHOD

The connection between probability and rates is the foundation of the kinetic Monte Carlo method. This connection enables a one-to-one correspondence between a step in a kinetic Monte Carlo calculation and time. The following discussion is based on the original paper by Bortz *et al.* [38].

The basic idea behind kinetic Monte Carlo is that the probability that an event will occur is proportional to its rate. We can understand it best by considering a simple system with only one event, for example an atom diffusing on a surface. Suppose the atom makes *on average* one jump every 10 seconds – its rate is $r = 0.1$/second. The average probability that the atom will jump in a 10 second time interval is thus 1. This does not mean that all atoms jump once every 10 seconds, but that on average they do. Since jumps are stochastic, there is nothing special about any specific time interval $\delta t$. Thus, the average probability that an atom will jump in the time interval $\delta t$ is just $r\,\delta t$. For example, in a 1 second interval, the probability that an atom will jump is just $1/10$.[3]

This simple example shows how the *rate* of an event can be associated with the *probability* of an event. An essential part of this argument is that we are talking about average properties, not individual jumps. These are stochastic processes and unpredictable except in an average sense.

Now suppose there are two atoms jumping on a surface, with rates $r_1$ and $r_2$. Assuming that the jumps are independent, then the probability that atom 1 will jump in $\delta t$ is $r_1\,\delta t$ and the probability that atom 2 will jump is $r_2\,\delta t$. From Eq. (C.45), the probability that either atom 1 or atom 2 would jump in a time $\delta t$ is just the sum of the probabilities that each can jump, i.e.,

$$P(1 \text{ or } 2) = r_1\delta t + r_2\delta t = (r_1 + r_2)\delta t . \tag{9.4}$$

If there were three independent atoms that could jump, the probability that a jump could occur in $\delta t$ is $(r_1 + r_2 + r_3)\delta t$, and so on. We emphasize again that this argument is based on probabilities – it holds *on average*, not for individual stochastic events.

Now consider a system of events such as described in the previous section. The sum of the rates for all possible events at time $t$ is $A(t)$ from Eq. (9.1). Thus, the probability, $f$, of some event occurring in a differential time $dt$ at time $t$ is

$$f = A(t)\,dt . \tag{9.5}$$

Consider the steps in a kinetic Monte Carlo calculation as outlined at the end of the previous section. In item (f) we advanced time by $\Delta t$, which is the finite time associated with having some event take place. We can use the connection between rates and probability to establish an equation for $\Delta t$ in terms of the activity $A$.

---

[3] If $\delta t > r$ then the probability is greater than 1, which is equivalent to a probability of 1 – it just says that on average a jump will occur in that time period.

We start by introducing the probability that *no* event takes place in a differential time $dt$, which we can write as[4]

$$g(dt) = 1 - f = 1 - A\,dt.$$ (9.6)

Now consider the probability that no event takes place over the time $\Delta t + dt$, where $dt$ is a differential change in time. We can evaluate $g$ with a Taylor series with the expansion parameter being the small variable, $dt$. Truncating at the linear term, we have

$$g(\Delta t + dt) = g(\Delta t) + \frac{dg(\Delta t)}{dt}dt.$$ (9.7)

From Eq. (C.46), since $g$ is a probability,

$$g(a + b) = g(a)g(b),$$ (9.8)

so $g(\Delta t + dt)$ can also be written as

$$g(\Delta t + dt) = g(\Delta t)g(dt).$$ (9.9)

Equating Eq. (9.7) and Eq. (9.9), we have

$$g(\Delta t)g(dt) = g(\Delta t) + \frac{dg(\Delta t)}{dt}dt,$$ (9.10)

or, by rearranging and dividing by $g(\Delta t)$,

$$\frac{1}{g(\Delta t)}\frac{dg(\Delta t)}{dt} = \frac{1}{dt}(g(dt) - 1).$$ (9.11)

From Eq. (9.6), $g(dt) - 1 = -A\,dt$, so we are left with

$$\frac{1}{g(\Delta t)}\frac{dg(\Delta t)}{dt} = -A.$$ (9.12)

Substituting $(1/g)dg/dt = d\ln(g)/dt$ in Eq. (9.12) yields the simple differential equation

$$\frac{d\ln[g(\Delta t)]}{dt} = -A,$$ (9.13)

with the solution

$$\Delta t = -\frac{1}{A}\ln(g(\Delta t)).$$ (9.14)

Thus, we have expressed the time associated with a kinetic Monte Carlo step in terms of the activity at that time and the probability that no event takes place in a time $\Delta t$. Since we do not know $g(\Delta t)$, this result may not seem like much progress.

$g(\Delta t)$ represents the probability that no event takes place in a time $\Delta t$. As we discussed, however, we do not know a priori when an event actually will take place. These are stochastic processes. We are, however, only interested in the average dynamics of the system. Thus, we do not need to know any specifics about $g(\Delta t)$. It is enough to know that it is a stochastic

---

[4] If $f$ is the probability of an event in time $dt$, then $1 - f$ is the probability of no event in that time period.

variable and, since it is a probability, is restricted to values between 0 and 1. Since we want the average dynamics, we can thus replace $g(\Delta t)$ by another stochastic variable that exhibits the same average properties. In this case, we can replace $g(\Delta t)$ by random number $\Re$ that is uniformly distributed between 0 and 1.[5] Thus, time associated with a kinetic Monte Carlo step whose current activity is $A$ is taken to be

$$\Delta t = -\frac{1}{A(t)} \ln(\Re). \tag{9.15}$$

In a calculation, $A(t)$ has already been calculated to determine the probabilities in each state. Once an event is chosen, as in Figure 9.2, a random number $\Re$ is chosen and the time incremented by Eq. (9.15).

## 9.3 KINETIC MONTE CARLO CALCULATIONS

To perform a kinetic Monte Carlo calculation, one must first identify the events that could take place in the system. Once the events are determined, the rates of each of those events must be obtained, from whatever source they are available.

Kinetic Monte Carlo is highly dependent on having a good list of the possible events and their rates of the system under study. If a critical event, for example a reaction in a chemical system, is left out, or if a supplied rate is incorrect, then the kinetic Monte Carlo results are not likely to be a good model for dynamic evolution of the system. However, for systems whose dynamics are determined by thermally activated events, kinetic Monte Carlo is often the only way to proceed.

### 9.3.1 Example I: diffusion on a surface

Voter was interested in the dynamics of atoms diffusing on free surfaces of other materials. His interests were on the formation and dynamics of clusters of atoms, rather than on individual atoms. Specifically, he examined the movement of clusters of rhodium atoms on a Rh(100) surface. He used the results to determine the diffusion constants for clusters of different sizes. His calculation was based on atoms interacting with the surface and with each other with short-range potentials. An atom sits in a potential well that depends on the local configuration of atoms around it. The barrier for an atom to move to another site depends on that local configuration of atoms.

Figure 9.3 shows the local environment used in Voter's calculation. Atoms sat on a surface that formed a square lattice and could jump to their nearest neighbor sites: left, right, up or down. In Figure 9.3, the local environment is shown for an atom hopping to the right. Note that by symmetry, all other directions of hopping (left, down, and up) are equivalent and can be

---

[5] This procedure is discussed in more detail in Appendix I.2. In Figure I.2 we show the probability distribution of $-\ln(\Re)$, where $\Re$ are random numbers on $(0,1)$. The mean of this distribution is 1.

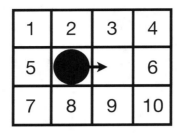

**Figure 9.3** The 10-atom local environment for an adatom jumping to the right on a square lattice. Each of the 10 binding sites may be empty or occupied, leading to $2^{10}$ possible environments [327].

modeled with an equivalent local environment to that shown in Figure 9.3. The numbered sites in Figure 9.3 may or may not be occupied. The rate for an atom to hop to the right in Figure 9.3 will depend on which of the numbered sites are occupied or not.

There are 10 sites that may or may not be occupied, so there are $2^{10} = 1024$ possible configurations of the surrounding atoms, many of which being equivalent by symmetry. For each of the 1024 configurations, the rate constants were calculated for a jump of an atom to its adjacent site as in Figure 9.3. Transition-state theory was used, with corrections that incorporated the possibility that secondary adatoms would make hops in response to the hopping of the primary atom. The net escape rate is $k_{esc}$, which is the sum of the rate constants for an atom in given configuration. All 1024 values of $k_{esc}$ were put in a table that also included a list of probabilities of transitions to various final states, i.e., any possible secondary motion. At each step in a simulation, the local environment of each atom was determined and a rate assigned from the tabulated list of values to the hopping of that atom to its adjacent site, as shown in Figure 9.3.

The calculation proceeded in the following steps:

(a) The catalog of hopping rates was used to look up the value of $k_{esc}$ for each of the $4N$ possible hops of each atom (each atom can hop in four possible directions). For hops that are blocked by an adatom in an adjacent binding site, $k_{esc} = 0$.

(b) Choose one of the $4N$ possible hops based on $k_{esc}$. Randomly select a final state weighted by the probability of jumping to that state as shown in Figure 9.2. Modify the position of the chosen adatom and any others that move in the transition.

(c) Increment the clock by

$$\Delta t_{hop} = -\left[ \sum_{i=1}^{4N} k_{esc}(i) \right]^{-1} \ln(\Re) . \tag{9.16}$$

(d) Repeat.

By ignoring all but the activated events, kinetic Monte Carlo avoids the time that atoms sit in their potential wells waiting for a jump and only includes the jumps themselves. Thus, kinetic Monte Carlo can describe systems over very long times. For example, Voter's calculations corresponded to over 1.5 microseconds of real time, far greater than feasible with standard molecular dynamics simulations with their very short time steps.

There are a number of limitations to kinetic Monte Carlo. Voter had to use physical intuition to make the list of possible jumps. Given the complexities of jumps that include more than a few atoms, that list was necessarily quite limited relative to what could possibly happen. Indeed, Voter later found from more sophisticated simulations that cluster diffusion involves concerted moves of many atoms [292]. The advantage of the accelerated dynamics methods discussed in Section 6.5 over kinetic Monte Carlo is that no assumptions are made about how the atoms behave.

### 9.3.2 Example II: chemical vapor deposition

One of the great strengths of the kinetic Monte Carlo method is that it is not restricted to the movement of atoms. Indeed, it can be used to model any system for which we have events and their rates. In Chapter 10, we will apply it to spin models. In this section, we show how it can be applied to complex chemical reactions on surfaces, leading to the deposition and growth of materials. Such problems have length and time scales beyond normal atomistic methods, and kinetic Monte Carlo is often the optimal approach to model them.

Our example is a simulation of the development of diamond films through chemical vapor deposition (CVD) based on the kinetic Monte Carlo method. The growth of diamond films by CVD occurs through the evolution and incorporation of chemisorbed species on a surface. There is a set of possible chemical reactions that can occur, depending on the current state of the surface. Each of these reactions occurs at different rates, which depend on temperature, vapor pressure in the gas, etc. Battaile and Srolovitz [25, 26] developed a model for the chemistry that incorporated known reactions and their reaction rates and used a kinetic Monte Carlo method to model the growth. They did not attempt to model all the details of the dynamics reactions, just their rates and outcome.

Battaile and Srolovitz used a complex set of gas-phase and surface reactions with a total of 18 reactions (forward and back). Their results showed impressive agreement with experiment for the growth of various surfaces. Their model was three-dimensional, showing growth patterns along both directions on the surface. Rather than describe their detailed calculations, here we discuss a simplified model presented by Battaile and Srolovitz as an illustration of the method [25]. The complete model is richer, but the basic ideas of kinetic Monte Carlo can be understood from the simple model.

The basic structure of the model:

(a) Establish the topology of the model, which in this example calculation is a one-dimensional surface along the $x$ axis with growth in the $y$ direction. For simplicity, the occupied sites will be displayed as filled squares and the unoccupied sites (in the growth direction) shown as empty squares. Thus, the basic simulation cell is a square lattice in two dimensions, as shown in Figure 9.4.

(b) Identify the possible species involved in the chemical reactions on or above the surface.

(c) Determine the reaction rates for each chemical reaction. For many reactions, especially in the gas phase, these rates are available. In the diamond CVD study, reactions (and rearrangements) on the surface were also included, the rates of which were calculated using atomistic simulations.

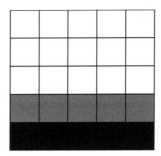

**Figure 9.4** Initial surface on a square lattice. Black squares are substrate (A) atoms and dark gray squares are adsorbed (B) atoms.

In this simple example, there are atoms on or below the surface, indicated by an ($s$), and atoms in the gas phase, indicated by ($g$). Atoms in the substrate are indicated by A, while the adsorbate atoms in contact with the vapor are B atoms. When a substrate atom is not covered by an adsorbate atom, there is a dangling bond (radical) at the surface, which is denoted by *. On the surface there are four types of species: A($s$), B($s$), AB$_2$($s$), and *($s$). The vapor consists of reacting species, B($g$) atoms and AB$_2$($g$) radicals, and two inert species that are not involved in the chemistry but provide the overall pressure in the system, B$_2$($g$) molecules and an AB$_3$($g$) precursor.

The possible chemical reactions in this simple model are

$$*(s) + B(g) \underset{k_f^1}{\overset{k_r^1}{\rightleftharpoons}} B(s) \tag{9.17a}$$

$$*(s) + AB_2(g) \underset{k_f^2}{\overset{k_r^2}{\rightleftharpoons}} AB_2(s) \tag{9.17b}$$

$$AB_2(s) \underset{k_f^3}{\rightarrow} A(s) + B(s) + B(g) . \tag{9.17c}$$

In Eq. (9.17a) and Eq. (9.17b), a surface radical can react with a gas-phase B atom or a gas-phase AB$_2$ molecule, respectively, adding the gas-phase species to the surface in a reversible way. In Eq. (9.17c), however, the AB$_2$($s$) molecule added to the surface in Eq. (9.17b) decomposes irreversibly to form an extra layer of A($s$) capped by a B($s$), releasing the other B atom to the gas phase. There are at least two steps that take place in this one reaction: chemical decomposition and diffusion of atoms A and B. One of the great strengths of the kinetic Monte Carlo method is that as long as the overall rate is known, one does not have to resolve all the individual steps in a reaction if no other species are involved. Note that if AB$_2$ is ever added to the surface, it either dissociates (Eq. (9.17c)) or it leaves the surface as itself (the reverse of Eq. (9.17b)). To illustrate their method, Battaile and Srolovitz came up with fictitious rates (in arbitrary units) for each of these reactions, which are given in Table 9.1. Note that in this simplified model there are no reactions that include more than one surface site at a time. Thus, each column is essentially independent of the others. In the accurate model, a number of reactions that involved more than one lattice site were included, modeling the effects of surface chemistry and diffusion.

In Figure 9.5 we show the first eight time steps for a five-site model based on the reactions and rates in Table 9.1. Per step, we have:

Table 9.1 **Fictitious reaction rates for the simple surface chemistry model**

| Reaction | | Forward rate | Reverse rate |
|---|---|---|---|
| 1 | Eq. (9.17a) | 100 | 200 |
| 2 | Eq. (9.17b) | 100 | 400 |
| 3 | Eq. (9.17c) | 400 | |

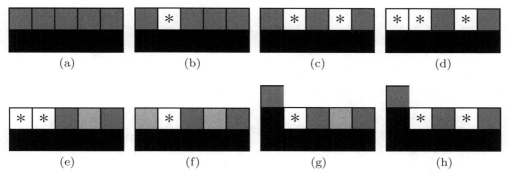

Figure 9.5 Steps of the growth process for a five-site model where black = A(s), dark gray = B(s), a "*" = *(s), and light gray = AB$_2$(s).

(a) The top surface only contains sites with a B($s$), so the only possible reaction for any of the sites is the reverse of reaction 1 (denoted $-1$) in Eq. (9.17a). All reactions are equally probable. The net activity for this step is $A = \sum_{i=1}^{5} r_i = 5(200) = 1000$, with the rates from Table 9.1. A random number between 0 and 1 was chosen and the selection process described in Figure 9.2 was followed, which led to site 2 being chosen, at which B($s$) is replaced by *($s$). Time was updated by $\Delta t = -\log(\Re)/1000$.

(b) The four B sites can only undergo reaction $-1$, but site 2 can now undergo either reaction 1 (by adding a B($g$)) or reaction 2 (by adding an AB$_2$($g$)). The net activity is $A = 4(200) + 100 + 100 = 1000$. A list of all reactions with each entry being $r_i/A$ in length was created. A random number was generated and the reaction at site 4 was chosen, replacing B($s$) by *($s$). Time was incremented by $\Delta t = -\ln(\Re)/1000$.

(c) The three B sites can only undergo reaction $-1$, but sites 2 and 4 can now undergo either reaction 1 or reaction 2. The net activity is $A = 3(200) + 2(100 + 100) = 1000$. Site 1 was randomly chosen to undergo reaction $(-1)$, replacing B($s$) by *($s$). Time was incremented by $\Delta t = -\ln(\Re)/1000$.

(d) The two B sites can only undergo reaction $-1$ and sites 1, 2 and 4 can undergo either reaction 1 or reaction 2. The net activity is $A = 2(200) + 3(100 + 100) = 1000$. From the list of all possible reactions, reaction 2 at site 4 was randomly chosen, adding AB$_2 s$ to the surface. Time was incremented by $\Delta t = -\ln(\Re)/1000$.

(e) Two B($s$) sites can undergo reaction $-1$, two *($s$) sites can undergo reaction 1 or 2, and one AB$_2$($s$) site can undergo either reaction $-2$ or reaction 3. The net activity is $A = 2(200) + 2(100 + 100) + 1(400 + 400) = 1600$. From the list of the eight possible reactions, reaction 2 at site 1 was randomly chosen, adding AB$_2$s to the surface. Time was incremented by $\Delta t = -\ln(\Re)/1600$.

(f) There are two B($s$) sites that can undergo reaction $-1$, one *($s$) site that can undergo reaction 1 or 2, and two AB$_2$($s$) site that undergo either the reverse of reaction 2 or reaction 3. The net activity is $A = 2(200) + 1(100 + 100) + 2(400 + 400) = 2200$. From the list of the eight possible reactions, reaction 2 at site 1 was randomly chosen, adding A($s$) + B($s$) to the surface and releasing a B($g$) to the vapor. Time is incremented by $\Delta t = -\ln(\Re)/2200$.

(g) There are three B($s$) sites that can undergo reaction $-1$, one *($s$) site that can undergo either reaction 1 or either 2, and one AB$_2$($s$) site that undergoes either reaction $-2$ or reaction 3. The net activity is $A = 3(200) + 1(100 + 100) + 1(400 + 400) = 1600$. From the list of the eight possible reactions, reaction $-2$ at site 4 was randomly chosen, adding *($s$) to the surface and releasing an AB2($g$) to the vapor. Time is incremented by $\Delta t = -\ln(\Re)/1600$.

Following the above discussion shows clearly how a kinetic Monte Carlo simulation should work. At each step, the possible events are identified and the reaction rates identified. The activity is calculated, an event is chosen, and time is incremented. Consider how the activity $A(t)$ varied during the steps: $A(t) = \{1000, 1000, 1000, 1000, 1600, 2200, 1600\}$. Thus, the maximum time for this series of events varies over time, with some events (Figure 9.5f) with smaller a-priori probability than others (e.g., Figure 9.5a).

Consider in a bit more detail the detailed model. For the forward reactions, the authors used the Arrhenius expression from Eq. (G.52), taking the appropriate data from the literature for most of the reactions. For the reverse reactions, Eq. (G.53) was used, combined with known thermodynamic data. They used results from detailed atomistic simulations to map out surface reconstruction reactions and to determine their rates. They had a full three-dimensional model, with the surface in two dimensions and the growth in the third. They mapped out different surface structures as a function of processing conditions (pressure and temperature).

As discussed above, kinetic Monte Carlo is not necessarily a good choice for modeling all systems. There must be discrete events and known rates. Kinetic Monte Carlo depends critically on the assumed mechanisms. For example, if in the study of CVD there was an important reaction path that was not included, then the results would be inaccurate (or just wrong). Application of kinetic Monte Carlo is thus often difficult, because the possible paths are not known or, if known, quite complex. One must also have good values for the rates. However, for many problems, kinetic Monte Carlo is an excellent choice to match the longer time scales of experimental studies.

## 9.4 APPLICATIONS

In this chapter we have given two examples of kinetic Monte Carlo applications, one focused on the modeling of diffusion and the other on the atomic processes of chemical vapor deposition.

These two examples are good representatives of applications of the kinetic Monte Carlo method in materials research, in which diffusion and chemistry are common themes. Here we list just a few other applications that indicate the wide spread use of this method.

- Radiation damage: one area of research that has had many applications of kinetic Monte Carlo is the study of radiation damage of materials. It is a classic problem for which kinetic Monte Carlo is suited, with long time dynamics associated with damage cascades and subsequent relaxation. A discussion of the application of kinetic Monte Carlo as well as other methods such as molecular dynamics to radiation damage is available [28].
- Diffusion: studies with kinetic Monte Carlo of diffusion and its consequences are common, including work on strain aging [115] and examining the effects of segregation on grain boundary diffusion [164]. A range of studies of processing methods that are diffusion dominated have been made, from the development of islands during epitaxial growth [19] to the effects of sample rotation on the deposition of thermal-barrier coatings [67].
- Dislocations: Lin and Chrzan introduced a kinetic Monte Carlo method for dislocation dynamics in [201]. The effects of solutes on dislocation motion were studied with kinetic Monte Carlo in [343].

From these, and other, studies, it is clear that the kinetic Monte Carlo method is very useful for describing systems with competing rates. Indeed, at times it is the only way to model on an atomistic basis systems whose time scales are long relative to molecular dynamics. The challenge for kinetic Monte Carlo is that one needs to have a very complete description of the system under study. If important processes are neglected or if the rates are incorrect, the kinetic Monte Carlo method can give misleading, or wrong, results. Despite these caveats, however, the kinetic Monte Carlo method is an important tool in the computational materials tool chest.

## 9.5 SUMMARY

In this chapter we introduced the basic ideas behind the kinetic Monte Carlo method for determining long-time dynamics of materials behavior. The key idea is the association of the rate of an event with the probability of that event. The basic methodology was introduced and two examples, showing different aspects of the method, were discussed in detail. Limitations of the method, based on the necessity for identifying the dominant events in the system and their rates prior to a calculation, were discussed.

# 10 Monte Carlo methods at the mesoscale

In Chapter 7, the basic ideas of the Monte Carlo method are presented and demonstrated on two types of problems, the Ising model and simulations of atomic systems. In Chapter 8, applications to molecular systems, including polymers and biomolecules, are discussed. The Ising model does not really fit into the theme of Chapter 7, which is included in the part of the text focused on atomic and molecular scale simulations. The Ising model really belongs in the current chapter, which focuses on the use of the Monte Carlo method to simulate models of materials behavior at the mesoscale. However, the Ising model is just too good an example of the Metropolis Monte Carlo method, so we included it in Chapter 7. Consider its natural home, however, here.

There are many types of Monte Carlo applications that have been developed to model material properties and response. We shall examine only one of them in detail, the Q-state Potts model as applied to grain growth. Not only has the Potts model had wide applicability and impact in materials research, it reflects most of the issues faced in any mesoscale modeling based on the Monte Carlo method.

## 10.1 MODELING GRAIN GROWTH

The goal of this section is to devise a model for grain growth by focusing on the migration of grain boundaries to lower the overall energy of the system. In Appendix B.6, we provide a short description of grains and grain growth. Details in that section will provide a guide for the development of the model.

The driving force for grain growth is minimization of the free energy of the system. Each boundary has a positive energy, so a model is needed in which the energy is lowered if a boundary is eliminated or reduced. We have already been introduced to a model that has that property – the Ising model of Chapter 7. Imagine a system of spins in an Ising model on a square lattice, with two regions, one in which all the spins have a value of $+1$ and another in which all spins have the value of $-1$, as shown in Figure 10.1. There is a boundary parallel to the $x$ axis between the two regions. From Eq. (7.12), the energy of the Ising model (with zero applied magnetic field) is

$$E = -\frac{J}{2} \sum_{i=1}^{N} \sum_{j \in Z} s_i s_j . \tag{10.1}$$

| +1 | +1 | +1 | +1 | +1 | +1 | +1 | +1 |
|----|----|----|----|----|----|----|----|
| +1 | +1 | +1 | +1 | +1 | +1 | +1 | +1 |
| +1 | +1 | +1 | +1 | +1 | +1 | +1 | +1 |
| +1 | +1 | +1 | +1 | +1 | +1 | +1 | +1 |
| -1 | -1 | -1 | -1 | -1 | -1 | -1 | -1 |
| -1 | -1 | -1 | -1 | -1 | -1 | -1 | -1 |
| -1 | -1 | -1 | -1 | -1 | -1 | -1 | -1 |
| -1 | -1 | -1 | -1 | -1 | -1 | -1 | -1 |

**Figure 10.1** Ising model with a boundary between a region with all spins equal to +1 and a region with all spins equal to −1. The boundary is indicated by the heavy fuzzy line. The interaction regions for two spins, one in the bulk −1 region and one for a spin on the boundary, are indicated with heavy lines.

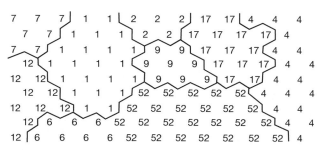

**Figure 10.2** Example of the discretization procedure in a two-dimensional Monte Carlo Potts model for grain growth. A lattice (in this case triangular) is mapped on to a two-dimensional slice of a material. Each site is marked with an index that indicates which grain that site is in. Adapted from [142].

Within the top and bottom half planes, the interaction energy of a spin with its neighbors is $-4J$, as indicated in Figure 10.1. The interaction energy for a spin on the boundary, however, is $(-1 - 1 - 1 + 1)J = -2J$. Thus, for each atom on the boundary, there is a net positive energy of $2J$. At low temperatures, the Ising model will evolve to eliminate the boundary to lower the energy of the system. It thus exhibits behavior that might make it a candidate for a model of grain growth.

The Ising model has been used to model an important materials problem – the growth of domains in magnetic systems. If one takes the two regions shown in Figure 10.1 as domains with different orientations, then the motion of the boundary between them can be monitored with the Monte Carlo methods discussed in this chapter.

The Ising model cannot, however, be employed to model grain growth in materials in which there is a distribution of grain orientations. Consider the schematic view of a two-dimensional microstructure shown in Figure 10.2. A triangular lattice is overlaid on the microstructure and then each site is marked by a number that indicates the specific grain the site is in. Each lattice site represents an area of material. Grain boundaries are shown as the dark lines between grains.

The problem with the Ising model as a model for a microstructural evolution is clear from Figure 10.2. The Ising model has only two possible "orientations" and thus has too few degrees of freedom to model realistic grain growth. The question of whether a variant of the Ising model that includes more degrees of freedom could serve to model grain growth was answered in a seminal series of papers in the mid 1980s, when the Potts model for grain growth was introduced [9, 294, 295]. The Potts model continues to play an important role in our understanding of grain growth.[1]

---

[1] Much of the discussion in this chapter is based on the work and papers of E. A. Holm.

The Q-state Potts model is a variant of the Ising model, in which instead of having just two allowed spin values per site, $Q$ possible values are allowed. These values are usually taken to be $\{1, 2, \ldots, Q\}$ and are usually referred to as spins in analogy with the Ising model. In the interpretation of the Potts model for grain growth, the spin value identifies which grain a site is in, as shown in Figure 10.2.

Since grains evolve to lower the energy of the system by removing grain boundary area, any model must incorporate an energy penalty for the existence of a grain boundary, which will lead to the removal of boundaries over time as the system evolves to lower its energy. The Potts model employs a very simple form of the energy,

$$E = \frac{1}{2} \sum_{i=1}^{N} \sum_{j \in Z_i} E_{s_i, s_j} \left[1 - \delta_{s_i, s_j}\right] + \sum_{i=1}^{N} F_i(s_i), \tag{10.2}$$

where the spin value on site $i$ is $s_i$, the sum over $j$ includes only the prescribed neighbors of $i$, $Z_i$, and $\delta_{s_i, s_j} = 1$ if $s_i = s_j$ and 0 otherwise.[2] Thus, the interaction between sites with the same spin value is 0 (within a grain) and that between sites with different spin values is $E_{s_i, s_j} > 0$, which represents a positive grain-boundary energy.

As written, the grain-boundary energy $E_{s_i, s_j}$ depends on the values of the spins on either side of a grain boundary. If the spin values were associated with specific orientations, then a grain-boundary energy that reflects the known relationship between energy and misorientation can be included. $F_i(s_i)$ is an energy associated with site $i$ and gives flexibility to the model, enabling it to be applied to such phenomena as recrystallization and other forms of abnormal grain growth. The simplest version of the Potts model is for a system with all grain-boundary energies being the same and with $F = 0$,

$$H = \frac{E_o}{2} \sum_{i=1}^{N} \sum_{j \in Z_i} \left[1 - \delta_{s_i, s_j}\right]. \tag{10.3}$$

We will use this form to describe the model.

From Eq. (10.3), the ground state of the Potts model (lowest energy at $T = 0$) is one in which all the sites have the same spin value, i.e., the total energy is zero. The system is thus $Q$-fold degenerate. As $T$ increases, more and more disorder will enter the system until a critical point is reached, beyond which the thermal fluctuations are large enough for the system to disorder, following the same basic physics as the Ising model.

## 10.2 THE MONTE CARLO POTTS MODEL

The evolution of the spins in the Potts model can be simulated with a Metropolis Monte Carlo method similar to that used in the study of the Ising model in Chapter 7. A trial consists of

---

[2] $\delta_{s_i, s_j}$ is the Kronecker delta discussed in Appendix C.5.1.

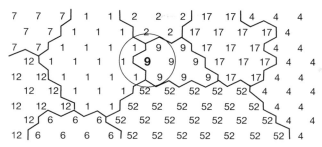

**Figure 10.3** Energy changes in a $T = 0$ Potts model based on flipping the site marked **9** from the microstructure in Figure 10.2, as described in the text.

picking a site at random and then changing the spin at that site to one of the $Q - 1$ other possible spins. The change in energy is calculated and a standard Metropolis procedure is employed to determine whether the trial move is accepted. If it is, then the new state is added to the list of configurations. If not, then the system stays in the old state, which is repeated in the list of configurations.

While the Potts model for grain growth can be studied at finite temperatures ($T > 0$), it is quite often used at $T = 0$. The basic system dynamics is unchanged – the system evolves to lower the energy by removing grain boundaries. As a reminder, in the Metropolis algorithm, a system changes to a new state in a trial if the energy change $\Delta E \leq 0$ or if $\exp(-\Delta E/k_B T) < \Re$, where $\Re$ is a random number on $(0, 1)$. If $T = 0$, then the trial would be accepted only if $\Delta E \leq 0$.

In systems that consist of discrete events, one can associate time with a Monte Carlo trial. Here we will define a Monte Carlo time step $\Delta t_{MC} = 1$ MCS/$N$, where $N$ is the total number of sites in the system. A Monte Carlo step (MCS) includes $N$ trials and, on average, after a long run of $X$ MCS, each site has been tried $X$ times. We define the net real time per MCS as $\tau$. The connection of $\tau$ to a real time is not prescribed.

Suppose we are doing a $T = 0$ Potts model simulation based on the isotropic energy in Eq. (10.3) and have the instantaneous spin distribution from Figure 10.3. As part of the Monte Carlo procedure, a site has been randomly chosen for a trial change in configuration, in this case the site marked **9**. Sites interact only with their six nearest neighbors, which are marked by a circle around site **9**. The energy associated with site **9** is $3E_o$ with the spin configuration shown in the figure.[3] In the Monte Carlo trial move, any of the $Q - 1$ other possible spin values could be picked for site **9**. Of those only one could lead to a move that would have $\Delta E \leq 0$, which is a change of the "9" to a "1". Any other choice would introduce a spin that would be different from all of its neighbors and thus have an energy of $6E_o$. Thus, most of the possible trials for the selected spin would be rejected. Suppose, however, that the "9" was flipped to a "1". The new energy would also be $3E_o$ and, based on the Metropolis scheme, the trial would be accepted, the spin at the selected site updated, and the procedure repeated.

The Potts model is inherently a lattice-based method. A poor choice of the lattice topology can, however, change the simulation results considerably, leading to incorrect and even nonsensical results. For example, the Potts model on a square lattice with nearest-neighbor interactions is

---

[3] There is an energy of 1 $E_o$ for every unlike neighboring spin.

Table 10.1 **Anisotropy in two-dimensional lattices.** $Z$ **is the number of neighbors included in the interaction energy from Eq. (10.3). Table from [150].**

| Lattice | Wulff shape | $Z$ | Anisotropy |
|---------|-------------|-----|------------|
| sq(1)   | Square      | 4   | 1.414      |
| tr(1)   | Hexagon     | 6   | 1.154      |
| sq(1,2) | Octagon     | 8   | 1.116      |
| tr(1,2) | Dodecagon   | 18  | 1.057      |

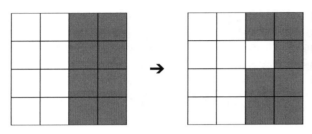

**Figure 10.4** Energy changes during grain-boundary motion on a square lattice, as described in the text. White squares indicate a specific spin value while gray squares indicate a different spin value.

notorious for yielding square grains, which are not very physical. Why this is so can be seen in Figure 10.4, where we indicate a lattice site with one spin value as white and with a different spin value as gray. With a nearest-neighbor square lattice, the energy of a gray site on the boundary is $1E_o$. If we flip it to a white site, as shown in the figure, the energy of that site is $3E_o$, with an energy change of $\Delta E = 2E_o$, representing a move that cannot occur at $T = 0$ and that would be energetically unfavorable even at relatively high temperatures. Thus, a straight boundary on a square lattice with nearest neighbors is stable, inhibiting further growth. Contrasting the behavior of the square with the situation of a boundary site shown on a triangular lattice in Figure 10.3, in which a boundary site can change, shows why a triangular lattice would be a much more physically reasonable model for grain growth.

The general effects of lattice topology on growth in two dimensions can be understood by comparing the energy anisotropy of the underlying lattice, where the anisotropy is defined as the ratio of the maximum to minimum grain boundary energy in the Wulff plot (which is a plot of surface energy versus surface orientation) [150, 152]. Table 10.1 summarizes the anisotropy for four lattice types: the nearest- and next-nearest-neighbor square lattice, designated sq(1) and sq(1,2), respectively, and the nearest- and next-nearest-neighbor triangular lattice, designated by tr(1) and tr(1,2), respectively. The square lattice with nearest-neighbor interactions had the largest anisotropy, which would lead to anisotropic growth, as demonstrated in Figure 10.4, with the other lattices showing much more isotropic surface energy distributions. Similar lattice effects have been seen in three dimensions [150], with simple cubic lattices with nearest and

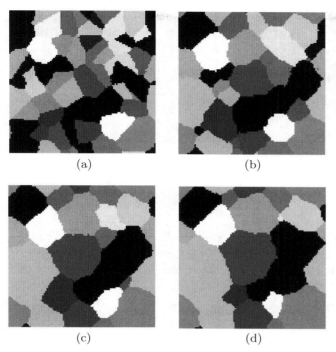

(a)  (b)

(c)  (d)

**Figure 10.5** Three microstructures from a MC Potts model study of normal "grain growth". (a) $t = 1$, (b) $t = 350$, (c) $t = 700$, and (d) $t = 1000$. The data for the figures are courtesy of A. D. Rollett. Visualization by the ParaView graphics package [265].

with next nearest interactions (sc(1) and sc(1,2)) showing inhibited growth. Inclusion of the third nearest neighbors, sc(1,2,3), leads to normal growth. fcc(1), fcc(1,2), and hcp(1) all show inhibited growth. Finite temperature simulations also can serve to roughen the boundaries, reducing lattice dependencies. More complete details are given in [151].

In Figure 10.5 we show results from a Potts model simulation of grain growth on a triangular lattice at $T = 0$. Three microstructures are shown at different times, showing *coarsening* of the microstructure with time. The grain boundary area (length of the lines in two dimensions) decreases as the grain size increases. In Figure 10.6, the grain area with time is shown for calculations in both two dimensions and three [150]. We see that the grain area increases roughly linearly with time, in agreement with experiment as discussed in Appendix B.6.2.

While the Potts model is very flexible and easy to implement, there are calculational issues that limit its effectiveness as a model. Consider the Potts model with the standard Metropolis algorithm at 0 K. Only those moves that lower the energy (or leave it unchanged) will be accepted, i.e., only those sites that are on a boundary can change their spins to a new value (they are "active"). It is easy to see why this is so. Consider a site in the center of a grain in Figure 10.3. Any change in the spin at that site has an energy of $6E_o$ and will not be allowed. Consider the physical meaning of such a flip. It corresponds to the nucleation of a new orientation right in the middle of a grain, which we expect to be vanishingly unlikely.

Figure 10.6 shows that the average grain size $\langle A_g \rangle \sim t$, from which we can show that the Monte Carlo Potts model will become increasingly inefficient as time increases. Since the only

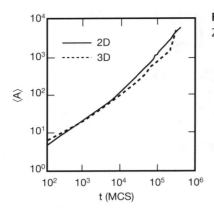

**Figure 10.6** Area of "grains" from normal grain growth studies in 2D and 3D with the Potts model. Adapted from [150].

sites that can change their spin values at low temperatures are along the boundaries, the ratio of those active sites to sites that cannot switch their spins is, in two dimensions, proportional to the average boundary length of a grain, $\langle S_g \rangle$, divided by the mean grain area, $\langle A_g \rangle$. An approximate value for the grain boundary length can be derived by considering the grains as circles, in which case the equivalent grain radius is $R_g = (\langle A_g \rangle / \pi)^{1/2}$ in two dimensions, with the circumference being $\langle S_g \rangle = 2\pi R_g$. Thus the approximate ratio of grain boundary length to mean grain area is $\langle S_g \rangle / \langle A_g \rangle \sim 1/R_g$. Since $\langle A_g \rangle \sim R_g^2 \sim t$, we have the ratio of active to inactive sites as $\langle S_g \rangle / \langle A_g \rangle \sim 1/t^{1/2}$ in two dimensions.[4]

The consequence of the change in surface to volume ratio is that a standard Monte Carlo sampling becomes very inefficient over time, with the fraction of sites that could flip decreasing as $1/t^{1/2}$. The problem is actually worse than just described. When doing a standard Monte Carlo simulation, there are $Q - 1$ possible values for the new spin value when a spin is flipped. If that spin is on a boundary with only one other type of spin adjacent to it, then there is only a $1/(Q - 1)$ chance of flipping to a spin that will actually lower the energy (i.e., to the other type already on the boundary). The net probability of making a successful flip thus goes as $1/(t^{1/2}(Q - 1))$. Making a random choice of sites and spin flips is thus very inefficient and, indeed, almost useless at long times.

All is not lost, however. In its original description, the kinetic Monte Carlo method of Chapter 9 was not developed to study reaction rates or diffusion, but rather to overcome the same types of inefficient sampling as seen at long times in the Potts model [38]. That approach, called the N-fold way, was developed for the Ising model and can be modified for use with the Potts model.

## 10.3 THE N-FOLD WAY

The *N-fold way* [142] is a specific example of a kinetic Monte Carlo (KMC) method, offering a way around standard Metropolis Monte Carlo's inefficiency at long times for some systems.

---

[4] In three dimensions, a similar analysis gives $\langle S_g \rangle / \langle A_g \rangle \sim 1/t^{1/3}$.

In the N-fold way, every trial is accepted, yet it still yields, on average, the structure, properties and dynamics that would be found in regular Metropolis Monte Carlo.

We will couch this discussion in terms of the isotropic $Q$-state Potts model of Eq. (10.2). We start by defining the activity of site $i$ as

$$a_i = \sum_{s_i^* \neq s_i} P(\Delta E[s_i \rightarrow s_i^*]), \qquad (10.4)$$

where $P(\Delta E[s_i \rightarrow s_i^*])$ is the Metropolis probability of changing the spin value on site $i$ from $s_i$ to $s_i^*$. $a_i$ is summed over all $Q - 1$ possible values of the spin that differ from $s_i$.

Consider a system at $T = 0$ K at a stage in the growth at which there are distinct grains and grain boundaries. The probability of a successful trial is 1 if the energy of the system stays the same or decreases during that trial and is 0 if the energy increases. Consider first a spin that is not on a boundary, but that is completely surrounded by spins of the same type. Changing that spin to any other value would increase the energy. At 0 K, the activity for that spin would be $a_i = 0$, i.e., there are no spin flips that could be successful. Now consider the highlighted site in Figure 10.3. There is only one possible flip that does not increase the energy, $s_i = 9 \rightarrow s_i = 1$. Thus the activity for that site is $a_i = 1$. If there were two possible flips (e.g., perhaps at a junction), then the activity would be $a_i = 2$, etc.

The total activity is derived from the list of the probabilities of all possible spin flips in the system and is given by

$$A = \sum_{i=1}^{N} a_i . \qquad (10.5)$$

At $T = 0$, $A$ is just the total number of possible flips. Note that as the system evolves, $A$ will evolve as well. Since the number of possible spin flips goes as the surface area (which $\sim 1/t^{1/2}$ in two dimensions), the total activity also decreases over time as $A \sim 1/t^{1/2}$ in two dimensions.

An N-fold way simulation proceeds in the same manner as the kinetic Monte Carlo calculations described in Chapter 9. At the beginning of the calculation, activities for all sites ($a_i$) are calculated. For each step in a $T = 0$ simulation:

(a) pick a site with probability $a_i$
(b) flip $s_i$ to one of the $s_i$ energetically possible values (with equal probability)
(c) recalculate activities
(d) advance time (as described next)
(e) repeat.

As in any kinetic Monte Carlo method, we must advance time at each step. For the Potts model, we also need to connect that time to the standard Monte Carlo time $\tau$, which we can do as follows. This discussion is based on those in [38, 142].

Since $A$ is the total number of possible flips (at $T = 0$), the average probability of a potentially successful flip per site on the simulation lattice is $\langle a \rangle = A/N$, where $N$ is the number of sites in the lattice. In the Monte Carlo Potts model, the number of trials per Monte Carlo step is $N$/MCS. The Potts model does not have a specific time scale. However, we can associate some time $\tau$ with a Monte Carlo step, in which case the number of flips per time is $N/\tau$, as discussed in Chapter 7 for the Ising model. The probability of a successful flip in an MCS is $N/\tau \times \langle a \rangle$. However, only $1/(Q-1)$ of those flips actually is successful, so the average number of successful flips per MCS is

$$\frac{N \langle a \rangle}{(Q-1)\tau} = \frac{A}{(Q-1)\tau}. \tag{10.6}$$

Equation (10.6) is the rate of successful flips.

The average time associated with a successful flip is the inverse of the rate, Eq. (10.6),

$$\langle \Delta t \rangle = \frac{(Q-1)\tau}{A}. \tag{10.7}$$

We can make use of the fact that these are stochastic processes to find an expression for $\Delta t$ for each time step that is consistent with $\langle \Delta t \rangle$. We follow the same type of thinking found in Section 9.2 for the time in a kinetic Monte Carlo step. We want an expression that is stochastic and whose average is $(Q-1)\tau/A$. Consider a random number $\Re$ on $(0,1)$. Averaging over many such random numbers, we find that $\langle -\ln(\Re) \rangle = 1$, as discussed in Appendix I.2. By writing

$$\Delta t_{NFW} = -\frac{(Q-1)\tau}{A} \ln(\Re), \tag{10.8}$$

we have a time for each step in the N-fold way that preserves the correct average time from Eq. (10.7).

The conventional Monte Carlo method and the N-fold way defined time in different, but equivalent, ways. The relationship between them can be easily understood by examining the steps in each method. Consider the microstructure in Figure 10.3. Most of the sites would have no activity, i.e., there are no flips that could be successful at $T = 0$. A typical Monte Carlo run would then have many trials in which nothing happens. Time is increasing by $\Delta t_{MCS} = \tau/N$ during each of those trials, so there will be considerable lengths of time in which time is progressing but the system is not changing. In the N-fold way, however, each trial leads to a successful move. The time step for a move is $\Delta t_{NFW} = -((Q-1)\tau/A) \ln(\Re)$, which increases as the microstructure evolves (and $A$ decreases). Thus, each time step in the N-fold way corresponds to the many time steps in an equivalent Monte Carlo calculation, during which nothing changes.

The N-fold way yields the same answers as conventional Monte Carlo, but in a fraction of the computer time at long Monte Carlo time as shown in Figure 10.7. For simulations starting from random initial conditions, at short times no grains have started to develop. Conventional Monte Carlo is more efficient for that situation. Typically, simulations are started with the conventional method and then switched to the N-fold way after a hundred MCS or so.

The N-fold way is equally applicable at finite temperatures ($T > 0$) or for anisotropic interactions (Eq. (10.2)), with the appropriate probabilities used in the definition of the activities. For $T > 0$, the probability in Eq. (10.4) is still 1 if $\Delta E \leq 0$, but is $\exp(-\Delta E/k_B T)$ for $\Delta E > 0$. The basic advantages of the N-fold way remain, however, as discussed elsewhere [142].

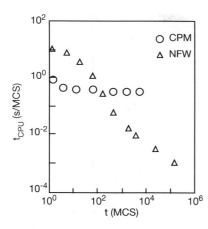

**Figure 10.7** Comparison of computational time (measured in cpu seconds per Monte Carlo step) versus Monte Carlo step (MCS) for the N-fold way (NFW) and conventional Monte Carlo (CPM) (adapted from [142]).

## 10.4 EXAMPLE APPLICATIONS OF THE POTTS MODEL

The Potts model has proven very successful in describing a number of important materials phenomena, including Zener pinning and recrystallization. The basic simulation methods are the same as that described for normal grain growth.

### 10.4.1 Zener pinning

Zener pinning refers to the effects of small particles in a material that can inhibit or prevent the motion of grain boundaries. The particles exert a pressure on the boundary that counteracts the driving force for boundary motion. Zener pinning is an important tool to modify materials behavior. For example, adding particles to a material to fix grain size is a common way to alter material properties. Since the strength of a material goes roughly as $1/D$, where $D$ is the grain diameter, fixing the maximum grain size determines the minimum strength, which is especially important when considering materials for high-temperature applications. If one can retard or eliminate grain growth, desired material properties are more likely to be maintained at high temperature. The basic physics of how particles pin moving grain boundaries is well understood and described elsewhere [263].

The Potts model is well suited to model Zener pinning and numerous studies have been reported in the literature. A set concentration of second-phase particles is chosen. Random sites are selected from the Potts model lattice consistent with that concentration. Clusters of spins are centered at those sites, with spin values set to a value that is different than any other in the system (e.g., $s_i = 0$) and then fixed for the duration of the simulation. These sites represent the pinning particles. The size of the pinning clusters is set by the problem under consideration. A moving boundary stops at a pinning particle because the spins in the particle cannot change. An example can be found in Figure 10.8c, in which all the boundaries can be seen to be pinned at second-phase particles.

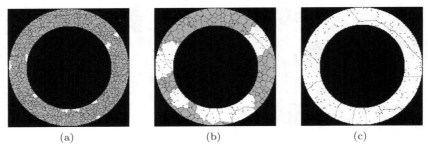

(a)                              (b)                              (c)

**Figure 10.8** Two-dimensional, 0 K, Potts model for recrystallization in a toroidal wire. Free surfaces surround the wire. A square lattice with first and second neighbors was used. The simulation started with a small number of recrystallized (low energy) sites at the free surface. The system contains 0.5% inert particles, shown as dark squares. Recrystallized (low strain energy) grains are white and unrecrystallized grains are gray. (a) 1000 MCS. (b) 10 000 MCS. (c) 100 000 MCS. Courtesy of E. A. Holm, Sandia National Laboratories.

Miodownik *et al.* point out that simulations at 0 K of Zener pinning are not completely correct owing to an artificial faceting of the boundaries, which leads to discrepancies between detailed predictions of the model and experiments [231]. They then show that using a simulation with sufficient thermal energy eliminates faceting and leads to results in agreement with experiment. A useful comparison of various methods used to simulate Zener pinning is given in [141].

## 10.4.2 Recrystallization

Recrystallization is the process by which undeformed grains grow into deformed grains, with the driving force being the reduction in strain energy associated with the deformation. If we assume that a deformed material with high dislocation density is replaced by material with few dislocations, an additional driving force on the interface between the undeformed and deformed grain is difference in strain energy, $\Delta E_{rxn}$, which is given approximately by, in energy per unit volume [218],

$$\Delta E_{rxn} = \mu b^2 \delta \rho \,, \tag{10.9}$$

where $\delta \rho$ is the dislocation density difference between the two grains, $\mu$ is the shear modulus, and $b$ the magnitude of the Burgers vector (see Appendix B.5). If this driving force is large enough, grains with lower strain energy will grow into the higher-strained grains, sweeping up dislocations and lowering the energy. Recrystallization is often an undesirable process. For example, suppose the goal is a fine-grained, and thus strong, material that is produced through a deformation process. Recrystallization occurs when a grain has a lower stored energy from deformation than other grains, which could arise because a grain (or, more likely, a set of grains) is oriented in such a way that an applied stress does not lead to significant deformation. If such grains exist, then recrystallization can occur, leading to large grains with little deformation, defeating the original goal of the processing.

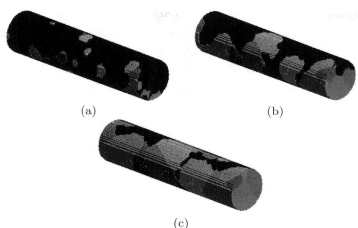

(a)

(b)

(c)

**Figure 10.9** 3D Potts model for recrystallization. The system is periodic in the axial direction with free surfaces in the radial directions. Simulations assumed random nucleation at the free surfaces. Recrystallized grains are light gray with unrecrystallized grains dark gray. (a) 10 000 MCS. (b) 30 000 MCS. (c) 50 000 MCS. Courtesy of E. A. Holm, Sandia National Laboratories.

A simple model for recrystallization can be developed based on Eq. (10.2), which for an isotropic grain boundary energy can be written as[5]

$$H = \frac{E_o}{2} \sum_{i=1}^{N} \sum_{j \in Z_i} \left[1 - \delta_{s_i, s_j}\right] + \sum_{i=1}^{N} E_i(s_i). \tag{10.10}$$

In this case, we include an extra term, $E_i(s_i)$ representing the strain energy stored in a site, which depends on the orientation (spin) of the grain at that site. For example, if a grain is deformed, the spins would have a high value of $E_i$, corresponding to a high strain energy, while undeformed grains would have a low (or zero) value for $E_i$. There is thus an extra driver for grain growth to eliminate grains with high stored energy.[6]

The results from a 0 K Potts model of recrystallization in a toroidal wire are shown in Figure 10.8,[7] with details of the simulation given in the caption. The goal of the study was to model recrystallization of toroidal wires used in incandescent lamps. At small times (Figure 10.8a), the recrystallized nuclei have started to grow, engulfing nearby grains. In Figure 10.8b, those grains have continued to grow and by Figure 10.8c all grains have recrystallized. Note that the grain boundaries are pinned by the second-phase particles.

In Figure 10.9 we show results from a three-dimensional simulation of recrystallization in a wire. The system was a simple cubic lattice with first, second, and third neighbors in the

---

[5] For other models of recrystallization, please see Chapter 11.

[6] Consider the two-dimensional square lattice in Figure 10.4 at $T = 0$. Suppose that each grain type had a stored strain energy arising from some deformation process, $E_g$ and $E_w$ for the gray and white grains, respectively. The energy change if one of the gray grain boundary sites flips to a white site is $\Delta E = 2E_o + E_w - E_g$. If a white square represents a grain with low strain energy and a gray square one with high strain energy, then as long as $E_g - E_w \geq 2E_o$, the move shown becomes favorable and grain growth can occur, serving as a model for recrystallization. Compare to the analysis given above without stored energy, in which no growth could occur.

[7] Figure 10.8–Figure 10.11 are courtesy of Sandia National Laboratories, which is a multi-program laboratory operated by Sandia Corporation, a wholly owned subsidiary of Lockheed Martin Corporation, for the US Department of Energy's National Nuclear Security Administration under contract DE-AC04-94AL85000. Support for this work was provided by the US Department of Energy, Office of Basic Energy Sciences.

interactions. It was periodic in the axial direction, with free surfaces in the radial direction. An initial set of recrystallized grains was created by randomly selecting grains at the surface and assigning them a zero stored energy, while all other grains were assigned a high stored energy. A 0 K Potts simulation was then performed. Note how the recrystallized grains (light gray) eventually replace the unrecrystallized grains (dark gray). Many other recrystallization studies using the Potts model have been performed, as in, for example [232].

### 10.4.3 Effects of boundary mobility anisotropy on grain growth

As another example of the utility of the Potts model, Holm and coworkers, building on earlier work [276], used it to examine how variations in the mobility of a boundary can promote abnormal grain growth, in which a few grains grow much faster than the others, eventually dominating the microstructure [153]. They assume a set of orientations for the grains, with a range of misorientation angles between grains. The grain boundary energy is isotropic, but the boundary mobility was a strongly varying function of grain boundary misorientation, as discussed in the above paper. In the Monte Carlo trials, only spin flips in which the boundary moved were considered – there was no grain nucleation. In practice this meant that in a trial a site was selected at random and the only possible change of the spin at that site was to a spin value from one of its neighboring sites (examine Figure 10.3). The Monte Carlo probability of switching a spin was taken to be

$$
\begin{aligned}
P(\Delta E_{i,j}) &= p_{i,j} & \Delta E \leq 0 \\
&= p_o e^{-\Delta E / k_B T} & \Delta E > 0,
\end{aligned}
\tag{10.11}
$$

where $p_o$ was a measure of the relative mobility of the boundary between the two selected grains. In Figure 10.10 we show a series of times from a simulation based on this model. Because grains that are outliers in the orientation distribution tend to have high misorientation boundaries with most other grains, they also have high mobility boundaries. It is these grains that grow abnormally. A similar study in three dimensions is shown in Figure 10.11.

## 10.5 APPLICATIONS IN MATERIALS SCIENCE AND ENGINEERING

Other applications of the Potts model in materials research include

- Biphasic materials: as discussed in this chapter, the Potts model is restricted to systems of a single phase. It has also been extended to the study of biphasic systems [29].
- Sintering: sintering of ceramics was modeled with the standard Potts model with the spin at each site reflecting the state of the grain or a pore [312].
- Magnetocaloric effect: a variation of the Potts model was used to model the magnetocaloric effect in a set of Heusler alloys[8] and compared to experimental results. The interaction

---

[8] A Heusler alloy is a metal alloy based on a Heusler phase, which is an intermetallic in the face-centered cubic crystal structure that consists of a certain set of elements. The Heusler alloys are ferromagnetic, even if the constituent elements are not.

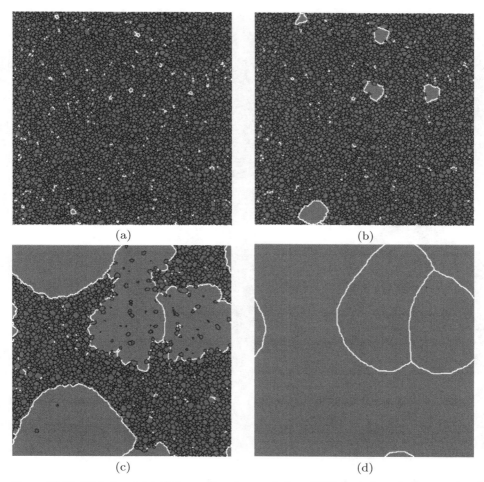

**Figure 10.10** 2D Potts model of abnormal grain growth from [153]. Grain boundaries are colored according to their mobilities. The abnormal grains clearly have white boundaries (high mobility), while most other grains have black boundaries (low mobility). (a) 0 MCS. (b) 500 MCS. (c) 1000 MCS. (d) 2000 MCS. Figures courtesy of E. A. Holm, Sandia National Laboratories.

energies were mapped onto a Potts-like Hamiltonian, with the spins corresponding to a structural parameter that differentiates between two crystal structures [44].

- Ferroelectric materials: the order-disorder properties of ferroelectrics have been studied using a three-dimensional Potts model in which the spins map onto the atomic displacements. In an application to $Cd_2Nb_2O_7$, a 12-site model on a pyrochlore lattice was used. The energy included a term coupling the electric field to the displacements (spins) [215].

- Biological materials: one of the more interesting applications of the Potts model has been to biological materials, as reviewed in [160]. One model, referred to as the cellular Potts model, modifies the Potts model to describe the contact energies between neighboring cells, keeping the cell areas constant (since cells do not disappear as do "grains" in microstructural evolution) [122, 128]. This method has been applied to a variety of phenomena, including the rearrangement of biological cells [122] and tumor growth [289].

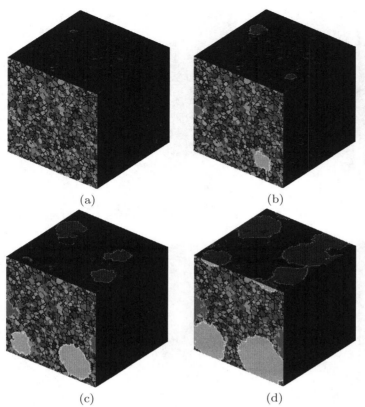

(a)　　　　　　(b)

(c)　　　　　　(d)

**Figure 10.11** 3D Potts model for abnormal grain growth owing to boundary mobility variations. Grain boundaries are colored according to their misorientation. The abnormal grains have light boundaries (high misorientation), while most other grains have dark boundaries (low misorientation). Details are given in [153]. Courtesy of E. A. Holm, Sandia National Laboratories.

## 10.6 SUMMARY

In this chapter we discuss how simple, energetically driven, models can be used to accurately model many materials phenomena, especially microstructural evolution. For example, in a model of a system in which local variables are used to indicate the local orientation and have a description of the energy in terms of those variables, the Monte Carlo method can almost always be employed to simulate the response of that system. The Potts model of grain growth is just one such method, showing the flexibility in describing with a simple model what is a very complex physical phenomenon. One does, perhaps, lose the ability to predict accurate response for a specific material, but gains an ability to examine phenomenological behavior.

### Suggested reading

Some books that you may find useful are listed here.

- D. P. Landau and K. Binder, *A Guide to Monte Carlo Simulations in Statistical Physics* [188]. One of a series of books by Kurt Binder and coworkers on Monte Carlo methods.

# 11 Cellular automata

An automaton is defined as a "a mechanism that is relatively self-operating; especially: robot" or as a "machine or control mechanism designed to follow automatically a predetermined sequence of operations or respond to encoded instructions" [226]. A classic cellular automaton is like an algorithmic robot. The cellular automata method describes the evolution of a discrete system of variables by applying a set of deterministic rules that depend on the values of the variables as well as those in the nearby cells of a regular lattice. Despite this simplicity, cellular automata show a remarkable complexity in their behavior.

Cellular automata have been used to model a number of effects in materials, mostly recrystallization, corrosion, and surface phenomena, with other applications ranging from hydration in cement to friction and wear, many of these applications being discussed below. Numerous applications extend classic cellular automata to include probabilistic rules, more complex lattice geometries, and longer-ranged rules. With the use of probabilistic rules, the distinction between cellular automata methods and Monte Carlo methods become a bit blurred, as will be discussed below.

In this chapter we will introduce the basic ideas behind cellular automata, using as examples some of the classic applications of the method. We will then go through a few applications of the methods to materials issues, highlighting the power of the method to model complex behavior. Much more detail about a range of applications, both in materials research and elsewhere, can be found elsewhere [68, 268, 269].

## 11.1 BASICS OF CELLULAR AUTOMATA

The idea behind cellular automata (CA) dates back to von Neumann and Ulam in the 1940s. von Neumann was seeking to show that phenomena as complex as life (survival, reproduction, evolution) could be reduced to the dynamics of many identical, very simple, primitive entities capable of interacting and maintaining their identity.[1] Von Neumann adopted a fully discrete method, i.e., space, time and the dynamical variables are all discrete. The result of these efforts was cellular-automaton theory, in which the world consists of cells that are endowed with a finite number of states. Those states evolve in discrete time steps according to some rules that

---

[1] A nice summary of the history of cellular automata is given in [68].

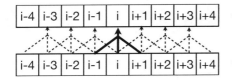

**Figure 11.1** Rules for one-dimensional cellular automata. Schematic drawing showing how the values of a site at a time $t + \delta t$ depend only on its value and those of its nearest neighbor sites at time $t$ ($r = 1$).

depend on the states of the cell (and perhaps its neighbors) at the previous time step. The cells all move to their new states simultaneously.

We will start the discussion with the original, classic, definition of cellular automata. New applications have broadened the original definition, but the basic ideas are the same. The general defining characteristics of classic cellular automata are:

(a) discrete in space: they consist of a discrete grid of spatial cells or sites
(b) homogeneous: each cell is identical and is located in a regular array
(c) discrete states: each cell is a state with a finite number of discrete values
(d) discrete in time: the value in each cell is updated in a series of discrete time steps
(e) synchronous updating: all cell values are updated at the same time
(f) deterministic rules: each cell value is updated according to a fixed, deterministic rule
(g) spatially local rule: the rule at each site depends only on the values of a local neighborhood around the site
(h) temporally local rule: the rule for the new value depends only on the values for a fixed number of preceding steps (usually one).

These basic characteristics of cellular automata can be understood by a simple one-dimensional example, shown in Figure 11.1. Each site is identical and labeled sequentially from left to right. As an example, assume each site can have only one of two possible values, 0 or 1. In Figure 11.1, the value of $a_i$ of site $i$ at a time $t + \delta t$ depends only on the values of $a_{i-1}$, $a_i$ and $a_{i+1}$ at time $t$, i.e., only the values of the nearest neighbors of site $i$ influence its evolution. Note that all sites move from time $t$ to time $t + \delta t$ simultaneously in the figure.

Since sites can only have the values of 0 or 1 and the rule includes only a site's nearest neighbors, there are eight possible configurations of the previous time step that can determine the new value of a site. For example, if site $i$ has a value of 0, then one possible configuration for site $i$ and its neighbors is 101, where we have ordered the values from left to right, $a_{i-1}\, a_i\, a_{i+1}$. All possible values are

$$111 \quad 110 \quad 101 \quad 100 \quad 011 \quad 010 \quad 001 \quad 000 \,. \tag{11.1}$$

We have ordered the possibilities in the way shown in Eq. (11.1) for a reason. If we consider each trio of numbers as a binary number, in decimal numbers, they are 7 6 5 4 3 2 1 0 in the order shown.[2] Eq. (11.1) is the standard way used by practitioners of cellular automata to order the possible configurations.

---

[2] For example, 111 as a binary number equals $(1 \times 2^2) + (1 \times 2) + 1 = 7$ in decimal numbers.

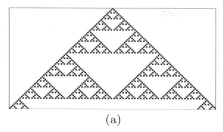 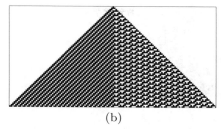

(a)                                               (b)

**Figure 11.2** Behavior of one-dimensional rules, with time $\ll$ increasing towards the bottom of the figures. Both figures were started from an initial state of a single site with a value 1 and the rest with a value of 0. The dots indicate a value of 1 and blanks indicate a value of 0. (a) Results of Rule 90. (b) Results of Rule 62.

To implement a cellular automaton, we need a rule that shows how each of the configurations in Eq. (11.1) leads to a new value of site $i$ at the next time step. Consider the simple rule,

$$a_i(t + \delta t) = \big(a_{i-1}(t) + a_{i+1}(t)\big)\mathrm{modulo}(2), \tag{11.2}$$

where $(b)\mathrm{modulo}(2)$ yields the remainder when $b$ is divided by 2. This rule is simple: if both $a_{i-1}$ and $a_{i+1}$ are 0, or both are 1, then the remainder is 0, otherwise it is 1. For example, if at time $t$ site $i$ and its neighbors have the values 101, the value of site $i$ at time $t + \delta t$ would be $(1 + 1)\mathrm{mod}(2) = 0$, etc.

Using Eq. (11.2) and the order for the configurations from Eq. (11.1), the possible changes from time $t$ to $t + \delta t$ are

$$
\begin{array}{ccccccccccc}
111 & 110 & 101 & 100 & 011 & 010 & 001 & 000 & & t & (11.3) \\
0 & 1 & 0 & 1 & 1 & 0 & 1 & 0 & & t + \delta t &
\end{array}
$$

From Eq. (11.3), we can identify the rule in Eq. (11.2) by the consequences of that rule, which are: 01011010. This sequence of numbers is cumbersome and difficult to remember. However, if we think of this sequence as a binary number and then convert it to a decimal number, then this becomes Rule 90.[3] It is common for all nearest-neighbor, one-dimensional rules to be numbered in this way.

Since any 8-digit binary number can specify a rule, there are $2^8 = 256$ possible rules for the one-dimensional case with nearest-neighbor interactions. However, it is generally assumed that 000 must always lead to a 0 and that symmetric pairs, 001/100 and 110/011, yield the same value by symmetry. With these assumptions, then there are 32 possible one-dimensional, nearest-neighbor rules.

Let us look more closely at Rule 90 (01011010). If we start with a single site with the value 1 and the rest set to 0, we see the patterns shown in Figure 11.2a, where the dots indicate a value of 1 and the blanks a value of 0. Note the very specific pattern that evolves from this very simple rule, with a set of nested triangles with varying, and growing sizes. Changing the rule

---

[3] $01011010 = (0 \times 2^7) + (1 \times 2^6) + (0 \times 2^5) + (1 \times 2^4) + (1 \times 2^3) + (0 \times 2^2) + (1 \times 2) + 0$, or
$0 + 64 + 0 + 16 + 8 + 0 + 2 + 0 = 90$.

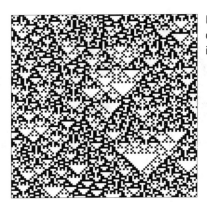

**Figure 11.3** Results of Rule 90 with a random initial conditions. The dots indicate a value of 1 and blanks indicate a value of 0. Time increases towards the bottom of the figure.

changes the behavior of the automaton. For example, the behavior of Rule 62 (00111110) from an initial state with a single 1 and the rest zeroes is shown in Figure 11.2b.

In Figure 11.3 we show a similar calculation for Rule 90, started not from a single 1 in a row of zeroes, but rather from a random sequence of 0s and 1s. While the pattern is far more complicated, we see the triangular pattern emerging from the random initial condition. If we started with a different initial state, then the system would develop different, but similar, patterns.

Cellular automata in one dimension are not restricted to having only two states per site, as in $a_i = 0$ or 1. Nor are they restricted to rules that only include the neighboring sites. A rule for the evolution of site $i$ could depend on neighbors up to $r$ sites away,

$$a_i(t + \delta t) = F[a_{i-r}(t), a_{i-r+1}(t), \dots, a_i(t), \dots, a_{i+r-1}(t), a_{i+r}(t)], \qquad (11.4)$$

where $F$ is some function that defines the rule. The parameter $r$ is said to define the *range* of the rule, i.e., how many sites to either side of site $i$ have an influence on the value at site $i$. There are a huge number of possible rules if $r > 1$, displaying the richness of cellular automata. That richness is even more on display in two and three dimensions, which offer almost limitless choices of rules, initial conditions, etc. and their consequent behavior.

The interest in cellular automata arises from the complex behavior that emerges from very simple rules, as seen in Figure 11.2 and Figure 11.3. Indeed, in the 1980s, there were many people who devoted years to understand this *complexity* and there are those who claim that cellular automata offer an entirely new way of looking at almost any aspect of science [359]. In materials research, we are often faced with very complex behavior based on physical mechanisms that we either do not know or cannot model directly. With cellular automata, we can model such complex behavior with very simple rules and computations, which can yield behavior that is very similar to known materials phenomena. As we will discuss below, however, it is relatively easy to come up with a cellular automaton that "looks like" the phenomena of interest. Verifying that the model correctly captures the physics and chemistry of underlying processes is not always so obvious.

Traditional cellular automata are deterministic, in which the configuration at a time step is completely determined by the environment of the previous time step. In probabilistic cellular

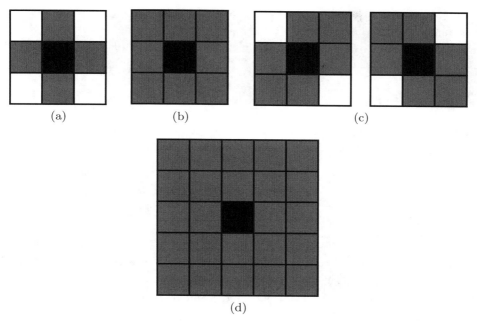

**Figure 11.4** Some of the more common local environments used in two-dimensional cellular automata. In these pictures, the state of the central site (in black) depends on rules that involve the neighbors shown in gray in the drawing. (a) The 5-neighbor environment, also known as the von Neumann environment. (b) The 9-neighbor or Moore environment. (c) Two versions of the antisymmetric 7-neighbor environments. (d) The 25-neighbor environment. Note that the central cell is included in the neighborhood, as the new state will depend on its value.

automata, deterministic rules are replaced by rules that depend on the *probability* of changing a state from one configuration to another. Examples of probabilistic cellular automata are common in applications in materials research.

## 11.2 EXAMPLES OF CELLULAR AUTOMATA IN TWO DIMENSIONS

Two-dimensional cellular automata have frequently been used to model materials. One of the first things that must be decided is the set of local sites that form the basis for the rules and thus determine the behavior of the automata. In one dimension, this was referred to as the range of the rules. In two dimensions, it is generally referred to as the local *environment*.

There are many ways to define a local environment in two dimensions. For simulations on a square lattice, for example, there are two common choices, which are shown in Figure 11.4. The *von Neuman* environment is shown in Figure 11.4a  and consists of a site and the four sites nearest to it. The *Moore* environment in Figure 11.4b includes the next-nearest-neighbor sites as well. The Moore environment is more symmetric and is more commonly used. Also shown are the asymmetric 7-neighbor and the symmetric 25-neighbor environments in Figure 11.4c and Figure 11.4d, respectively.

Before going on to some materials examples, we would be remiss if we did not mention the most famous cellular automaton, Conway's so-called "The Game of Life" [73, 113]. Perhaps more than any other example, the Game of Life brought to light how very simple rules could yield quite complex behavior.

### 11.2.1 The Game of Life

The Game of Life is based on a two-dimensional square lattice in which each lattice site has a site value (called the "activity", $a_i$) of 1 if the site is "alive" or 0 if the site is "dead". The quantity

$$A_i = \sum_{j=1}^{8} a_j, \tag{11.5}$$

where the sum is over the eight neighboring cells in the "Moore" environment from Figure 11.4b, is calculated for each site at time $t$ and then used to determine the value of $a_i$ at time $t + \delta t$. Since each site has a value of 0 or 1, $A_i$ is a count of how many of the cells adjacent to $i$ are "alive". The rule for the game of life is

$$\begin{aligned}
a_i(t + \delta t) &= 0 && \text{if } A_i(t) > 3 \tag{11.6}\\
a_i(t + \delta t) &= 1 && \text{if } A_i(t) = 3\\
a_i(t + \delta t) &= a_i(t) && \text{if } A_i(t) = 2\\
a_i(t + \delta t) &= 0 && \text{if } A_i(t) < 2
\end{aligned}$$

The basic idea is that if the neighborhood is too crowded around site $i$ (high $A$ value) then $i$ dies. If there are too few alive sites around $i$ (low $A$ values) then $i$ dies. Only if there is an optimum number of organisms nearby does the site flourish.

While this rule is very simple, it leads to characteristic patterns that oscillate with time (i.e., they cycle through a specific set of configurations). An example of this behavior is shown in Figure 11.5, in which four sequential time steps show an oscillation between an unconnected box of the form ▪▪ and a cross-like structure ▪▪▪. There are also patterns that "move" with time. For example, the simple pattern ▟ is called a glider and it moves in straight lines across the system, as shown for a series of times in Figure 11.6. Indeed, there are many very complex interactions and behaviors such that it is essentially impossible to anticipate behavior of the system based on the initial conditions.

The Game of Life was very influential in establishing interest in cellular automata. It shows that simple rules can lead to complex and unpredictable behavior.

### 11.2.2 "Solidification"

In an early survey on two-dimensional cellular automata, Packard and Wolfram discussed a range of rules and characterized their behavior [248]. One such rule yielded structures that look similar to the dendritic structures seen in snowflakes, which is surprising in that the cellular automaton includes no explicit description of the physics of solidification.

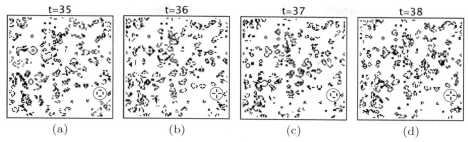

**Figure 11.5** Four consecutive snapshots from a $100 \times 100$ Game of Life. Note the pulsating "star" near the right edge about 1/4 the way up from the bottom of the cell and surrounded by a circle, which looks like an unconnected box ⬚ in (a), a cross-like structure in (b), an unconnected box in (c), and a cross-like structure in (d). This structure was stable for a few more time steps, but was eventually "consumed" by neighboring structures.

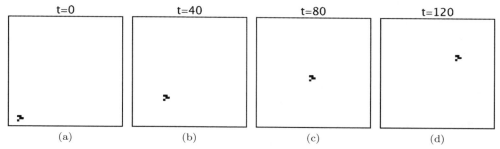

**Figure 11.6** (a) Pattern for a "glider". The four pictures are the position of the glider at time: (a) $t = 0$, (b) $t = 40$, (c) $t = 80$, and (d) $t = 120$.

In Packard's model, a two-dimensional triangular lattice was used, in which the activity of a site was defined as $a_i = 1$ for a solid and $a_i = 0$ for a fluid. The rules for the dynamic evolution of the activity are

$$a_i(t) = f[A_i(t-1)],\qquad(11.7)$$

where $f$ is some function and

$$A_i(t) = \sum_{j=1}^{6} a_j(t).\qquad(11.8)$$

The sum in Eq. (11.8) is over the six nearest neighbors on the triangular lattice. Packard identified four types of growth depending on $f[A]$:

(a) no growth: $f(A) = 0$
(b) plate-like growth: $f(A) = 1$ for $A > 0$ and $= 0$ otherwise
(c) dendritic growth: $f(A) = 1$ for $A = 1$ and $= 0$ otherwise
(d) amorphous growth: $f(A) = 1$ for $A = 2$ and $= 0$ otherwise.

In Figure 11.7 we show steps 1–5 of a cellular automaton on a triangular lattice based on rule c. Note that at step 1 and step 3, the growing crystal "closes in" on itself – it grows into a plate.

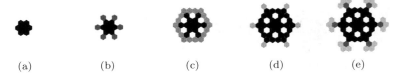

(a)  (b)  (c)  (d)  (e)

**Figure 11.7** Snapshots from the rule c for solidification. Growth from a single-site initial condition under the action of the solidification cellular automaton rule item c on a triangular lattice. The last set of sites added to the growing solid are shown in gray. The figure is based on calculations using a variant of a Mathematica® code described in [117].

The system repeats a cycle every $2^n$ time steps, where $n$ is an integer, in which the growing seed forms a plate. It displays self-similarity of the growth process, as the growing seed oscillates between dendrite and plate forms. Arms grow from the corners of the plate, side branches grow, and eventually the side branches grow into a new plate. The behavior is reminiscent of dendritic growth, which led to further exploration of the role that cellular automata could play in the study of solidification and microstructural evolution.

## 11.3 LATTICE-GAS METHODS

The flow of fluids is of critical interest in many aspects of engineering, from the development of casting processes to controlling flow through porous media. The continuum, macroscopic, hydrodynamic, flow of a liquid is described by the Navier-Stokes equation. Many computer codes can numerically solve the Navier-Stokes equation and fluid flow can be calculated very accurately. There are cases, however, in which the complex geometry of a problem can make it very difficult to numerically solve the Navier-Stokes equation. In these cases, and others that will be discussed below, the *lattice-gas* cellular automata offer a practical and useful alternative for modeling fluid flow.

In 1986, Frisch, Hasslacher and Pomeau (FHP) [110] showed that one can derive the Navier-Stokes equation from a microdynamics based on an artificial set of rules for the collision and propagation of identical particles constrained to move on a regular lattice with discrete time steps and with only a few allowed velocities. Their approach, called the lattice-gas method, is a simple cellular automaton. What FHP found is remarkable. The number of degrees of freedom in their system is small, yet on a scale much bigger than the lattice on which the particles move and at a time much longer than the time step of their simulations, their simple rules asymptotically simulate the incompressible Navier-Stokes equation.

The FHP rules are:

(a) particles are on a triangular lattice
(b) each particle moves with unit speed, with directions given by the directions to the neighboring sites
(c) no more than one particle with the same velocity can occupy a lattice site at any time
(d) each time step of the lattice gas consists of two steps:

Input State   Output State

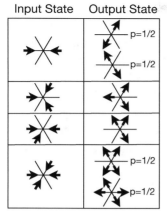

**Figure 11.8** Collision rules in the FHP model [158]. The particles arrive at a point as shown on the left and leave that point with velocities as shown on the right. In the two cases with two possible outcomes, the outcomes are equally probable.

(1) each particle hops to a neighboring site in the direction given by its velocity
(2) if particles collide, their velocities change according to specific rules shown in Figure 11.8.

Consider the top case in the rule list of Figure 11.8. Two particles come directly at each other. After they collide, they leave with a 50% probability of going along one diagonal and 50% along the other. The other collision rules are interpreted the same way. There are mathematical representations of these rules given elsewhere [278]. One can show that mass and momentum are preserved in these collisions and that these simple rules and geometries yield, asymptotically, the Navier-Stokes equation [278].

In spite of its successes, there are some deficiencies with the lattice-gas model, which have been addressed in a variant of the model, called the lattice-Boltzman method. In the lattice-Boltzmann method, the single particles that move around the lattice are replaced with single particle distribution functions, $f(\langle n \rangle)$, where $\langle \rangle$ indicates a local ensemble average and $n_i$ is the occupation number of site $i$. $n_i$ is not actually a single number, but rather a set of six numbers indicating whether there is a particle with velocity in the six possible directions.

The lattice-gas and lattice-Boltzmann methods are quite powerful and offer new opportunities to model flows in quite complex geometries. That such simple rules yield the same results as the Navier-Stokes equation is quite remarkable. These methods are widely used to model flows through porous media, where traditional computational fluid dynamics cannot be used (the boundary conditions are too complicated). Three-dimensional versions have also been developed, which enables application to very complex flows. Some applications of these methods to materials processing have been made (e.g., [175, 269, 270]), though their use is not extensive.

## 11.4 EXAMPLES OF CELLULAR AUTOMATA IN MATERIALS RESEARCH

In this section we highlight two of the many applications of cellular automata in materials research. These examples are chosen in part because they are representative of the broad class of cellular automata, but also because they build on the basic concepts and geometries of Game of Life, extending it to the study of microstructural evolution. One of the challenges of cellular automata is that as the complexity of the physical phenomena increases, the rule set to accurately describe that complexity grows. For example, one of the most successful models based on cellular automata is the Cellular Automata – Finite Element (CAFE) model of Gandin and Rappaz and co-workers [112, 272]. This model, which is a coupling of cellular automata and a finite-element method, describes solidification and the growth of dendritic structures. It is a relatively simple model, but too complicated for discussion in this text. We thus recommend other sources, such as [268, 269], for a more complete review of the use of cellular automata in materials research.

### 11.4.1 A simple model for recrystallization

Hesselbarth and Göbel [143] presented a very straightforward application of cellular automata to recrystallization.[4] We will describe a version of their simulation briefly here, as it highlights the basic way that models of physical phenomena can be implemented as a cellular automaton.

The authors focused on capturing three phenomena:

(a) nucleation of grains
(b) growth of grains
(c) the slowing of growth owing to the impingement of grains.

The basic model used by Hesselbarth and Göbel included, as in all cellular automata, the following generic issues, followed by their choices,

(a) the geometry of the cells
  - a two-dimensional square lattice
(b) the number and the kind of states a cell can possess
  - there were two states per site, recrystallized or not recrystallized
(c) the definition of the neighborhood of a cell
  - the authors considered each of the two-dimensional neighborhoods in Figure 11.4
(d) the rules that determine the state of each cell in the next time step
  - the authors developed rules for nucleation of new grains, growth of grains, and the impingement of grains.

At the beginning of the simulation, all sites were set to zero (not recrystallized) and then $N_{embryo}$ grain "embryos" were placed in the system by assigning non-zero values to randomly selected sites on the lattice. While not specified by the authors, a likely strategy would be to assign to each embryo a label to identify the grain that arose from that embryo.

Defining the activity $A$ as the sum of recrystallized neighbors of the central site based on the neighborhood in Figure 11.4, they applied a simple rule to describe growth: if $A \geq 1$ at time step $t$, then the central site would be considered recrystallized at time step $t + 1$ and take

---

[4] For a discussion of recrystallization, please see Section 10.4.2.

on the identity of the grain that extends into its neighborhood. Near impingement, when more than one grain might be growing into a neighborhood, some model for choosing between the identities of the impinging grains would be needed. There are a number of modifications that could be included to control the growth rate of the grains. One could decide to include a growth step only every $n$ time steps, where $n$ is an input parameter. Or, one could say that the growth happens not with a probability $P = 1$, but with some lower probability.[5]

The authors explicitly assumed a constant nucleation rate, $\dot{N}_{embryo}$, such that at each time step new embryos would be created by choosing a set of random positions among the non-recrystallized lattice sites. To reflect the loss of available sites as recrystallization advances, they assumed the form

$$\frac{\Delta N_{embryo}}{\Delta t} = \dot{N}_{embryo}(1 - x),\qquad(11.9)$$

where $x$ is the fraction of the total number of recrystallized sites, $N_{rxn}$ to the total number of sites, $N_s - x = N_{rxn}/N_s$. $\dot{N}_{embryo}$ is a parameter set to model different behaviors.

To model the slowing down of growth as the growing grains approach each other (impingement), the authors implemented one other rule. The number of recrystallized sites within the 25-neighbor environment of Figure 11.4d is $A_{25}$. If $A_{25} > 11$, then the probability $P$ of growth was reduced and a probabilistic model was used.

The authors varied the environment, the frequency of growth steps $n$, the nucleation rate $\dot{N}_{embryo}$, and the probability $P$ of growth under impingement conditions. They generated microstructures based on a variation of these conditions. To examine the dynamics of growth, they compared their results to the Johnson-Mehl-Avrami-Kolmogrov (JMAK) growth equation, showing good agreement.[6]

In Figure 11.9 we show results of two calculations based on this model. The program that implements this model is very similar to that used in the Game of Life. In these simulations, a set number of sites, $N_i$, were chosen initially as embryos. They were numbered sequentially. The growth rule was then employed. At each time step, the creation of another set of $N_d$ embryos was attempted. $N_d$ sites were chosen at random and if a site was not recrystallized, an embryo was created there, also with a sequential number. Thus all growing grains have their own unique identification number. For these examples, we did not include the authors's final rule for impingement.

The dependence of the growth characteristics on the environment of the growth rule can be seen in Figure 11.9. The Moore environment was employed in Figure 11.9a and Figure 11.9b while the von Neuman environment was used in Figure 11.9c and Figure 11.9d. In each case, a $400 \times 400$ square lattice was employed, with $N_i = 20$ and $N_d = 5$. In Figure 11.9, the plots

---

[5] As in much of the modeling discussed in this text, one can choose an event with probability $P$ by choosing a random number $\Re$ on $(0,1)$ and invoking the event if $\Re < P$.

[6] The JMAK growth law relates the growth fraction $f$ to time by

$$f = 1 - e^{-kt^n},\qquad(11.10)$$

where $k$ is a constant and $n$ depends on the type of growth. A good discussion of the JMAK equation is given in [263].

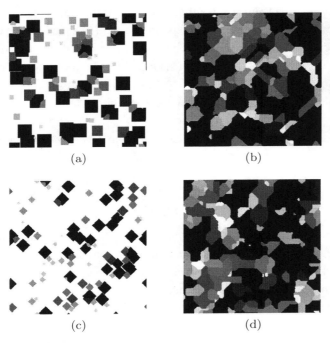

(a)  (b)

(c)  (d)

**Figure 11.9** Results of simulations of a cellular automaton model for recrystallization based on [143]. Panels (a) and (b) were based on the Moore environment while (c) and (d) were modeled with the von Neumann environment. Panels (a) and (c) correspond to $t = 20$ while (b) and (d) are $t = 70$. Details are given in the text.

on the left correspond to $t = 20$ and those on the right to $t = 70$. Note the variation in the sizes of the growing embryos at $t = 20$, reflecting the nucleation of new sites. Generally, the simulations based on the von Neumann environment grew a bit more slowly, though the final microstructures are quite similar.

## 11.4.2 A cellular automaton model for spinodal decomposition

Binary alloy systems can undergo a nonequilibrium phase separation in which domains of two stable phases grow from a thermodynamically unstable homogeneous phase as a result of quenching below a critical temperature. This phenomenon is known as *spinodal decomposition*.

Most modeling of spinodal decomposition has been done with methods such as the phase-field method, described in the following chapter. Here we describe an alternative approach based on a cellular automaton. Details of the model are given in the original paper by Oono and Puri [246]. Note that while they refer to their model as a "cell dynamical system", it is a cellular automaton.

There are three main features that Oono and Puri try to represent in their model. Suppose a system has two species, $A$ and $B$. They assumed:

(a) there is a tendency on a local level to segregate, i.e., to increase the local concentration difference between the two species;

(b) the pure bulk phases are stable;

(c) matter is conserved. By this we mean that the total concentrations of the two species is fixed. The concentrations can change locally, but not overall. Thus, if one species is added in one place, it must be reduced somewhere else. Since materials move by diffusion, that decrease must occur locally.

The first step is to define the state of the system at each site. In this case, we introduce a parameter, which varies in space and time, that defines the local state of the system, i.e., whether the system is species $a$, species $b$, or some mixture of the two. That parameter is referred to as an *order parameter*. For this problem, the order parameter, $\eta$, is the concentration difference between the two components in our binary mixture, i.e., $\eta = \Delta c = c_a - c_b$. In this calculation, we are not going to focus on the actual behavior of a binary material system, but rather will look at the general phenomenology. Thus, it does not matter what the concentrations are exactly, but only that they have the right basic trends. To simplify the model, we normalize the order parameter such that it takes on values between $-1$ and $1$, i.e., when the order parameter at a site is $\eta = -1$, the system is made up of one component and when it is $\eta = 1$ at a site, the system is locally the other component. For other values of $\eta$, the system consists of a mixture of the components. The goal of the cellular automaton is to compute the spatial and temporal dependence of this order parameter.

The driving force for this phase separation is the reduction of free energy. The lowest-energy state has regions of pure $A$ and regions of pure $B$, separated by an interface. In the phase-field model, those free energies are included explicitly to determine the behavior of the system. For a cellular automaton, we need a set of rules to drive the same behavior.

The model is a two-dimensional cellular automaton with the Moore environment of Figure 11.4b, as in the Game of Life. Below we will differentiate the sites in the Moore neighborhood, referring to the nearest-neighbor sites (up, down, left, right) and the sites along the diagonals (right up, left up, left down, right down).

The rule that guides the behavior of $\eta$ of a site, for example site $i$, is written as

$$\eta_i(t + 1) = f\big(\eta_i(t)\big) \tag{11.11}$$

where $f$ is the function

$$f(\eta) = A \tanh(\eta) \tag{11.12}$$

and $A > 0$.

Consider the effects of $f$ on $\eta$. As shown in Figure 11.10, $\tanh(x)$ asymptotes to $-1$ at increasingly negative $x$ and to $+1$ at increasingly positive $x$. So, if $\eta < 0$, $f(\eta)$ drives the value of $\eta$ towards $-A$. If $\eta > 0$, $f(\eta)$ goes towards $+A$. Thus, the rule in Eq. (11.11) tends to drive the system to segregate to one phase or the other, i.e., it mimics the effects of lowering the free energy of the system by creating pure phases.[7] Note that this rule does not include any influence by the neighborhood of a site – it only represents the driving force of thermodynamics.

The local environment is, however, very important and has a strong effect on the evolution of the order parameters $\eta$. Consider the Moore environment. The only way to change $\eta$ in site

[7] As we shall see in Chapter 12, $f$ mimics the structure of the interface between two regions of pure material.

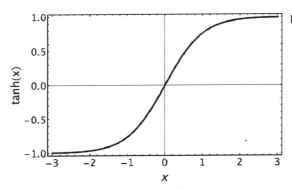

**Figure 11.10** Plot of tanh($x$).

$i$ is for material to diffuse into or out of site $i$ from the neighboring sites. From the theory of diffusion, discussed in Appendix B.7, we have the relation (see Eq. (B.46))

$$\frac{\partial c_a}{\partial t} = -D_a \nabla^2 c_a , \qquad (11.13)$$

where $D_a$ is the diffusion coefficient of species $a$. This equation states that the time rate of change in concentration goes as the local Laplacian of the concentration.[8] To include that relation here, we need an estimate for the Laplacian on a square lattice. This is an example of a two-dimensional numerical derivative, which is discussed in Appendix I.4.2. Using the approximation from the Oono and Puri article, the diffusion term is

$$\frac{\partial c_a}{\partial t} = -D_a\big(\langle\langle\eta\rangle\rangle_i - \eta_i\big), \qquad (11.14)$$

where

$$\langle\langle\eta\rangle\rangle_i = \frac{1}{6} \sum_{j\in r,u,l,d} \eta_j + \frac{1}{12} \sum_{j\in ru,lu,ld,rd} \eta_j . \qquad (11.15)$$

The first sum is over the nearest-neighbor sites in the Moore environment, right, up, left and down, while the second sum is over the next-nearest neighbor sites along the diagonals. The total rule is thus given by

$$\eta_i(t+1) = \mathfrak{F}[\eta_i] = A \tanh\big(\eta_i(t)\big) + D\big(\langle\langle\eta\rangle\rangle_i - \eta_i\big), \qquad (11.16)$$

where we introduce the notation $\mathfrak{F}$ to indicate a rule with both thermodynamics and diffusion.

Using Eq. (11.16) would not be quite correct, however, because the concentration is not conserved, i.e., concentration changes in a site are not reflected by opposite changes in the adjacent sites. Oono and Puri modified Eq. (11.16) to approximate those changes by

$$\eta_i(t+1) = \mathfrak{F}\big[\eta_i(t)\big] - \big\langle\big\langle\mathfrak{F}\big[\eta_i(t)\big] - \eta_i(t)\big\rangle\big\rangle, \qquad (11.17)$$

where the notation $\langle\langle\rangle\rangle$ is defined in Eq. (11.15). What this rule does is to calculate the change at each site and then to subtract the average of that change from the concentrations at the neighboring sites, thus conserving the concentration.

---

[8] The Laplacian, $\nabla^2$, is defined in Eq. (C.17).

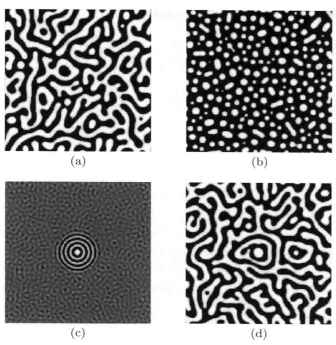

(a)

(b)

(c)

(d)

**Figure 11.11** Results of simulations of a cellular automaton model for spinodal decomposition. All simulations had $D = 0.7$ and $A = 1.3$ in Eq. (11.16). (a) Random initial conditions with $-0.1 < \eta < 0.1$ at $t = 2000$, (b) Random initial conditions with $-0.4 < \eta < -0.2$ at $t = 2000$, (c) and (d) Snapshots of a simulation with random initial conditions with $-0.0001 < \eta < 0.0001$, except for a $5 \times 5$ block in the center of the simulation cell at which $\eta = 0.5$. (c) $t = 500$, (d) $t = 2000$.

The program that implements this model is very similar to that used in the Game of Life, except that the order parameters can be non-integers. The system can be started with many initial conditions, with a consequent variation in behavior. The calculation of the evolution of the conserved order parameter requires two passes through the lattice at each step: the change in each site is calculated by applying Eq. (11.16) and then the two-dimensional average of that change is subtracted from its neighboring sites based on Eq. (11.17).

In Figure 11.11, examples of simulations based on a variety of starting conditions are displayed. All simulations had $D = 0.7$ and $A = 1.3$ in Eq. (11.16). In Figure 11.11a, all sites were initialized with values between $-0.1$ and $0.1$, leading to a roughly equal mixture of the two components. Consequently, the microstructure at a late time is evenly distributed between the two types of material, showing that matter was indeed conserved. In contrast, Figure 11.11b shows results from a system in which the sites were initialized with values between $-0.4$ and $-0.2$, starting the system with more of one component than the other. The late-time microstructure shows inclusions of the minority material into the majority host material. Finally, in Figure 11.11c and Figure 11.11d, we show two times from a simulation in which an initial seed was placed in the middle of the cell while the other sites were initialized with values between $-0.001$ and $0.001$. The dominant growth at early times shows circular bands of the two phases, though eventually the random initial conditions lead to a less symmetric growth pattern.

This model shows the power of a cellular automaton – simple rules can lead to complex behavior. Moreover, by building rules that mimic the physical processes in a real material, realistic descriptions of evolving material properties can be found.

### 11.4.3 Other applications of cellular automata

There are many other applications of cellular automata in materials science. Here we list a few examples, without detail, to show the breadth of these applications. While many of these examples focus on microstructural evolution of one sort or another, a wide range of other phenomena have also modeled with automata.

- Grain growth: grain growth has been a prime focus of cellular automaton models, and many new and interesting variants have been proposed, of which just a few will be mentioned. Modeling normal curvature-driven grain growth using a simple set of rules and a von Neumann environment was one of the earlier examples of cellular automata in materials research [204]. Since then, a variety of alternative methods have been proposed, including expector cellular automata (in which cells of the lattice represent locations in space and the automata are entities that occupy those cells) [22], and frontal cellular automata, in which, among other differences, only cells that sit on boundaries are considered in each time step [300]. An energy-based cellular automaton was used to model grain growth in alloys [8].
- Recrystallization: recrystallization modeling, such as described in Section 11.4.1, continues to be a common goal for cellular automaton modeling. Kroc, for example, constructed a model that includes, in an approximate way, the evolution of the dislocation density during dynamic recrystallization [182]. At each step, the change in dislocation density is calculated from a simple rate equation, with recrystallization being driven by changes in dislocation density, and thus strain energy, between neighboring grains. Other applications include an extension of the Hesselbarth and Göbel model described in Section 11.4.1 to model dynamic recrystallization [125] and a three-dimensional model of recrystallization in steel [39].
- Solidification: in addition to the CAFE model mentioned above [272, 112] a number of other applications of cellular automata to solidification have been made. One of the most interesting is a coupling of the Potts model of Chapter 10 with the lattice-gas method of Section 11.3 [95]. In the Potts lattice-gas model, there are solute and solvent particles as in the lattice-gas model, but the interactions also depend on the relative orientations of nearest neighbors, as in the Potts model, which enabled the author to include complicated nucleation pathways.
- Surface evolution: there have also been a number of applications of cellular automata to the evolution of surfaces. One example was a study of the evolution of the deposition of inert gas monolayers, in which surface diffusion was ignored, but a cellular automaton included the surface flux, a desorption process, and an adsorption process. It was an atomic-based stochastic method using the energy of interaction between the atoms in the monolayer and the surface atoms [365]. Taking an entirely different approach is a study of deposition and etching, in which there are no physically based models governing the rule set, but rather a set of probabilities that depend strictly on the population of the local environment. In this two-dimensional probabilistic cellular automaton, transition probabilities were adjusted to match experiment [134].

- Corrosion: two articles developed cellular automata for corrosion [203, 305]. One of these articles focused mostly on the propagation of the corrosion front, with the chemistry being treated in an abstract way [305]. The other study was much more detailed in how it treated specific chemical reactions, and emphasized the propagation of intergranular corrosion through the material [203]. The models had similar intents, but the details are very different, showing the extreme flexibility inherent in cellular automaton modeling.

These are just a few of the many applications of cellular automata to materials. Many of these methods are quite simple, while others are extremely detailed. All, however, build on the basic ideas outlined in Section 11.1.

## 11.5 RELATION TO MONTE CARLO

In some sense, the Monte Carlo method of Chapter 7 looks like a probabilistic cellular automaton – the system at one "time" is determined by finding the energy change from the configuration at the previous time to the current time, which is, for systems like the Ising model, just a rule. To see that, consider an Ising model on a two-dimensional lattice at 0 K. The only moves accepted are those that lower the energy. For example, a configuration of a 1 surrounded by four 0's will flip to 0. That is a rule. At finite temperature it becomes a probabilistic rule, with the probability given by the Boltzmann factor.

The difference is how the states are chosen. In the regular Metropolis Monte Carlo method, states are sampled uniformly, with one change at a time. At late stages, we argued that the N-fold way gave the same results on average as Metropolis Monte Carlo when one chooses a spin to flip proportional to the probability of that flip. In a regular cellular automaton, on the other hand, every state is updated at the same time. There is no guarantee that the system evolves within a specific ensemble and the dynamics are not necessarily related to real times. Thus, a cellular automaton is not the same as Monte Carlo. For every cellular automaton, its connection to the real world must be examined.

## 11.6 SUMMARY

A cellular automaton is a lattice-based method in which the evolution of values at each site of the lattice is governed by rules. If the rules are chosen with care, then cellular automata can be used to mimic real-world behavior. Cellular automata are interesting for many reasons, not the least of which is that very complex behavior can be obtained from very simple rules.

Applications to materials behavior have been made in a number of contexts, with microstructural evolution receiving the most attention. Fine-tuning the rules has led to a number of robust and useful methods. Care must be taken, however, to ensure that while the results may *look* reasonable, they must also have meaning. Cellular automata can create results that may be nothing more than pretty pictures.

## Suggested reading

Some references on cellular automata that I find useful are:

- B. Chopard and M. Droz, *Cellular Automata Modeling of Physical Systems* [68].
- S. Wolfram, editor, *Theory and Applications of Cellular Automata*, especially the outline and Chapter 1.1 [358].
- While some find this a bit excessive in its claims, the work by S. Wolfram, *A New Kind of Science* [359], has some fascinating discussions of the power of cellular automata.

# 12 Phase-field methods

The *phase-field* method is a thermodynamics-based approach most often employed to model phase changes and evolving microstructures in materials. It is a mesoscopic method, in which the variables may be abstract non-conserved quantities measuring whether a system is in a given phase (e.g., solid, liquid, etc.) or a conserved quantity, such as a concentration. Interfaces are described by the smooth variation of those quantities from one phase to another and are diffuse, not sharp.

The phase-field method is increasingly being used in materials science and engineering because of its flexibility and utility. We discuss the basic method here, but researchers are continually creating new features and new approaches within the basic phase-field framework.

We first introduce the basic mathematical formalism, followed by some simple examples of the phase field in one and two dimensions. Implementation of the phase field requires some new computational methods, which will be discussed in the regular text and an appendix. Finally, we will discuss some applications of the phase-field method in materials research.

## 12.1 CONSERVED AND NON-CONSERVED ORDER PARAMETERS

In phase-field modeling, the state of a system is described by a function of position and time. This function could be a specific property of the system such as concentration or it could be a parameter that indicates what phase the system is in, e.g., solid or liquid. This function is generally referred to as an *order parameter*.

Suppose, for example, we could model solidification by considering a parameter $\phi(\mathbf{r}, t)$ that is a function of the position in the system ($\mathbf{r}$) and time ($t$) and whose value determines the thermodynamic phase of the material. $\phi = 1$ could indicate a solid phase and $\phi = -1$ a liquid.[1] As in all thermodynamic systems, the phase at each position, $\phi(\mathbf{r}, t)$, is determined from the free energy of the system. There are no local or global constraints on its value other than the energy. Suppose, for example, that a certain volume of system is a solid. For the volume of solid to transform to a liquid does not require any transport of material into or out of it so, except for energetic considerations; what happens in that volume is independent of its neighboring

---

[1] Any other values indicating solid or fluid would work as well in that they serve only as indicators of the state of the system.

volumes. In this case, there are no conservation laws that must be maintained and the order parameter is referred to as being *non-conserved*.

In other situations, the order parameter may be a conserved quantity. For example, suppose the order parameter at a point $\mathbf{r}$ represents the local concentration of a species $C(\mathbf{r}, t)$. For the concentration at $\mathbf{r}$ to increase, atoms must be transferred to that position from other regions in the material. Thus growth of a concentration at one point requires a drop in value at another. The order parameter at $\mathbf{r}$ cannot show unbounded growth since the total concentration is conserved. The equations governing the behavior of such systems must reflect the conserved quantities and the order parameters must also be conserved.

## 12.2 GOVERNING EQUATIONS

There are two equations that underlie the general phase-field method, one that describes the dynamics of systems with non-conserved order parameters and one for systems with conserved order parameters.

### 12.2.1 The Allen-Cahn equation for non-conserved order parameters

The phase-field method is based on a thermodynamic description of a system, so the first step is to establish an expression for the total free energy. For a system with non-conserved order parameters, the free energy of a two-phase system includes three types of terms: the energy of the volume of the system that contains one of the phases, the energy of the volume of the system that contains the other phase, and the volume of that system that corresponds to the *interfaces* between the phases. Given the free energy, expressions that describe how the system will evolve to lower the free energy can be found. In the phase-field method, the change from one phase to another will be monitored by the change in the order parameters. For non-conserved order parameters, the equation that describes the evolution of the order parameters is called the *Allen-Cahn* equation.

Suppose we have a system that has two phases, which we indicate by different values for $\phi$ in each phase (for example, $+1$ for a solid and $-1$ for a liquid). Ignoring, for now, any interfaces between the two phases, we can write the total free energy of the system as a sum of the volume of each phase multiplied by the free-energy density (i.e., the free energy per volume) of that phase. The free-energy density, which we will call $f$, is a local variable that depends only on the state of the system at each point and represents the free energy for a phase at the current thermodynamic conditions, e.g., at some $T$ and $P$. Since the state of the system is specified by the order parameter $\phi$, which is a function of the position $\mathbf{r}$ and time $t$, we can specify the local free energy at any point of the system by $f[\phi(\mathbf{r}, t)]$.

The total free energy $\mathfrak{F}$ is an integral over the volume of the system $\Omega$ of the free-energy density $f[\phi(\mathbf{r}, t)]$,

$$\mathfrak{F} = \int_{\Omega} f\big[\phi(\mathbf{r}, t)\big] d\mathbf{r}. \tag{12.1}$$

Note that $\mathfrak{F}$ is a function of $\phi$, which is a function of position and time. Thus, $\mathfrak{F}$ is a *functional* of $\phi$, which will change how we approach the derivation below.

Eq. (12.1) does not include the interfaces between phases, which have positive energy and lead to phenomena such as grain growth. Interfaces within the phase-field method are indicated by a change in the value of the order parameter. For example, if $\phi = +1$ means solid and $\phi = -1$ means liquid, then the interface between a solid and liquid is indicated by a change in $\phi$ from $+1$ to $-1$. The width of the interface, and thus its energy, will depend on how quickly $\phi$ changes with distance. Thus, the free-energy density depends not only on the value for $\phi$ at a point but also on the *change* in $\phi$ at that point. It is reasonable to assume that this change in energy depends on the gradient of $\phi$, $\nabla\phi(\mathbf{r}, t)$. The total free-energy density is $f^{total}[\phi(\mathbf{r}, t), \nabla\phi(\mathbf{r}, t)]$ and the total free energy is

$$\mathfrak{F} = \int_\Omega f^{total}[\phi(\mathbf{r}, t), \nabla\phi(\mathbf{r}, t)]d\mathbf{r}. \tag{12.2}$$

In Appendix 12.9.1 we present the derivation of equations for $\mathfrak{F}$ as well as for time rate of change of the order parameter $\phi$. The key assumptions in that derivation are:

(a) the interfaces are diffuse
(b) Eq. (12.2) can be expanded in a Taylor series in $\nabla\phi$ that can be truncated at second order in $(\nabla\phi)^2$
(c) $\phi$ will change with time to decrease the free energy
(d) in analogy with classical mechanics, the dynamics of $\phi$ are governed by a thermodynamic "force", which is defined as the negative of the derivative of the free energy, $\mathfrak{F}$, with respect to $\phi$.

The outcome of these assumptions (and a little calculus and algebra) is the Allen–Cahn equation. In one dimension, the free energy is

$$\mathfrak{F} = \int_\Omega \left[ f[\phi(x)] + \frac{\alpha}{2}\left(\frac{\partial\phi}{\partial x}\right)^2 \right]dx \tag{12.3}$$

and the time derivative of $\phi$ is

$$\frac{\partial\phi(x, t)}{\partial t} = -L_\phi \frac{\delta\mathfrak{F}}{\delta\phi(x, t)} = -L_\phi \left( \frac{\partial f(\phi)}{\partial\phi} - \alpha\frac{\partial^2\phi(x)}{\partial x^2} \right). \tag{12.4}$$

The free energy in Eq. (12.3), $\mathfrak{F}$, consists of two terms. The first is the thermodynamic free-energy density of the phases. The second is the energy of the interfaces between phases, with $\alpha$ being a constant that sets the overall energy of the interface. The interfacial energy is proportional to $(\partial\phi/\partial x)^2$, the square of the slope of the variation of $\phi$ through the interface. Sharper interfaces in this description lead to higher interfacial energies.

Equation (12.4) states that the time derivative of $\phi$ is simply a constant, $L_\phi$, multiplied by the negative of the functional derivative of the free energy with respect to $\phi$, which consists of two terms. The first term is the simple derivative of the free-energy density function $f(\phi)$ with respect to $\phi$ and the second term arises from the change in interfacial energy, which is proportional to $\partial^2\phi/\partial x^2$, the $x$-component of the Laplacian operator $\nabla^2\phi$ from Eq. (12.38).[2]

From Eq. (12.5) and Eq. (12.6), in three dimensions

$$\mathfrak{F} = \int_\Omega \left( f[\phi] + \frac{\alpha}{2}|\nabla\phi|^2 \right) d\mathbf{r} \tag{12.5}$$

and

$$\frac{\partial\phi(\mathbf{r}, t)}{\partial t} = -L_\phi \left( \frac{\partial f(\phi)}{\partial\phi} - \alpha\nabla^2\phi(\mathbf{r}, t) \right). \tag{12.6}$$

Equation (12.6) is a statement that the *dynamics* of the variation of the order parameter $\phi_i$ are driven by the system's approach to its minimum free energy. If $\mathfrak{F}$ decreases with an increase in $\phi_i$, then $\phi_i$ increases with time; if $\mathfrak{F}$ increases with an increase in $\phi_i$, then $d\phi_i/dt$ is negative and $\phi_i$ decreases. When the right hand side is zero, $\mathfrak{F}$ is at its minimum and $\phi_i$ is stationary. Below we shall give examples of how Eq. (12.6) can be used to study microstructural evolution, solidification, etc.

## 12.2.2 The Cahn-Hilliard equation for conserved order parameters

The previous section described the time evolution of non-conserved order parameters. This section focuses on situations in which conservation laws come into play.

Consider a binary alloy with $A$ and $B$ type atoms in which there is only one phase. For this problem, it is convenient to take as the order parameter the concentration of one of the species, for example species $A$. We define the order parameter $C$ such that $C(\mathbf{r}, t) = C_A(\mathbf{r}, t) = 1 - C_B(\mathbf{r}, t)$. $C$ is *conserved* since the total concentration is fixed and if any region has an increase in one of the species, the amount of the other species in that region is reduced. The conservation of the order parameter leads to an equation for the dynamics that is somewhat more complicated than in the non-conserved case. This equation is referred to as the *Cahn-Hilliard equation* [52, 53].

We describe in detail the derivation of the Cahn-Hilliard equation in Appendix 12.9.2. The derivation is a bit more complicated than that for the Allen-Cahn equation of Section 12.2.1, but not more difficult. In one dimension, we have

$$\begin{aligned}
\frac{\partial C}{\partial t} &= \frac{\partial}{\partial x}\left( M\frac{\partial\mu}{\partial x} \right) \\
&= \frac{\partial}{\partial x}\left( M\frac{\partial}{\partial x}\left[ \frac{\partial f}{\partial C} - \alpha\frac{\partial^2 C(x)}{\partial x^2} \right] \right),
\end{aligned} \tag{12.7}$$

where $M$ is a concentration-dependent mobility. $M$ could be anisotropic (i.e., dependent on the direction of diffusion in the lattice), in which case it would be a second-rank tensor.

---

[2] Functional derivatives require some special techniques, as described in Appendix C.6.

If $M$ is independent of concentration (and therefore $x$),

$$\frac{\partial C}{\partial t} = M\left(\frac{\partial^2}{\partial x^2}\frac{\partial f}{\partial C} - \alpha\frac{\partial^4 C(x)}{\partial x^4}\right). \tag{12.8}$$

Equation (12.7) is the one-dimensional version of the Cahn-Hilliard equation [52, 53].

In three dimensions, the Cahn-Hilliard equation is

$$\frac{\partial C}{\partial t} = \nabla \cdot (M\nabla\mu)$$

$$= \nabla \cdot \left(M\nabla\left(\frac{\partial f}{\partial C} - \alpha\nabla^2 C\right)\right). \tag{12.9}$$

### 12.2.3 Systems with both conserved and non-conserved order parameters

For a system described by a set of conserved $\{c_i\}$ and unconserved $\{\phi_i\}$ order parameters, the free energy can be written as

$$\mathfrak{F}[c_1, c_2, ...c_p; \phi_1, \phi_2, ..., \phi_k] = \int \left[ f(c_1, c_2, ...c_p; \phi_1, \phi_2, ..., \phi_k) \right. \tag{12.10}$$

$$\left. + \sum_{i=1}^{p}\alpha_i(\nabla c_i)^2 + \sum_{i=1}^{3}\sum_{j=1}^{3}\sum_{l=1}^{k}\beta_{ij}\nabla_i\phi_l\nabla_j\phi_l \right]d\mathbf{r},$$

where $f$ is a free-energy density that is a function of $\{c_i\}$ and $\{\phi_i\}$ and $\alpha_i$ and $\beta_{ij}$ are coefficients. The integral is over the volume. $\mathfrak{F}$ is thus a functional of the variables $\{c_i\}$ and $\{\phi_i\}$. The dynamics for the order parameters will be of the form shown in Eq. (12.9) for the $\{c_i\}$ and Eq. (12.6) for the $\{\phi_i\}$.

## 12.3 A ONE-DIMENSIONAL PHASE-FIELD CALCULATION

The phase-field model describes the evolution of systems described by equations like the Allen-Cahn and Cahn-Hilliard equations. The first step is to develop a description of the local free energy in terms of order parameters that indicate the material's phase (i.e., various solid phases, liquid) or some other material property, such as concentration.

A principal assumption in the derivation of the phase-field equations is that the order parameters vary continuously and smoothly throughout the system. Consequently, interfaces are smooth and much broader than their physical counterparts. In Figure 12.1a is shown a schematic view of such an interface, where the variation of $\phi$ through an interface is plotted. The integrand of the free energy in Eq. (12.5) is $f[\phi] + (\alpha/2)(\partial\phi/\partial x)^2$. A schematic view of this function is shown in Figure 12.1b, showing the peak in free-energy density in the interfacial region. This figure highlights two of the essential features of the physics that governs the phase-field model – the diffuse interfaces and the excess interfacial free energy.

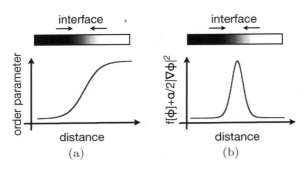

**Figure 12.1** (a) A schematic view of the one-dimensional diffuse interface. Plotted is the variation of the order parameter through an interface, $\phi(x)$. (b) Schematic drawing of the integrand of the energy expression from Eq. (12.5), showing the peak in free-energy density in the interface.

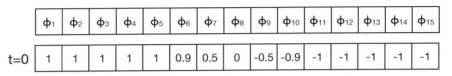

**Figure 12.2** A one-dimensional model showing the values of the order parameter at $t = 0$.

To make the discussion more specific, consider the one-dimensional system in Figure 12.2, in which we divide the volume into a grid. Each grid point is associated with a certain volume of material $v = a^3$. The order parameter has the same value throughout the individual grid volumes. This procedure converts the problem from a continuous system to a discrete one. For the purposes of this example, we assume fixed boundary conditions at the edges, i.e., sites to the left are fixed with a value of $+1$ and those on the right to $-1$.[3] Assume that $+1$ represents a solid and $-1$ a liquid, in which case this system is solid to the left and liquid to the right with an interface in between.

Suppose that we have a solid and liquid in equilibrium such that the free energies of each phase are the same. In that case, the free energy as a function of the phase might look like

$$f[\phi(x, t)] = 4\Delta\left( -\frac{1}{2}\phi^2 + \frac{1}{4}\phi^4 \right), \tag{12.11}$$

which is shown in Figure 12.3. This is not a thermodynamic function that corresponds to a real material. This function is chosen to model the phenomenological behavior of a system with two phases of equal energy. The values of $f$ for $\phi$ in the interface (i.e., between $-1$ and $1$) are not related to any physics, but are meant to reflect the approximate way such systems might behave.

The expression for $f[\phi(x)]$ goes into the total energy expression in Eq. (12.3). For calculating the time rate of change of $\phi(x)$ as in Eq. (12.4), we need

$$\frac{\partial f[\phi(x, t)]}{\partial \phi} = 4\Delta\left( -\phi + \phi^3 \right), \tag{12.12}$$

which is shown in Figure 12.3.

---

[3] Boundary conditions depend on the problem of interest. They may be fixed, as in this example, or periodic or whatever the problem requires.

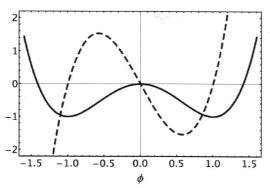

**Figure 12.3** Free-energy density functional $f(\phi(x, t))$ from Eq. (12.11) (solid curve) and $df/d\phi$ from Eq. (12.12) (dashed curve) for $\Delta = 1$.

To evaluate the energy in Eq. (12.29) and the time derivative in Eq. (12.4) we need values for $d\phi/dx$ and $d^2\phi/dx^2$ at site $i$ on the grid. We evaluate those terms numerically, using the finite difference equations presented in Appendix I.4. Simple one-dimensional expressions for the first and second derivatives on a mesh yield (from Eq. (I.10) and Eq. (I.12), respectively)

$$\frac{d\phi_i}{dx} = \frac{\phi_{i+1} - \phi_{i-1}}{2a}$$

$$\frac{d^2\phi_i}{dx^2} = \frac{\phi_{i+1} + \phi_{i-1} - 2\phi_i}{a^2}, \tag{12.13}$$

respectively, where $a$ is the grid spacing.

One of the assumptions is that the values of the functions are constant across each grid volume, which leads to a simplification in evaluating the integral needed in the evaluation of the total free energy, $\mathfrak{F}$ in Eq. (12.3) – it is just the value of the integrand at each grid point multiplied by the volume, $v$, of the grid. Combined with the expressions for the derivatives, the energy and time derivative become

$$\mathfrak{F} = \sum_{i=1}^{N_{grid}} v \left[ 4\Delta \left( -\frac{1}{2}\phi_i^2 + \frac{1}{4}\phi_i^4 \right) + \frac{\alpha}{2} \left( \frac{\phi_{i+1} - \phi_{i-1}}{2a} \right)^2 \right] \tag{12.14}$$

and

$$\frac{\partial \phi_i}{\partial t} = -L \left[ 4\Delta \left( -\phi_i + \phi_i^3 \right) - \alpha \left( \frac{\phi_{i+1} + \phi_{i-1} - 2\phi_i}{a^2} \right) \right], \tag{12.15}$$

respectively. Thus, knowing the values of $\phi_i$ in each grid volume, we can calculate the total energy and $d\phi_i/dt$.

In standard molecular dynamics, it is essential to have an accurate integration of the equations of motion to ensure that the total energy was conserved. In contrast, in the phase-field method the driving force is the *decrease* in the total free energy. The simplest approach to integrating the equation of motion is just to assume a first-order Taylor expansion, such as

$$\phi_i(t + \delta t) = \phi_i(t) + \frac{\partial \phi}{\partial t} \delta t . \tag{12.16}$$

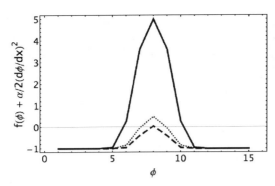

**Figure 12.4** $f[\phi_i] + \frac{\alpha}{2}\left(\frac{d\phi_i}{dx}\right)^2$ evaluated at the $t = 0$ values for $\phi_i$ in Figure 12.2 for $\Delta = 1$ and $\alpha = 10$ (solid), $\alpha = 1$ (dotted) and $\alpha = 0.1$ (dashed).

This approach is known as an Euler equation. The magnitude of $\delta t$ is set to achieve a balance between numerical accuracy (which requires a small value for $\delta t$) and a rapid evaluation of the equations (large $\delta t$).

For this simple model, the only parameters are $\Delta$, which governs the local free-energy density (i.e., the thermodynamics), and $\alpha$, which reflects the interfacial energy. The kinetic parameter $L$ also must be prescribed to set the time scale. For convenience, $L = 1$ for the rest of this discussion.

Consider the *energy* of the configuration at $t = 0$ shown in Figure 12.2. As we noted above, we assume that $\phi$ is constant over each grid volume. Taking $a = 1$ (and therefore $v = 1$), from Eq. (12.14) the energy for each grid point equals

$$f[\phi_i] + \frac{\alpha}{2}\left(\frac{d\phi_i}{dx}\right)^2 = 4\Delta\left(-\frac{1}{2}\phi_i^2 + \frac{1}{4}\phi_i^4\right) + \frac{\alpha}{2}\left(\frac{\phi_{i+1} - \phi_{i-1}}{2a}\right)^2. \qquad (12.17)$$

We plot this energy for differing values of $\Delta$ and $\alpha$ in Figure 12.4. Note that as $\alpha/\Delta$ increases, the interfacial energy becomes much larger in magnitude. We thus expect that simulations of systems with high $\alpha/\Delta$ ratios should show fewer boundaries than those with low values of $\alpha/\Delta$.

Consider now a one-dimensional system with periodic boundary conditions with the energy expressions just described. A phase-field calculation of this model would proceed as follows:

(a) choose an order parameter, $\phi(x)$, at each mesh site for $t = 0$
(b) from the $\phi$, calculate $\mathfrak{F}$ and $d\phi/dt$ for each grid point at time $t$
(c) use Eq. (12.16) to calculate the new values of $\phi$ at $t + \delta t$
(d) go to item b and repeat until the free energy converges to a minimum.

The initial configuration involves the choice of mesh and the initial values for the order parameter. In this case, the mesh size is $a = 1$ and an initial configuration of $\phi_i$ values was chosen randomly over a range of values between $-0.1$ and $0.1$. The dynamic parameter in Eq. (12.16) was taken as $L = 1$. The procedure in items a–d was followed, varying the time step, $\delta t$ in Eq. (12.16), until a stable solution was found. In Figure 12.5 we show results from a series of long-time configurations for different values of the ratio $\alpha/\Delta$, i.e., the ratio between the coefficient of the interface penalty and the magnitude of the local free-energy term. For the three simulations shown, $\Delta$ was fixed (at 1) and $\alpha$ was varied. In all cases, the calculations were run until a stable configuration was found, i.e., until $d\phi_i/dt \approx 0$. Note that the equilibrium

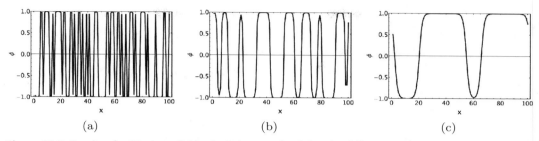

**Figure 12.5** Results of a 1D phase-field calculation at a fixed time for different starting configurations, as described in the text. (a) $\alpha/\Delta = 0.1$, (b) $\alpha/\Delta = 1$, (c) $\alpha/\Delta = 10$.

configuration should have no interfaces – the cases shown in Figure 12.5 are at long times at which the energy change with time is essentially zero, but they are not at equilibrium.

Since the system has the same local free energy whether $\phi = 1$ or $-1$, if there were no interfacial penalty ($\alpha = 0$), the order parameters would take on random values of $+1$ or $-1$ depending on their initial state. Having a positive interfacial energy drives the system to minimize the interfaces. In Figure 12.5a we show the values of $\phi_i$ for the case where $\alpha/\Delta = 0.1$, i.e., the system is dominated by the local free energy. Note that the interfaces between regions are sharp. (In the next section we show that the width of an interface should scale as $\sqrt{\alpha/\Delta}$ and its energy as $\sqrt{\alpha\Delta}$.) Figure 12.5b is an intermediate case with $\alpha/\Delta = 1$. We see that the system still has many interfaces, which are somewhat broader, and there are states intermediate between the order parameters. Finally, in Figure 12.5c we see that for $\alpha/\Delta = 100$, the number of interfaces is reduced and that they are broader, all consistent with the energetics seen in Figure 12.4.

## 12.4 FREE ENERGY OF AN INTERFACE

Take a system described by the free-energy function in Eq. (12.11) with an interface located in the $yz$ plane at $x = 0$. The free energy per area is given by

$$\frac{\mathfrak{F}}{A} = \int dx \left( 4\Delta \left( -\frac{1}{2}\phi(x)^2 + \frac{1}{4}\phi(x)^4 \right) + \frac{\alpha}{2}\left( \frac{\partial \phi(x)}{\partial x} \right)^2 \right) \tag{12.18}$$

and the minimum free energy is the solution to

$$\frac{\delta \mathfrak{F}/A}{\delta \phi(x)} = 0 = 4\Delta \left( -\phi(x) + \phi(x)^3 \right) - \alpha \left( \frac{\partial^2 \phi(x)}{\partial x^2} \right). \tag{12.19}$$

We assume the profile is given by $\phi(x) = \tanh(\gamma x)$, as shown in Figure 11.10, which has the correct general shape. Inserting this form of $\phi$ in Eq. (12.19) yields, after simplification,[4]

$$\gamma = \sqrt{\frac{2\Delta}{\alpha}}. \tag{12.20}$$

[4] In Figure 12.1b the integrand of the expression for the energy in Eq. (12.18) is plotted for this value of $\gamma$.

If we define the interface to be the distance between the points where $\phi(x)$ reaches 96% of its asymptotic values, then the width (thickness) of the interface is given approximately by $\delta = 4/\gamma$. The width of the interface thus goes as

$$\delta \sim \sqrt{\frac{\alpha}{\Delta}} \,.$$ 

(12.21)

Using the value for $\gamma$ from Eq. (12.20) (i.e., the minimum energy) in the expression for the free energy per area in Eq. (12.18) and integrating over $x$ from $-\infty$ to $\infty$ yields the free energy per area of the (minimum-energy) interface to be

$$\frac{\mathfrak{F}}{A} = \frac{4\sqrt{2}}{3}\sqrt{\alpha\Delta} \,.$$ 

(12.22)

We see in Eq. (12.21) and Eq. (12.22) the interplay between the energy of the homogeneous phases (controlled by $\Delta$) and the interfacial energy (controlled by $\alpha$). The width of the interface, $\delta$, increases with increasing $\alpha$ and decreasing $\Delta$. On the other hand, the energy associated with the interface increases with increasing $\alpha$ and $\Delta$. Thus, there is a competition between width and energy, which is seen in the simulations shown in Figure 12.5.

## 12.5 LOCAL FREE-ENERGY FUNCTIONS

Much of the materials science in the phase-field method goes into the definition of the free-energy density function $f$. Here we discuss some very simple forms for the free energy, first with a small number of order parameters and then two examples with many order parameters.

### 12.5.1 Systems with one order parameter

In Eq. (12.11) we introduced a simple form for the free-energy density function $f[\phi]$ that describes a material with two, equally energetic, phases. The form of the function was chosen to have minima at $\phi = \pm 1$ with a value of $-\Delta$ at the minima. This function could be appropriate, for example, as a model for the generic features of spinodal decomposition.[5] If the system were started with a set of values for $\phi$ not at $\pm 1$, then it would spontaneously order into regions with phases characterized by $\phi = 1$ or $\phi = -1$, with interfaces between them. The positive interfacial energy, given by the gradient terms in Eq. (12.10), would, at low temperatures, drive the system to one phase or another. We note that the prefactors of each term in Eq. (12.11) are chosen to make the derivatives simple, as seen in Eq. (12.12). The forms of these functions are shown in Figure 12.3.

One could imagine that instead of having two phases of equal energy, one phase might be more stable than the other at some temperatures and less stable at others. A simple form that

---

[5] See Section 11.4.2.

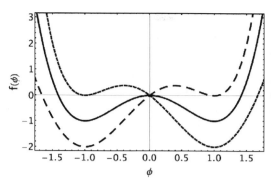

**Figure 12.6** Plot of the two-state free-energy function of Eq. (12.23). The stable states are at $\phi = 1$ and $\phi = -1$. The solid curve is for $T = T_m$, the long dash is with $T < T_m$ and the short dash is for $T > T_m$. We thus see that $\phi = 1$ with a solid phase and $\phi = -1$ with a liquid.

shows such behavior is

$$f(\phi, T) = 4\Delta\left(-\frac{1}{2}\phi^2 + \frac{1}{4}\phi^4\right) + \frac{15\gamma}{8}\left(\phi - \frac{2}{3}\phi^3 + \frac{1}{5}\phi^5\right)(T - T_m), \tag{12.23}$$

where $\gamma$ is a positive constant and $T_m$ is a "melting" temperature [37]. In Figure 12.6 we show this potential for three values of $T$. This function is still not an actual thermodynamic model; the forms were chosen to mimic the behavior of real systems, not to match them. Accurate free-energy expressions can also be found for each phase, for example from the CALPHAD method [208], and then interpolated together as discussed in the next section.

### 12.5.2 Systems with more than one order parameter

In this section we discuss two types of problems in which more than one order parameter is necessary. Both of these problems serve as examples of many of the methods in use today. The first is one in which there is more than one component and more than one phase. This type of model could be used, for example, to model alloy solidification. The second example has only one component, but many "phases", which will represent different orientations of grains in a model of grain growth.

#### Free-energy function for two-phase, two-component, systems

Suppose the goal is to model the solidification of a binary alloy. We thus need a phase-field model that can capture the solidification as well as the equilibration of the two alloy species. Constructing a model for this two-component, two-phase system thus requires two order parameters, a non-conserved order parameter $\phi$ to indicate the phase of the system at a given location and time and a conserved order parameter $C$ to indicate the local concentration of the two species. The free-energy functional must depend on both of these order parameters, for example as seen in Eq. (12.10).

Regular solution theory gives us a way to construct a two-component free-energy function by starting with the free energies of each of the pure systems.[6] We could use a simple expression

---

[6] This section is based on a model discussed in [37].

such as given in Eq. (12.23), adjusting the parameters for each phase. However, that function will not provide an adequate representation of the real thermodynamics of an alloy system. In that case, more complex thermodynamic functions can be used.[7]

Suppose we have experimental free-energy curves for both the solid and liquid phases ($S$ and $L$) for each of the pure components ($A$ and $B$), determined from some source of data, e.g., $f_A^S(T)$, $f_A^L(T)$, $f_B^S(T)$, and $f_B^L(T)$. The net free energy for component $A$ would look something like

$$f_A(T) = \left(1 - p(\phi)\right) f_A^S(T) + p(\phi) f_A^L(T),  \tag{12.24}$$

where the function $p(\phi)$ is chosen so that the system goes smoothly from pure liquid ($p(\phi) = 1$) to pure solid ($p(\phi) = 0$) as a function of the order parameter, $\phi$.

Having expressions for $f_A(T)$ and $f_B(T)$, we can use regular solution theory to construct the free energy of the $AB$ solution,[8] which takes the form

$$
\begin{aligned}
f(\phi, C, T) = {} & (1 - C) f_A(\phi, T) + C f_B(\phi, T) \\
& + RT \left[ (1 - C) \ln(1 - C) + C \ln(C) \right] \\
& + C(1 - C) \left[ \Omega_S (1 - p(\phi)) + \Omega_L p(\phi) \right],
\end{aligned}
\tag{12.25}
$$

where $C$ is the concentration of $B$ and $\Omega_{S(L)}$ are the regular solution parameters for the solid (liquid).

## Free-energy function for multi-phase systems

For a phase-field model of grain growth, Tikare *et al.* [311] employed a free-energy function that had one component but many "phases", with each phase representing a grain with a different orientation. They chose a free-energy function that has an arbitrary number of order parameters, with an energy expression that had distinct minima, each with only one order parameter that is non-zero. In this way, a grain (or any distinct phase) would consist of contiguous sites with one non-zero order parameter.

Interfaces between grains (or phases) would show a gradual change of order parameter between values in each grain, as shown in Figure 12.7 for a simple case with four order parameters. At the left, the system corresponds to grain "1", with $\phi_1 = 1$ with all other order parameters being 0. At the right, the system is in grain "4", with $\phi_4 = 1$ and all other order parameters are 0. In between, there is a gradual decrease in $\phi_1$ and a gradual increase in $\phi_4$ from left to right.

Tikare *et al.* employed the following function, which has a set of $P$ order parameters $\{\phi\}$

$$f[\{\phi\}] = \frac{\gamma}{2} \sum_{i=1}^{P} \phi_i^2 + \frac{\beta}{4} \left( \sum_{i=1}^{P} \phi_i^2 \right)^2 + \left( \lambda - \frac{\beta}{2} \right) \sum_{i=1}^{P} \sum_{j \neq i=1}^{P} \phi_i^2 \phi_j^2,  \tag{12.26}$$

---

[7] For example, see [208].
[8] Regular solution theory couples an ideal entropy of mixing with a mean-field model for the interaction energy of randomly distributed alloy species, as described in [114].

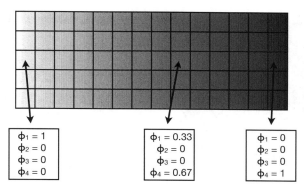

Figure 12.7 Variation of order parameters between grains in two dimensions. On the left, the system has a grain (or phase) with one non-zero order parameter, $\phi_1$. On the right, there is also one non-zero order parameter, $\phi_4$. Going from left to right in the interface region, there is a gradual decrease of $\phi_1$ to 0 and an increase of $\phi_4$ to 1.

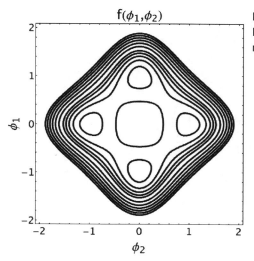

Figure 12.8 Contour plot of the free-energy function from Eq. (12.26) with $\alpha = \beta = \lambda = 1$ and $P = 2$. There are four minima, corresponding to the values $\pm 1$, 0 and 0, $\pm 1$.

where $\gamma$, $\beta$, and $\lambda$ are constants. This function has minima at $\phi_i = \pm 1$ with $\phi_{j \neq i} = 0$, so there are $2P$ minima, all with equal energy. A simple case with $P = 2$ is shown in Figure 12.8. The dynamics of a system described with this free-energy function will be discussed next.

## 12.6 TWO EXAMPLES

In this section are presented two examples of the phase-field method applied to materials phenomena. These are presented in some detail, while additional examples will be listed in a later section.

### 12.6.1 A 2D model of grain growth

In [311], Tikare *et al.* presented a very simple phase-field model for grain growth in which they employed the local free-energy functional shown in Eq. (12.26), with interfaces as in Figure 12.8. The model was implemented on a two-dimensional square mesh, with the Laplacian in Eq. (12.6) calculated as described in Appendix I.4 and shown in Eq. (I.17). Since the order

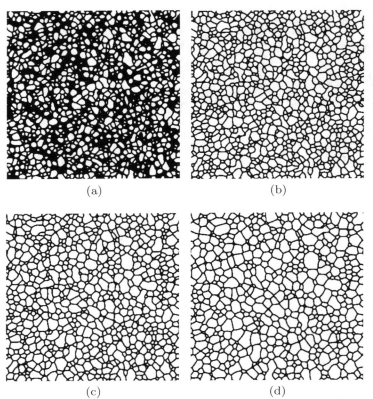

(a)                              (b)

(c)                              (d)

**Figure 12.9** Four snapshots of a 2D phase-field model of grain growth, based on the model of Tikare *et al.* [311]. The solid lines are the boundaries. (a) 20 000 time steps, (b) 40 000 time steps, (c) 60 000 time steps, (d) 100 000 time steps.

parameters are non-conserved, the dynamics was governed by the Allen-Cahn equation of Eq. (12.6).

The driving force for the evolution of the order parameters is the removal of the interfacial energy, just as in the Potts model discussed in Chapter 10. Indeed, the purpose of the paper by Tikare and coworkers was to compare the results for grain growth from a phase-field calculation to that from the Potts model. They found that the phase-field approach gives the same time dependence of the grain growth as does the Potts model.

Results from a calculation based on this model are shown in Figure 12.9, in which $P = 25$. The system was started with all 25 values of $\phi$ at each site taken as random numbers between $-0.1$ and $0.1$ and then followed using the non-conserved dynamics of Eq. (12.6), with $L = 1$. At each time step the derivative of Eq. (12.26) with respect to each $\phi_i$ was evaluated, as well as the Laplacian. Each lattice site was evolved concurrently. As time progresses, the values of $\phi$ coalesce into regions with only one non-zero order parameter, which then coarsen with time.

## 12.6.2 Solidification

The formation of complex microstructures during the solidification of metals and alloys is not only a challenging scientific problem, but has great significance in determining the quality of liquid-based processing, such as casting, welding, etc. Modeling solidification has been one of

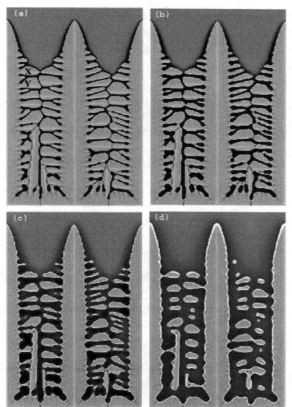

**Figure 12.10** Two-dimensional phase-field simulation of dendritic growth from [37]. The growth temperature was 1574 K in (a) and increased to 1589 K, shown at later times in (b–d). Light gray is solid and dark gray is liquid. Reprinted with permission from the Annual Review of Materials Research, Volume 32, copyright 2002 by Annual Reviews, http://www.annualreviews.org.

the great successes of the phase-field approach, describing with excellent success such processes as the growth of dendrites and eutectics.

The basic idea behind solidification modeling is, at its simplest, what we have already described. Suppose we have a binary alloy whose thermodynamics can be described by equations like those in Eq. (12.24) and Eq. (12.25). There are two order parameters, $\phi$ that specifies whether the system is solid or fluid and whose dynamics are described by Eq. (12.6) and $C$, the concentration, whose dynamics are given by the Cahn-Hilliard equation in Eq. (12.7). Application of the method has been done in two and three dimensions, with fluid flow and without. People have studied dendritic growth and a variety of other phenomena. An excellent review of these applications has been given by Boettinger *et al.* [37]. Not discussed in this text is the linkage of phase-field calculations to fluid flow.

In Figure 12.10 are shown snapshots from a two-dimensional phase-field calculation of dendritic growth based on the model described in Eq. (12.24) and Eq. (12.25) (from [37]). The temperature in Figure 12.10a is 1574 K, which was increased to 1589 K instantaneously. While this change may not realistically capture the fluctuations seen in physical systems, it can shed light on the effects of temperature excursions on the dendritic structure. Figure 12.10b–d are shown at later times, respectively. Note that the dendritic arms are progressively melting over time.

## 12.7 OTHER APPLICATIONS IN MATERIALS RESEARCH

A few other applications of the phase-field method in materials research include:

- Phase-field crystal model: one of the more intriguing advances in phase-field modeling has been the introduction of the phase-field crystal model [97]. In this approach, the atomic structure is kept, enabling the modeling of sharp interfaces, in contrast to the very broad interfaces of the standard phase-field model. There have been many applications of this method, for example to the dynamics of dislocations [61], the Kirkendall effect [98], melting at dislocations and grain boundaries [33], etc.

- Elastically driven phase transformations: we have seen how the phase-field model can be applied to thermodynamically driven phase transitions. It can also be applied to systems in which the driving forces are elastic interactions. Excellent reviews of this type of application of the phase field are in [64] and [65]. Since phase transitions involve some change in structure, there is an associated strain with those transformations. To use a phase-field approach, two things are needed – order parameters that describe the state of the system, and that go smoothly from one structure to the other, and a free-energy functional of those order parameters. The order parameters reflect the symmetry of the transition. The free energy is the strain energy described in Eq. (H.17). Applications of these methods have been made to numerous systems, including to martensitic transformations [186].

- Plastic deformation: the phase-field method has been used to model plastic deformation in two [177, 178, 179] and three dimensions [345]. An overview of this approach, including a comparison to the phase-field crystal model, is given in [342]. Numerous applications have been made, including to the deformation of nanocrystalline nickel and to dislocation shearing of precipitates in Ni-base superalloys [375].

## 12.8 SUMMARY

This chapter introduces the basic ideas and implementations of the phase-field method. Details of the steps needed in a phase-field calculation are discussed, with an emphasis on the free-energy functionals and the balance between the phase energy and the interfacial energy. Numerical implementation of the method in one, two and three dimensions is discussed. A few simple example calculations are presented.

## 12.9 APPENDIX

In this appendix we present the derivation of the Allen-Cahn and Cahn-Hilliard equations for systems with non-conserved and conserved order parameters, respectively.

## 12.9.1 Derivation of the Allen-Cahn equation

The derivation starts from Eq. (12.2), a statement that the total free energy depends on a functional of the local free-energy density $f[\phi(\mathbf{r}, t)]$, which is a function of some order parameter $\phi(\mathbf{r}, t)$, and an interfacial term that depends on the gradient of $\phi(\mathbf{r}, t)$.

For simplicity, consider a one-dimensional system. The total free-energy density is then $f^{total}[\phi(x, t), \partial\phi/\partial x]$. The phase-field method assumes that any interfaces are broad and diffuse and not sharp. It that case, the derivative $\partial\phi/\partial x$ is presumed to be small, which means that we can simplify the free-energy expression by expanding it in terms of $\partial\phi/\partial x$. To make the notation a bit easier, we introduce the quantity $g = \partial\phi/\partial x$, which is assumed to be small. Expanding $f^{total}$ around $g = 0$ in the usual way, we have

$$f^{total}[\phi(x, t), g] = f'[\phi(x, t), 0] + \left(\frac{\partial f'}{\partial g}\right)_{g=0} g + \frac{1}{2}\left(\frac{\partial^2 f'}{\partial g^2}\right)_{g=0} g^2 + \cdots . \tag{12.27}$$

$f^{total}[\phi(x, t), g = 0]$ is the equilibrium state – there is no interface and thus $f^{total}[\phi(x, t), g = 0] = f[\phi(x, t)]$. The next term depends on the derivative of $f^{total}$ with respect to $g$, evaluated at $g = 0$. Since interfacial energies are positive, $f^{total}[\phi(x, t), g = 0]$ must be the minimum energy, in which case the derivative with respect to $g$ evaluated at $g = 0$ must be zero. Since we do not actually know how to evaluate $(\partial^2 f^{total}/\partial g^2)_{g=0}$, we *assume* that it is some constant, which we will call $\alpha$. We know that $\alpha > 0$, since interfacial energies are positive.

Ignoring higher-order terms and remembering that $g = \partial\phi/\partial x$, Eq. (12.27) becomes

$$f^{total}[\phi(x, t), g] = f[\phi(x, t)] + \frac{\alpha}{2}\left(\frac{\partial\phi}{\partial x}\right)^2 \tag{12.28}$$

and the total free energy is

$$\mathfrak{F} = \int_\Omega \left[f[\phi(x)] + \frac{\alpha}{2}\left(\frac{\partial\phi}{\partial x}\right)^2\right] dx . \tag{12.29}$$

This is the basic expression for the free energy of a one-dimensional phase-field model. Note that it depends on $f[\phi]$ and $\partial\phi/\partial x$ – the thermodynamic free energy and the slope of the change of $\phi$ with distance.

Over time, the order parameter $\phi(x, t)$ will evolve to minimize the total free energy of the system, $\mathfrak{F}$, reaching the lowest-energy state when the derivative of $\mathfrak{F}$ with respect to $\phi(x, t)$ is zero. Since $\mathfrak{F}$ is a functional of $\phi(x, t)$, not a function, taking its derivative is a bit different than in normal calculus. A functional derivative is denoted by this somewhat different notation,

$$\frac{\delta\mathfrak{F}[\phi(\mathbf{r}, t)]}{\delta\phi(\mathbf{r}, t)}, \tag{12.30}$$

and the condition for the minimum is

$$\frac{\delta\mathfrak{F}[\phi(x)]}{\delta\phi(x)} = 0 = \frac{\delta}{\delta\phi(x)}\int_\Omega\left[f[\phi(x)] + \frac{\alpha}{2}\left(\frac{\partial\phi}{\partial x}\right)^2\right] dx . \tag{12.31}$$

After evaluating the various terms, the functional derivative of $\mathfrak{F}$ is[9]

$$\frac{\delta \mathfrak{F}}{\delta \phi} = \frac{\partial f(\phi)}{\partial \phi} - \alpha \frac{\partial^2 \phi(x)}{\partial x^2} . \qquad (12.32)$$

In classical mechanics, the time-dependent position of an atom changes in response to the force acting on that atom. When the atom is at rest, the net force is zero. That force is related to the potential energy $U$ of the system by the negative of the gradient of $U$, in one dimension as $F = -\partial U / \partial x$. Thus, for the atom at rest, $U$ must be at least at a local minimum.

By analogy, we can consider the negative derivative of the free energy with respect to $\phi$ as an effective "force" acting on $\phi$, leading to a reduction of the free energy. Given this force, we can solve an "equation of motion" for $\phi$ to calculate how it evolves with time. A major difference between the time-dependence of $\phi$ and that of an atom is that there is no inertia associated with $\phi$ and thus no acceleration term. The response of $\phi$ to the "force" acting on it is similar to an object's response to a frictional force, a topic that is discussed in detail in Section 13.1. Such systems have a linear relation between the force and the velocity of the object. We assume a similar behavior for $\phi$, where its "velocity" is $\partial \phi(x, t) / \partial t$. Putting it all together, we have an expression called the Allen-Cahn equation, which in one dimension can be written as

$$\frac{\partial \phi(x, t)}{\partial t} = -L_\phi \frac{\delta \mathfrak{F}}{\delta \phi(x, t)} = -L_\phi \left( \frac{\partial f(\phi)}{\partial \phi} - \alpha \frac{\partial^2 \phi(x)}{\partial x^2} \right) . \qquad (12.33)$$

$L_\phi$ serves as a coupling coefficient that sets the time scale [6]. It is a parameter that must be determined separately. The Allen-Cahn equation is an example of what is often called a Ginzburg-Landau theory, in the sense that the dynamics are written as functional derivatives of a free energy with respect to an order parameter.[10]

In three dimensions

$$\mathfrak{F} = \int_\Omega \left( f[\phi] + \frac{\alpha}{2} |\nabla \phi|^2 \right) d\mathbf{r} , \qquad (12.34)$$

where

$$|\nabla \phi|^2 = \nabla \phi \cdot \nabla \phi = \left( \frac{\partial \phi}{\partial x} \right)^2 + \left( \frac{\partial \phi}{\partial y} \right)^2 + \left( \frac{\partial \phi}{\partial z} \right)^2 , \qquad (12.35)$$

---

[9] Evaluating functional derivatives is described in Appendix C.6. From those results we can see that

$$\frac{\delta}{\delta \phi(x)} \int_\Omega f[\phi(x)] dx = \frac{\partial f[\phi]}{\partial \phi}$$
$$\frac{\delta}{\delta \phi(x)} \int_\Omega \frac{\alpha}{2} \left( \frac{\partial \phi}{\partial x} \right)^2 = -\alpha \frac{\partial^2 \phi(x)}{\partial x^2} , \qquad (12.36)$$

where $\partial f / \partial \phi$ is a regular derivative since $f$ is a simple function of the order parameters. Note that the interfacial energy term (the square of the first derivative of $\phi$ with respect to $x$) becomes a second derivative of $\phi$ with respect to $x$.

[10] Ginzburg and Landau were interested in superconductivity and developed this phenomenological approach [121].

and the Allen-Cahn equation is

$$\frac{\partial \phi(\mathbf{r}, t)}{\partial t} = -L_\phi \left( \frac{\partial f(\phi)}{\partial \phi} - \alpha \nabla^2 \phi(\mathbf{r}, t) \right), \tag{12.37}$$

where

$$\nabla^2 \phi = \frac{\partial^2 \phi}{\partial x^2} + \frac{\partial^2 \phi}{\partial y^2} + \frac{\partial^2 \phi}{\partial z^2}. \tag{12.38}$$

Note again that the energy depends on the squares of the first derivatives of $\phi$ with respect to the coordinates but that the driving force to reach equilibrium goes as the second derivatives of $\phi$ with respect to the coordinates.

For systems with more than one order parameter, there can be a coupling of the dynamics between the order parameters, in which case the more general expression, for $N$ order parameters,

$$\frac{\partial \phi_i(\mathbf{r}, t)}{\partial t} = -\sum_{j=1}^{N} L_{ij} \frac{\delta F}{\delta \phi_j(\mathbf{r}, t)}, \tag{12.39}$$

governs the evolution of the system.

## 12.9.2 Derivation of the Cahn-Hilliard equation

In this section we derive the basic expressions for the phase-field method when there are conserved order parameters, as discussed in Section 12.2.2. We start with the results of Section 12.2.1 and add terms that take into account the conservation requirements. In one dimension we have,

$$\mathfrak{F} = \int_\Omega \left[ f[C(x)] + \frac{\alpha}{2} \left( \frac{\partial C(x)}{\partial x} \right)^2 \right] dx. \tag{12.40}$$

From basic thermodynamics, the derivative of the free energy with respect to the number of particles of a given type of atom is the chemical potential $\mu$ of that atom. Thus, we can define as the chemical potential the functional derivative of $\mathfrak{F}$ with respect to the order parameter $C(x)$,

$$\mu = \frac{\delta \mathfrak{F}}{\delta C(x)} = \frac{\partial f}{\partial C} - \alpha \frac{\partial^2 C(x)}{\partial x^2}, \tag{12.41}$$

following the same procedure for functional derivatives as presented in Section 12.2.1.

We now need to build into the dynamics the local conservation of $C$, which we do by considering how atoms must diffuse from one region to another to change $C$. Since $C$ is a concentration, we assume its time rate of change is given by Fick's second law, which in one dimension is

$$\frac{\partial C}{\partial t} = -\frac{\partial J}{\partial x}, \tag{12.42}$$

where $J$ is the flux. From our interpretation of $\mu$ as a chemical potential, we expect the flux to be given by an expression of the form (in one dimension)

$$\mathbf{J} = -M \frac{\partial \mu}{\partial x},$$ (12.43)

where $M$ is a concentration-dependent mobility. $M$ could be anisotropic (i.e., dependent on the direction of diffusion in the lattice), in which case it would be a second-rank tensor. Putting together Eq. (12.42) and Eq. (12.43) and then Eq. (12.41) yields

$$\frac{\partial C}{\partial t} = \frac{\partial}{\partial x}\left( M \frac{\partial \mu}{\partial x} \right)$$

$$= \frac{\partial}{\partial x}\left( M \frac{\partial}{\partial x}\left[ \frac{\partial f}{\partial C} - \alpha \frac{\partial^2 C(x)}{\partial x^2} \right] \right).$$ (12.44)

If $M$ is independent of concentration (and therefore $x$),

$$\frac{\partial C}{\partial t} = M\left( \frac{\partial^2}{\partial x^2}\frac{\partial f}{\partial C} - \alpha \frac{\partial^4 C(x)}{\partial x^4} \right).$$ (12.45)

Equation (12.44) is the one-dimensional version of the Cahn-Hilliard equation [52, 53]. In three dimensions, it is

$$\frac{\partial C}{\partial t} = \nabla \cdot (M \nabla \mu)$$

$$= \nabla \cdot \left( M \nabla \left( \frac{\partial f}{\partial C} - \alpha \nabla^2 C \right) \right).$$ (12.46)

If $M$ is independent of position, then Eq. (12.46) can be simplified as

$$\frac{\partial C}{\partial t} = M\left( \nabla^2 (\frac{\partial f}{\partial C}) - \alpha \nabla^2 \nabla^2 C \right),$$ (12.47)

where

$$\nabla^2 \nabla^2 C = \left( \frac{\partial^2}{\partial x^2} + \frac{\partial^2}{\partial y^2} + \frac{\partial^2}{\partial z^2} \right)\left( \frac{\partial^2}{\partial x^2} + \frac{\partial^2}{\partial y^2} + \frac{\partial^2}{\partial z^2} \right)C$$

$$= \left( \frac{\partial^4}{\partial x^4} + \frac{\partial^4}{\partial y^4} + \frac{\partial^4}{\partial z^4} + 2\frac{\partial^4}{\partial x^2 \partial y^2} + 2\frac{\partial^4}{\partial x^2 \partial z^2} + 2\frac{\partial^4}{\partial y^2 \partial z^2} \right)C.$$ (12.48)

# 13 Mesoscale dynamics

Molecular dynamics provides a way to model the dynamical motion of atoms and molecules by calculating the force on each atom and solving the equations of motion. In this chapter, we apply the same approach to the motion of entities other than atoms. These entities will typically be collected groups of atoms, such as dislocations or other extended defects. The first step will be to identify the entities of interest, to determine their properties, and then to calculate the forces acting on them. By following similar procedures as in molecular dynamics, the equations of motion can then be solved and the dynamics of the entities determined.

The principal focus of these types of simulations is the mesoscale, that region between atomistics and the continuum, and the goal is often the determination of the microstructure. These extended defect structures are typically many $\mu$m in scale and are thus beyond what can generally be studied atomistically. It is not just the length scale that limits the applicability of atomistic simulations to microstructural evolution. The time scales for microstructural evolution are also much much longer than the nanoseconds of typical molecular dynamics simulations. The defects in question could be grains and the questions of interest could be the growth of those grains and their final morphology. One could also be interested in determining the development of dislocation microstructure and its relation to deformation properties. There, the dislocations might be the entities of interest.

In this chapter, we outline some of the issues of importance in mesoscale dynamics and give some brief examples.

## 13.1 DAMPED DYNAMICS

In molecular dynamics, we solved Newton's equations of the form

$$m\frac{d^2\mathbf{r}_i}{dt^2} = \mathbf{F}_i ,$$ (13.1)

where the force $\mathbf{F}_i$ on particle $i$ arises from inter-particle interactions, external forces, etc. We will solve similar equations when modeling the dynamics at the mesoscale, with one major difference. Eq. (13.1) will often be replaced by equations that reflect the presence of *dissipative* forces[1]. Suppose, for example, that the entities of interest are dislocations. While we can treat

---

[1] Dissipative forces are those forces that act like friction, in which energy is lost from a system to heat.

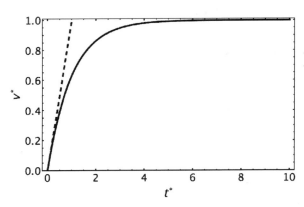

**Figure 13.1** Solid curve: damped velocity (in reduced units) versus reduced time. Dashed curve: velocity under same constant force but with no damping.

dislocations as "particles", they are made up of collective groups of atoms that when moving interact with the lattice to produce phonons, etc. In the absence of any new forces, a moving dislocation will slowly lose energy to the lattice (hence the word dissipative) and will stop.

We start by adding to the force in Eq. (13.1) a dissipative force $\mathbf{F}^{diss}$, the form of which depends on the type of system that one is studying. Here, we commonly discuss systems in which we can use the form for viscous drag,

$$\mathbf{F}_i^{diss}(\mathbf{r}) = -\gamma \mathbf{v}_i \,, \tag{13.2}$$

where $\mathbf{v}_i$ is the velocity of particle $i$ and $\gamma$ is a "friction coefficient". The physical origins and magnitude of $\gamma$ will depend on the physics of the modeled system. Including the viscous drag term, the equation of motion is

$$m \frac{d\mathbf{v}_i}{dt} = \mathbf{F}_i - \gamma \mathbf{v}_i \,. \tag{13.3}$$

The presence of a dissipative force changes the dynamics considerably. As a simple example, consider a system under a constant force in one dimension. The equation of motion in Eq. (13.3) can be solved to find[2]

$$v(t) = \frac{F}{\gamma} \left( 1 - e^{-\gamma t/m} \right), \tag{13.4}$$

which is plotted in Figure 13.1 as the reduced velocity ($v^* = \gamma v(t)/F$) versus the reduced time, $t^* = \gamma t/m$. We also plot the velocity in the limit of no damping (the dashed curve). Note that the damped velocity asymptotes to a constant, steady-state, value, referred to as the *terminal* velocity, with a value $v = F/\gamma$. The time for reaching 95% of the terminal velocity is $t_{term} \approx 3m/\gamma$.

This type of behavior is quite familiar. A marble dropped through a vacuum under the constant gravitational force speeds up until it hits the ground. Dropping the same marble into a viscous

---

[2] Let $t^* = (\gamma/m)t$. We then have that $dv/dt = (\gamma/m)dv/dt^*$, which leads to $dv/dt^* + v = F/\gamma$. Now consider that $d(e^{t^*}v)/dt^* = e^{t^*}(dv/dt^* + v) = e^{t^*}F/\gamma$. Integrating both sides from 0 to $t^*$ and rearranging, we find Eq. (13.4).

fluid leads to, after an initial acceleration, a constant velocity. If the damping coefficient is very large, the time to reach the steady-state velocity can be quite short. If the time to reach steady state is small relative to the important time scales in a problem, then a common approximation is to ignore the acceleration to terminal velocity entirely and to assume

$$v_{overdamped} = \frac{F}{\gamma} = MF, \qquad (13.5)$$

where $M = 1/\gamma$ is the *mobility*. $M$ could be a constant or could depend on direction, as in a lattice. The linear force-velocity relation in Eq. (13.5) is generally referred to as the *overdamped* limit, since it is a good approximation when the damping is high.

## 13.2 LANGEVIN DYNAMICS

The motion of bits of dust suspended in air, or small particles in a liquid, seems random, as was first described in detail by Thomas Brown (in 1827). This type of motion is, appropriately enough, called *Brownian motion*. In a classic paper, Paul Langevin [189] showed that this random motion could be described by the following equation of motion

$$m\frac{d^2\mathbf{r}}{dt^2} = -\gamma\frac{d\mathbf{r}}{dt} + \mathfrak{F}_{rand}(t), \qquad (13.6)$$

where $\mathbf{r}$ is the position of a particle, $\gamma$ is a drag (friction) coefficient, and $m$ is the mass of the particle. $\mathfrak{F}_{rand}(t)$ is a random force describing the impulsive forces from collisions with molecules in the surrounding fluid. It is usually referred to as *noise*. We can write

$$\mathfrak{F}_{rand}(t) = \sqrt{2\gamma k_B T}\,\mathfrak{R}(t), \qquad (13.7)$$

where $\mathfrak{R}(t)$ is a random variable that is assumed to be uncorrelated in time, i.e., the random force at any time is unaffected by that force at any other time. The form of the random force in Eq. (13.7) arises from an important result from nonequilibrium statistical mechanics, the fluctuation-dissipation theorem.[3] Note that the magnitude of the force depends on the temperature and the magnitude of the frictional force. $\mathfrak{R}(t)$ is usually chosen as a random number from a normalized Gaussian distribution.[4]

Eq. (13.6) is an interesting equation. It says that the motion of a Brownian particle is only affected by random collisions from molecules in the surrounding fluid. It ignores any specific interactions by treating the effects of the fluid in a continuum, macroscopic, way, i.e., it ignores the atoms of the fluid and treats their effects collectively. It is an excellent example of a method that bridges length and time scales from small (in this case atoms and molecules) to large (a macroscopic system) by defining an averaged quantity (the noise) that represents the smaller scale physics.

---

[3] For readers wanting a detailed description of nonequilibrium statistical mechanics, a useful, if not easy, book is [376].

[4] In Appendix I.2.1, we show how to pick random numbers from a normal (Gaussian) distribution.

Based on the ideas expressed in Eq. (13.6), we can also account for energy dissipation through a frictional term and an impulsive energy through a random force term. The Langevin equation is thus

$$m\frac{d^2\mathbf{r}}{dt^2} = \mathbf{F} - \gamma\frac{d\mathbf{r}}{dt} + \mathfrak{F}_{rand}(t),$$ (13.8)

where $\mathbf{F}$ is the usual interatomic force term used in the Newtonian formalism of classical molecular dynamics ($\mathbf{F} = -\nabla U$). $\mathfrak{F}_{rand}(t)$ is the same random force used in Eq. (13.6).

The Langevin equation in Eq. (13.8) is a stochastic differential equation that approximates the dynamics of a system. We can use this to ignore certain degrees of freedom. For example, the effects of solvent molecules on the dynamics of a solute can be modeled in terms of a frictional drag on the solute molecules and random impulsive forces that are associated with the thermal motions of the solvent molecules, ignoring the actual details of the molecular interactions. In mesoscale simulations, it enables the implicit inclusion of the effects of temperature in the simulation of dissipative systems.

## 13.3 SIMULATION "ENTITIES" AT THE MESOSCALE

Dynamic simulations at the mesoscale start with the definition of the "entities" of the simulation. These entities are variables that define the physics of interest. Typically, these entities are *collective variables*, in which the actions of many smaller-scale entities are treated as one. For example, in mesoscale simulations of dislocations, the entities are the dislocations and not the underlying atoms that make them up. As the dislocations move, they represent the collective motion and displacements of the atoms, but the simulations do not capture individual atomic-level events. Thus, any aspect of the simulation that depends directly on atomic-level events must be captured through models of that behavior.

The advantage to mesoscale dynamic simulations is that they extend length and time scales over atomistic-based simulations. The quality of a simulation depends on the degree to which the models upon which the simulation is based capture the desired physical phenomena. The types of problems that can be modeled successfully with such techniques are ones in which there is a clear separation into collective variables.

There are a number of issues faced by all mesoscale simulations. After the variables have been identified, they must be represented in some way. For example, in three-dimensional dislocation simulations, the dislocation lines must be described in such a way that is tractable in a numerical simulation. Setting time scales for a dynamic simulation depends on the effective mass of the collective variables. Unfortunately, the mass of such variables is not often well defined.

It is important to remember that the success of any calculation depends on the degree to which the models upon which it is based represent the real system. All models are an approximation to reality and thus have inherent errors associated with them. It is up to the modeler to validate the model by comparing calculated results with experimental data.

In the next few sections we will briefly describe a few examples of dynamic simulations at the mesoscale, focusing on simulations of the collective properties of defects.

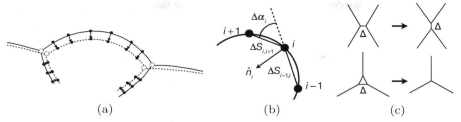

**Figure 13.2** (a) Discrete model of grain boundaries showing motion of nodal points in one grain. (b) Geometry of the discrete points used to determine curvature and the normal. (c) Models of recombination (top) and annihilation (bottom).

## 13.4 DYNAMIC MODELS OF GRAIN GROWTH

Two mesoscale approaches for modeling grain growth are described elsewhere in this text, the Potts model in Chapter 10 and the phase-field method in Chapter 12. Both of these approaches are energy-based methods; the driving force is a lowering of the interfacial energy of the system. Here we describe two approaches that convert the energy into forces and solve for the grain growth explicitly as a dynamic process.

Suppose we want to develop a dynamic method to model grain growth. The first step is to decide what we actually want to model. We assume that within the grains the system is a perfect crystal and that all the interesting physics occurs where two grains with different orientations meet. One approach would be to treat the grain boundaries as the "entities" in the model. For clarity, we will describe a two-dimensional simulation, in which the normal plane that defines a grain boundary becomes a curved line.

Having chosen the simulation entities to be the grain boundaries, we need to find some way to represent them. Here we will consider two approaches, as they illustrate how increasing levels of abstraction can lead to greater simplification, if somewhat less accuracy, in the calculation. In each case, we will start by defining a set of variables that define the location of the grain boundaries, followed by expressions for the forces that act on those variables. Note the difference between Potts model and phase-field method, in which the system was defined in such a way that the boundaries naturally developed and did not have to be explicitly incorporated into a simulation.

### 13.4.1 A boundary model of grain growth

Frost and Thompson [111] introduced a simple 2D model based on the following assumptions:

(a) The boundary is discretized by a simple set of nodal points, as shown in Figure 13.2a.
(b) Only trijunctions are allowed. Note that the intersection of three grain boundaries is a point in two dimensions (a curved line in three, as seen in Figure B.16).
(c) Each segment of a boundary moves according to $\Delta x = M\kappa \, \Delta t$ (curvature-driven growth), where the mobility $M$ and the time step $\Delta t$ are input parameters and the curvature $\kappa$

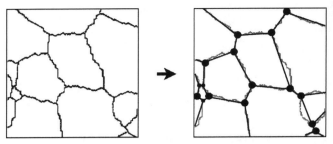

**Figure 13.3** (a) Results from a Potts model simulation of grain growth, as discussed in Chapter 10. (b) Straight lines overlaid on Potts model grain boundaries.

is calculated at each segment point by a simple geometric formula based on the nodal positions.[5]

(d) The positions of the trijunctions are constrained such that a 120° angle between the boundaries is maintained, which corresponds to the case in which all three grain boundary energies are the same.

(e) There is a set of recombination and annihilation rules, as shown in Figure 13.2c, in which junctions that are moving together recombine and move off in a new direction when they are closer than a distance $\Delta$ apart (top) and small grains are annihilated when they are smaller than $\Delta$ (bottom).

Numerical errors arise from the finite number of points on each boundary, the definition of curvature, the finite time step, and the algorithm for moving the trijunctions. Despite these errors, results are quite good and the method is computationally fast. For example, von Neumann's rule, $dA/dt = -M\pi(6 - n)/3$, is obeyed to better than 3% and the average grain area increases as $A \sim t$ at long times, as expected. The method reproduces normal grain growth kinetics and it matches experimental grain size and shape distributions. However, the approach depends on an input model for the dynamics and for junction angles. It is also difficult to implement in 3D and contains no finite temperature effects.

### 13.4.2 A vertex model of grain growth

Another approach to modeling grain growth illustrates a different choice of the fundamental "entity" of the simulation. Developed by Kawasaki [166], the approach was motivated by two observations: (1) for normal grain growth, there are only trijunctions with 120° angles (assuming isotropic grain boundary energies) and (2) if we examine microstructures, we see that, at least at long times, the grain boundaries connecting the trijunctions are approximately linear, as shown in Figure 13.3. If they are indeed taken as straight lines, then we could imagine developing

---

[5] The curvature is given by $\kappa = d\alpha/ds$, the derivative of an angle $\alpha$ (relative to some reference state) with respect to the arc length $s$. For discrete segments as shown in Figure 13.2b, we approximate the derivative at point $i$ by $\Delta\alpha_i/\Delta s_i$, where $\Delta s_i = (\Delta S_{i-1,i} + \Delta S_{i,i+1})/2$, $\Delta\alpha = \cos^{-1}((\Delta x_{i-1,i}\Delta x_{i,i+1} + \Delta y_{i-1,i}\Delta y_{i,i+1})/(\Delta S_{i-1,i}\Delta S_{i,i+1}))$ and $\Delta S_{i,i+1} = (\Delta x_{i,i+1}^2 + \Delta y_{i,i+1}^2)^{1/2}$. The normal vector is $\hat{n} = (m/\sqrt{m^2 + 1}, -1/\sqrt{m^2 + 1})$ where $m$ is the slope of the line connecting the point $i - 1$ to point $i + 1$, $m = (y_{i+1} - y_{i-1})/(x_{i+1} - x_{i-1})$.

equations of motion for the junctions, or vertices, and connecting them with straight lines, i.e., the entities are the vertices themselves. This choice would be a great simplification of the modeling of grain growth, as we have far fewer vertices than grain boundary sites.

Their general approach was to derive damped equations of motion for the vertices, to simulate their motion, and then to create a microstructure, by connecting the vertices with straight lines. They then characterized the grain growth in the usual ways (measuring area, etc. with time).

The derivation of the equations of motion of the vertices, even for isotropic growth, is complicated, so we will skip the details and just outline the general approach. The main point is that this method is a form of coarse-graining, in which most of the details are averaged out of the problem, leading to a simple set of equations.

The key assumptions made by Kawasaki are:

(a) The total energy of the boundaries is $H = \sigma A$, where $A$ is the total area of the boundaries in the system and $\sigma$ is the grain-boundary free energy per unit area. In two dimensions, $A$ is just the total length of the boundaries.
(b) Each point of the boundary moves with velocity $v(a) = M\kappa_i$, where $M$ is the mobility and $\kappa$ is the curvature at point $i$.
(c) The boundary motion is dissipative. Kawasaki [166] derived an expression for the average dissipation $R$ such that the steady-state velocity is given by $v_i = M\kappa_i$.
(d) To determine the energy, Kawasaki assumed that the boundaries are linear, i.e., $H = \sigma \sum_{ij} r_{ij}$, where $r_{ij}$ is the distance between vertices $i$ and $j$ and the sum goes over all vertices and the three boundaries connected to them.
(e) The velocities along a boundary are assumed to be the weighted average of the vertex velocities.

With these assumptions, Kawasaki derived a number of models, the simplest of which (his Model II) results in an equation of motion

$$D_i \mathbf{v}_i = -\sigma \sum_{j}^{(i)} \frac{\mathbf{r}_{ij}}{r_{ij}}, \tag{13.9}$$

with

$$D_i = \frac{\sigma}{6M} \sum_{j}^{(i)} r_{ij}. \tag{13.10}$$

The notation $\sum_{j}^{(i)}$ indicates a sum over the three boundaries connected to $i$. The equations of motion in Eq. (13.9)–Eq. (13.10) are really rather simple. The dynamics of the vertices are described only in terms of the vectors connecting the vertices to each other and some material parameters.

The method works very well, even with the simplifications included in Eq. (13.9)–Eq. (13.10). In comparisons of the Kawasaki complete model, his model II (described here), and the Potts model for the normalized grain distribution all compare favorably to each other (and to experiment). A number of versions of the vertex model in three dimensions have been introduced in which the computational speed, as well as reasonable accuracy, have been demonstrated, for example in [233, 194, 304].

The real advantage to this approach is that it has converted a very complicated problem, grain growth, to one in which we track a relatively few number of variables with a relatively simple equation of motion. Since we have fewer grains at long time, the number of computations required for the simulation decreases with increasing time. Perhaps the biggest advantage to this approach is that it is dynamic and can be linked to other methods (such as the finite-element method) to examine very large numbers of boundaries, examining the coupling of grain-boundary motion to such properties as external stress, diffusion, etc. The weaknesses of this approach are that it is based on many assumptions about growth. It is certainly not as flexible as the Potts model and, as described, has no finite temperature effects. There is no restriction that keeps the vertices from deviating from $120°$ angles.

## 13.5 DISCRETE DISLOCATION DYNAMICS SIMULATIONS

The plastic deformation of crystals involves the creation and motion of dislocations.[6] The dislocations organize into structures, which are typically microns to tens of microns in scale, far beyond what atomistic simulations can describe. While it is often useful to assume a phenomenological model for plasticity, in this section we describe how to model the dislocations directly by treating them as the entities in the simulation. This approach is generally referred to as dislocation dynamics and has become a popular method for studying plasticity. We first describe simplified simulations in two dimensions and then briefly discuss extensions to three dimensions. Many more details are given in [49].

### 13.5.1 Two-dimensional simulations

The computational simulation of dislocations started many years ago, with the pioneering simulations of Foreman and Makin [107, 108]. They employed a very simple line-tension model and calculated the effects of random obstacles on the motion of a single dislocation on its slip plane. While simplistic, their model provides a reasonable explanation of the role of obstacles on dislocation movement, as well as showing the power of simulations to shed light on dislocation behavior.

The next significant advance in dislocation simulations was the use of a two-dimensional model of parallel dislocations [133, 193]. Consider a set of $N$ parallel edge dislocations, all with Burgers vectors in the $\pm\hat{x}$ direction with their line directions in the $\hat{z}$ direction as shown in Figure 13.4a. If the dislocations do not change line direction, then their positions are completely determined by their coordinates in the $xy$ plane, which we indicate in Figure 13.4b, using $\perp$ to represent dislocations with $\mathbf{b} = +b\hat{x}$ and $\top$ to represent dislocations with $\mathbf{b} = -b\hat{x}$. A simulation of this system, while representative of three-dimensional dislocations, is two-dimensional, with only the coordinates in the plane shown in Figure 13.4b being relevant.

---

[6] A brief review of dislocations and their properties and interactions is given in Appendix B.5.

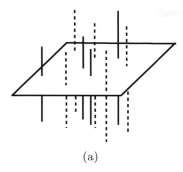
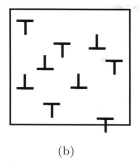

(a)                                    (b)

**Figure 13.4** Model based on two-dimensional set of edge dislocations. (a) A set of infinite, parallel edge dislocations can be viewed as a two-dimensional system of points. The dislocations in (a) denoted as dashed lines have a Burgers vector of $\mathbf{b} = -b\hat{x}$ while the solid lines are dislocations with $\mathbf{b} = b\hat{x}$. (b) The dislocations in (a) correspond to those in (b) with dislocations with $\mathbf{b} = -b\hat{x}$ denoted as $\top$ and dislocations with $\mathbf{b} = b\hat{x}$ as $\perp$.

The dynamics of the dislocations in Figure 13.4b are governed by dissipative equations such as Eq. (13.3), with the dissipation arising from a number of sources, including the generation of phonons as the dislocations move through the lattice. The force $\mathbf{F}$ on a dislocation is given by the Peach-Koehler equation of Eq. (B.29). For the parallel dislocations in Figure 13.4b, the Burgers vectors are $\hat{b} = \pm\hat{x}$ and the line direction of all the dislocations is $\hat{\xi} = \hat{z}$. If we assume that only glide is important, then the movement is restricted to the $\pm\hat{x}$ direction, so that only the $x$-component of the force is needed. The Peach-Koehler force on dislocation $i$ along the $x$ direction is

$$F_i = b_i \sigma_{xy}, \tag{13.11}$$

where $b_i = \pm b$ and $b$ is the magnitude of the Burgers vector. $\sigma_{xy}$ is the shear component of the stress.

The net stress at a dislocation arises from an externally applied stress $\sigma_{xy}^{app}$ plus the stress due to other dislocations (and any other defects) in the system. If there are only edge dislocations as in Figure 13.4b, then the net stress at dislocation $i$ is

$$\sigma_i = \sigma_{xy}^{app} + \sum_{j \neq i} \sigma_{xy}(ij), \tag{13.12}$$

where $\sigma_{xy}(ij)$ is the $xy$ component of the stress from dislocation $j$ evaluated at dislocation $i$ and we sum over all dislocations in the system.

The $\sigma_{xy}$ component of the stress from an edge dislocation with Burgers vector along $\hat{x}$ and line direction along $\hat{z}$ is given in Eq. (B.32). The net force on dislocation $i$ from Eq. (13.11) is

$$\mathbf{F}_i = b_i \sigma_{xy}^{app} + \sum_{j \neq i} \frac{\mu b_i b_j}{2\pi(1-\nu)} \frac{x_{ij}(x_{ij}^2 - y_{ij}^2)}{(x_{ij}^2 + y_{ij}^2)^2}, \tag{13.13}$$

with $b_i(j) = \pm b$.

As in [133], a number of dislocations could be distributed on random slip planes in a simulation cell with periodic boundaries. Examination of Eq. (13.13) shows that the force decreases as the inverse of the inter-dislocation separations, i.e., it is very long ranged. Thus, much as for ionic materials as discussed in Section 3.6, the interactions cannot be truncated with a cutoff. Approaches to deal with the long-range interactions must be used.

One method for dealing with the long-ranged nature of the interactions for parallel dislocations in periodic cells is to recognize that a dislocation and its periodic images in the $y$ direction form a low-angle grain boundary with periodic repeat distance that is the size of the simulation cell [133]. Suppose the cell size is $D$, then the vertical images of a dislocation located at a position $\mathbf{r}$ in the simulation cell are located at $\mathbf{r} + nD\hat{y}$, where $n$ is an integer ranging from $-\infty < n < \infty$. The vertical lines are periodic along $\hat{x}$ and are located at $\mathbf{r} + mD\hat{x}$, where $-\infty < m < \infty$. The dislocation text by Hirth and Lothe [147] gives expressions for the stress field of a periodic line of edge dislocations, which falls off exponentially with distance away from the line. Thus, the net stress arising from a dislocation and all its images is just a sum of the stresses arising from the lines of dislocations along $\hat{x}$, which converges quickly.[7] Note that this method only works for systems with parallel dislocations and periodic boundaries.

Another approach for dealing with the long-range interactions is through the use of the fast-multipole method described in Appendix 3.8.2. This approach was developed for parallel dislocations in two dimensions in [338] and could be extended to arbitrary sets of dislocations in three dimensions.

Given the force on the dislocations, their positions evolve by solving the equations of motion. Assuming overdamped dynamics as in Eq. (13.5), we can ignore the inertial terms and have that the velocity is proportional to the force. Thus, the equation of motion becomes

$$\mathbf{r}_i(t + \Delta t) = \mathbf{r}_i(t) + \mathbf{v}_i \Delta t = \mathbf{r}_i(t) + M\mathbf{F}_i \Delta t. \tag{13.14}$$

There is a computational issue with this simple equation of motion. When dislocations are close to each other, the stresses between them can become very large. Assuming a fixed time step, $\Delta t$, could lead to unreasonably large changes in position, $\mathbf{r}_i(t + \Delta t) - \mathbf{r}_i(t)$. Very small time steps for the entire simulation would eliminate this problem, but would be inefficient. A more efficient approach would be to use a dynamic time step, in which, for example, a maximum distance that a dislocation can travel in a time step, $\Delta r_{max}$, is set as an input parameter. At each time step, after the forces on each dislocation, $\mathbf{F}_i$, are calculated, the maximum of the

---

[7] For dislocations periodically located along the $y$ axis with a separation of $D$ between the dislocations, and with a Burgers vector of $\mathbf{b} = \pm b\hat{x}$ and line direction along $\hat{z}$, the stress field is

$$\sigma_{12}(xy) = \frac{\mu b_j}{2D_y(1-\nu)} \sum_{n=-m}^{m} \frac{t_n\left(\cosh t_n \cos u - 1\right)}{\left(\cosh t_n - \cos u\right)^2}, \tag{13.15}$$

where

$$u = 2\pi y_{ij}/D$$
$$t_n = 2\pi(x_{ij} - nD)/D. \tag{13.16}$$

In this case, $D$ is the simulation cell size. The parameter $m$ is the number of lines used in the sum. At large $|x|$, the stress from a line of dislocations falls off in the direction perpendicular to the line approximately as $e^{-2\pi x/D}$. Thus, by $m = 3$, the relative contribution is only about $10^{-8}$ so that only a few terms are required in the sum in Eq. (13.15).

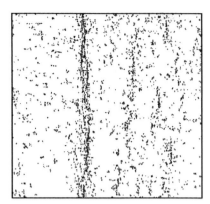

**Figure 13.5** Dislocation microstructure from a simulation of parallel edge dislocations with Burgers vectors along the $\pm \hat{x}$ directions. The system was loaded under an increasing stress, with a simple nucleation model to ensure that the density followed Taylor's law, as described in the text.

magnitudes of those forces is determined, i.e., $F_{max} = \max\{|\mathbf{F}_i|\}$. The magnitude of that time step is then set to $\Delta t = \Delta r_{max}/F_{max}$.

A typical simulation proceeds as follows:

(a) Dislocations are placed at random positions in the cell.

(b) The forces are calculated and the equation of motion is solved, allowing the positions to evolve with time.

(c) If dislocations of opposite sign on the same slip plane are within a small distance of each other, then they could annihilate, as shown in Figure B.13a. If they are close, but on separate slip planes, dislocation dipoles would form.[8]

(d) External stress could be applied, various obstacles representing other defects could be included, etc.

(e) If there is no external stress, the simulation would typically be run until the dislocation motion stops, i.e., if the force on each dislocation is less than a Peierls stress (an input parameter).[9]

(f) Standard analysis tools could then be employed to analyze the structures, for example distribution functions, cluster analysis, etc., as described in [340].

In Figure 13.5 we show results from a calculation of a system under an increasing stress. As discussed in Appendix B.5.1, the density of dislocations, $\rho$, increases with the stress, often following Taylor's relation in which $\rho \propto \tau^2$, where $\tau$ is the stress. In Figure 13.5, a simple nucleation model was used in which new dislocations were created to ensure that the dislocation density followed Taylor's relation. Note the formation of a distinct dislocation microstructure, including a band of increased density, and therefore increased slip. Many useful calculations have been done with this simple two-dimensional model. For example, a good review of two-dimensional dislocation dynamics simulations of fracture is given in [318]. Many other applications of 2D dislocation models such as the one described here have been made. A few of

---

[8] A dislocation dipole is one in which two dislocations, on different slip planes and with opposite signs, come close together, forming a stable structure owing to the high interaction stresses.

[9] The Peierls stress is the minimum force needed to move a dislocation along its slip plane.

these include: the development of the plastic zone in Mode III fracture [366, 367], viscoplastic deformation [230], effects of grain size on strengthening [34], and the coupling of discrete dislocation modeling and continuum plasticity [336].

While useful, this type of two-dimensional dislocation simulation is not very realistic – dislocations are in general curved and not parallel. In three dimensions they can form quite complex structures, with entanglements and various connections, none of which being possible in a two-dimensional simulation. In the next section, we will show how to rectify those limitations by introducing the basics of three-dimensional simulations.

## 13.5.2 Three-dimensional simulations

Three-dimensional dislocation simulations present a number of challenges that must be overcome, including

(a) representation of three-dimensional dislocation loops
(b) efficient calculation of the forces
(c) models or rules to handle short-range interactions
(d) imposition of boundary conditions
(e) ....

In this text, we can give only a brief introduction. More complete details are given in the useful book on dislocation dynamics simulations by Bulutov and Cai listed in the Suggested reading section. We now will discuss items (a)–(d) in turn.

### Representation of dislocations in three dimensions

The first requirement of a three-dimensional discrete dislocation simulation is to find a representation of the curved dislocations in space, which is usually chosen as a series of nodes along the dislocation line.[10] To create the dislocation, the nodes must be connected. One approach is to connect them as a series of straight dislocation segments of mixed edge and screw character, as shown in Figure 13.6a [368], but using segments that are restricted to being pure edge or screw is common in some methods [87]. One advantage to this approach is that the stress tensor arising from a straight segment is known analytically [147]. Thus, the total stress on a node is just a sum of contributions from each segment of the dislocations in the system. From the stress, the Burgers vector of the dislocation, and the line direction of the segment, the force can be evaluated using the Peach-Koehler force in Eq. (B.29).

Another approach to representing dislocations was developed by Ghoniem and collaborators [119, 347] and uses a cubic spline representation of the dislocation lines. Evaluation of the stress from a dislocation in this formulation requires a line integral over the dislocation line, as shown in Eq. (B.33). This integration is done numerically using a Gauss-Legendre quadrature [264]. The examples given in this text are based on the approach by Ghoniem.

---

[10] Please see the discussion of dislocation types in Appendix B.5.2.

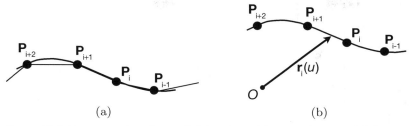

**Figure 13.6** Representation of curved dislocations. (a) Representation of dislocation line with discrete straight segments. Note the error in the representation in regions of high curvature. (b) Parametric representation with cubic splines. The tangents are determined to ensure continuous curvature. The curve is mapped out from node $P_i$ to $P_{i+1}$.

Since the interactions are long ranged, they should be calculated using methods such as described in Appendix 3.8. The fast-multipole method described in that Appendix is particularly suited to discrete dislocation simulations and has been developed for two-dimensional simulations [338]. Multipole expansions of the three-dimensional stresses are available [346], as well as a description of a methodology for evaluating these stresses in a simulation on parallel computers [347].

## Models and rules

While the equations above describe the interactions and forces acting on dislocations, they do not yet incorporate any information about processes that are inherently atomistic in nature, for example climb (Figure B.10a), cross slip (Figure B.10b), or such phenomena as junction formation and annihilation, as described in Figure B.13a. Including these phenomena thus requires models that reflect the underlying atomistic processes.

Climb is relatively easy to incorporate by a modification to the basic equation of motion, which, in the over-damped limit of Eq. (B.37), is

$$\mathbf{v} = \frac{1}{\gamma}\mathbf{F} = M\mathbf{F},\tag{13.17}$$

where $M$ is a mobility. As written, the mobility is independent of direction, and would, in a calculation without climb, be restricted to movement along the glide planes. To include climb, an anisotropic mobility would be used, with a temperature-dependent mobility along the climb direction, reflecting that climb is a diffusive, and thus activated, process.

One of the more important phenomena in dislocation motion is cross slip, in which a dislocation moves from its slip plane to another plane, changing the direction of its motion in the lattice, as shown in Figure B.10b. Cross slip is an activated, and thus temperature-dependent, process that depends on the resolved stress along the cross-slip direction. If the planes have the correct relative orientations for cross slip to occur, then one approach to modeling a cross-slip event is to employ a simple Arrhenius-type expression that reflects the effects of both temperature and applied stress. There have been many models developed for cross slip, with a variety

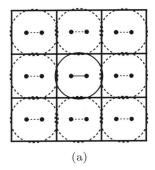 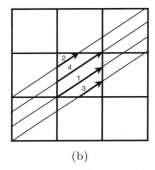

(a)                    (b)

**Figure 13.7** Impact of periodic boundary conditions on dislocation simulations. (a) A Frank-Read source in the central simulation cell, marked as a solid line, with its periodic replicas shown as dashed lines. (b) Glide planes shown perpendicular to the dislocation motion. When a dislocation bowing out along glide plane 1 crosses the boundary, its image returns along glide plane 2, with its image returning along glide plane 3, and so on.

of functional forms [266]. Perhaps the most used in dislocation dynamics simulations is that developed by Kubin and coworkers [184], in which the probability of cross slip is proportional to an exponential that is a function of the resolved stress along the new plane.

Junctions occur when dislocations moving on their slip planes encounter dislocations on nearby slip planes, forming stable structures owing to high interaction stresses [157]. They are typically modeled in dislocation simulations by rules based on the inter-dislocation distances. Annihilation, in which two oppositely oriented dislocations react, leaving behind a perfect lattice (Figure B.13a), is also generally modeled with rules. More details are given in [49].

## Boundary conditions

Finding the correct boundary conditions for the simulation of bulk samples is problematic. The easiest to implement are periodic boundaries, though there are a number of issues that arise from the periodicity as dislocations cross a boundary. One issue is shown in Figure 13.7a, which shows a Frank-Read source, from Figure B.14, in the central simulation cell, marked as a solid line. Its replicas in the neighboring cells are shown as dashed lines. As the dislocation bows out in the central cell, its replicas bow out in the neighboring cells. Where the dislocation and its replicas meet at a cell edge, their line directions are opposite. Since they are replicas of each other, they have the same Burgers vector. Thus, the dislocations annihilate at the cell edges, leaving behind a series of circular dislocations at the cell corners, an entirely unphysical structure that arises from the boundary conditions.

Another issue with the use of periodic boundaries is shown in Figure 13.7b, in which the glide planes are seen edge on so they appear as lines. Dislocations are moving on those planes. Imagine a case in which a dislocation is moving along slip plane 1 with no obstacles to stop its motion. In a real system, it would continue along that slip plane until it runs into another dislocation or a boundary or some other defect. With periodic boundaries, the situation is rather different. When a dislocation bowing out along glide plane 1 crosses the boundary, its image returns into the cell along glide plane 2, which does not describe the actual motion of the dislocation. When that dislocation on plane 2 crosses the boundary, its image enters the cell along glide plane 3, which is again unphysical. The effect would continue. Consider the outcome of this simulation. In the real crystal, after the dislocation leaves the volume associated with the simulation cell and if no other dislocations enter that volume, the central cell would be empty of dislocations. With periodic boundary conditions, however, there will always be a dislocation in the cell. If the dislocation is extended, such that it leaves behind a segment in the cell, then when the dislocation crosses the boundary into the central cell, interactions with the remaining segment will affect its motion, leading to a dislocation density that is greater than in the real crystal. The physical reasonableness of this approach can certainly be questioned. A detailed discussion of the effects of periodic boundary conditions on dislocation dynamics simulations as well as some simple steps to alleviate some of the problems are given in [211].

For small samples, in which the periodic boundaries are replaced by free surfaces, the image forces must be taken into account [147]. There are a number of ways to include these forces, including a simple analytical approximation introduced by Lothe and coworkers [206]. A more sophisticated approach is to use the boundary element method (BEM), in which an extra field is introduced along the surface to ensure a stressless surface [96]. The BEM was used, for example, in successful calculations on the plasticity of nano- and microscale pillars [372, 373, 374].

### 13.5.3 Limitations and assessment

There are many approximations inherent in discrete dislocation dynamics simulations. Some of these issues are numerical in nature, while some arise from the basic approximations of the dislocation model upon which the simulations are based.

Numerical errors are introduced in many ways. For example, the choice of discretization in Figure 13.6, as well as the density of nodes along the dislocation, affects the accuracy of the force determinations. As dislocations grow in length, adding more nodes becomes necessary, with a balance between numerical accuracy and computational time.

The equations of motion are generally solved as in Eq. (13.14). As discussed with regard to that equation, often a dynamic time step is employed to avoid unreasonably large dislocation motion. Another approach would be to solve the full dynamic equations of motion in Eq. (13.3). For dislocations, this is complicated by the effective dislocation mass being dependent on the dislocation velocity [148]. The importance of including the full dynamical equations of motion for systems at high-strain rates ($> 10^3$) is discussed in [348].

As noted above, atomistic properties are included with models, not all of which are well developed. In addition to those mentioned above, a major drawback to dislocation simulations is that partial dislocations, which occur readily in many crystal systems, as described in Appendix B.5.11, are rarely incorporated within a dislocation simulation. While methods that can include them have been suggested [220], their complexity limits their utility.

In addition to the inherent errors associated with the numerics of dislocation dynamics simulations as well as the uncertainties in many of the underlying models, there are a number of other limitations to these methods. The time and length scales, for example, while much larger than those in atomistic simulations, are still small relative to many problems of interest. The limitations of time scales are related to the need for small time steps discussed above.

The length scale problem is easy to understand. It is well known that the dislocation density increases with stress. For $fcc$ materials, the dislocation density often follows the Taylor law, increasing with stress as $\rho \propto \tau^2$, the flow stress [306]. This increase in dislocation density has a number of effects on the ability to do calculations. First, an increase in dislocation density means an increase in dislocation length, which in turn leads to an increase in the number of dislocation segments needed to achieve an accurate representation of the dislocation. The increased number of segments in turn causes an increase in computational time. For example, if the density goes as $L$, the total dislocation length, then the calculational time goes approximately as $L^2$ for a direct summation of the interaction terms. Thus, the calculational time for a direct summation goes as $\tau^4$, a prohibitive expense as the stress increases.[11] The high density also leads to dislocations being very close to each other, with high stresses and a consequent decrease in the time step. Efficient remeshing and time-step optimization are thus extremely important. Even with large-scale parallel computing, the difficulties in modeling high dislocation densities limit the ability of these methods to do very large systems [347].

While having limitations, the importance of discrete dislocation simulations in helping developing a deeper understanding and predictive power for plasticity has been growing. While many articles along these lines have been written, a few summarize the field quite nicely, including the connection between plasticity and the simulations in [88], describing transitions in plasticity [86], and a recent review that shows the power of connecting experiment and simulation to understand the fundamental scaling laws for dislocation microstructures in the deformation of $fcc$ metals [281].

Dislocation simulations also offer a unique opportunity to examine the development of local ordering in dislocation microstructures. Their role in understanding plasticity, however, is dependent on the development of the theory of the behavior of ensembles of dislocations. Unlike the case for atomistics, there is as yet no well-accepted statistical mechanics of dislocations that can be used to connect the simulations to macroscopic behavior.

## 13.5.4 Applications

Numerous simulations based on three-dimensional dislocation dynamics have been performed, examining a wide range of phenomena. Some representative examples include:

- Plasticity at high-strain rates: 3D dislocation dynamics simulations based on the discretization in Figure 13.6b were employed to study the plastic anisotropy of a single crystal of copper under a wide range of high strain rates, $10^4$–$10^6$ s$^{-1}$ [350]. The simulations included the model for cross slip from [184] as well as the full dynamical equation of motion, both of

---

[11] Use of an $O(N)$ method, such as the fast multipole method, reduces that to $\tau^2$.

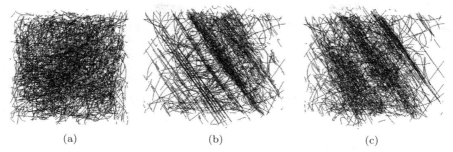

(a)　　　　　　　　　(b)　　　　　　　　　(c)

**Figure 13.8** Three-dimensional simulation of slip band formation. Shown are three views of the same microstructure, rotated relative to each other. Note the populated slip planes in (b), connected by dislocations that have cross slipped between the two planes, as described in [350].

which having been shown to have a large impact on the way dislocations interact and move at high strain rates [348, 349].

- Slip band formation: 3D simulations were used to examine development of slip band structures, which were also shown to be highly dependent on the loading orientation [349]. In Figure 13.8 we show a structure formed with a loading direction of [111] with a $10^5$ s$^{-1}$ strain rate.

- Strain hardening: Kubin and coworkers used 3D dislocation simulations to map out aspects of Stage II strain hardening in $fcc$ crystals,[12] showing how the simulations could yield parameters in constitutive laws that yielded stress-strain curves in good agreement with experiment [89, 185].

- Small-scale plasticity: 3D dislocation dynamics simulations have done much to elucidate the mechanisms of plasticity in small samples, including micropillars [372, 373, 374] and thin films [370, 371].

- Fatigue: Fivel and coworkers have applied 3D dislocation simulations to the early stages of fatigue,[13] with applications to fatigue in: steel [83, 84], $fcc$ materials [85] and precipitation hardened materials [288].

## 13.6 SUMMARY

In this chapter we summarized some examples of methods which involve solving the equations of motion for coarse-grained variables at the mesoscale. We discussed damped dynamics and introduced the Langevin equation. Two methods for grain growth were introduced, one in which the grain boundaries served as the entities of the simulation and one in which the vertices were

---

[12] Stage II strain hardening corresponds to a second linear part of the stress-strain curve in single-crystal $fcc$ materials [171].

[13] Fatigue is the cumulative structural damage that occurs when a material is subjected to cyclic loading.

the entities. Finally, we introduced dislocation dynamics simulations, in which dislocations are the simulation entities. Both two- and three-dimensional dislocation simulations were discussed.

## Suggested reading

There are many books and articles available that discuss this topic, with a wide range of readability. I like:

- The basics of the Langevin equation are described in Zwanzig's excellent text on *Nonequilibrium Statistical Mechanics* [376], while details on the application of the method are given in Coffey *et al.*, *The Langevin Equation* [71].
- Reviews and papers of grain growth simulations abound. One interesting source for more information on the basics of microstructural evolution is Phillips's excellent text, *Crystals, Defects, and Microstructures* [260].
- The book by Bulatov and Cai, *Computer Simulations of Dislocations*, is an excellent source for more details of dislocation simulations [49].

# Part Four

# Some final words

# 14 Materials selection and design

Engineered designs are generally based on the use of a constrained, and fixed, set of materials. Because materials development is slow, the role of the materials engineer is generally one of materials selection, i.e., choosing a material from a restricted list to fit a specific need in a product design process. Traditionally, the optimal material was a balance between best meeting the product performance goals and minimizing the cost of the material. In recent years, an increased focus has been on the life cycle of the material, with an eye towards recycling and reuse.

The selection of the best material for an application begins with an understanding of the properties needed for the design as well as a way to display and access the properties of candidate materials. If the design is based on a single criterion for the material, such as density, for example, then the choice of a material is usually pretty simple. If multiple criteria must be met, then a way to compare multiple properties of a set of materials with each other is needed. A common way to do that is through an "Ashby plot", a scatter plot that displays one or more properties of many materials or classes of materials [13, 14]. For example, suppose one needs a material that is both stiff and light. Stiffness is measured in Young's modulus, while knowing the density of a material will enable one to pick the lightest material for a specific volume. Thus, one could plot the Young's modulus along one axis and the density along the other axis, with each material represented by a point on that plot. Materials can be grouped together depending on the type of material, i.e., whether a metal, ceramic, etc. Given the needs of the design, a set of candidate materials can often be easily identified from such plots.

The ultimate goal in materials science and engineering is, however, the discovery and development of new materials. Historically, materials development has been Edisonian in nature, involving numerous trials and incremental gains. This process is by its nature slow and inefficient and creates limitations for the timely development and application of new materials. Reducing the time, and cost, of materials development is thus a major goal.

## 14.1 INTEGRATED COMPUTATIONAL MATERIALS ENGINEERING

The integration of the types of modeling and simulation discussed in this text with experiment offers a path to accelerate materials development. Referred to as *Integrated Computational Materials Engineering* (ICME), the goal is to accelerate the development, certification, and insertion of new materials into both existing and new technologies. A recent report [72] described

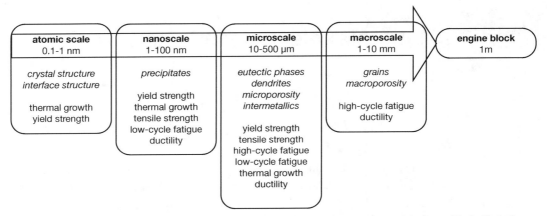

**Figure 14.1** Scales in an ICME process. At each scale, the dominant structures are indicated in italics. Also listed are the materials properties at each scale of importance to the product under development. Adapted from Figure 3 in [7].

the ICME process in some detail and provided a few examples that showed strong economic benefits for companies that implemented at least partial versions of an ICME strategy [7, 17]. With those examples in mind, the report concluded that widespread utilization of ICME could enable sweeping advances in our ability not only to improve materials design and development, but also to include materials design directly in the product design process, which we will discuss below as *concurrent design*.

The idea behind ICME can perhaps best be understood by considering an example, which we will take from a paper describing a program to develop a casting process for manufacturing aluminum drive-train components [7]. Specifically, they were developing a new process for casting of engine blocks with the goal of reducing manufacturing defects and optimizing properties. They started by considering the dominant length scales of the materials structures and properties in developing their product with the right overall properties, which are shown schematically in Figure 14.1. In their program, which was performed by a number of different groups, they combined casting simulations and experiment with research at each of the scales in Figure 14.1 to refine the structures to yield the desired properties. What makes this an ICME process is that they coupled experiment and modeling at each scale, taking advantage of the strengths of both approaches to accelerate the process development.

Each of the scales in Figure 14.1 required different modeling approaches. For example, electronic structure calculations were employed to determine such thermodynamic properties as the volume and energy differences between precipitate phases (at the atomic scale). At the nanoscale, they linked the calculated thermodynamic quantities to thermodynamic phase equilibria calculations coupled with a microstructural evolution model (phase field) for precipitate growth. To predict the microstructure, they coupled models of solid-state diffusion with approximate analytical models of dendrite formation, though they stated that they were examining the use of phase-field calculations to replace the phenomenological analytical models. At each scale, different levels of modeling were used as both an adjunct to the experiment, providing additional information, and as an integrator of the data into empirical models that could be coupled to other simulations.

The project described in [7] was successful, with the cost of the development of the final manufacturing process, including the modeling, being significantly less than the more traditional approach of design, build, and test. Estimates of the return on investment for the ICME process versus their traditional approach was 1:3 – the traditional approach was three times as expensive [72]. Similar returns on investment were seen by many companies that employed an ICME process [72].

There are, however, many challenges that remain before this approach can come into more common use. In almost all regimes, for example, models are limited in their ability to predict materials behavior without at least some experimental validation. One of the greatest challenges, inherent in the basic idea of ICME, is how to best capture the wide range of disparate, yet connected, information that results from modeling and experiment. In this chapter we introduce the basic methodologies by which we can integrate that information – *materials informatics* – as well as some more detailed comments about the challenges inherent in ICME and materials design.

## 14.2 CONCURRENT MATERIALS DESIGN

Engineering has been described as "design under constraint" [361], with an emphasis on the word "design". Engineers design complex technological objects, from integrated circuits to large bridges. The development paths of these objects have at least two things in common: a design process and the requirement of the use of materials. What has been lacking in the design process, however, is that materials have only been included in a static way, creating a design based on a fixed, and often limited, set of materials. An untapped potential would be a design process in which the materials are designed concurrently with the product, an approach that goes beyond the materials selection described earlier, with the goal of *designing* the material of each part with properties optimized as part of the overall product design cycle. The net design could then include materials whose properties are *tailored* to meet specific design goals and whose properties vary throughout that component. Referred to as *concurrent engineering* [242, 222], it could provide enormous freedom in the design of new processes and products.

Figure 14.2 shows a schematic view of concurrent design. The top right shows the current approach to designing products in which information flows down from the assembled system level to provide constraints on the design of individual parts, and up from the parts to determine the properties of the assembled system. As Figure 14.2 indicates, current design processes typically stop above the diagonal, dashed, line, with the materials for the parts selected from a known set of possibilities, i.e., materials selection. In concurrent design, the materials themselves will be included in the design process. The challenge is that the same type of information exchange that is available in our current design processes is not generally available for materials design. Materials modeling, in which we include modeling derived from experiment, may include a linkage of scales, a building up of details and accuracy from one scale to

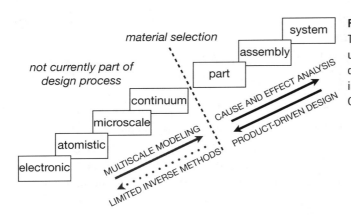

**Figure 14.2** Scales in concurrent design. The current design process, shown in the upper right, is coupled to materials design on the bottom left, as described in the text. Adapted from McDowell and Olson [222].

the next. We generally do not have, however, a way to inform the materials modeling of the overall design, which originates in the limited availability of inverse models in which one can describe properties at a finer scale based on information from a larger scale. This inability to exchange information back and forth between different scales is a major challenge for concurrent design.

There are a number of challenges to creating a concurrent design process. In addition to the development of inverse models, there is a need for an increased ability to computationally design the materials independently of design. Computational materials design faces numerous theoretical and computational challenges that arise from the extreme range of spatial and temporal scales inherent in materials response.

Consider the various length scales in Figure 1.1. The physics of each scale is dominated by a different set of fundamental microvariables, which typically describe some sort of collective behavior of sets of microvariables from smaller scales. To model each scale generally requires unique methods and is usually done by different research groups, using many of the methods described in this text. While much progress has been made at modeling at each of the scales shown in Figure 1.1 (and equivalent diagrams for other material properties), advancing those models is still a major thrust.

One approach has been to average properties at one scale (from either experiment or modeling) and to use those averages to develop models for behavior at larger scales. There are a number of limitations to this approach. The averaged models tend to be limited in range and applicability, they are often time-consuming to produce, they typically are limited in the quality of physics they can represent, and they cannot represent abnormal or rare events, which are stochastic in nature and not generally amenable to an averaged description. This "message passing" approach is, however, the most common way people attempt to link simulations from one scale to another. An alternative is to embed one simulation method within another, i.e., different methods are employed to model different regions or times of the material. The most common example in materials is the incorporation of atomistic simulations within finite-element calculations, which has been accomplished by a number of groups. What has been done to date has, however, been limited. Moreover, the design of the computational strategy is generally static, with fixed interfaces between the simulations. Thus, they do not provide flexibility to handle the information inherent in a multiscale description of material properties nor to describe the range of physical states different parts of a material might have in use.

The design of multiscale materials is unique in several ways – by working across scales, additional degrees of freedom are introduced into the design space and small changes at one scale can have a significant impact on other scales relevant to the intended purpose of the design. In this way multiscale materials and processes have the potential to be built piece by piece at one scale and then experienced or used at larger scales, in effect creating a canvas on which a designer can work to create new materials and processes. The methodologies and tools to do this type of creative engineering design do not, however, currently exist.

## 14.3 METHODS

The beginnings of material design are the computational models used in each of the boxes in Figure 14.2. Even for models of the same phenomenon, each scale may require a different set of methods. However, a common feature of modeling mesoscale properties is that, often, many methods are available, each with different capabilities in terms of length and time scales, accuracy, etc. The choice of methods is based on many requirements, with none of the methods actually being clearly "best". The choice of what model to use will depend on a variety of factors, including computational resources, the methods at other scales to which they may be linked,[1] the questions under consideration, and so on.

The goal of this section is not to prescribe what methods should be used, but rather to show, for a single example, how bringing the range of methods to a single problem sheds light on both the physics of the problem as well as the strengths and limitations of the methods. Grain growth was chosen as being illustrative of the types of challenges faced in all multiscale modeling.

### 14.3.1 Grain growth

In this text we have discussed six different ways that could be used to model grain growth: atomistic simulations, the Monte Carlo Potts model, phase-field methods, cellular automata, boundary dynamics, and vertex dynamics. All have strengths and limitations, and can answer different sets of questions, though all should yield the same fundamental properties of grain growth as discussed in Appendix B.6.

(a) The atomistic-level simulations of energy minimization (Chapter 3), molecular dynamics (Chapter 6), and Monte Carlo (Chapter 7) provide the most fundamental description of grain-boundary properties and motion. In these methods, no assumptions are made about

---

[1] The choice of model may be dictated by what information is needed by modeling at the next scale up the chain in Figure 14.2.

grain-boundary structures or how grain-boundary motion occurs. The only input is the potentials (Chapter 5),[2] and the simulation methods automatically take into account the structures, temperature effects, etc.[3] The limitations with atomistic-level simulations are that the length and time scales are much too small to examine multiple grain boundaries or long-time dynamics.

(b) The Monte Carlo Potts model of Chapter 10 is very flexible and makes no direct assumptions about the dynamics of grain growth, other than it is driven by energy reduction. It is easily extendable to three dimensions and finite temperature, and, within the limits owing to the choice of the lattice upon which the model is implemented, a boundary can take on any configuration. It is a very flexible method, and has been used to study not just normal grain growth but phenomena such as recrystallization as well. When implemented as a kinetic Monte Carlo method (Chapter 9), called for the Potts and similar spin models the N-fold way, the Potts model is relatively fast computationally, though it slows at finite temperatures. While the Potts model gives overall dynamics that match experiment, its weaknesses are that it depends on the underlying lattice and there is no absolute length scale.

(c) In Chapter 11 we introduced the cellular automata method, discussing some examples that include the basic physics of grain growth. As in all classical cellular automata, the dynamics arise from local rules, so the methods are computationally fast, as well as being very flexible. A weakness in a traditional cellular automaton is that the model is deterministic, and thus cannot model systems at finite temperatures. That restriction can be relaxed by allowing for probabilistic rules to mimic the Boltzmann factor of the Monte Carlo method. Just as for the Potts model, lattice effects are very important. It is important to recognize that a cellular automaton may have very little physics involved. There is no absolute length scale.

(d) The phase-field method of Chapter 12 is a thermodynamics-based method that has been widely applied to the study of microstructural evolution in both two and three dimensions. The phase-field method is very powerful, enabling the inclusion of a variety of thermodynamic drivers, which enables the study of many phenomena in microstructural evolution. Because it is a thermodynamic model, the phase-field can be readily connected to the real energetics of a material, which has led to its common use within an ICME framework [7]. A limitation to the phase-field method is that, as described in Chapter 12, the interfaces are diffuse, which limits its ability to describe details of interfacial structures. This method also suffers from a lack of a well-defined length scale.

(e) The boundary dynamic method (Chapter 13) tracks the dynamics of mesh points along the boundary. The dynamics arise from a local velocity relation based on curvature-driven growth. Boundaries have a reasonable shape and there are no lattice effects. The method is fast and reasonably flexible. Its weaknesses are the assumption of curvature-driven

---

[2] While calculations have been done on grain-boundary structures using the electronic structure methods of Chapter 4, calculations of motion are limited by system size, so essentially all studies at the atomistic level have been done using potential-based methods.

[3] A caveat is that the structure of the simulation cell and the boundary conditions must be consistent with the symmetry of the grain boundary.

dynamics, its restriction to zero temperature, and its lack of flexibility relative to the Potts model. This method also suffers from a lack of a well-defined length scale.

(f) The vertex dynamics method of Chapter 13 is very fast, as it focuses only on the motion of the relatively few vertices in the grain structure. It assumes the dynamics is curvature driven, has straight grain boundaries, and is restricted (as described) to zero temperature. While it is not as flexible as the Potts model in the incorporation of new physics, it can be directly connected to other dynamical methods, such as the finite-element method. It has been implemented in two and three dimensions. This method also suffers from a lack of a well-defined length scale.

Consider the range of methods that we could use to model microstructural evolution. By going from atomistics to the Potts model, to a cellular automaton, to a phase-field method, to boundary dynamics, and finally to vertex dynamics, is to steadily reduce the dimensionality of the problem. Clearly, we have sacrificed details as we went from small to large scales; however, we have gained computational speed, which means we can do more complicated problems. Modeling grain growth thus serves as a paradigm for materials modeling – as we move from smaller to larger length scales, we gain in systems size and time but give up detail by incorporating information from the smaller scales in some average way.

## 14.4 MATERIALS INFORMATICS

Materials informatics is a computation-based method that is beginning to play an essential role in materials design and development. Materials informatics is what its name applies – the application of the ideas of informatics to materials science and engineering. Informatics is a field of research in which information science, computer processing, and systems thinking combine to examine the structure and behavior of information, enabling new ways to access and explore that information. In materials research, information comes in many forms, as does the possible use of that information. It should thus not be surprising that the potential applications of informatics to materials are extremely diverse.

The basic idea of informatics applied to materials and their properties is not new. Materials development, while a trial-and-error process, was guided by cumulative knowledge, which often was captured in the form of rules. These rules were generally empirical relationships based on a combination of data and observation. Indeed, they were an early form of materials informatics. A simple example is the so-called Hume-Rothery rules, which describe the conditions under which an element can dissolve in a metal to form a solid solution [221]. These rules take simple forms based on the relative ionic radii of the solute and solvent atoms and correlate experimental observations (degree of solubility) with simple characteristics of the material (ionic radii). They are empirical, based on fits to data, but yield important information that can guide materials development. Such rules are inherently limited, however, in the complexity of data that they can describe, with few examples in which correlations between more than a few variables have been identified.

Modern informatics takes advantage of recent advances in computational data manipulation and analysis to enable the material scientist to greatly expand both the range and complexity of the data being studied as well as the quality and utility of information retrieved from that data.

## 14.4.1 Data mining

The fundamental task of materials informatics is to extract the maximum information from data, which it does using the basic ideas from a field often referred to as *data mining*. While the details are beyond the scope of this text, it is useful to summarize the basic tasks of data mining, which include (details are given in [137]):

(a) exploratory data analysis
- to visualize what is in the data
(b) descriptive modeling
- probability distributions, cluster analysis, etc.
(c) predictive modeling
- classification and regression
(d) discovering patterns and rules
- spotting behavior, outlier identification
(e) retrieval by content
- match desired patterns.

All these tasks employ a suite of methods. Deciding what method and how it can be best used is often the "art" in informatics.

One of the most challenging problems in data mining is to retrieve information from the content of the data. An example is the determination of critical descriptions of complex structures from, for example, images. A classic problem of this form was the development of the fingerprint database, a complicated problem that is not totally solved [66]. In materials research, a similar, but much more complicated, problem is to extract information from microstructures as part of the development of more informative correlations between structures and properties. Some of the challenges of describing microstructures can be seen, for example, in [163, 198, 200], as will be discussed in somewhat more detail in the next section.

## 14.4.2 Applications

The application of informatics in materials research is, at least conceptually, straightforward [100]. As an example, consider a number of applications of informatics that have employed electronic structure methods such as described in Chapter 4 to create databases of energy and structures. From these databases, correlations and predictions can be extracted using standard data-mining methods [58, 297]. This approach greatly aids in identifying new alloy systems for specific applications, a process that has traditionally been time-consuming and expensive owing to the large multi-component design space.

Informatics enables the materials developer to combine data from many forms (e.g., experiment and modeling) and with a range of uncertainty. Common characteristics between potential alloy systems can be isolated, and trends in the data that might normally be overlooked can be identified. These can then be employed to identify promising classes of materials. For example, Ceder and colleagues have developed a database of free energies (determined from calculations) and structures for a wide range of binary and ternary intermetallic systems that can be employed to determine lowest-energy structures and to help construct their phase diagrams [103]. When a user requests information for a system not in the database, statistical methods are applied to the database to identify likely structures, the choice of which is then refined by a minimal series of *ab initio* calculations. These types of studies offer a new way to use modeling to develop trends and to guide materials selection.

The behavior of materials is, however, often complicated by the presence of complex structures that manifest themselves in physical responses with a wide range of length and scales, as shown in Figure 1.1. Classic structure-property relations are dependent on the distributions of defects at the mesoscale. Characterization of those structures in three dimensions is challenging, as described for the distribution of grains (microstructure) in a series of articles [310]. Developing ways to represent those structures for incorporation in an informatics framework is also a challenging task. One promising approach employs serial sectioning combined with optical microscopy and electron backscattered diffraction to yield three-dimensional microstructures represented with voxels [198, 199]. These voxels then serve as a finite-element mesh. One study employed this approach combined with materials informatics to extract correlations between mechanical response and microstructure in a commercial steel [200]. Linking materials informatics with experiment in this way can greatly enhance the information from, and thus impact of, the data [271].

Informatics also offers a way to help make progress in one of the most fundamental challenges in materials – the linkage across length and time scales. In most materials modeling, methods are used that are only appropriate for a relatively limited set of lengths and times. Atomistic simulations based on interatomic potentials, for example, can be routinely applied for systems with linear sizes in the hundred or so nanometers [279] – much too small to model the many grains in a polycrystalline sample. Time scales are limited to the nanosecond regime (except for some problems for which accelerated methods are available [333]). Other methods, such as the phase-field model, can, however, be used to model the dynamics of microstructural evolution [64]. The challenge is how to couple information across the scales. Informatics offers a promising approach to that problem, as described in detail elsewhere [205].

As noted above, integrated computational materials engineering and concurrent engineering employ a combination of experimental data and computational modeling to accelerate the development of materials for specific engineered applications. It is in this regime that materials informatics may play its most critical role. In a materials development process, data come in many forms, from the basic information about how to create and process a material to the analysis of the properties of both the individual materials and the engineered product. We use the term "data" in this case to refer to all information about the material, whether it comes from experiment, modeling or simulation. The data are often of mixed origin, with varying degrees of certainty. The analysis is often highly dependent on computer modeling. For all of these components to be integrated with each other requires a robust, flexible approach to extract the maximum information possible. Materials informatics seems to fit that requirement.

## 14.5 SUMMARY

In this chapter we discuss an important use of many of the models introduced in this text – the discovery and development of materials. Based on ideas from a field now being called integrated computational materials engineering, the goal is to base materials development on all information available, be it from experiment or modeling, the linkage between the two often requiring methodologies from materials informatics. The ultimate goal is concurrent engineering, in which the materials are designed concurrently with the desired structures. Key to all of these advances is a robust set of modeling tools. Increasing the recognition of and role that these methods play in the materials community is one of the reasons behind the writing of this text.

### Suggested reading

- Many of the issues raised in this chapter are discussed in detail in the National Academy report on *Integrated Computational Materials Engineering* [72].
- Many books on data mining are available. A useful one is *Principles of Data Mining*, by Hand *et al.* [137].

# Part Five

# Appendices

# A Energy units, fundamental constants, and conversions

## A.1 FUNDAMENTAL CONSTANTS

Table A.1 **Some useful fundamental constants**[a]

| unit | cgs | MKS | name |
|---|---|---|---|
| $k_B$ | $1.3806 \times 10^{-16}$ erg/°K | $1.3806 \times 10^{-23}$ J/°K | Boltzmann's constant |
| $\hbar$ | $1.05457 \times 10^{-27}$ erg sec | $1.05459 \times 10^{-34}$ J sec | Planck's constant |
| $e$ | $4.80320 \times 10^{-10}$ esu | $1.60219 \times 10^{-19}$ coulomb | electron charge |
| $m_e$ | $9.10938 \times 10^{-28}$ g | $9.10938 \times 10^{-31}$ kg | electron mass |
| $m_p$ | $1.67262 \times 10^{-24}$ g | $1.67262 \times 10^{-27}$ kg | proton mass |
| $a_o$ | $5.299172 \times 10^{-9}$ cm | $5.29177 \times 10^{-11}$ m | Bohr radius |
| $c$ | $2.99792 \times 10^{10}$ cm/sec | $2.99792 \times 10^{8}$ m/sec | speed of light |
| $N_A$ | $6.02214 \times 10^{23}$ mol$^{-1}$ | $6.02214 \times 10^{23}$ mol$^{-1}$ | Avagadro's number |

[a] From http://physics.nist.gov/cuu/Constants/index.html.

## A.2 UNITS AND ENERGY CONVERSIONS

In MKS units, the position is measured in meters (m), the mass in kilograms (kg), and time in seconds (sec). The acceleration thus has units of m/sec$^2$ and the force has units of kg m/sec$^2$. The MKS unit of force is called a *Newton* (N), with 1 N = 1 kg m/sec$^2$. In cgs units, the length is in centimeters (cm) and mass is in grams (g). The force has units of 1 *dyne* = 1 cm g/sec$^2$. The conversion between MKS and cgs units of force is: 1 N = (1000 g )(100 cm)/sec$^2$ = $10^5$ dynes. The unit of energy in MKS units is 1 *Joule* (J) which is defined as 1 N m = 1 kg m$^2$/sec$^2$ whereas in cgs units 1 *erg* = 1 dyne cm = 1 g cm$^2$/sec$^2$. We thus have 1 J = (1000 g)(100 cm/sec)$^2$ = $10^7$ erg.

Pressure (or stress) is defined as force per area (or energy per volume). The unit for pressure (or stress) in MKS units is the *Pascal* (Pa), with 1 Pa = 1 N/m$^2$. The pressure (stress) unit in

Table A.2 **Energy relationships**[a]

| | J | eV | K | kcal/mol |
|---|---|---|---|---|
| 1 J | 1 | $6.2415 \times 10^{18}$ | $7.2430 \times 10^{22}$ | $1.4393 \times 10^{20}$ |
| 1 eV | $1.6022 \times 10^{-19}$ | 1 | $1.1605 \times 10^{4}$ | 23.0609 |
| 1 K | $1.3807 \times 10^{-23}$ | $8.6173 \times 10^{-5}$ | 1 | $1.9872 \times 10^{-3}$ |
| 1 kcal/mol | $6.9477 \times 10^{-21}$ | 0.04336 | 503.228 | 1 |

[a] From http://physics.nist.gov/cuu/Constants/energy.html.

cgs units is 1 *bayre* (ba) = 1 dyne/cm$^2$ = 0.1 Pa. A more common unit of pressure is the *bar* = $10^5$ Pa or $10^6$ ba. A bar is approximately equal to a standard *atmosphere* (atm), with 1 bar = 0.98692 atm.

The units of electrostatics require some attention. In MKS units, the unit of charge is the Coulomb (C). In cgs units, however, the unit of charge is the *statcoulomb* (statC), which is also called the *electrostatic unit of charge* (esu); 1 esu = $3.33564 \times 10^{-10}$ C. Please see Appendix E.1 for a more detailed description of units in electrostatics.

In Table A.2 we give conversions between the standard MKS units of energy and two other units that are commonly used in materials research, the electron volt (eV) and the thermal energy (K). These are defined as:

(a) 1 eV is the amount of energy gained by a single unbound electron when it is accelerated in vacuum through an electrostatic potential difference of 1 volt, where the volt is defined (MKS units) as 1 volt = 1 J/C, where charge is measured in *coulomb* (C). The charge on an electron is $e = 1.6022 \times 10^{-19}$ C. We thus have that 1 eV = $1.6022 \times 10^{-19}$ J = $1.6022 \times 10^{-12}$ erg.

(b) The thermal energy of a system is given by $k_B T$, where $k_B$ is Boltzmann's constant from Table A.1 and $T$ is the temperature measured in Kelvin (K). An energy of 1 K corresponds to the amount of thermal energy in 1 K, which has the value of 1K $\times$ $k_B$, i.e., 1 K $\times$ $1.3806504 \times 10^{-23}$ J/K = $1.3806 \times 10^{-23}$ J = $1.3806 \times 10^{-16}$ erg.

We also introduce the chemical units *kcal/mol*. These convert the standard energies per atom to the equivalent for a mole of atoms, using the basic conversion 1 cal = 4.184 J. Conversions to *kJ/mol* can be found by dividing those for kcal/mol by 4.184.

# B    A brief introduction to materials

This text is focused on the modeling of materials structure and properties. The language and choice of problems and methods reflects the interests of the materials science and engineering (MSE) community. We realize, however, that there is increased interest in these problems from people in fields outside MSE. The purpose of this chapter is to give a rapid overview of materials science strictly from the point of view of what is covered elsewhere in the text. It is certainly not a comprehensive introduction to materials.

## B.1    INTRODUCTION

Materials in use are solids and most, but certainly not all, are crystals, by which we mean systems of atoms that have a regular, periodic structure. Few materials in actual use, however, are perfect crystals. Most have defects, imperfections in their lattices that have a profound effect on the overall properties of those materials. These defects may be point defects, such as vacancies, line defects (typically dislocations), or planar defects, such as surfaces or interfaces between two crystals. The distribution of those defects is referred to as a materials *microstructure*. Understanding the evolution of the microstructure as well as its role in determining overall properties is a major thrust of materials modeling and simulation.

In this chapter, we introduce basic crystallography of simple crystals, as well as how to represent that crystallography in calculations. We then discuss the defects of those materials and the ramification of those defects on materials properties. We also emphasize the role of dynamic processes, such as diffusion, on materials.

The focus of the discussion is on metals, in part because of their importance in technology, but more so because they tend to have simple crystal structures. We emphasize, however, that the phenomena discussed here are ubiquitous across all types of materials. For example, phase transitions occur in all materials types, and the evolution of microstructure is thus important in most, if not all, materials systems. Likewise, dislocations are essential for understanding plasticity not only in metals, but also in ceramics, semiconductors, and molecular systems. Our focus is on crystalline systems, though many of the methodologies are equally applicable to liquids, amorphous solids, and quasicrystals.[1]

---

[1] Quasicrystals were discovered in the 1980s by Dan Shechtman, who won the Nobel Prize in Chemistry in 2011 for his discovery. They are structures that are ordered, but not periodic.

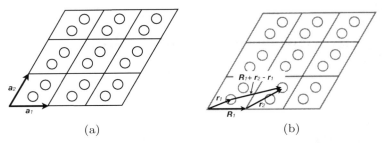

**Figure B.1** Two-dimensional lattice with a basis. (a) There are two atoms per unit cell. (b) Vectors defining atomic positions and interatomic distances. Position vectors within the unit cell are denoted by $r_i$. Shown is the vector connecting atom 1 in one cell to atom 2 in a cell separated from the first by the lattice vector **R**.

## B.2   CRYSTALLOGRAPHY

A periodic crystal is a solid in which its constituents, be they atoms, ions, or molecules, are arranged in an ordered, periodic, pattern extending in all spatial directions and completely filling space (i.e., no voids).[2] Because of the restriction that they must be space-filling, the periodic structures are restricted by symmetry to be of a very few types, called Bravais lattices. There are are 14 Bravais lattices, arising from seven basic lattice systems. All periodic crystal structures are derivable from these basic lattices.

Consider, as an example, the two-dimensional lattice shown in Figure B.1a. The basic crystal structure can be defined in a variety of ways. As shown, the system is made up of repeated units (the solid lines) that are defined by the two vectors $\mathbf{a}_1$ and $\mathbf{a}_2$, which may or may not be orthogonal. The fundamental unit is called the *unit cell*.

There are two atoms in the unit cell for the example shown, which is referred to as the *basis* of the cell. In this text the positions of the basis atoms are denoted by **r**. We could have chosen the unit cell with its origin at the site of one of the basis atoms or, indeed, any number of other ways as long as the fundamental unit can be repeated to fill space with no overlaps.

Any one of the repeated cells can be connected to another through a direct lattice vector, which throughout this text is denoted by the vector **R**, which has the general form (in three dimensions)

$$\mathbf{R} = n_1\mathbf{a}_1 + n_2\mathbf{a}_2 + n_3\mathbf{a}_3 , \tag{B.1}$$

where $n_1$, $n_2$, and $n_3$ are integers that span from $-\infty$ to $\infty$. One such lattice vector is shown in Figure B.1b.

The position of the $j$th basis atom in a cell defined by the lattice vector $\mathbf{R}_1$ (relative to the origin) is $\mathbf{R}_1 + \mathbf{r}_j$, as shown in Figure B.1b. Thus the vector connecting the $i$th basis atom in the central cell ($\mathbf{R} = \mathbf{0}$) with the $j$th atom in a cell located at $\mathbf{R}_1$ is

$$\mathbf{R}_1 + \mathbf{r}_j - \mathbf{r}_i . \tag{B.2}$$

---

[2] After the discovery of quasicrystals, the International Union of Crystallography redefined crystals to be "any solid having an essentially discrete diffraction diagram".

Table B.1  **Basic crystal systems**

| Crystal system | Lattice lengths | Lattice angles |
| --- | --- | --- |
| Triclinic | $a \neq b \neq c$ | $\alpha \neq \beta \neq \gamma$ |
| Monoclinic | $a \neq b \neq c$ | $\alpha \neq \beta = \gamma = 90°$ |
| Orthothomic | $a \neq b \neq c$ | $\alpha = \beta = \gamma = 90°$ |
| Tetragonal | $a = b \neq c$ | $\alpha = \beta = \gamma = 90°$ |
| Rhombohedral | $a = b \neq c$ | $\alpha = \beta = \gamma \neq 90°$ |
| Hexagonal | $a = b \neq c$ | $\alpha = \beta = 90°, \gamma = 120°$ (or $60°$) |
| Cubic | $a = b = c$ | $\alpha = \beta = \gamma = 90°$ |

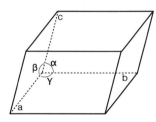

**Figure B.2**  A generic unit cell, showing the lattice lengths, $a$, $b$, and $c$, and the lattice angles, $\alpha$, $\beta$, and $\gamma$.

## B.2.1  Basic crystal structures

A unit cell in three dimensions can be defined by the *lattice parameters* shown in Figure B.2. The basic crystal systems are defined by the relationships between the lattice parameters indicated in that figure. These are summarized in Table B.1. Specific crystal structures are based on the seven lattice systems in Table B.1, often with a basis of atoms as shown in two dimensions in Figure B.1. A good description of simple crystal structures can be found in many introductory texts.

The simplest crystal structure is the *simple cubic structure*, which has a single atom at each corner of a cube. The basis of this system is 1, with the atom at the origin "belonging" to the central unit cell, and all other atoms being associated with other unit cells, as indicated in Figure B.1.

In Figure B.3 we show three basic cubic structures, from which many other structures can be derived, as shall be discussed a bit later in this section. Numerous elements crystallize into these basic structures. In Figure B.3a, we show the *body-centered cubic* lattice, which is often abbreviated as *bcc*. There are two atoms per unit cell, with one at the origin $(0,0,0)$ and the other at the center of the cubic cell at a position $(\frac{1}{2},\frac{1}{2},\frac{1}{2})a$, where $a$ is the lattice parameter. Many important elements have a *bcc* structure, including iron (Fe), chromium (Cr), molybdenum (Mo), cesium (Cs), and tungsten (W).

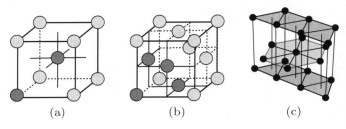

**Figure B.3** Basic cubic crystal structures as described in the text. (a) Body-centered cubic (*bcc*). The atoms in the basis are shown in darker gray. (b) Face-centered cubic (*fcc*). The atoms in the basis are shown in darker gray. (c) Hexagonal-closest-packed (*hcp*). Figures adapted from [15].

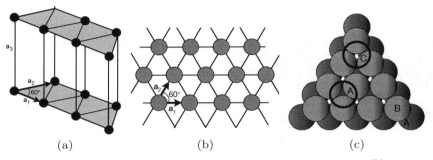

**Figure B.4** (a) Simple hexagonal structure, showing the lattice parameters. (b) Hexagonal planes in the hexagonal structure. (c) Stacking sequence of close-packed balls, showing two triangular layers, marked *A* and *B*. Indicated in that picture are the two options for the next layer, as discussed in the text. Figures adapted from [15].

In Figure B.3b is shown the *face-centered cubic* lattice (*fcc*). There are four atoms per unit cell, located at $(0,0,0)$, $(\frac{1}{2},\frac{1}{2},0)a$, $(\frac{1}{2},0,\frac{1}{2})a$, and $(0,\frac{1}{2},\frac{1}{2})a$. The face-centered-cubic lattice is the stable structure in silver (Ag), gold (Cu), copper (Cu), nickel (Ni), argon (Ar), and many other elements.

We show the *hexagonal-closest packed* (*hcp*) structure in Figure B.3c. It is closely related to the *fcc* structure, but a bit more complicated to describe. In Figure B.4a we show the simple hexagonal lattice. It consists of planes of triangular nets as shown in Figure B.4b. The *hcp* lattice of Figure B.3c consists of two interpenetrating simple hexagonal lattices that are displaced by each other by a vector $\mathbf{a}_1/3 + \mathbf{a}_2/3 + \mathbf{a}_3/2$, where the lattice vectors are indicated in Figure B.4a.

The *hcp* structure consists of stacks of triangular nets. An easy way to understand this structure is to think of a close-packed stacking of balls, in which all the balls touch, as shown in Figure B.4c. Shown in that figure are two triangular layers, marked *A* and *B*. Also shown are the two options for the next layer. If a ball in the next layer sits directly over an atom in the *A* layer, then it will be another *A* layer. If a ball in the next layer sits over an empty space in the *A* layer, then it will be different than either *A* or *B* and is indicated by *C*. *hcp* structures have an ...*ABABABAB*... stacking sequence. For a perfect *hcp* structure, the ratio of the *c* axis (indicated in Figure B.4a by $\mathbf{a}_3$) to the length of the *a* axis ($\mathbf{a}_1 = \mathbf{a}_2$) is $c/a = \sqrt{8/3}$, which can be understood from the simple packing of spheres. Elements that have the *hcp* structure include beryllium (Be), cadmium (Cd), titanium (Ti), zirconium (Zr), and numerous others.

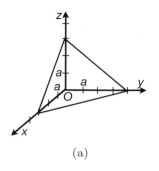

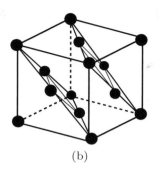

(a)  (b)

**Figure B.5** (a) Definition of the nomenclature for lattice planes. The ticks are measured in units of the lattice parameter *a*. The plane intersects the *x, y, z* axes at 3*a*, 4*a*, 3*a*, respectively. (b) (111) lattice planes in a face-centered cubic structure. Figures adapted from [15].

The *fcc* structure has an ...*ABCABCABCABC*... stacking sequence. To see how the *fcc* structure has this stacking sequence requires a bit closer look at crystallography. We start by introducing some nomenclature used to describe the orientation of a plane in a crystal. In Figure B.5a we show the intersection of a plane intersecting the principal axes of a cubic crystal. The standard notation for such a plane is to use the *reciprocal intercepts* of the plane on the axes relative to the lattice parameters, i.e., the plane shown in Figure B.5a has the reciprocal intercepts of the three axes of $\frac{a}{3a}$, $\frac{a}{4a}$, and $\frac{a}{3a}$. These are usually written as the three smallest integers in these ratios, i.e., 4, 3, and 4. The standard notation is that a plane is indicated by parentheses, i.e., (*hkl*), where $h, k, l$ are the *Miller indices*. The plane in Figure B.5a is thus denoted by (434).

Suppose that a plane is parallel to one of the axes, such as a plane that intersects the *y* axis at 1 and the *z* axis at 1. The convention is to say that the plane intersects the *x* axis at ∞. Taking the reciprocal yields 0, so that plane would be denoted by (011). Families of planes are related by symmetry. For example the (111), $(1\bar{1}1)$, $(11\bar{1})$, and $(\bar{1}11)$ planes are all equivalent in the *fcc* structure (the notation $\bar{k}$ indicates that the plane intersects the axis on the negative side of the origin). Families of planes are indicated with braces, i.e., {*hkl*}. Directions in the lattice are denoted by square brackets and are given in terms of the unit cell parameters. For example, the [111] direction in a cubic lattice points along the body diagonal. Families of symmetry-related directions are indicated by angle brackets, such as ⟨111⟩. Note that the direction [*hkl*] is the normal to the plane indicated by (*hkl*).

In Figure B.5b, we show a sequence of (111) *lattice planes* in an *fcc* structure. While it might not be apparent in that figure, the {111} planes are close-packed (all nearest-neighbor atoms are at the minimum separation in the crystal). Analysis of the stacking of the (111) planes shows that the face-centered cubic lattice has the *ABC* stacking sequence.

We have described all of these structures based on a cubic lattice with a basis. For example, the face-centered cubic structure was written as a cube with four atoms per unit cell. The *fcc* structure can, however, also be described in a *primitive* unit cell that has one atom per cell. In this case, the three lattice parameters are not along the *x, y,* and *z* directions, but point towards the nearest-neighbor atoms, which are on the faces of the cube. The primitive lattice vectors, in Cartesian coordinates, can be written as

$$\mathbf{a}_1 = \frac{a}{2}(\hat{y} + \hat{z}) \quad \mathbf{a}_2 = \frac{a}{2}(\hat{x} + \hat{z}) \quad \mathbf{a}_3 = \frac{a}{2}(\hat{x} + \hat{y}). \tag{B.3}$$

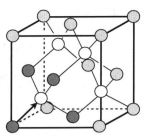

**Figure B.6** The unit cell for diamond, which is represented by two interpenetrating face-centered cubic structures offset by the vector $\frac{a}{4}(\hat{x} + \hat{y} + \hat{z})$, which is indicated as an arrow. The dark circles represent the basis for one of the $fcc$ lattices, while the white circles represent the other $fcc$ lattice. The lines indicate nearest neighbors.

As noted above, the hexagonal-closest-packed structure can be represented as two interpenetrating simple hexagonal lattices. Other structures can be created in the same way based on different basic crystal lattices. Sodium chloride (NaCl) consists of interpenetrating simple cubic lattices offset by $\frac{a}{2}(\hat{x} + \hat{y} + \hat{z})$. One of the lattices consists of the $Na^+$ ions and the other has the $Cl^-$ ions. The cesium chloride (CsCl) structure can be described as two interpenetrating body-centered cubic lattices offset by $\frac{a}{2}(\hat{x} + \hat{y} + \hat{z})$, with one lattice of $Cs^+$ ions and another of $Cl^-$ ions.

An important example is shown in Figure B.6, which is the diamond (C) structure. Diamond shares this structure with a number of important technological materials, including silicon (Si), germanium (Ge), and $\alpha$-tin (Sn). The diamond lattice can be represented as two interpenetrating face-centered cubic lattices offset by $\frac{a}{4}(\hat{x} + \hat{y} + \hat{z})$. In Figure B.6, the lines indicate the four nearest neighbors for each atom, which are located at the corner of a tetrahedron, with bond angles of $109.5°$.

We close this brief discussion of crystallography by noting that while many materials have crystal structures that may be much more complicated than the simple structures shown here, the basic structure of all crystals is the same. One must specify the six lattice parameters, $a$, $b$, $c$, $\alpha$, $\beta$, and $\gamma$, as well as the positions of all the atoms in the basis within the unit cell. There are many good discussions of these structures, as well as of the basic symmetries and crystallography of them, given elsewhere.

## B.2.2 The direct lattice

In Eq. (B.2) we show an expression for the vector connecting atoms within the central unit cell and that in a unit cell located a lattice vector **R** away. The separation between such atom pairs is commonly needed in calculations with periodic boundary conditions, for example. Here we summarize a few results that are used throughout the text.

It is often convenient to write the position of the basis atoms as *fractions* of the unit cell lattice vectors, for example as

$$\mathbf{r}_i = s_{i1}\mathbf{a}_1 + s_{i2}\mathbf{a}_2 + s_{i3}\mathbf{a}_3 . \tag{B.4}$$

We can thus write the position of an atom ($i$) within the cell as

$$\mathbf{s}_i = (s_{i1}, s_{i2}, s_{i3}) . \tag{B.5}$$

With this notation, the vector connecting two atoms in different cells (as in Figure B.1b) is

$$\mathbf{R} + \mathbf{r}_j - \mathbf{r}_i = (n_1 + s_{j1} - s_{i1})\mathbf{a}_1 + (n_2 + s_{j2} - s_{i2})\mathbf{a}_2 + (n_3 + s_{j3} - s_{i3})\mathbf{a}_3 , \quad \text{(B.6)}$$

where $\mathbf{R}$ is given in Eq. (B.1). Equation (B.6) has the general form

$$\mathbf{X} = x_1\mathbf{a}_1 + x_2\mathbf{a}_2 + x_3\mathbf{a}_3 . \quad \text{(B.7)}$$

with $(x_1, x_2, x_3) = (n_1 + s_{j1} - s_{i1}, n_2 + s_{j2} - s_{i2}, n_3 + s_{j3} - s_{i3})$. The distance between the two atoms is the length of $X$,

$$X = \sqrt{\mathbf{X} \cdot \mathbf{X}} = |\mathbf{X}| . \quad \text{(B.8)}$$

If we use the standard lattice lengths and angles: $a_1 = a$, $a_2 = b$, $a_3 = c$, $\angle(a_1, a_2) = \gamma$, $\angle(a_1, a_3) = \beta$ and $\angle a_2 a_3 = \alpha$, then the length between two atoms is

$$X^2 = x_1^2 a^2 + x_2^2 b^2 + x_3^2 c^2 + 2x_1 x_2 ab \cos\gamma + 2x_1 x_3 ac \cos\beta + 2x_2 x_3 bc \cos\alpha , \quad \text{(B.9)}$$

which for cubic systems is

$$X^2 = a^2\left(x_1^2 + x_2^2 + x_3^2\right) . \quad \text{(B.10)}$$

## B.2.3 The reciprocal lattice

It is sometimes necessary to consider the *reciprocal lattice*, which is a transform of the direct lattice. A reciprocal lattice vector $\mathbf{K}$ is defined by the relation[3]

$$e^{i\mathbf{K}\cdot\mathbf{R}} = 1 , \quad \text{(B.11)}$$

where $\mathbf{R}$ is a direct lattice vector. It is straightforward to show that $\mathbf{K}$ must be of the form

$$\mathbf{K} = k_1\mathbf{b}_1 + k_2\mathbf{b}_2 + k_3\mathbf{b}_3 \quad (k_1, k_2, k_3) \text{ are integers} , \quad \text{(B.12)}$$

where the reciprocal lattice is defined by the vectors

$$
\begin{aligned}
\mathbf{b}_1 &= 2\pi \frac{\mathbf{a}_2 \times \mathbf{a}_3}{\mathbf{a}_1 \cdot (\mathbf{a}_2 \times \mathbf{a}_3)} \\
\mathbf{b}_2 &= 2\pi \frac{\mathbf{a}_3 \times \mathbf{a}_1}{\mathbf{a}_1 \cdot (\mathbf{a}_2 \times \mathbf{a}_3)} \\
\mathbf{b}_3 &= 2\pi \frac{\mathbf{a}_1 \times \mathbf{a}_2}{\mathbf{a}_1 \cdot (\mathbf{a}_2 \times \mathbf{a}_3)} ,
\end{aligned}
\quad \text{(B.13)}
$$

where $\mathbf{a}_1 \cdot (\mathbf{a}_2 \times \mathbf{a}_3) = V$, the volume of the direct lattice cell; $\times$ indicates a cross product, as defined in Eq. (C.12). Distances in the reciprocal lattice are described in an equivalent way to those in the direct lattice.

---

[3] From Eq. (C.37) we have that $\exp(i2\pi n) = 1$ for $n$ equal to an integer.

## B.2.4 A more formal description of non-cubic lattices

In this section we discuss a general way to describe any cubic or non-cubic lattice type. It is equivalent to the discussion of the direct lattice, but there are times when this notation and structure are a bit more useful.

We start by writing the position of the atoms as fractions of the simulation cell as in Eq. (B.4). We then create a matrix $\mathbf{h}$, whose columns are made up of the lattice vectors of the simulation cell in cartesian coordinates, i.e.,[4]

$$\mathbf{h} = (\mathbf{a}, \mathbf{b}, \mathbf{c}). \tag{B.14}$$

The volume of the cell is given by the formula

$$V = \det \mathbf{h} = \mathbf{a} \cdot (\mathbf{b} \times \mathbf{c}), \tag{B.15}$$

in which $\det \mathbf{h}$ is the determinant of $\mathbf{h}$.

If the fractional coordinates of an atom $i$ are given by the vector $\mathbf{s}_i$, as in Eq. (B.5), then the *absolute* coordinates are given in terms of $\mathbf{h}$ by

$$\mathbf{r}_i = \mathbf{h}\,\mathbf{s}_i, \tag{B.16}$$

which is just a compact way of writing Eq. (B.4).

Calculating lattice sums using this representation is straightforward – a general position is

$$\mathbf{r} + \mathbf{R} = \mathbf{h}\,(\mathbf{s} + \mathbf{S}), \tag{B.17}$$

where the scaled lattice vectors are of the form $\mathbf{S} = (n_1, n_2, n_3)$ as in Eq. (B.1).

Distances between atoms in the simulation cell are easily related to the fractional coordinates. If $\mathbf{r}_{ij} = \mathbf{r}_j - \mathbf{r}_i$, then

$$\begin{aligned}
r_{ij} = |\mathbf{r}_{ij}| &= \left\{ \mathbf{h}\,(\mathbf{s}_j - \mathbf{s}_i) \right\}^T \left( \mathbf{h}\,(\mathbf{s}_j - \mathbf{s}_i) \right) \\
&= (\mathbf{s}_j - \mathbf{s}_i)^T\, \mathbf{h}^T\, \mathbf{h}\,(\mathbf{s}_j - \mathbf{s}_i) \\
&= \mathbf{s}_{ij}^T\, \mathbf{G}\, \mathbf{s}_{ij},
\end{aligned} \tag{B.18}$$

where the superscript $T$ indicates a transpose of a vector or a matrix as defined in Eq. (C.21). The *metric tensor* is defined as

$$\mathbf{G} = \mathbf{h}^T\, \mathbf{h}. \tag{B.19}$$

Equation (B.19) arises from writing a dot product as $\mathbf{a} \cdot \mathbf{b} = \mathbf{a}^T \mathbf{b}$.

---

[4] For a cubic lattice, $\mathbf{h}$ is given by

$$\mathbf{h} = \begin{pmatrix} a & 0 & 0 \\ 0 & a & 0 \\ 0 & 0 & a \end{pmatrix}, \tag{B.20}$$

while for the simple hexagonal lattice in Figure B.4a

$$\mathbf{h} = \begin{pmatrix} a & a\cos(60°) & 0 \\ 0 & a\sin(60°) & 0 \\ 0 & 0 & c \end{pmatrix}. \tag{B.21}$$

The metric tensor $\mathbf{G}$ is given in terms of the standard lattice parameters $a$, $b$, $c$, $\alpha$, $\beta$, and $\gamma$ by

$$\mathbf{G} = \begin{pmatrix} a^2 & ab\cos\gamma & ac\cos\beta \\ ab\cos\gamma & b^2 & bc\cos\alpha \\ ac\cos\beta & bc\cos\alpha & c^2 \end{pmatrix}. \tag{B.22}$$

Calculations of interactions as described in Chapter 3 for non-cubic simulation cells thus present no special challenges. The atomic positions and lattice vectors are all monitored within the scaled coordinate system. Distances, $s_{ij}$, are calculated in the scaled lattice, then converted to the actual lattice, using the metric tensor as in Eq. (B.18). The real distances are used to determine interaction energies and forces.

## B.3   DEFECTS

While the basis for the properties of materials is the crystal structures described in the previous section, few technological materials are based on single crystals, and none have crystals that are "perfect", i.e., without flaws. Often, it is the distribution of these flaws that dominates the behavior of a material. Indeed, many processing methods are designed to create certain defect structures to optimize a specific macroscopic property.

There are three main types of defects in a material. *Point defects* are things such as vacancies, in which a lattice site is empty, or interstitial atoms, in which atoms are located in sites that are not part of the regular lattice. The most important *line defects* are dislocations, the movement of which leads to plasticity, the permanent deformation of a material. Finally, *planar defects* include any planar interface, which could be a free surface or a boundary between crystallites of either the same or another type of material. Some examples of these defects are described in the following sections.

## B.4   POINT DEFECTS

A vacancy is formed by removing an atom from a lattice site, which requires energy, appropriately enough called the *vacancy formation energy*. In metals, typical vacancy formation energies are of the order of about 1 eV/atom. In ordered alloyed systems in which one atom is smaller than the other, the smaller atom can at times force itself into a hole between the lattice sites, forming an *interstitial* defect, which has an energy associated with it that depends on the types of atoms in the system and the crystal structure. Since no material is 100% pure, impurities can also be located in interstitial sites. *Substitutional* impurities are created by replacing an atom of one type with a different type, for example by inserting a Ge atom on to a silicon lattice. All of these processes have energies associated with them.

Vacancies are always present in a material, with a fraction of vacant sites that is approximately given by [263]

$$X_v^e = e^{\Delta S_v/R} e^{-\Delta H_v/RT} , \qquad (B.23)$$

where $\Delta S_v$ and $\Delta H_v$ are the entropy associated with the change in vibrational properties of atoms near the vacancy and the vacancy formation energy, respectively. In metals, typical vacancy concentrations near the melting point are $10^{-4}$–$10^{-3}$. Vacancies play an important role in diffusion in solids, as discussed in Section 2.3 and Appendix B.7.

## B.5  DISLOCATIONS

Dislocations are curvilinear defects that play the dominant role in plasticity, the process that creates a permanent deformation in a material. Here we introduce the basics of dislocation-based plasticity, with the goal of providing sufficient information to understand the simulations on dislocations discussed throughout this text. For more information, please see the introductory books by Weertman or Hull and Bacon listed in the Suggested readings. For an advanced treatment, see the definitive book by Hirth and Lothe.

### B.5.1  Plastic deformation

Plastic deformation in crystalline metals is a consequence of the collective motion of large numbers of curvilinear defects called dislocations. Dislocations as topological objects were first described by Volterra in 1905, long before their application to crystal deformation. In 1934, G. I. Taylor [306], E. Orowan [247], and M. Polanyi [261] independently suggested that dislocations should serve as a way to explain why the strength of materials was so much less than expected.[5]

The mobility of dislocations gives rise to plastic flow at relatively low stress levels compared to the theoretical strength. In a typical metal, dislocation densities, $\rho$, range from $10^{10}$ to $10^{15}/m^2$, i.e., $10^{10}$ to $10^{15}$ meters of dislocations in a cubic meter of material, values that typically increase rapidly under applied stress (or strain).[6] Dislocations form organized structures such as walls, cells and pile-ups. Differences owing to the topological constraints placed by crystallography greatly add to the complexity of describing dislocation evolution and dynamics.

We want to emphasize the importance of dislocations in essentially all types of materials systems. Dislocations are found in ceramics, leading to plasticity in some systems at high temperature. They are extremely important in determining the properties of thin films of silicon, where they affect not only the mechanical properties, but also the diffusive properties of the

---

[5] If no dislocations existed, the strength would be determined by the force necessary to move an entire plane of atoms one step over an adjacent plane.

[6] *fcc* materials often obey the Taylor law, in which the applied stress is related to the dislocation density by $\rho = \alpha \tau^2$, where $\alpha$ is a constant.

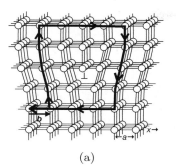

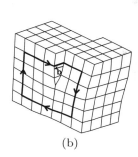

(a)  (b)

**Figure B.7** Types of dislocations. (a) Pure edge dislocation, showing the Burgers loop and the Burgers vector, **b**. The Burgers vector is perpendicular to the line direction and the glide plane is along the Burgers vector. (b) Pure screw dislocation, showing the Burgers loop and the Burgers vector, **b**. The Burgers vector is parallel to the line direction. Figures adapted from [157].

films. They are found in molecular crystals as well, including materials used in high explosives, in which the dislocations may play a critical role in creating hot spots for ignition.

## B.5.2 Structure of dislocations

Dislocations can be easiest visualized in their limiting "pure" states of edge dislocations and screw dislocations, which are shown in Figure B.7.

Consider first an edge dislocation, as shown in Figure B.7a. It can be described as an extra half plane of atoms inserted into the lattice. The extra plane of atoms introduces distortions in the lattice, which leads to an elastic stress and strain field. The dislocation can be characterized by two parameters, the Burgers vector, **b**, and the line direction, $\hat{\xi}$. The Burgers vector is a measure of the lattice distortion caused by the dislocation, and can be found by creating a Burgers circuit, which is shown as the heavy black line in Figure B.7a. One starts in the bottom half of the figure and traces a line vertically a set number of lattice positions, say $m$, continuing the line to the right another $m$ lattice sites, then down the lattice for $m$ sites, and then left $m$ sites. The difference between the starting position and the final position after the Burgers circuit is the Burgers vector, as shown in Figure B.7a. The Burgers vector thus is a measure of the amount of lattice displacement associated with the dislocation. The line direction, $\hat{\xi}$, is along the direction of the extra half plane, which is into the paper in Figure B.7a. Thus for edge dislocations, $\hat{b} \perp \hat{\xi}$. Note the organization of atoms at the center of the dislocation line (into the page in Figure B.7a). The displacements of the atoms in this region are large and can only be described using detailed atomistic simulations. This region is called the *dislocation core* and has an excess energy associated with it, called the core energy and given in units of energy per length of the dislocation.

A pure screw dislocation in shown in Figure B.7b. In this case, drawing a Burgers circuit shows that screw dislocations have their Burgers vector parallel to the line direction, $\hat{b} \parallel \hat{\xi}$.

In general, however, dislocations have mixed character. In Figure B.8 we show a dislocation loop, with the variation of dislocation character around the loop. This loop was created by a shift of atoms in the indicated $ABCD$ plane in Figure B.8a. In Figure B.8b we show the atomic positions in the plane indicated by $E$. Note that the deformation was in the $x$ direction, i.e., the Burgers vector of the deformation is along the $x$ axis. The lines of the two edge dislocations

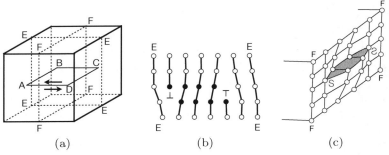

**Figure B.8** Dislocation loop showing mixed dislocation. (a) Small block of material is deformed as shown with the arrows. (b) Atom positions in the two-dimensional plane described by $E$. The atoms that have moved are indicated in black. Note the two edge dislocations, with the one on the right pointing down and the one on the right pointing up. These dislocations mark the $DC$ and $AB$ boundaries of the displaced atoms in (a) as it intersects the $E$ plane. (c) Atom positions in the two-dimensional plane described by $F$. The extra plane caused by the movement of the atoms is indicated as a shaded region and is bounded by $AD$ and $BC$ in (a). Adapted from Weertman and Weertman [351].

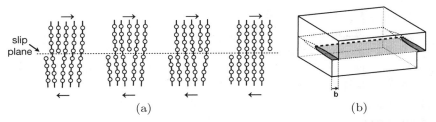

**Figure B.9** Dislocation motion. Figures adapted from [157]. (a) Edge movement along its glide direction, showing that movement of dislocation leads to deformation. Moving the dislocation one lattice space requires relatively small atomic motions. (b) Motion of a screw dislocation. The heavy dashed line is the dislocation, the light shaded region is the dislocation plane, and the heavy shaded region is the deformed material. As the dislocation moves to the back surface of the drawing, the dislocation plane will grow, as will the deformed region.

are into the paper and are thus perpendicular to the deformation. In Figure B.8c, we show the screw dislocation along the deformation in the plane $F$. The Burgers vector must be the same as for the edge dislocation in Figure B.8b, as it is the same change in atom positions. Thus, the Burgers vector is the same for the entire dislocation loop. In a general loop, the edge component at a point is given by $\mathbf{b}_e = \hat{\xi} \times (\mathbf{b} \times \hat{\xi})$, where $\hat{\xi}$ is the line direction (the tangent to the loop) at that point, and the screw component by $\mathbf{b}_s = (\mathbf{b} \cdot \hat{\xi})\hat{\xi}$ [147].

## B.5.3 Movement of dislocations

The importance of dislocations is that they move relatively easily through a lattice and that their motion "carries" the deformation. In Figure B.9, we show schematically how edge and screw dislocations move. Figure B.9a shows how the movement of an edge dislocation requires only

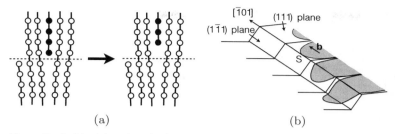

(a)                            (b)

**Figure B.10** (a) Dislocation climb in edge dislocations. Note the planes of atoms (extending into the paper) indicated in black. Movement of the edge dislocation off its slip plane requires diffusion of atoms either from (in this case) or to the dislocation. (b) Cross slip of dislocation onto a new, related, slip plane, followed by cross slip back to the original plane (double cross slip), as described in the text.

the shifting of one plane of atoms relative to the lattice. It also shows how the movement of the dislocation leads to deformation, with the extra step at the surface on the right as the dislocation leaves the system. Note that there is a minimum stress required to move a dislocation from one lattice site to another. Called the *Peierls-Nabarro stress*, it ranges in size from $\leq 10^{-6}-10^{-5}\,\mu$ for *fcc* systems to $10^{-2}\,\mu$ in covalent systems like silicon, where $\mu$ is the shear modulus, with body-centered cubic and hexagonal materials somewhere in between [157]. Edge dislocations are thus restricted to movement along a plane, called the slip or glide plane, which is indicated as a dashed line in Figure B.9a, remembering that the lattice extends into the plane of the paper. Movement off the slip plane requires physical mechanisms covered in the next section.

In Figure B.9b, the movement of a screw dislocation is shown, and explained. Note that as the screw dislocation moves, there is an extension of the dislocation step on the surface of that small block of material. That extension is a measure of the deformation caused by the dislocation movement.

## B.5.4 Movement of edge dislocations off their slip plane

There are two primary mechanisms by which dislocations can move off their slip planes: climb and cross slip. The movement of an edge dislocation perpendicular to its slip plane is called *climb*, and is shown in Figure B.10a. Climb is a diffusive process, requiring movement of material into or out of the dislocation core. Thus, climb is thermally activated and is usually insignificant at low temperatures.

Screw dislocations tend to move along certain crystallographic planes. In face-centered cubic materials, for example, screw dislocations move in {111} planes. Screw dislocations can switch from one {111} type plane to another, as long as the new plane contains the direction of the Burgers vector **b**. This behavior is shown in Figure B.10b. In that figure, a dislocation with Burgers vector $\mathbf{b} = 1/2[\bar{1}01]$ moves along a (111) plane. The leading screw component switches to a $(1\bar{1}1)$, with the rest of the expanding dislocation loop following, including all the edge components. This is an example of an important process called *cross slip*. In enables edge dislocations to move off their glide planes. In Figure B.10b, the dislocation switches back to its original (111) plane, which is an example of a phenomenon called double cross slip.

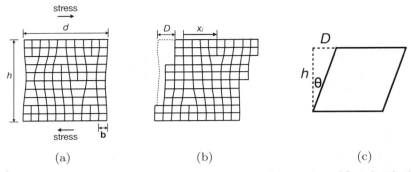

**Figure B.11** Relation of dislocation motion to strain. Figures adapted from [157]. The lattice planes extend into the paper. (a) System of edge dislocations before deformation. (b) System of dislocations from Figure B.11a after dislocations have moved under stress. Total displacement is $D$.

### B.5.5 Relation of dislocation motion to plastic strain

In Figure B.9 we show how the movement of a dislocation leads to deformation of a lattice. It should not be surprising, then, that the net *plastic strain* can be related to the movement of dislocations.[7] Hull and Bacon, in their excellent introductory text on dislocations, present a simple expression for the relation of plastic strain to dislocation movement. This discussion is based on that text.

Consider the simple system of parallel edge dislocations in Figure B.11a. If a stress is applied to that system, the dislocations will move and slip will occur, as shown in Figure B.11b. Dislocations of opposite sign will move in opposite directions, as discussed above. The macroscopic displacement, $D$, can be written in terms of the dislocation movement as

$$D = \frac{b}{d} \sum_{i=1}^{N} x_i \,,$$ (B.24)

where $N$ dislocations move in the system, $x$ is the change in position of a moving dislocation and $b$ is the magnitude of the Burgers vector.

The shear strain is shown in Figure B.11a, with

$$\epsilon_P = \theta = \tan^{-1}\left(\frac{D}{h}\right) \approx \frac{D}{h} \,,$$ (B.25)

where the right-hand term arises from a small-angle approximation. Thus, the macroscopic shear plastic strain is

$$\epsilon^p = \frac{D}{h} = \frac{b}{dh} \sum_{i=1}^{N} x_i \,.$$ (B.26)

[7] Elastic and plastic strain is discussed in Appendix H.

If $\langle x \rangle$ is the average distance a dislocation moves, then $\sum_{i=1}^{N} x_i = N \langle x \rangle$ and

$$\epsilon^p = \frac{b}{dh} \sum_{i=1}^{N} x_i = b \frac{N}{dh} \langle x \rangle = b\rho \langle x \rangle , \tag{B.27}$$

where $\rho$ is the density of dislocations. While the figures look two-dimensional, it is important to remember that dislocations are three-dimensional. The dimensions of our system are, in two dimensions, $h$ by $d$. Suppose we extend that in the third dimension by a length $L$, then the volume is $V = hdL$. The total length of dislocations is $NL$ (they are straight), so the density is $\rho = NL/(hdL) = N/(hd)$. Note that the dislocation density in Eq. (B.27) only refers to the *mobile* dislocations, i.e., if dislocations are trapped by defects or other dislocations, then they cannot move and cannot add to the deformation.

Similar expressions hold for screw and mixed dislocations. Thus, if we can calculate the total dislocation movement, we can determine macroscopic strain from a simulation of dislocations. From the applied stress, we can calculate the elastic part of the strain, $\epsilon^e$, from Eq. (H.16) and, thus, the total strain, as discussed in Appendix H.5, is given by

$$\epsilon = \epsilon^e + \epsilon^p . \tag{B.28}$$

## B.5.6 Stresses and forces

The force per unit length on a dislocation is determined using the Peach-Koehler relation [147, 255].[8]

$$\frac{\mathbf{F}}{L} = \left( \mathbf{b} \cdot \sigma_T \right) \times \hat{\xi} , \tag{B.29}$$

where $\mathbf{b}$ is the Burgers vector, $\hat{\xi}$ is the line direction of the dislocation and $\sigma_T$ is the total stress tensor evaluated at the position of the dislocation. The total stress is in general a sum of contributions from various sources

$$\sigma_T = \sigma_{applied} + \sigma_\perp + \sigma_{other} , \tag{B.30}$$

where $\sigma_{applied}$ is the applied (external) stress, $\sigma_\perp$ is the stress from other dislocations (including the self stresses) in the system, and $\sigma_{other}$ is the stress from other defects (grain boundaries, vacancies, etc.).

We note that one usually assumes the results from linear elasticity in Appendix H.4 to solve for the stress field. It is also usual to assume isotropic elasticity. The expressions for fully anisotropic elasticity are available [147], but are considerably more complicated and take more computational time. Here we give only expressions for the isotropic case.

The total stress at a point on a dislocation, $a$, arises from all other dislocations in the system, $b$, and the self-stress of dislocation $a$, i.e.,

$$\sigma_\perp^{(a)} = \sum_b \sigma_\perp^{(b)} + \sigma_{self}^{(a)} , \tag{B.31}$$

The self stress is discussed in Appendix B.5.7.

---

[8] The level of detail in this section is needed for the discussion in Section 13.5.

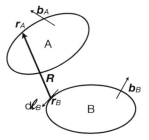

**Figure B.12** Schematic view of the evaluation of the stress at a point on dislocation loop $A$ that arises from loop $B$. A line integral, Eq. (B.33), is performed along the differential line direction, $d\ell_B$, along dislocation $B$. $\mathbf{R}$ is the vector from the integration point on loop $B$, $\mathbf{r}_B$ to a point, $\mathbf{r}_A$, on dislocation $A$, $R = |\mathbf{r}_A - \mathbf{r}_B|$. $\mathbf{b}_A$ and $\mathbf{b}_B$ are the Burgers vectors of $A$ and $B$, respectively.

The evaluation of the stress at a point on a dislocation from another dislocation is shown schematically in Figure B.12. The stress from dislocation $B$ at a point on dislocation $A$ can be evaluated by a line integral along dislocation $B$ of a function of the separation of the dislocations as well as the Burgers vector on dislocation $B$ and the differential line direction, $d\ell_B$.[9] Once the stress is determined at $\mathbf{r}_A$, the force on that point is found using the Peach-Koehler force in Eq. (B.29), with Burgers vector $\mathbf{b}_A$ and noting that the line direction $\boldsymbol{\xi}$ is the tangent to the dislocation line at $\mathbf{r}_a$. Much more detail is given in [147] and [82].

The integrals in Eq. (B.33) for a straight dislocation, either infinitely long or a finite segment, have been worked out analytically and are summarized in the book by Hirth and Lothe [147]. We will not repeat all of those relations here, but want to emphasize their availability, as they form the basis for a number of simulation methods. As an example, the $xy$ component of the stress from an infinite dislocation oriented such that its Burgers vector is along $\hat{x}$ and its line direction is along $\hat{z}$ is [147]

$$\sigma_{xy} = \frac{\mu b}{2\pi(1-\nu)} \frac{x(x^2 - y^2)}{(x^2 + y^2)^2},$$ (B.32)

where $b$ is the Burgers vector. Note that the stress is long-ranged, going approximately as $1/r$ for large $r$, and thus the force is as well.[10] Accurate evaluation of these forces requires numerical methods similar to those discussed in Appendix 3.8.

---

[9] From a very useful, if a bit mathematical, paper by deWit [82], the components of the stress tensor arising from a dislocation loop

$$\sigma_{ij}^B = \frac{\mu b_n^B}{8\pi} \oint \left[ R_{,mpp}\left(\epsilon_{jmn}d\ell_i + \epsilon_{imn}d\ell_i\right) + \frac{2}{1-\nu}\epsilon_{kmn}\left(R_{,ijm} - \delta_{ij}R_{,ppm}\right)d\ell_k \right],$$ (B.33)

in which the notation that the $x$ component of a vector is indicated by a subscript 1, the $y$ component by a subscript 2, and the $z$ component by a subscript 3. In any expression in which an index is repeated, it implies a sum. For example, $x_i x_i = \sum_{i=1}^3 x_i^2 = x_1^2 + x_2^2 + x_3^2 = x^2 + y^2 + z^2$. $\delta_{ij}$ is the Kronecker delta ($\delta_{ij} = 1$ if $i = j$ and 0 otherwise) and $\epsilon_{ijk}$ is the Levi-Civita tensor defined in Eq. (C.52). $R$ is the distance between the point at which the stress is being evaluated and the point along the integration of the dislocation curve. The notation $R_{,i}$ indicates the derivatives of $R$ with respect to $x_i$. For example, $R_{,mkn}$ indicates the derivative $\partial^3 R/\partial x_m^B \partial x_k^B \partial x_n^B$, where $R = |\mathbf{r}_A - \mathbf{r}_B|$. To evaluate the stress at a point on dislocation $A$, the integral around the whole line of $B$ must be performed. Finally, $\mu$ is the shear modulus and $\nu$ is Poisson's ratio.

[10] Please see the discussion in Section 3.6.

There are a number of approximations in Eq. (B.33). First, as noted above, Eq. (B.33) was derived assuming isotropic linear elasticity. Also, when two dislocations are sufficiently close to each other so that the dislocation cores begin to overlap, Eq. (B.33) is no longer valid.

### B.5.7 Self energies and forces

An important feature of dislocations is that they have a *self energy*, i.e., there are terms in the energy that arise from this dislocation itself. One of these terms is the energy arising from the structure of the dislocation cores. As noted above, the total core energy is proportional to the length of the dislocation and scales with the Burgers vector as $b^2$. Thus, any change in length of a dislocation is accompanied by a change in energy and thus there is a force that works against creating new dislocations and dislocation length. An easy way to think of this term is as a *line tension*. An approximate form for a dislocation line tension, $T$, which equals the self energy of the dislocation core, $E_{dis}$ when both are measured as an energy per unit length of dislocation, is [157]

$$T = E_{dis} = \alpha \mu b^2 , \tag{B.34}$$

where $\alpha \approx 0.5 - 1.0$. Based on this expression, the stress, $\tau_o$, required to form a curved dislocation with a radius of curvature equal to $R$ is [157]

$$\tau_o = \frac{\alpha \mu b}{R} . \tag{B.35}$$

There is another term in the self energy (and thus force) of a dislocation that arises from the interaction of one part of the dislocation with the elastic stress field of another part of the same dislocation. The self stress would have the same form as in Eq. (B.33) and the self force would be evaluated by the Peach-Koehler force in Eq. (B.29).

### B.5.8 Dislocation non-linear interactions and reactions

When dislocations are close to each other, the cores of the dislocations begin to overlap and linear elasticity no longer holds. Such small separations correspond to a realm in which the details of the atomic nature of dislocation dominate. Molecular dynamics simulations indicate that for separations beyond ten Burgers vectors or so, linear elasticity holds, at least for interacting screw dislocations in face-centered cubic materials [301]. For most discrete dislocation simulations, that distance is so small that the details of the short-range interactions can be ignored – the dislocation simulations are typically on the scale of microns and the core size is on the scale of Å.

Short-range interactions between dislocations can have important consequences on dislocation behavior, as both a source of new dislocations and a sink of existing dislocation. A classic example is shown in Figure B.13a, in which two edge dislocations of opposite sign move along the same slip plane until they combine, eliminating the defect, leaving behind a perfect lattice. This process is called annihilation and serves to reduce the density of dislocations. In

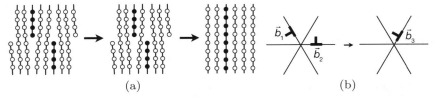

(a)                                                                          (b)

**Figure B.13** Dislocation reactions. (a) Dislocation annihilation. Two dislocations of opposite sign combine to eliminate the dislocation. (b) Dislocation reactions, in which two dislocations combine to make another. The total Burgers vector must be conserved in the reaction, i.e., $\mathbf{b}_3 = \mathbf{b}_1 + \mathbf{b}_2$.

Figure B.13b, two dislocations combine to form a new dislocation, with the restriction that the total Burgers vector after the "reaction" must equal that from before the reaction, $\mathbf{b}_3 = \mathbf{b}_1 + \mathbf{b}_2$. In discrete dislocation simulations, described in Chapter 13, such interactions are usually handled by rules and models.

Two other, related, types of short-range interaction are *intersections* and *junctions*. Intersections occur when non-parallel dislocations glide into and through each other, introducing jogs into the dislocations, which affect subsequent motion. Junctions form when dislocations interact at longer range, forming stable structures. Junctions may move easily (glissile) or be hindered in their movement (sessile). In any case, intersections and junctions serve as important barriers for dislocation motion, leading to pile-ups and the development of the *forest* of dislocations that forms at high dislocation densities. Details are beyond this text, but are covered very clearly in the book by Hull and Bacon [157].

### B.5.9 Dislocation sources

As noted above, the density of dislocations, measured in units of length per volume, increases under plastic load. Some increase in density arises from the elongation of dislocations during movement. However, they are mechanisms that serve as sources of dislocations. The most well-known of these is the Frank-Read source, shown schematically in Figure B.14 [273]. A dislocation segment is pinned at both ends, perhaps by an intersection with another dislocation, at a solute particle, etc. A stress acting in the appropriate direction causes the dislocation to bow out on its slip plane away from the pinning points. The shape of the bow out of the Frank-Read source in Figure B.14 can be accounted for by a simple line-tension model. The dislocation will continue to bow, until it eventually bows back around the original line between the pinning points. As it continues to bow, eventually the two lines of the dislocation will be close to each other. Since the Burgers vector is the same, but the line directions are opposite, the dislocations can annihilate as in Figure B.13a, forming a dislocation loop, which continues to grow under stress. The residual dislocation moves back to the original line between pinning points, from which the process begins again. A single Frank-Read source can thus generate many dislocations. Other sources of dislocations include nucleation at other defects, such as surfaces or grain boundaries.

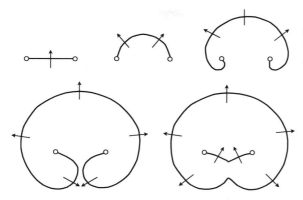

**Figure B.14** Schematic view of a Frank-Read source, as discussed in the text. The arrows indicate the direction of the dislocation movement. Adapted from [273].

## B.5.10 Dynamics

Dislocation motion is damped owing to interactions with the lattice, interactions with other defects, etc. We can approximate that effect by assuming that there is an added frictional force, $\gamma\mathbf{v}$, with the net force on the dislocation being

$$\mathbf{F}_{net} = \mathbf{F} - \gamma\mathbf{v}, \tag{B.36}$$

where $\gamma$ is a damping term and $\mathbf{v}$ is the dislocation velocity.

Dislocation drag arises from a number of sources. For example, the phonon drag coefficient varies with temperature as $\gamma_{phonon} = C\mu b(T/\Theta)/C_t$, where $C$ is a constant, $C_t$ is the sound speed, $\Theta$ is the Debye temperature, and $g$ is a function that (ignoring electron scattering) varies from 0 to 1 at $T = \Theta$, with a $T^{-3}$ dependence at low $T$ [147]. Climb (Figure B.10a) can be described as diffusive motion by assuming a drag coefficient for climb given by $\gamma_{climb} = b_e^2 k_B T/D_s V_a$, where $b_e$ is the edge component of the Burgers vector, $D_s$ is the self-diffusion coefficient and $V_a$ is the volume of a vacancy. Note that this is valid only at high temperatures.

In the limit that the drag term is large, we can ignore inertial effects in solving the equations of motion (i.e., the acceleration) and use the steady-state velocity ($\mathbf{F}_{net} = 0$ in Eq. (B.36)), yielding a linear force-velocity equation of motion

$$\mathbf{v} = \frac{1}{\gamma}\mathbf{F} = M\mathbf{F}, \tag{B.37}$$

where $M$ is a mobility. This relation is often referred to as the overdamped limit, as discussed in Section 13.1. Note that for some crystal structures, $M$ may be anisotropic (different mobilities along different directions) and a tensor form may be required. While assuming overlapped dynamics seems to work well in many cases, in some circumstances, such as high-strain rates, the full equation of motion must be employed [348].

## B.5.11 Crystallography of dislocations

The behavior of dislocations is highly dependent on the crystallography of the material. For example, the available slip systems in a body-centered cubic system differ from those in a

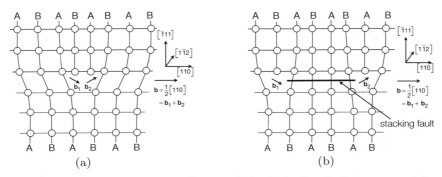

**Figure B.15** Partial dislocations in a face-centered cubic lattice. the Burgers vector is $\mathbf{b} = \frac{1}{2}[110]$, the normal to the $(\bar{1}11)$ slip plane is $[\bar{1}11]$ and the direction into the paper is $[1\bar{1}2]$. (a) Formation of $\mathbf{b}_1 = \frac{1}{6}[\bar{2}11]$ and $\mathbf{b}_2 = \frac{1}{6}[\bar{1}2\bar{1}]$ partials, with $\mathbf{b} = \mathbf{b}_1 + \mathbf{b}_2$. (b) Partial dislocations separated by a stacking fault extending into the plane.

face-centered cubic material, which also are different from those in a hexagonal close-packed system. These differences are quite important and lead to deformation properties that vary widely between crystal structures. We redirect the interested reader to other sources for more information, in particular [157], in which the discussion is clear and concise.

In the face-centered cubic system, for example, the smallest interatomic separation is along the {111} planes shown in Figure B.5b. As noted above in Eq. (B.34), the self energy of a dislocation goes as the Burgers vector squared, $E_{dis}$ $b^2$. Thus, the dominant Burgers vector in an *fcc* solid is the smallest vector between lattice sites, which is the vector connecting the nearest neighbors. Thus, the dominant Burgers vector in an *fcc* solid is of the type $\frac{1}{2}\langle 110 \rangle$, with the slip being on the {111} planes. We show this for the $(1\bar{1}1)$ slip plane in Figure B.15.

Because of the structure of the atomic lattices, it is energetically favorable for the $\mathbf{b} = \frac{1}{2}[110]$ to split into two dislocations, called *partial dislocations*, as shown in Figure B.15a. The dislocations are of the form

$$\mathbf{b} \rightarrow \mathbf{b}_1 + \mathbf{b}_2 \qquad (B.38)$$

or

$$\frac{1}{2}[110] \rightarrow \frac{1}{6}\langle 211 \rangle + \frac{1}{6}\langle 12\bar{1} \rangle,$$

where we use the $\langle \rangle$ notation to indicate families of directions. The partials do not have to stay together. They can separate, forming a stacking fault plane as shown in Figure B.15b. The separation between partials depends on the Burgers vectors and the stacking fault energy. The formation of partials in *fcc* systems affects their deformation properties, especially in promoting cross slip, interactions with grain boundaries, nucleation of twin boundaries, etc.

# B.6 POLYCRYSTALLINE MATERIALS

Planar defects include any interfaces between volumes of a material that differ in some way. These regions may be the same phases, but with different crystallographic orientations. They may be different phases, in which the composition or thermodynamic phase changes across the interface. The surfaces of the material are also interfaces, in this case between the material and the atmosphere, which may play a critical role in bulk materials because of corrosion, as well as in nano-sized materials, where surfaces can dominate behavior.

To simplify the discussion, we will focus on one type of boundary, the boundary between the crystallites that make up a polycrystalline material. The crystallites are called *grains* and the interfaces between them are *grain boundaries*. The grains have varying size and orientation. The distribution of these orientations is called the *texture*. The orientations may be random, appropriately called random texture, or directed, which largely arises from the processing of the material. Almost all metals, and most ceramics, are polycrystalline.

The distribution of grain boundaries in a material is referred to as its *microstructure*. The microstructure of a material can have a profound effect on many of its properties. Grain boundaries typically have larger local volumes and thus more space between atoms, enhancing diffusion. This enhanced interfacial diffusion plays an important role in many phenomena, such as sintering.[11] The distribution of grains also has a major impact on the macroscopic properties of the material. Thus, it should be no surprise that understanding the development, evolution, and consequence of microstructures is a major thrust in the modeling and simulation of materials.

While we describe grain boundaries and grain growth in detail, we emphasize that there are many types of planar interfaces that are important in materials. Inclusions, which are solid particles embedded within a material, can influence material properties through, for example, hindering the movement of dislocations and pinning of grain boundaries. Interphase boundaries in multiphase systems play similar roles. The evolution and impact of such boundaries are covered in detail in other sources. For example, in metals they are described in many texts, including [263].

## B.6.1 Grain boundaries

The boundaries between two grains are sheets; they are two-dimensional, but not necessarily flat, as shown in Figure B.16. Grain edges occur where three or more grains meet along a curve, again as shown in Figure B.16. Grain vertices are the points where four or more grains meet.

A grain boundary has five macroscopic degrees of freedom describing the relative orientations of the two grains: the orientation of the boundary to the two grains and the three-dimensional rotation required to bring the grains into coincidence. To simplify matters, boundaries are usually characterized by the misorientation angle between the grains. In Figure B.17a, we show one type of boundary, the tilt boundary, in which one grain is tilted relative to the other. In Figure B.17b a twist boundary is shown. Low-angle boundaries, in which the misorientation angle is less than $11°$ or so, can be modeled as a set of dislocations. Larger-angle boundaries are generally more complicated, consisting of structural units that depend on the grain disorientation and the plane of the interface.

---

[11] Sintering is a process by which powders are consolidated into a material. It is diffusion based, and thus highly temperature dependent. It can be used for any powders, and is commonly used in ceramics and metals.

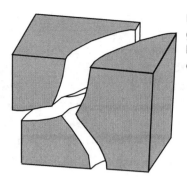

**Figure B.16** Three-dimensional view of grain boundaries between three grains, each grain having a different orientation. Note the grain edge between the three grains, which is a curved line in space. Figure courtesy of Professor Greg Rohrer, Carnegie Mellon University.

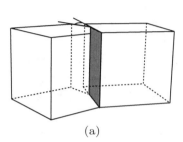
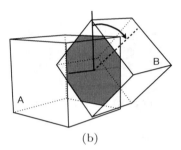

(a)                                    (b)

**Figure B.17** (a) Schematic view of a tilt boundary. The two grains share the same horizontal and vertical planes, but are rotated relative to one another. (b) Schematic view of a twist boundary. Grain *B* joins the back plane of grain *A*. *B*'s vertical axis has been rotated as shown relative to the vertical axis of *A*.

Because atomic bonds are altered at grain boundaries from what they would be in a perfect, equilibrium crystal, boundaries contribute a finite, positive, free energy per unit boundary area to the system energy. Low-angle boundaries have an energy described by the *Reed-Shockley equation*

$$\gamma_s = \gamma_o \theta \left( A - \ln \theta \right), \tag{B.39}$$

where the disorientation angle is $\theta = b/h$, where $b$ is the Burgers vector of the dislocations making up the boundary and $h$ is the separation between dislocations in the boundary. $\gamma_o = \mu b/(4\pi(1-v))$, where $\mu$ is the shear modulus and $v$ is Poisson's ratio.[12] $A = 1 + \ln(b/2\pi r)$ and $r_o$ is the size of the dislocation core.

The energy of high-angle boundaries is more complicated and no single equation exists to describe it. The variation of grain-boundary energy with misorientation angle is periodic in the misorientation, since rotation of one grain relative to another eventually returns the two grains back into registry. Note also that there are "special boundaries" with low grain-boundary energy. These are boundaries in which an ordered structure forms.

In three dimensions, a trijunction is formed when three grains join in a line, as shown in Figure B.18. In two dimensions, that line is a point, with the angles between the boundaries

---

[12] Appendix H.

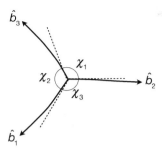

**Figure B.18** Angles at a trijunction in two dimensions. The vectors $b$ indicate the boundary. When all grain-boundary energies are equal, the dihedral angles $\theta_1 = \theta_2 = \theta_3 = 120°$.

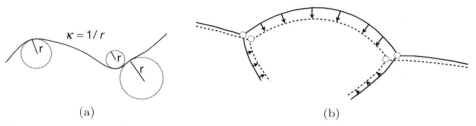

(a)  (b)

**Figure B.19** Curvature-driven movement of a grain boundary in two dimensions. (a) The curvature at a point is equal to the inverse of the radius of the circle that best approximates the curve at that point (a sphere in three dimensions). (b) Movement towards the center of curvature.

being related to the interfacial energies by the equation

$$\frac{\gamma_1}{\sin \chi_1} = \frac{\gamma_2}{\sin \chi_2} = \frac{\gamma_3}{\sin \chi_3}, \tag{B.40}$$

where the $\gamma$s are the interfacial free energies. If the interfacial energies are equal, then the equilibrium angles are equal to 120°. In real systems, the boundary energies are not necessarily equal, as discussed, so the trijunctions will not always have 120° angles. Many simulations, however, assume isotropic boundaries, so they should exhibit 120° angles.

Because of the positive grain-boundary energy, a polycrystalline material is *metastable*, the lowest energy of the material being a crystal without boundaries. It should not be surprising that when you heat a material, the grains evolve to lower the total grain-boundary area, and thus the energy associated with the boundaries. Thus, the mean grain size increases over time. This process is referred to as grain growth.

## B.6.2  Grain growth

It has been observed from experiment that grain growth occurs by grain boundaries migrating toward their centers of curvature, as shown in Figure B.19a. It has also been observed that, in general, grains with few sides shrink, and grains with many sides grow. The grain structure is often observed to be self similar – while the average sizes increases with time, the shapes of the grains and the distribution of grain shapes do not change. Any successful model of grain growth must reproduce these features.

Burke and Turnbull [50] employed simple approximations of the grain morphology and a simple rate theory for movement of atoms across the grain boundaries. They showed that a spherical boundary moves with a speed

$$v = M\kappa \,, \tag{B.41}$$

where $M$ is the atomic mobility and $\kappa$ is the mean curvature of the boundary. They assumed a constant activation energy for moving an atom across a boundary, $\Delta E_A$, which leads to a mobility of the form

$$M = \left(\frac{\gamma}{k_B T}\right) v \lambda e^{-\Delta E_A / k_B T} \,, \tag{B.42}$$

which has dimensions of a diffusion coefficient. $\gamma$ is the grain boundary interfacial energy, $\lambda$ is the jump distance, $v$ is a jump frequency, and $k_B$ is Boltzmann's constant. Assuming spherical grains of radius $R$, so that the curvature goes as $1/R$, they showed that the mean grain size (in two dimensions the area $A$) should approach $\langle A \rangle \sim t$ at long times. While this relation comes after many approximations, it seems to match experiments.

von Neumann derived a simple expression for the rate of growth in two dimensions [326],

$$\frac{dA}{dt} = -M \frac{\pi}{3}(6 - n) \,, \tag{B.43}$$

where $A$ is the area of the grain, $M$ is the mobility, and $n$ is the number of sides of the grain. He based this expression on some simple geometric assumptions, including that grain boundaries are (more or less) straight and that for isotropic grain growth (i.e., all boundaries have the same energy), grains meet only at trijunctions (three grains in two dimensions meeting at a point) with an equilibrium angle of 120°, which suggests that grains with $n < 6$ shrink, those with $n > 6$ will grow and $n = 6$ is stable. This result matches many experimental observations. Work by MacPherson and Srolovitz has extended this analysis to describe three-dimensional grains as well [210]. The results are more complicated, but the basic ideas are the same.

We will not discuss in detail in this text such important phenomena as abnormal grain growth, or recrystallization, or any effects of three dimensions. Many of these phenomena have, however, been successfully modeled by methods described in this book.

## B.7 DIFFUSION

Diffusion is a process of great importance in essentially all materials systems.[13] It is a principal mechanism for transporting atoms through a material, which manifests itself in many important phenomena, such as phase transformations, corrosion, etc., and has an impact on many technological materials and products, including the thin films in semiconductor devices.

---

[13] Please see one of the many texts that describe diffusion in detail, e.g., [263, 287].

In Section 2.3 we discuss the relation of vacancy diffusion to the diffusion of atoms in a lattice. A similar analysis can be done for interstitial diffusion, which, for some systems, dominates the diffusion in a bulk material.[14] Diffusion depends on a number of variables, including initial concentration, surface concentration, diffusivity, time, and temperature.

## B.7.1 Continuum descriptions of diffusion

At the continuum level, the fundamental equations governing diffusion of a species ($i$) are Fick's first law, in which the flux is written in terms of the concentration of $i$,

$$\mathbf{J}_i = -D_i \nabla c_i , \tag{B.44}$$

and Fick's second law, in which the time rate of change of the concentration is written in terms of the flux,

$$\frac{\partial c_i}{\partial t} = -\nabla \cdot J_i . \tag{B.45}$$

$D_i$ is the diffusion coefficient of species $i$ and is a fundamental property of the species and the material through which it is diffusing. If $D_i$ is independent of position (as it would be for bulk diffusion), then the second law becomes

$$\frac{\partial c_i}{\partial t} = -D_i \nabla^2 c_i . \tag{B.46}$$

Equation (B.46) is a differential equation that, subject to boundary conditions that define the problem, can be solved to yield the concentration as a function of position $\mathbf{r}$ and time $t$, i.e., $c(\mathbf{t}, t)$. For example, if we assume that the diffusing material is initially localized at the origin with a concentration $c_o = 1$, then concentration profile at time $t$ is [287]

$$c(\mathbf{r}, t) = \frac{1}{8(\pi Dt)^{3/2}} e^{-r^2/4Dt} , \tag{B.47}$$

where we have written $c(\mathbf{r}, t)$ in spherical polar coordinates; $c(\mathbf{r}, t)$ gives the concentration over space at time $t$. Since material is conserved, then summing the concentration at each point must yield the same total concentration that the system had initially, in this case 1. $c(\mathbf{r}, t)$ is a continuous function, so we need to integrate $c(\mathbf{r}, t)$ over all space. Assuming that the diffusion is isotropic (the same in all directions), then the integral in spherical polar coordinates becomes

$$\int_0^\infty c(\mathbf{r}, t) 4\pi r^2 dr = 1 . \tag{B.48}$$

Using $c(\mathbf{r}, t)$ we can also relate the mean square displacement of the diffusing species to the diffusion coefficient. The mean square displacement is defined for a system with $N$

---

[14] Interstitial diffusion occurs when the diffusing atom is small enough to move between the atoms in the lattice and requires no vacancy defects to operate.

particles as

$$\langle r^2(t) \rangle = \langle |\mathbf{r}_i(t) - \mathbf{r}_i(0)|^2 \rangle, \tag{B.49}$$

where the average is taken over all particles. Since we started all the particles at the origin, $\mathbf{r}(0) = (0, 0, 0)$. $c(\mathbf{r}, t)$ is the distribution of all the diffusing species in space at a given time $t$. The average of $r^2$ is then just the integral over that distribution, as discussed in Appendix C.4. The mean square displacement becomes

$$\langle r^2 \rangle = \int_0^\infty r^2 c(\mathbf{r}, t) 4\pi r^2 dr = 6Dt, \tag{B.50}$$

or

$$D = \frac{1}{6t} \langle r^2 \rangle. \tag{B.51}$$

Equation (B.51) is valid at long times ($t$ large).

On the right-hand side of Eq. (B.51) we have an average (denoted by $\langle \rangle$) of the square of the distance between particle $i$ at time $t$ and its position at some arbitrary initial time, which we set to $t = 0$. The average is taken over all diffusing particles in the system. Since we expect $D$ to be a constant, we can infer from Eq. (B.51) that $\langle r^2 \rangle \propto t$ (so that $t$ cancels out).

The relation in Eq. (B.51) relates an important material property, the diffusion coefficient, to quantities that we can, in principle, calculate. For example, if we could calculate the movement of atoms through a solid, we could determine the mean square displacement to directly calculate $D$, using methods discussed in Chapter 6. There are also other ways to calculate $D$ that are somewhat more amenable to calculation. As we see in Appendix G, the diffusion coefficient is also related to an integral over a quantity called the velocity autocorrelation function, which is sometimes easier to calculate. That method of calculating $D$ is also discussed in Chapter 6.

Diffusion is an activated process involving the jumping of atoms from site to site and thus follows an Arrhenius law, i.e., it follows a relation of the form $D \sim e^{\Delta E/k_B T}$, where $\Delta E$ is the energy of the barrier to jump from site to site, $k_B$ is Boltzmann's constant and $T$ is the temperature. More details are given in Appendix G.8.

### B.7.2 Diffusion along defects

As noted above, defects such as dislocations and grain boundaries have structures that typically have a larger local volume than in the lattice. Diffusion is typically enhanced along those interfaces. Diffusion can also occur along defects, for example along grain boundaries, dislocations, or on free surfaces. We note that the overall effect of defects on the diffusivity of a bulk material is highly dependent on their concentration. Typically, diffusion coefficients follow the following relation: $D_{lattice} < D_{dislocation} < D_{grain\ boundary} < D_{surface}$ [263]. Calculation of diffusion along interfaces has been of great interest to the modeling community, as simulations can easily differentiate between lattice and defect diffusion, while that is not always so clear experimentally.

## Suggested reading

Many texts provide details neglected in this chapter.

- An excellent summary of many of these ideas is in the book by Rob Phillips, *Crystals, Defects, and Microstructures: Modeling Across Scales* [260].
- For dislocations, the introductory text, *Dislocations*, by Hull and Bacon [157], is an excellent place to start. A simpler text, by Weertman and Weertman [351], has excellent sketches, which make dislocation structures quite understandable. For a much fuller treatment, the classic text by Hirth and Lothe, *Theory of Dislocations* [147], is the best reference.
- A good treatment of the basics of diffusion is given in *Diffusion in Solids* by Shewmon [287]. Another excellent book on diffusion is Glicksman's *Diffusion in Solids: Field Theory, Solid-State Principles, and Applications* [123].

# C Mathematical backgound

The intent of this section is to review some mathematical topics. If these topics are completely unfamiliar to you, please consult the references at the end of the section. We note that this discussion is by no means complete – we only present topics that arise within the main body of the text.

## C.1 VECTORS AND TENSORS

Here we review the notation used in this text as well as some useful results for vectors and tensors.

### C.1.1 Basic operations

A vector $\mathbf{r}$ that identifies a point in space can be written in Cartesian coordinates as

$$\mathbf{r} = x\hat{x} + y\hat{y} + z\hat{z}, \tag{C.1}$$

where vectors are denoted by **bold** type and unit vectors (with magnitude equal to 1) by a $\hat{\ }$, i.e., $\hat{x}$. As shown in Figure C.1, vectors have length and direction.

If two points in space are denoted by $\mathbf{r}_1 = x_1\hat{x} + y_1\hat{y} + z_2\hat{z}$ and $\mathbf{r}_2 = x_2\hat{x} + y_2\hat{y} + z_2\hat{z}$, then the vector connecting point 1 to point 2 (i.e., pointing from 1 to 2) is

$$\mathbf{r}_{12} = \mathbf{r}_2 - \mathbf{r}_1 = (x_2 - x_1)\hat{x} + (y_2 - y_1)\hat{y} + (z_2 - z_1)\hat{z}. \tag{C.2}$$

The *length* of the vector $\mathbf{r}_{12}$ is just the distance between the points and is often denoted as $r_{12}$. The length is written as

$$r_{12} = |\mathbf{r}_2 - \mathbf{r}_1| = ((x_2 - x_1)^2 + (y_2 - y_1)^2 + (z_2 - z_1)^2)^{1/2}. \tag{C.3}$$

The *unit vector* $\hat{r}_{12}$ is the vector with unit length pointing along the line that connects point 1 to point 2 and is $\hat{r}_{12} = \mathbf{r}_{12}/r_{12}$, i.e., the vector divided by its length.

We can write a general vector in three dimensions as

$$\mathbf{a} = a_x\hat{x} + a_y\hat{y} + a_z\hat{z}, \tag{C.4}$$

where the components of the vector (e.g., $a_x$) give the value of the projection of the vector quantity in the $x$ direction. An example of such a vector could be the force of gravity, which varies in direction and magnitude depending on where it is being measured.

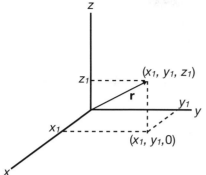

**Figure C.1** Vector in three-dimensional Cartesian coordinates.

Suppose we have two vectors **a** and **b**. The *dot product* of the two vectors is defined as

$$\mathbf{a} \cdot \mathbf{b} = a_x b_x + a_y b_y + a_z b_z . \tag{C.5}$$

The dot product is a scalar (a number) whose value is

$$\mathbf{a} \cdot \mathbf{b} = ab \cos\theta , \tag{C.6}$$

where $a$ is the magnitude of **a**, $b$ is the magnitude of **b**. $\theta$ is the angle between **a** and **b**. The magnitude of a vector is its dot product with itself, for example,

$$a = |\mathbf{a}| = \sqrt{\mathbf{a} \cdot \mathbf{a}} = \sqrt{a_x^2 + a_y^2 + a_z^2} . \tag{C.7}$$

The dot product of vectors that are perpendicular to each other ($\theta = 90^o$) is zero.

For notational convenience, the book sometimes uses a more convenient way to write the Cartesian coordinates $x$, $y$, and $z$, using $\hat{x}_1 = \hat{x}$, $\hat{x}_2 = \hat{y}$, and $\hat{x}_3 = \hat{z}$. A vector then takes the form $\mathbf{a} = a_1 \hat{x}_1 + a_2 \hat{x}_2 + a_3 \hat{x}_3$ and the dot product between **a** and **b** is given by

$$\mathbf{a} \cdot \mathbf{b} = a_1 b_1 + a_2 b_2 + a_3 b_3 , \tag{C.8}$$

which is equivalent to Eq. (C.5). Using the notation

$$\sum_{i=1}^{m} a_i = a_1 + a_2 + a_3 + \cdots + a_{m-1} + a_m . \tag{C.9}$$

we have

$$\mathbf{a} \cdot \mathbf{b} = \sum_{i=1}^{3} a_i b_i . \tag{C.10}$$

This notation is very useful for writing compact expressions, especially when one invokes the so-called *repeated index convention*, which states that whenever two indices are the same in an expression, it implies a sum over those indices. For example, the dot product becomes simply $\mathbf{a} \cdot \mathbf{b} = a_i b_i$. We will not (generally) employ the repeated index convention in this text.

The *cross product* (denoted by $\times$) of two vectors yields a third vector that is mutually perpendicular to the first two. Consider the cross product of **a** and **b**

$$\mathbf{a} \times \mathbf{b} = \mathbf{c} \tag{C.11}$$

An easy way to remember how to evaluate the cross product is by evaluating the determinant[1]

$$\mathbf{c} = \begin{vmatrix} \hat{x} & \hat{y} & \hat{z} \\ a_x & a_y & a_z \\ b_x & b_y & b_z \end{vmatrix} = (a_y b_z - a_z b_y)\hat{x} + (a_z b_x - a_x b_z)\hat{y} + (a_x b_y - a_y b_x)\hat{z}, \tag{C.12}$$

which can be written in index form as

$$\mathbf{a} \times \mathbf{b} = (a_2 b_3 - a_3 b_2)\hat{x}_1 + (a_3 b_1 - a_1 b_3)\hat{x}_2 + (a_1 b_2 - a_2 b_1)\hat{x}_3. \tag{C.13}$$

The *magnitude* of the cross product is

$$c = |\mathbf{c}| = ab \sin\theta, \tag{C.14}$$

where $\theta$ is the angle between **a** and **b**. Note that the cross product of vectors that are parallel to each other ($\theta = 0$) is zero.

## C.1.2 Differential vector operators

The *gradient* of a scalar field is a vector field that points in the direction with the largest slope and whose magnitude is that slope.[2] It is indicated by the operator $\nabla$ and is defined by

$$\nabla A(x, y, z) = \frac{\partial A}{\partial x}\hat{x} + \frac{\partial A}{\partial y}\hat{y} + \frac{\partial A}{\partial z}\hat{z}, \tag{C.15}$$

which can also be written as

$$\nabla A(x_1, x_2, x_3) = \frac{\partial A}{\partial x_1}\hat{x}_1 + \frac{\partial A}{\partial x_2}\hat{x}_2 + \frac{\partial A}{\partial x_3}\hat{x}_3 = \sum_{i=1}^{3} \frac{\partial A}{\partial x_i}\hat{x}_i. \tag{C.16}$$

We will encounter the gradient often as it defines how a function varies in space, i.e., its shape, where large gradients indicate a steep slope, small gradients indicate a shallow slope.

Another measure of the shape of a function is the second derivative (i.e., the rate of change of the slope), which is given by the *Laplacian*

$$\nabla^2 = \frac{\partial^2}{\partial x^2} + \frac{\partial^2}{\partial y^2} + \frac{\partial^2}{\partial z^2}, \tag{C.17}$$

---

[1] For a $3 \times 3$ matrix, the determinant is the sum of the products of the downward diagonal rows minus the upward diagonal rows.

[2] A scalar field consists of a single value at each point in space. A vector field has both magnitude and direction at each point.

or by[3]

$$\nabla^2 = \frac{\partial^2}{\partial x_1^2} + \frac{\partial^2}{\partial x_2^2} + \frac{\partial^2}{\partial x_3^2} \, . \tag{C.18}$$

The Laplacian operates on scalar fields.

## C.1.3 Matrices and tensors

Second-rank tensors occur often in the theory of elasticity. For the purposes of this text, these tensors take the form of $3 \times 3$ matrices.[4] There are many relationships and properties of tensors. The book by Malvern [216] has an excellent summary of tensor properties. The stress tensor, for example, can be written as

$$\boldsymbol{\sigma} = \begin{pmatrix} \sigma_{11} & \sigma_{12} & \sigma_{13} \\ \sigma_{21} & \sigma_{22} & \sigma_{23} \\ \sigma_{31} & \sigma_{32} & \sigma_{33} \end{pmatrix} \, . \tag{C.19}$$

If a matrix **h** is defined as

$$\mathbf{h} = \begin{pmatrix} h_{11} & h_{12} & h_{13} \\ h_{21} & h_{22} & h_{23} \\ h_{31} & h_{32} & h_{33} \end{pmatrix} \, , \tag{C.20}$$

then the *transpose* of **h** is

$$h_{ij}^T = h_{ji} \, , \tag{C.21}$$

where we have reflected the components of **h** across its diagonal.

The determinant of a square matrix arises in a number of applications, for example in the solving of linear equations. For the $3 \times 3$ matrix in Eq. (C.20), the determinant is

$$det(\mathbf{h}) = \begin{vmatrix} h_{11} & h_{12} & h_{13} \\ h_{21} & h_{22} & h_{23} \\ h_{31} & h_{32} & h_{33} \end{vmatrix}$$

$$= h_{11}h_{22}h_{33} - h_{13}h_{22}h_{31} - h_{12}h_{21}h_{33} - h_{11}h_{23}h_{32}$$
$$+ h_{13}h_{21}h_{32} + h_{12}h_{23}h_{31} \, . \tag{C.22}$$

---

[3] The derivatives take somewhat different forms in other coordinate systems. For example, the Laplacian in spherical polar coordinates, where

$$x = x_1 = r \sin\theta \cos\phi$$
$$y = x_2 = r \sin\theta \sin\phi$$
$$z = x_3 = r \cos\theta \, . \tag{C.23}$$

is

$$\nabla^2 = \frac{1}{r^2} \frac{\partial}{\partial r}\left(r^2 \frac{\partial}{\partial r}\right) + \frac{1}{r^2 \sin\theta} \frac{\partial}{\partial \theta}\left(\sin\theta \frac{\partial}{\partial \theta}\right) + \frac{1}{r^2 \sin^2\theta} \frac{\partial^2}{\partial \phi^2} \, . \tag{C.24}$$

[4] A scalar is a zeroth-rank tensor and a vector is a first-rank tensor.

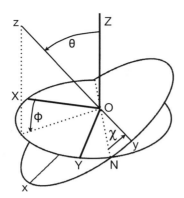

**Figure C.2** Definition of Euler angles. Adapted from [353].

## C.1.4 Euler angles

It is sometimes necessary to transform Cartesian coordinates in one system to that in another, rotated, system. In this text, this occurs most frequently in the discussion of the interactions between molecules. The most common way to perform that transform is through the use of the Euler angles.

With the angles defined as in Figure C.2 (note that there is more than one convention for how these are defined), the relation between a vector in a new, rotated, frame, defined by the vector $\mathbf{x} = (x, y, z)$ and that in the old frame, $\mathbf{X} = (X, Y, Z)$, is

$$\mathbf{x} = \mathbf{R}\mathbf{X}, \tag{C.25}$$

where the rotation matrix $\mathbf{R}$ is [353]

$$\begin{pmatrix} \cos\theta\cos\phi\cos\chi - \sin\phi\sin\chi & \cos\theta\sin\phi\cos\chi + \cos\phi\sin\chi & -\sin\theta\cos\chi \\ -\cos\theta\cos\phi\sin\chi - \sin\phi\cos\chi & -\cos\theta\sin\phi\sin\chi + \cos\phi\cos\chi & \sin\theta\sin\chi \\ \sin\theta\cos\phi & \sin\theta\sin\phi & \cos\theta \end{pmatrix} \tag{C.26}$$

## C.2 TAYLOR SERIES

The Taylor series gives us a way to *estimate* the value of a function at the point $x_o + a$ when we know the value of the function and at least some of its derivatives at $x_o$. Suppose one has a function $f(x)$. One can then *expand* $f$ in a Taylor series in powers of $a$

$$f(x_o + a) = f(x_o) + \left(\frac{\partial f}{\partial x}\right)_{x_o} a + \frac{1}{2}\left(\frac{\partial^2 f}{\partial x^2}\right)_{x_o} a^2 + \cdots . \tag{C.27}$$

The notation $(\partial f/\partial x)_{x_o}$ indicates a partial derivative of $f$ with respect to $x$ evaluated at $x = x_o$. The $n$th-order term in Eq. (C.27) has the form $(\partial^n f/\partial x^n)_{x_o} a^n/n!$ .

We illustrate a Taylor series in Figure C.3. We know the value of a function $f$ and its derivatives (e.g., $(df/dx)_{x_o}$, $(d^2 f/d^2 x)_{x_o}, \cdots$) at a position $x_o$. We want to know its value at

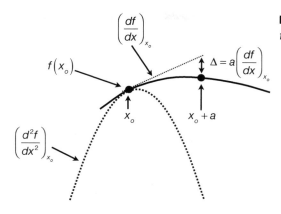

**Figure C.3** Expansion in a Taylor series. $f(x_o + a) \approx f(x_o) + (df/dx)_{x_o} a + (1/2)(d^2f/dx^2)_{x_o} a^2 + \cdots$.

$x_o + a$. Our first guess is that $f(x_o + a) \approx f(x_o)$, which clearly underestimates the real value in this example. We find the slope of the tangent to $f$ at $x_o$, which of course is just the value of the first derivative. The change in $f$ along that tangent line as we move a distance $a$ along the $x$ axis is just $\Delta = (df/dx)_{x_o} a$, as shown in the figure. The next guess to $f(x_o + a)$ is then $f(x_o + a) \approx f(x_o) + (df/dx)_{x_o} a$, which overestimates the actual value in this example. The second derivative is negative ($f$ curves downward here), so adding the next term, which goes as $(1/2)(d^2f/d^2x)_{x_o} a^2$, will bring the result closer to the true value. Additional terms involving higher derivatives might be needed to obtain convergence to the actual value within a desired accuracy.[5]

For functions of three variables we have [10] (using the index form of Cartesian coordinates)

$$f(\mathbf{r}_o + \mathbf{a}) = f(\mathbf{r}_o) + \sum_{\alpha=1}^{3} \left( \frac{\partial f}{\partial x_\alpha} \right)_{\mathbf{r}_o} a_\alpha + \frac{1}{2} \sum_{\alpha=1}^{3} \sum_{\beta=1}^{3} \left( \frac{\partial^2 f}{\partial x_\alpha \partial x_\beta} \right)_{\mathbf{r}_o} a_\alpha a_\beta + \cdots, \qquad (C.28)$$

where the derivatives are evaluated at $\mathbf{r}_o = (x_{o1}, x_{o2}, x_{o3})$.

## C.3 COMPLEX NUMBERS

Complex numbers have both real and imaginary parts and can be written in the form

$$z = a + ib, \qquad (C.29)$$

---

[5] Consider a simple example. Suppose we want to estimate a square root, i.e., $f(x) = \sqrt{x}$ where we know its value at a point $x + o = x - a$. To second order we have

$$f(x_o + a) \approx f(x_o) + \frac{1}{2\sqrt{x_o}} a - \frac{1}{8x_o^{3/2}} a^2 + \cdots, \qquad (C.30)$$

where we took the derivative of $\sqrt{x}$ with respect to $x$ and then evaluated them at $x = x_o$. If you want the square root of some number $x$, then find the nearest number $x_o$ for which you know the square root (e.g., 9, 16, 25, 36, ...). For example, suppose we want to estimate $\sqrt{90}$. Using the first-order term only, taking $x_o = 81$ and $a = 9$, we can estimate $\sqrt{90} \approx 9 + 9/18 = 9.5$. The actual value is $\sqrt{90} = 9.486$. The error grows as $a$ increases. For example, the first-order answer for $\sqrt{95}$ is off by 0.3%, for $\sqrt{100}$ by 0.6%, and for $\sqrt{110}$ by 1.2 %. It is instructive to evaluate the Taylor series to higher terms to see how well the series converges to the actual value.

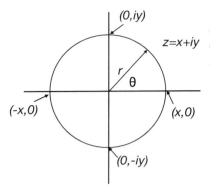

**Figure C.4** Representation of a complex number on a cartesian coordinate system with the *x* axis being real and the *y* axis being complex. General complex numbers can be written as $z = x + iy$ or $z = re^{i\theta}$ as described in the text

where $a$ and $b$ are real and $i = \sqrt{-1}$. In Eq. (C.29) the real part of $z$ ($\Re[z]$) is $a$ and the imaginary part ($\Im[z]$) is $b$. The complex conjugate of $z$ is denoted by $z^*$ and equals $z$ with $i$ replaced by $-i$, i.e.,

$$z^* = a - ib. \tag{C.31}$$

The magnitude of a complex number is defined as

$$|z| = (z^* z)^{1/2} = ((a - ib)(a + ib))^{1/2} = (a^2 + b^2)^{1/2}. \tag{C.32}$$

In Figure C.4 we show how a complex number can be represented on a simple coordinate system, where the real part is plotted along $x$ and the imaginary part along $y$, with the imaginary number being represented as

$$z = x + iy. \tag{C.33}$$

Using polar coordinates $x = r \cos\theta$ and $y = r \sin\theta$, where $r$ is the magnitude of $z$ ($\sqrt{x^2 + y^2}$), we can rewrite Eq. (C.33) as

$$z = r(\cos\theta + i \sin\theta). \tag{C.34}$$

Equation (C.34) can be simplified using Euler's formula

$$e^{ix} = \cos x + i \sin x \tag{C.35}$$

to be

$$z = re^{i\theta}. \tag{C.36}$$

Note that when you see an expression of the form $e^{ia}$, $a$ is an *angle* measured in *radians*.[6] By examining Eq. (C.35), we see, for example, that[7]

$$e^{i2\pi n} = 1 \qquad \text{for integer values of } n. \tag{C.37}$$

Similarly $\exp(i(2n + 1)\pi) = -1$ for integer values of $n$.[8]

---

[6] $2\pi$ radian = 360°, i.e., 1 radian = $(180/\pi)°$.
[7] Consider the value of cosine and sine when the angle is an integer multiple of $2\pi$ radians (360°).
[8] For integer $n$, $2n + 1$ is odd and, thus, $(2n + 1)\pi$ corresponds to multiples of 180°.

We will see expressions of the type $e^{i\mathbf{k}\cdot\mathbf{r}}$ many times in this text, where $\mathbf{r}$ is a vector in space and $\mathbf{k}$ is a wave vector. $\mathbf{k}$ must have units of radians/length, since we know that $\mathbf{k} \cdot \mathbf{r}$ must be unitless (it is in an exponential function).[9]

## C.4   PROBABILITY

Probability is the likelihood that something is true or will happen. Here we will discuss just a few central ideas that are important in our other discussions.[10]

Suppose we have a system that is made up of a discrete number of states. For example, we could talk about a coin flip, in which there are two states, heads and tails, or we could talk about dice, in which there are six states, etc.[11] Suppose we roll a die many many times. We will find that we pick each state approximately 1/6 of the rolls. For example, if we roll a die 60 000 times, we will have a 1 approximately 10 000 times, etc. The *probability* that we roll a 1 is 1/6. The probability that we roll any one of the six numbers is 1 – each roll of the die gives us one of the states.

Now suppose we have some system with $n$ states and that we do $N$ trial selections of that system, seeing what state the system is in during each trial. If we make a list of those states as we do the $N$ trials, we can count the number of times we find each of the $n$ states on the final list. Suppose we find that state $i$ occurs $m_i$ times. Assuming that $N$ is large enough to have good statistics[12] the probability of being in state $i$ is

$$P(i) = \frac{m_i}{N}.$$ (C.38)

Summing $P(i)$ over all states we have

$$\sum_{i=1}^{n} P(i) = \sum_{i=1}^{n} \frac{m_i}{N} = \frac{1}{N} \sum_{i=1}^{n} m_i = 1,$$ (C.39)

i.e., we always pick at least one of the states in each of the $N$ trials.

---

[9] From Eq. (C.35) we can easily derive expressions for cosine and sine in terms of exponentials,

$$\cos x = \frac{1}{2}\left(e^{ix} + e^{-ix}\right).$$ (C.40)

Similarly,

$$\sin x = \frac{1}{2i}\left(e^{ix} - e^{-ix}\right).$$ (C.41)

[10] An excellent source for more information is [120], which as of the publication of this text is available as a free download.

[11] A coin typically has different symbology on each of its sides. One can be designated "heads" and one "tails", indicating the two possible states. A die (plural dice) is a small rollable object with multiple resting states. Typically a die is a cube with each of its faces being designated by a number from 1 to 6.

[12] Flipping a coin five times is not likely to give a good value for the probability of having heads or tails, where flipping it 10 000 times would.

Suppose that there is some quantity, $A$, that has a different value in each of the $n$ states, which for the $i$th state is $A_i$. After $N$ trials, the system has been in state $i$ $m_i$ times. The *average* value of $A$ is

$$\langle A \rangle = \frac{1}{N} \sum_{i=1}^{n} m_i A_i = \sum_{i=1}^{n} P(i) A_i \,. \tag{C.42}$$

Equation (C.42) is a very important result that occurs many times in this text. It says that the average of a function, $A$, is the sum over all states of the probability that the system is in each state multiplied by the value of $A$ in that state.

Suppose now that there is a continuous set of states over the range $(a, b)$. The sum becomes an integral and the normalization condition is

$$\int_{a}^{b} P(x) dx = 1 \,, \tag{C.43}$$

where the probability the particle is at point $x$ is $P(x)$. The average value of a quantity $A(x)$ is

$$\langle A \rangle = \int_{a}^{b} P(x) A(x) dx \,. \tag{C.44}$$

Now suppose we consider an experiment in which a number of events could occur. The probability that *either* event $a$ *or* event $b$ could occur in a single performance of an experiment is called the union of events $a$ and $b$ and is written as $P(a \cup b)$, with

$$P(a \cup b) = P(a) + P(b) \,. \tag{C.45}$$

Similarly, the probability that *both $a$ and $b$* occur is the intersection of the events $P(a \cap b)$, which is

$$P(a \cap b) = P(a) P(b) \,. \tag{C.46}$$

## C.5   COMMON FUNCTIONS

We will often run across the Gaussian, or normal, distribution defined as (in one dimension)

$$f(x) = \sqrt{\frac{\alpha}{\pi}} e^{-\alpha(x-x_o)^2} \,, \tag{C.47}$$

where the constants are chosen such that $\int_{-\infty}^{\infty} f(x) dx = 1$. The function is centered at $x = x_o$. Defining $\langle x^n \rangle = \int_{-\infty}^{\infty} x^n f(x) dx$, we find that $\langle x^2 \rangle - \langle x \rangle^2 = 1/(2\alpha)$. We thus can relate the parameter $\alpha$ to the standard deviation of the Gaussian distribution. We show an example in Figure C.5a. In three dimensions, the normalized Gaussian distribution is

$$f(x, y, z) = \left( \frac{\alpha}{\pi} \right)^{3/2} e^{-\alpha(x^2 + y^2 + z^2)} \,. \tag{C.48}$$

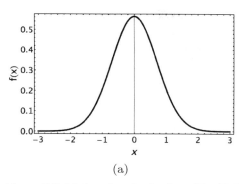 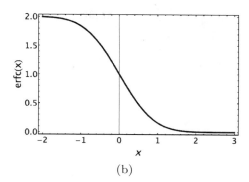

(a)  (b)

**Figure C.5** (a) Gaussian distribution of Eq. (C.47) centered at $x_o = 0$ with $\alpha = 1$. (b) The complementary error function of Eq. (C.50).

The *complementary error function* arises in the evaluation of the Coulomb energy by the Ewald method. With the error function defined as

$$\mathrm{erf}(x) = \frac{2}{\sqrt{\pi}} \int_0^x e^{-t^2} dt \,, \tag{C.49}$$

the complementary error function is

$$\mathrm{erfc}(x) = 1 - \mathrm{erf}(x) = \frac{2}{\sqrt{\pi}} \int_x^\infty e^{-t^2} dt \tag{C.50}$$

and is shown in Figure C.5. Note that the error function can be considered as an incomplete integral of the Gaussian distribution and is easily generalizable to more than one dimension.

## C.5.1 The Kronecker delta, Levi-Civita symbol, and the Dirac delta function

The Kronecker delta is defined as

$$\begin{aligned} \delta_{ij} &= 1 \quad \text{if} \quad i = j \\ &= 0 \quad \text{if} \quad i \neq j \,, \end{aligned} \tag{C.51}$$

where $i$ and $j$ are integers.

Another useful function is the Levi-Civita tensor $\epsilon_{ijk}$, which is

$$\begin{aligned} \epsilon_{123} &= \epsilon_{231} = \epsilon_{312} = 1 \\ \epsilon_{132} &= \epsilon_{213} = \epsilon_{321} = -1 \\ \epsilon_{ijk} &= 0 \quad \text{for all other } ijk \quad . \end{aligned} \tag{C.52}$$

The Levi-Civita tensor gives us a very handy, if rather opaque, way to write a cross product, where $(\mathbf{a} \times \mathbf{b})_i = \epsilon_{ijk} a_j b_k$.

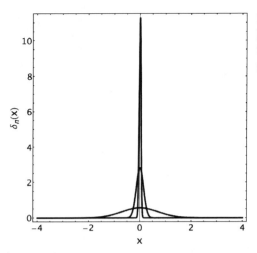

**Figure C.6** Graphical representation of a $\delta$ function from Eq. (C.54) centered at $a = 0$. As $n$ increases from 1 to 5 to 20, the Gaussian peaks become more and more highly peaked, while maintaining unit area.

The Dirac delta function is a continuous function that is extraordinarily useful. In one dimension, the Dirac delta function can be defined by its properties:

$$\delta(x - a) = 0, \quad x \neq a \tag{C.53}$$

$$\int_{-\infty}^{\infty} \delta(x - a)dx = 1,$$

$$\int_{-\infty}^{\infty} f(x)\delta(x - a)dx = f(a).$$

From these definitions, $\delta(x)$ must be a function that is an infinitely high, infinitely thin, spike centered at $x = 0$ with unit area. A delta function *selects* only one point out of all the numbers.

The problem in defining the delta function is that no analytic function actually exists. We can approximate the delta function with a variety of functional forms. For example,

$$\delta_n(x - a) = \frac{n}{\sqrt{\pi}}e^{-n^2(x-a)^2}, \tag{C.54}$$

where the delta function is better and better approximated as $n$ becomes large. We show $\delta_n(x)$ in Figure C.6 for $n = 1$, 5, and 20. As $n$ increases, $\delta_n(x)$ becomes more and more sharp and thin, keeping a unit area under the curve. However, the limit $\lim_{n \to \infty} \delta_n(x)$ does not exist in a mathematical sense.

Another representation of the delta function is based on $\exp(i\mathbf{k} \cdot \mathbf{r})$. In one dimension, we can define (setting $a = 0$ for convenience)

$$\delta(x) = \frac{1}{2\pi} \int_{-\infty}^{\infty} e^{ikx}dk. \tag{C.55}$$

Equation (C.55) deserves a few comments. The key feature of $\delta(x)$ is that it is zero everywhere but at $x = 0$. Consider the integral[10] (remembering Eq. (C.41))

$$\delta_n(x) = \frac{1}{2\pi} \int_{-n}^{n} e^{ikx}dk = \frac{\sin nx}{\pi x}. \tag{C.56}$$

Since the limit of $\sin nx/x$ as $x \to 0$ is $n$, as $n \to \infty$, the function becomes infinitely peaked at $x = 0$ (or $x = a$).

For three dimensions, we have $\delta(\mathbf{r})$, with

$$\int_{-\infty}^{\infty} f(\mathbf{r})\delta(\mathbf{t} - \mathbf{r})d\mathbf{r} = f(\mathbf{t}), \tag{C.57}$$

where $d\mathbf{r} = dx\, dy\, dz$.

## C.6  FUNCTIONALS

Suppose that $\phi(x)$ is a function of $x$ and

$$F[\phi(x)] = \int f\big(\phi(x)\big)dx. \tag{C.58}$$

$F[\phi(x)]$ is a *functional* of $\phi(x)$. Functionals arise many times in materials modeling. For example, the most common form of electronic structure calculation is based on density-functional theory, in which the electronic energy is a functional of the electronic density.

The calculus of functionals differs from the normal calculus. For example, the *differential* of a functional $F[\phi(x)]$ is defined as the linear term in $\delta f$ in the expression $\delta F[\phi(x)] = F[\phi(x) + \delta\phi(x)] - F[\phi(x)]$ and is given by

$$\delta F[\phi(x)] = \int \frac{\delta F[\phi(x)]}{\delta\phi(x)}\delta\phi(x)dx, \tag{C.59}$$

where $\delta F[\phi(x)]/\delta\phi(x)$ is the *functional derivative* of $F$ with respect to $\phi$ at the position $x$ [324].

There is another way to write the relation in Eq. (C.59) that is based on the formal definition of a derivative. Suppose we write $\delta\phi(x)$ explicitly as $\delta\phi(x) = \epsilon g(x)$, where $g(x)$ is an arbitrary function and $\epsilon$ is a small constant that will be sent to zero. We then have

$$\lim_{\epsilon \to 0}\left[\frac{F[\phi(x) + \epsilon g(x)] - F[\phi(x)]}{\epsilon}\right] = \left[\frac{d}{d\epsilon}F[\phi(x) + \epsilon g(x)]\right]_{\epsilon \to 0}$$
$$= \int \left(\frac{\delta F[\phi(x)]}{\delta\phi(x)}\right)_{\epsilon=0} g(x)dx. \tag{C.60}$$

Comparison with Eq. (C.59) shows that these expressions are identical. Since $g$ is arbitrary, it is often convenient to take it as the Dirac delta function, $\delta(x)$ of Eq. (C.53).[13]

Consider a simple example. Suppose we have an integral functional of the form

$$F[\phi(\mathbf{r})] = C \int \phi(\mathbf{r})^n d\mathbf{r}. \tag{C.61}$$

---

[13] $g(x) = \delta\phi(x)/\epsilon$.

Then

$$F\big[\phi(\mathbf{r}) + \delta\phi(\mathbf{r})\big] = C \int (\phi(\mathbf{r}) + \delta\phi)^n d\mathbf{r}$$

$$= C \int \left( \phi(\mathbf{r})^n + n\phi(\mathbf{r})^{n-1}\delta\phi(\mathbf{r}) + \cdots \right) d\mathbf{r}$$

$$= F\big[\phi(\mathbf{r})\big] + nC \int \phi(\mathbf{r})^{n-1}\delta\phi(\mathbf{r})d\mathbf{r} + \cdots , \tag{C.62}$$

where we used a Taylor series to expand $(\phi(\mathbf{r}) + \delta\phi(\mathbf{r}))^n$. Comparing with Eq. (C.59), we see that the functional derivative is the term linear in $\delta\phi(\mathbf{r})$, so

$$\frac{\delta F\big[\phi(\mathbf{r})\big]}{\delta\phi(\mathbf{r})} = nC\phi(\mathbf{r})^{n-1} . \tag{C.63}$$

It is a general result that for *local* functionals of the form

$$F\big[\phi(x)\big] = \int g\big(\phi(x)\big)dx , \tag{C.64}$$

we have

$$\frac{\delta F\big[\phi(x)\big]}{\delta\phi(x)} = \frac{\partial g\big(\phi(x)\big)}{\partial\phi(x)} . \tag{C.65}$$

Consider now another case that we will see in the discussion of phase fields

$$F\big[C(\mathbf{r})\big] = \frac{1}{2} \int \nabla C(\mathbf{r}) \cdot \nabla C(\mathbf{r})d\mathbf{r} . \tag{C.66}$$

The functional derivative is

$$\frac{\delta F\big[C(\mathbf{r})\big]}{\delta C(\mathbf{r}))} = -\nabla^2 C(\mathbf{r}) . \tag{C.67}$$

The relation in Eq. (C.67) is a bit tricky to derive. We start with the formal definition given in Eq. (C.60). For clarity we will show the result for functions of $x$ only – extension to three dimensions is straightforward. From Eq. (C.60),

$$\frac{\delta}{\delta C(x)}\frac{1}{2} \int \left( \frac{\partial C(x')}{\partial x} \right)^2 dx' = \lim_{\epsilon \to 0} \frac{1}{2\epsilon} \int \left\{ \left( \frac{\partial(C(x') + \epsilon g(x'))}{\partial x} \right)^2 - \left( \frac{\partial C(x')}{\partial x} \right)^2 \right\} dx'$$

$$= \lim_{\epsilon \to 0} \frac{1}{\epsilon} \int \frac{\partial C(x')}{\partial x}\frac{\partial \epsilon g(x')}{\partial x}dx'$$

$$= \frac{\partial C(x')}{\partial x}\bigg|_{-\infty}^{\infty} g(x') - \int g(x')\frac{\partial^2 C(x')}{\partial x^2}dx'$$

$$= -\frac{\partial^2 C(x)}{\partial x^2} . \tag{C.68}$$

In step 3, we did an integration by parts, assuming that the function $C$ vanishes at infinity, and in step four we substituted $g(x') = \delta(x' - x)$, which we are free to do since $g(x)$ is an arbitrary perturbation.

A more complete summary of the properties of functionals is given in [251].

## Suggested reading

- A good source for review of mathematical topics are the *Schaum's Outlines*. They are inexpensive yet provide a good guide to how to solve problems. There are many volumes, including advanced calculus, vector and tensor analysis, complex variables, etc.
- For those interested in a more advanced treatment, the book by Arfken and Weber *Mathematical Methods for Physicists* [10], is excellent.

# D A brief summary of classical mechanics

Classical mechanics is the study of the dynamical response of macroscopic objects under applied forces. The basic tenets of classical mechanics are familiar to all of us – it is the study of the motion and stability of everyday objects. We give a very brief summary here of topics needed in this text.

*Potential energy* is the energy that results from the position or configuration of an object (or system of objects). For example, an object has the capacity for doing work as a result of its position in a gravitational field, an electric field, or a magnetic field, etc. It may have elastic potential energy as a result of a stretched spring or other elastic deformation. Potential energy will generally be represented in this text by the symbol $U$.

An object in motion also has *kinetic energy*. There are many forms of kinetic energy – vibrational, rotational, and translational (the energy due to motion from one location to another). Kinetic energy is denoted in this text as $K$.

## D.1 NEWTON'S EQUATIONS

Suppose we want to determine the motion of an object. In the Newtonian formulation of classical mechanics, we have the quite familiar relation that force equals mass times acceleration (Newton's second law), or[1]

$$\mathbf{F} = m\mathbf{a}, \tag{D.1}$$

where $\mathbf{F}$ is the force on the object, $m$ the mass of the object, and $\mathbf{a}$ is the object's acceleration. The acceleration is the time rate of change of the velocity $\mathbf{v}$, which in turn is the time rate of change of the position $\mathbf{r}$ of the object, i.e.,

$$\mathbf{a} = \frac{d\mathbf{v}}{dt} = \frac{d^2\mathbf{r}}{dt^2}, \tag{D.2}$$

or, using the notation that a dot indicates a time derivative, $\mathbf{a} = \ddot{\mathbf{r}}$.[2] We can also write Newton's equations in terms of the momentum $\mathbf{p} = m\mathbf{v}$

$$\mathbf{F} = \frac{d\mathbf{p}}{dt} = \dot{\mathbf{p}} = m\mathbf{a}. \tag{D.3}$$

---

[1] For a discussion of the various units of forces and energies, please see Appendix A.2.

[2] $\frac{dx}{dt} = \dot{x}$, $\quad \frac{d^2x}{dt^2} = \ddot{x}$ ....

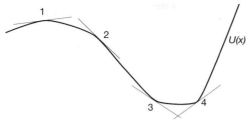

**Figure D.1** Shown is a one-dimensional potential energy surface $U(x)$. In 1D, the force is $F_x = -dU(x)/dx$. We indicate the slope in four places: at 1 and 4, the slope is positive so the force drives the system to the left. At positions 2 and 3, the slope is negative and the system is driven to the right.

If we know **F** as a function of position and knew the initial positions and velocities, we could solve the second-order partial differential equation in Eq. (D.1) for the positions and velocities as a function of time, thus completely characterizing the system. The force causes a free body to undergo a change in speed, orientation, shape, etc. It is a vector, with direction, and if $F_{12}$ indicates the force of atom 2 on atom 1, then $\mathbf{F}_{12} = -\mathbf{F}_{21}$, the force of atom 1 on atom 2.

Work (energy) is defined as the force multiplied by the distance over which it is applied. Suppose we have an object under a force, $F$, that is dependent on position only. The definition of the potential energy, $U$, is that it is equal to the work needed to move the object against that force from some reference position, $x_o$, with potential energy $U(x_o)$, to some final position, $x$. $U(x_o)$ is an arbitrary starting energy, which we can set to zero. The force exerted on the object to move against $F$ must have the same magnitude, but opposite direction, as $F$. If $F$ is a continuous function, then, in one dimension,

$$U(x) = -\int_{x_o}^{x} F(x)dx , \qquad (D.4)$$

remembering that $U(x_o) = 0$. By the definition of an integral, we can identify $F(x)$ as the negative derivative of $U$ with respect to $x$,

$$F = -\frac{\partial U(x)}{dx} . \qquad (D.5)$$

In three dimensions, the equivalent equation is

$$\mathbf{F} = -\nabla U(\mathbf{r}) , \qquad (D.6)$$

where the gradient, $\nabla$, is given in Eq. (C.15). The relation in Eq. (D.6) makes good sense – an object under a force moves to reduce the energy, with the steepness of the "cliff" being the force. We show this schematically in Figure D.1.

An important point is that there is no absolute energy scale, only *relative* energies. Suppose we have a system with potential energies based on some scale, i.e., $U$. If we add a constant to that energy $U' = U + C$, then the force that arises from that potential energy is unchanged, because the derivative of a constant is 0.

The kinetic energy $K$ is defined as

$$K(v_x, v_y, v_z) = \frac{1}{2}mv^2 = \frac{1}{2}m(v_x^2 + v_y^2 + v_z^2) = \frac{p_x^2 + p_y^2 + p_z^2}{2m} = \frac{p^2}{2m} , \qquad (D.7)$$

where we have given the form of the expression in Cartesian coordinates and have expressed it using both the velocity and the momentum.

It turns out that the Newtonian formalism is not always convenient for actually solving problems. There are other, more powerful, ways to do the work of classical mechanics, including use of the Lagrangian or Hamiltonian formalism. For the purpose of this text, however, we will restrict our discussion of those more advanced methods to only what we need.

## D.2 THE HAMILTONIAN

Newton's equations can be very difficult to solve for some problems that require non-Cartesian coordinates. An alternative approach defines an energy function, called the Hamiltonian, $\mathcal{H}$. The Hamiltonian is simplified in systems in which the potential energy depends only on the coordinates and the kinetic energy only on the momenta. Systems like these are referred to as Newtonian. In this case, the Hamiltonian is very simple; it is just the sum of the kinetic plus potential energies, which, for a system of $N$ particles, can be written as

$$\mathcal{H}(\mathbf{p}^N, \mathbf{r}^N) = K(\mathbf{p}^N) + U(\mathbf{r}^N),$$ (D.8)

where the notation $\mathbf{r}^N = \mathbf{r}_1, \mathbf{r}_2, \ldots, \mathbf{r}_N$, i.e., the set of the positions of the $N$ particles. Similarly $\mathbf{p}^N$ represents the set of $N$ momenta. The Hamiltonian is thus just a function of momenta and positions. The value of the Hamiltonian is the total internal energy of the system, which is usually denoted by $E$ (i.e., it is the $E$ of thermodynamics).

The Hamiltonian has an important property for conserved (Newtonian) systems – its value, $E$, is a constant, i.e., the energy is *conserved*, with

$$\frac{dE}{dt} = \frac{d\mathcal{H}}{dt} = 0.$$ (D.9)

The conservation of total energy is a very important property. It says that as long as the potential energy does not depend on velocity, an isolated system's energy remains constant. For systems in which there is a velocity-dependent potential energy, for example in systems with friction, the energy is not conserved, i.e., it changes with time.

Equation (D.9) is very easy to prove. In one dimension we have:

$$\mathcal{H} = \frac{1}{2}mv^2 + U(x)$$
$$\frac{d\mathcal{H}}{dt} = mv\frac{dv}{dt} + \frac{dU}{dt}$$
$$= mva + \frac{dU}{dx}\frac{dx}{dt} = mva - Fv$$
$$= mva - mva = 0,$$ (D.10)

where we used the chain rule and the relation of the force to the gradient of the potential. It is also easy to show in three dimensions.[3]

## D.3  EXAMPLE: THE HARMONIC OSCILLATOR

Consider a particle attached to a massless spring in one dimension (this problem can easily be extended to two or three dimensions). This problem is commonly called the *harmonic oscillator*. The force is just Hooke's law and is given by $F = -k(x - x_o) = ma$, where $x_o$ is the equilibrium distance of the spring, $k$ is the force constant, and $m$ is the mass of the particle. Letting $z = x - x_o$, the *equation of motion* becomes

$$m\frac{d^2z}{dt^2} = -kz.$$  (D.11)

The general solution to this differential equation is[4]

$$z(t) = A\cos(\omega t) + B\sin(\omega t),$$  (D.12)

where the *angular frequency* of the oscillator is $\omega = \sqrt{k/m}$. The velocity is

$$v(t) = \frac{dz(t)}{dt} = -\omega A\sin(\omega t) + \omega B\cos(\omega t).$$  (D.13)

The parameters $A$ and $B$ are determined from the *initial conditions*, i.e., the initial (i.e., at time $t = 0$) position and velocity.

There is often some confusion about the angular frequency $\omega$ of an oscillator and the frequency $\nu$. $\omega$ corresponds to the number of radians per second in the solution Eq. (D.12). There are $2\pi$ radians per one revolution, so

$$\omega = \frac{2\pi}{T},$$  (D.14)

where $T$ is the period of the oscillation (the number of seconds per one revolution). The frequency, however, is just

$$\nu = \frac{1}{T},$$  (D.15)

so

$$\nu = \frac{\omega}{2\pi}.$$  (D.16)

It is common to refer to both $\omega$ and $\nu$ as frequencies, using the notation to distinguish them.

---

[3]  Show that $d\mathcal{H}/dt = 0$ in three dimensions for systems in which the kinetic energy depends only on velocity and the potential only on positions. Hint: $dU/dt = \nabla U \cdot \mathbf{v}$.

[4]  Given Eq. (C.40) and Eq. (C.41), Eq. (D.12) can also be expressed as $z(t) = C\exp(i\omega t) + D\exp(-i\omega t)$, in which $C$ and $D$ are constants.

The potential energy is

$$U = \frac{1}{2}k(x - x_o)^2 = \frac{1}{2}kz^2 \tag{D.17}$$

and the kinetic energy is

$$K = \frac{1}{2}mv^2 = \frac{1}{2}m\left(\frac{dz}{dt}\right)^2, \tag{D.18}$$

where $v$ is the velocity $(dz/dt)$. The total energy is the value of the Hamiltonian $\mathcal{H} = U + K$. It is easy to show by inserting the solution in Eq. (D.12) into the Hamiltonian that the energy of the system is a constant in time (conserved).[5]

## D.4 CENTRAL-FORCE POTENTIALS

We will come across a number of systems in which the interaction potential between two particles is a function only of the distance between them – these are called *central-force potentials*.

For particles $i$ and $j$, the interaction energy is

$$U(\mathbf{r}_i, \mathbf{r}_j) = \phi(r_{ij}), \tag{D.19}$$

where $\phi$ is some function and $r_{ij}$ is the magnitude (Eq. (C.7)) of the vector $\mathbf{r}_{ij}$ extending from particle $i$ to particle $j$,

$$\mathbf{r}_{ij} = \mathbf{r}_j - \mathbf{r}_i = (x_j - x_i)\hat{x} + (y_j - y_i)\hat{y} + (z_j - z_i)\hat{z}. \tag{D.20}$$

To calculate the force on particle $i$ arising from particle $j$, we must determine

$$\mathbf{F}_i(\mathbf{r}_{ij}) = -\nabla_i\phi(r_{ij}) = -\left(\frac{\partial}{\partial x_i}\hat{x} + \frac{\partial}{\partial y_i}\hat{y} + \frac{\partial}{\partial z_i}\hat{z}\right)\phi(r_{ij}). \tag{D.21}$$

Similarly, to determine the force of $i$ on $j$, we take the gradient with respect to the coordinates of atom $j$.

Using the chain rule, we find

$$\frac{\partial\phi(r_{ij})}{\partial x_i} = \frac{\partial\phi(r_{ij})}{\partial r_{ij}}\frac{\partial r_{ij}}{\partial x_i} = -\frac{\partial\phi(r_{ij})}{\partial r_{ij}}\frac{x_j - x_i}{r_{ij}}. \tag{D.22}$$

Adding in the $y$ and $z$ terms, we find

$$\mathbf{F}_i(\mathbf{r}_{ij}) = \frac{\partial\phi(r_{ij})}{\partial r_{ij}}\left(\frac{x_j - x_i}{r_{ij}}\hat{x} + \frac{y_j - y_i}{r_{ij}}\hat{y} + \frac{z_j - z_i}{r_{ij}}\hat{z}\right) = \frac{\partial\phi(r_{ij})}{\partial r_{ij}}\frac{\mathbf{r}_{ij}}{r_{ij}}. \tag{D.23}$$

---

[5] A useful exercise is to verify the solution to the harmonic oscillator: (a) verify that Eq. (D.12) is a solution; (b) verify that $z(t) = A\exp i\omega t + B\exp -i\omega t$ is a solution; evaluate the energy of the harmonic oscillator and show that it is a constant with time $t$; determine $A$ and $B$ for the initial condition that at $t = 0$, $z = x - x_o = 1$. Repeat for the case of the velocity equal to 0.1 at $t = 0$. What is the total energy in each case?

Taking the derivative with respect to the coordinates of atom $j$ we find

$$\mathbf{F}_j(\mathbf{r}_{ij}) = -\frac{\partial \phi(r_{ij})}{\partial r_{ij}} \frac{\mathbf{r}_{ij}}{r_{ij}}, \tag{D.24}$$

i.e., the force on atom $j$ from atom $i$ is equal in magnitude but in the opposite direction to the force on atom $i$ from atom $j$. Note that $\mathbf{r}_{ij}/r_{ij} = \hat{r}_{ij}$ is just the *unit vector* from atom $i$ to atom $j$.

A common form for a central-force potential involves terms such as

$$\phi(r_{ij}) = \frac{C}{r^n}, \tag{D.25}$$

where $C$ is some constant that depends on the kind of atoms.[6] Using $\partial \phi/\partial r_{ij} = -nC r_{ij}^{-(n+1)}$ the force takes the very simple form

$$\mathbf{F}_i = -n\frac{C}{r^{(n+1)}}\hat{r}_{ij} \quad \text{and} \quad \mathbf{F}_j = n\frac{C}{r^{(n+1)}}\hat{r}_{ij}. \tag{D.26}$$

## Suggested reading

There are many good books available on classical mechanics. For a comprehensive text, I like the book *Classical Mechanics*, by Herbert Goldstein, Charles P. Poole, and John L. Safko [126].

---

[6] A classic example would be the Lennard-Jones potential of Eq. (5.6).

# E Electrostatics

## E.1 THE FORCE

The force between two charged particles with charges $q_1$ and $q_2$, respectively, (in the absence of any external electric fields) has the general form

$$\mathbf{F}(r) = k \frac{q_1 q_2}{r^2} \hat{r} , \tag{E.1}$$

where $r$ is the distance between the charges, $\hat{r}$ is the unit vector along the line connecting the charges, and $k$ is a constant that depends on the units. The force acts in a direction along $\hat{r}$ and is attractive when $q_1$ and $q_2$ have opposite signs and repulsive when they have the same sign.

In cgs units, $k = 1$ in Eq. (E.1) and the magnitude of the force between the two charged particles is simply

$$\mathbf{F}(r) = \frac{q_1 q_2}{r^2} \hat{r} . \tag{E.2}$$

The unit of charge is the *electrostatic unit of charge*, or esu (also called the statcoulomb), and the distance is measured in cm. The esu is defined such that if two objects, each carrying a charge of $+1$ esu, are 1 cm apart, then they repel each other with a force of 1 dyne. Thus, 1 esu $= (1 \text{ dyne cm}^2)^{1/2} = g^{1/2} \text{ cm}^{3/2}/\text{sec}$. In MKS (SI) units, however,

$$k = \frac{1}{4\pi \epsilon_o} , \tag{E.3}$$

where $\epsilon_o$ is the *permittivity constant*, with a value of $8.854 \times 10^{-10}$ C$^2$/(N m$^2$). The charges are measured in coulombs (C). This form arises from a long series of definitions that we will not go into here. The simplicity of the form in Eq. (E.2) is why cgs units are often preferred in physics or electrical engineering.

## E.2 ELECTROSTATIC POTENTIALS AND ENERGIES

The electrostatic potential from a charge $q$ is

$$\Phi(\mathbf{r}) = k \frac{q}{r} , \tag{E.4}$$

where $r$ is the distance from the charge.

The *energy* between two charges ($q_i$ and $q_j$) is just the $q_i$ multiplied by the potential from $q_j$ (and vice versa),

$$U(r) = k \frac{q_i q_j}{r} , \tag{E.5}$$

where $r$ is the distance between the charges.

From electromagnetism, the electrostatic potential at a point $\mathbf{r}$ from a group of $n$ charges is found by summing over the charges, i.e.,

$$\Phi(\mathbf{r}) = k \sum_{i=1}^{n} \frac{q_i}{|\mathbf{r} - \mathbf{r}_i|} , \tag{E.6}$$

where the distance between the point $\mathbf{r}$ and each charge is

$$|\mathbf{r} - \mathbf{r}_i| = \left( (\mathbf{r} - \mathbf{r}_i) \cdot (\mathbf{r} - \mathbf{r}_i) \right)^{1/2} . \tag{E.7}$$

The electrostatic energy between a group of $n$ charges is the sum of the interactions

$$U = k \frac{1}{2} \sum_{i=1}^{n} \sum_{j=1}^{n} {}' \frac{q_i q_j}{r_{ij}} , \tag{E.8}$$

where $r_{ij}$ is the distance between the charges and the $'$ indicates that the $i = j$ terms are not included in the sum.[1]

The potential of a *continuous* distribution of charge ($\rho(\mathbf{r})$) is found by changing the sum in Eq. (E.6) to an integral

$$\Phi(\mathbf{r}) = k \int \frac{\rho(\mathbf{r}')}{|\mathbf{r} - \mathbf{r}'|} d\mathbf{r} . \tag{E.9}$$

Note that the position at which the potential is evaluated is $\mathbf{r}$ while the integration is over the charge distribution's value at $\mathbf{r}'$.

The electrostatic energy of the charge distribution $\rho(\mathbf{r})$ is found by converting the sums in Eq. (E.8) to integrals

$$J[\rho] = k \frac{1}{2} \iint \frac{\rho(\mathbf{r}_1)\rho(\mathbf{r}_2)}{|\mathbf{r}_2 - \mathbf{r}_1|} d\mathbf{r}_1 \mathbf{r}_2 , \tag{E.10}$$

where we integrate over the values of the charge distribution at two points in space divided by the separation between those points. This term is often called the Coulomb integral $J$.

# E.3 DISTRIBUTION OF CHARGES: THE MULTIPOLE EXPANSION

Suppose we have a distribution of $n$ point charges in some region defined by its volume $V = L^3$, where $L$ is the linear size of the region, as shown in Figure E.1. The charges have positions

---

[1] This type of sum is discussed in detail in the section describing Eq. (3.2).

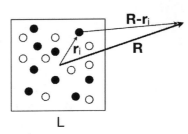

$\mathbf{r}_i$ relative to some origin in the distribution. The net electrostatic potential arising from these charges at some external point $\mathbf{R}$ (measured from the same origin) is

$$\Phi(\mathbf{R}) = k \sum_{i=1}^{n} \frac{q_i}{|\mathbf{R} - \mathbf{r}_i|} \,.$$  (E.11)

If we introduce a charge $q'$ at position $\mathbf{R}$, then the energy is

$$U = q' \, \Phi(\mathbf{R}) \,.$$  (E.12)

Suppose the magnitude of $\mathbf{R}$ is large relative to the size of the distribution $L$, for example if $R > 2L$. Instead of summing over all the charges individually, which can be computationally prohibitive if $n$ is large, it is possible to develop an expansion of the potential $\Phi$ in terms of moments of the distribution of charges. More specifically, we can rearrange the expression for $\Phi$ and do a Taylor expansion[2] in terms of $R/L$, which is a small quantity. The procedure is straightforward, but not sufficiently instructive to include here. Suffice it to say that the potential can be written as a series over the *multipole moments* of the distribution relative to its origin. The first three multipole moments are

$$Q = \sum_{i=1}^{N} q_i$$  (E.13a)

$$\boldsymbol{\mu} = \sum_{i=1}^{N} q_i \, \mathbf{r}_i$$  (E.13b)

$$\boldsymbol{\Theta}_{\alpha\beta} = \sum_{i=1}^{N} q_i (3\mathbf{r}_{i\alpha} \, \mathbf{r}_{i\beta} - r_i^2 \delta_{ij}) \,.$$  (E.13c)

$$\ldots$$  (E.13d)

where $Q$ is the total charge in $V$, $\boldsymbol{\mu}$ is the dipole moment (a vector), $\boldsymbol{\Theta}_{\alpha\beta}$ is the (traceless) quadrupole moment tensor, and so on. The notation $\mathbf{r}_{i\alpha}$ indicates the $\alpha$ component of the vector $\mathbf{r}_i$, i.e., the notation discussed in Appendix C.1.1.

[2] See Appendix C.2.

The potential in Eq. (E.11) can then be written as

$$\frac{1}{k}\Phi(\mathbf{R}) = \frac{Q}{R} + \frac{\boldsymbol{\mu} \cdot \mathbf{R}}{R^3} + \frac{1}{6R^5} \sum_{\alpha=x,y,z} \sum_{\beta=x,y,z} Q_{\alpha\beta}(3R_\alpha R_\beta - \delta_{\alpha\beta}R^2) + \cdots . \tag{E.14}$$

$R_\alpha$ is the $\alpha$ (x, y, or z) component of $\mathbf{R}$ and $\delta_{\alpha\beta}$ is the Kronecker delta defined in Appendix C.5.1.

The number of moments needed to describe the potential depends on the ratio $R/L$ – as $R/L$ increases, then fewer terms in Eq. (E.14) are needed to achieve an accurate value for $\Phi$. The advantage to the multipole expansion is that once the moments of a given charge distribution are determined, then the complicated sum in Eq. (E.11) is replaced by the much more efficient sum in Eq. (E.14) for all future calculations.

The electrostatic interaction between two well-separated charge distributions can be written as a sum of multipole-multipole interactions, as discussed in detail elsewhere (e.g., [190]). For example, suppose we are considering the interaction between two neutral molecules that have a dipole moment $\boldsymbol{\mu}$ (e.g., two water molecules). The leading term in the multipole moment expansion of the electrostatic energy is the well-known dipole-dipole energy expression

$$u_{dd} = \frac{1}{r_{ij}^3}\left(\boldsymbol{\mu}_i \cdot \boldsymbol{\mu}_j - 3(\boldsymbol{\mu}_i \cdot \hat{r}_{ij})(\boldsymbol{\mu}_j \cdot \hat{r}_{ij})\right), \tag{E.15}$$

where $r_{ij}$ and $\hat{r}_{ij}$ are the distance and the unit vector connecting the centers of the two molecules, respectively. Since the dipole moment reflects the charge distribution within the molecule, Eq. (E.15) will depend on the relative orientations of the molecules. General expressions for the interactions bewteen other moments are given in [45, 46, 47].

# Elements of quantum mechanics

In this appendix we review some of the basic ideas and methods behind quantum mechanics. This brief treatment is meant only to introduce the reader to this important subject. A number of elementary texts are listed in the Suggested reading for those who would like to go further into this fascinating field.

## F.1 HISTORY

Quantum mechanics arose from an attempt to understand discrepancies between predictions of classical mechanics and observed (experimental) behavior. Around 1900, there was increasing recognition that some phenomena could not be understood based on classical physics. One of these problems was *blackbody* radiation, i.e., the glow that is given off by a heated object which is an indicator of its temperature. Planck came up with an explanation for blackbody radiation in a cavity, but had to describe the energetics of the system as consisting of oscillators whose energy was quantized (i.e., integer multiples of some quantity). In 1905, Einstein took that idea one step farther and proposed that electromagnetic radiation (i.e., light) is itself quantized as an explanation of the photoelectric effect. We now call these quanta of light *photons*.

One of the other main failures of classical theory was its inability to explain the spectrum of hydrogen, which has distinct lines. One of the most important results from the quantum mechanical description of the H atom, and of all matter, is that quantum systems have states with discrete energy levels (not continuous as in classical mechanics). Transitions of electrons between these discrete levels lead to the observed spectra of the H atom and other atoms and molecules. They also found that quantum particles did not behave like classical particles – their behavior was like a wave, which has a profound effect on how they must be described, as will be discussed in the next section.

While we know that quantum mechanics is important at small scales, we have yet to identify how small a system has to be before quantum effects are important. One measure is found through the *Heisenberg Uncertainty Principle*, which states that we cannot know the precise position and momentum of a particle – there is an inherent uncertainty in their values, with the product of the uncertainty in position of a particle and the uncertainty in its momentum being related by $\Delta x \Delta p_x > \hbar$, where $\hbar$ is Planck's constant. The ramifications of the uncertainty principle are quite profound. It says that we cannot, for a quantum system, know exactly both the position and momentum of a particle – the better we know the position, the less well we know the momentum and vice versa.

Since $\hbar$ ($= 1.054 \times 10^{-27}$ erg sec) is very small, for the macroscopic systems familiar to us in everyday life any uncertainties are negligible and the system obeys classical mechanics. The uncertainties for an electron, however, which has a very small mass ($m_e = 0.911 10^{-27}$ g), can be quite large. Consider an electron in a hydrogen atom. In the ground state (i.e., the state with lowest energy), the average radius of the orbit is 1 bohr ($\approx 0.53$ Å). The root mean square momentum in the same state is about $\sqrt{\langle \rho^2 \rangle} = 2 \times 10^{-19}$ g cm/sec. Suppose we wanted to know the position to within 1% of the radius (0.005 Å). The uncertainty in momentum would be about $2 \times 10^{-17}$ g cm/sec, or a factor of 100 greater than its average value. Electrons are clearly quantum in nature.

## F.2  WAVE FUNCTIONS

It was a big conceptual leap to recognize that, at some small scale, objects can behave as both waves and particles. Thus light, which had always been described as a wave, acts like a particle, and electrons, which are clearly particles, also behave like waves, and thus are described by a wave function, usually written as $\psi(\mathbf{r}) = \psi(x, y, z)$. It is important to remember that we cannot measure a wave function. We can only see observable quantities, which we will discuss below.

Because of the uncertainty principle, we do not know exactly where a particle is, but we can calculate the *probability* that a particle will occupy a certain spot in space.[1] If we call the wave function of a particle $\psi(\mathbf{r})$ at a point $\mathbf{r}$ in space, then the probability for that particle to be at the point $\mathbf{r}$ is

$$p(\mathbf{r}) = \psi^*(\mathbf{r})\psi(\mathbf{r}) \,, \tag{F.1}$$

where the $^*$ indicates the complex conjugate.[2] As will become apparent, Eq. (F.1) is fundamental to all applications of quantum mechanics to the electronic structure of materials.

Much of the time, we will deal with wave functions that are *normalized*, i.e., that satisfy the relation that probabilities integrate to 1 over the volume of the system. For a quantum system, this would take the form (for normalized wave functions)

$$\int p(\mathbf{r})d\mathbf{r} = \int \psi^*(\mathbf{r})\psi(\mathbf{r})d\mathbf{r} = 1 \,, \tag{F.2}$$

where the notation $d\mathbf{r} = dxdydz$ and the integrals are taken over the volume of the system. Since $\psi^*\psi$ is a probability, Eq. (F.2) just states that a particle described by $\psi$ is somewhere in the system. There will be times when $\psi$ might not be normalized, in which case the probability that the particle is at $\mathbf{r}$ becomes

$$p(\mathbf{r}) = \frac{\psi^*(\mathbf{r})\psi(\mathbf{r})}{\int \psi^*(\mathbf{r})\psi(\mathbf{r})d\mathbf{r}} \,, \tag{F.3}$$

which of course integrates to 1 in Eq. (F.2).

---

[1] For a discussion of probability, see Appendix C.4.
[2] Complex numbers are discussed in Appendix C.3.

## F.3   THE SCHRÖDINGER EQUATION

Since we have said that the particles act like waves, it should not be surprising that the fundamental equation that describes the particles is similar to that seen in the classical description of waves. That equation was first proposed by Schrödinger, and is the quantum mechanical analogue to the Hamiltonian function in classical mechanics, $H = K + U$, where $K$ is the kinetic energy and $U$ the potential energy. As in classical mechanics, the value of the quantum Hamiltonian is just the total energy $E$, the sum of kinetic plus potential energies.

In quantum theory, we replace the familiar quantities of position and momentum that define a particle's state by linear operators that act on the particle's wave function.[3] Based on the analogue of position and momentum in waves, we can define the quantum kinetic energy operator, which is a function of the momentum, as

$$\mathcal{K}\Psi(\mathbf{r}) = \frac{p^2}{2m}\Psi(\mathbf{r}) = -\frac{\hbar^2}{2m}\nabla^2\Psi(\mathbf{r}),  \tag{F.4}$$

where the momentum operator is $i\hbar\nabla$ and the gradient ($\nabla$) and Laplacian ($\nabla^2$) are discussed in Appendix C.1.2. The potential energy operator, as a function of the position, simply multiplies the wave function,

$$\mathcal{U}(\mathbf{r})\Psi(\mathbf{r}).  \tag{F.5}$$

The net quantum Hamiltonian is $\mathcal{H} = \mathcal{K} + \mathcal{U}$, has the value $E$ and takes the form

$$\mathcal{H}\psi(\mathbf{r}) = \left(-\frac{\hbar^2}{2m}\nabla^2 + U(\mathbf{r})\right)\psi(\mathbf{r}) = E\psi(\mathbf{r}),  \tag{F.6}$$

which is the familiar time-independent Schrödinger equation. Solving this equation for a given potential energy yields the energies and wave function $\psi$, which defines the quantum state.

## F.4   OBSERVABLES

We cannot experimentally observe wave functions. However, quantum theory gives us a way to calculate observable quantities as *averages* over the probabilities $\psi^*\psi$. For example, suppose we had an operator $\mathcal{A}(\mathbf{r})$. From Appendix C.4, its *average* can be calculated as in Eq. (C.44), i.e., by integrating its value in a state multiplied by the probability it is in that state. Of course, being quantum mechanics, it becomes a bit more complex because $\mathcal{A}$ is an operator, not a simple function.

An average takes the form (for a normalized wave function)

$$\langle \mathcal{A} \rangle = \int \psi^*(\mathbf{r})\mathcal{A}(\mathbf{r})\psi(\mathbf{r})d\mathbf{r}.  \tag{F.7}$$

---

[3] An operator is a function that acts on another function, in this case a wave function. By definition, an operator acts on the function to its right. An operator, $\mathcal{L}$, is linear if, for a set of functions $f$ and $g$, $\mathcal{L}(f + g) = \mathcal{L}f + \mathcal{L}g$.

The order of the multiplication in Eq. (F.7) matters. Since $\mathcal{A}$ is an operator, it acts on the function to its right, in this case the wave function. Thus, the integrand in Eq. (F.7) is a product of the complex conjugate of the wave function and $\mathcal{A}$ operating on the wave function.

Suppose we wanted to know the average position of a particle. We would calculate $\langle \mathbf{r} \rangle$ by evaluating $\int \psi^*(\mathbf{r}) \, \mathbf{r} \, \psi(\mathbf{r}) d\mathbf{r}$. In this case, the order does not matter. However, if we want the average *momentum*, then we would need to evaluate $\int \psi^*(\mathbf{r}) i\hbar \nabla \, \psi(\mathbf{r}) d\mathbf{r}$. The steps would be to take the gradient of $\psi$, then multiply by $\psi^*$, and then integrate.

For *notational convenience*, integrals such as found in Eq. (F.7) are often written as

$$\langle \mathcal{A} \rangle = \langle \psi | \mathcal{A} | \psi \rangle = \int \psi^*(\mathbf{r}) \mathcal{A}(\mathbf{r}) \psi(\mathbf{r}) d\mathbf{r}. \tag{F.8}$$

This notation is called the *Dirac notation*, after its inventor Paul Dirac, and is very powerful in many ways that we will not discuss.

For unnormalized wave functions, we have

$$\langle \mathcal{A} \rangle = \frac{\int \psi^*(\mathbf{r}) \mathcal{A}(\mathbf{r}) \psi(\mathbf{r}) d\mathbf{r}}{\int \psi^*(\mathbf{r}) \psi(\mathbf{r}) d\mathbf{r}} = \frac{\langle \psi | \mathcal{A} | \psi \rangle}{\langle \psi | \psi \rangle}. \tag{F.9}$$

## F.5 SOME SOLVED PROBLEMS

### F.5.1 Particle in a box

The simplest problem, yet one that brings to light many features of quantum mechanics, is that of a quantum particle in a one-dimensional box defined by the potential

$$\begin{aligned} U(x) &= \infty & x &= 0 \\ &= 0 & 0 &< x < L \\ &= \infty & x &= L. \end{aligned} \tag{F.10}$$

Since the potential at the box wall is infinite, the wave function must vanish at $x = 0$ and $x = L$ (otherwise the Schrödinger equation would have infinite energies at those points).[4]

When the potential term is zero, the Schrödinger equation is

$$E\psi(x) = -\frac{\hbar^2}{2m} \frac{d^2 \psi(x)}{dx^2}. \tag{F.11}$$

A solution to Eq. (F.11) based on real functions is[5]

$$\psi(x) = A\cos(kx) + B\sin(kx), \tag{F.12}$$

---

[4] The potential term in the Schrödinger equation is $U\psi$. If $U = \infty$, then $\psi$ must be zero or there will be an infinite contribution to the energy.

[5] We could also have used a solution of the form $\psi(x) = Ae^{ikx} + Be^{-ikx}$. These are called *plane waves* and will be used in Chapter 4 and in Appendix F.6.3.

where $A$, $B$ and $k$ are determined from the boundary conditions that the wave function must vanish at the sides of the box, i.e., at $x = 0$ and $x = L$.

At $x = 0$, $\cos(kx) = 1$, which violates the boundary conditions. Thus, $A = 0$ and only the second term in Eq. (F.12) contributes (since $\sin(0) = 0$). $\sin(kx) = 0$, at $x = L$ only if $k = \pi n/L$, where $n$ is an integer. Thus, the normalized wave function is

$$\psi_n(x) = \sqrt{\frac{2}{L}} \sin\left(\frac{n\pi x}{L}\right) \qquad n \geq 1, \tag{F.13}$$

where $n$ is an integer – the *quantum number*, i.e., the solutions are discrete (quantized).[6]

In Eq. (F.13), we have ensured that the wave function is normalized so that

$$\int_0^L |\psi(x)|^2 dx = 1. \tag{F.14}$$

Plugging Eq. (F.13) into Eq. (F.11), we see that the energy takes the form

$$E_n = \frac{\hbar^2 \pi^2 n^2}{2mL^2} = \frac{h^2 n^2}{8mL^2} \qquad n \geq 1, \tag{F.15}$$

where the energy is written in terms of both $\hbar = h/(2\pi)$ and $h$, the two versions of Planck's constant. Note that $E_n \propto k^2$, where $k$ is the wave vector for the electron.

We can calculate many average properties using Eq. (F.8). For example, the mean position for the energy level defined by $\psi_n$ is

$$\langle x \rangle_n = \int_0^L \psi_n^2(x) \, x \, dx = \frac{L}{2}, \tag{F.16}$$

i.e., the average position of the particle is in the middle in the box, independent of the quantum number $n$.

In three dimensions, the energy takes the form

$$E_{n_1,n_2,n_3} = \frac{h^2}{8mL^2}\left(n_1^2 + n_2^2 + n_3^2\right) \qquad n_1 \geq 1,\ n_2 \geq 1,\ n_3 \geq 1. \tag{F.17}$$

The lowest-energy state has $n_1 = n_2 = n_3 = 1$. There are three states with the next lowest energy, $(n_1, n_2, n_3) = (2,1,1)$, $(1,2,1)$, and $(1,1,2)$. These states are referred to as being *degenerate*, i.e., they all have the same energy.

Consider Figure F.1, which shows the possible states for a particle in a two-dimensional box. Each point represents a choice of $n_1$ or $n_2$. Because of the restrictions on the $n_1$ and $n_2$ to be positive integers, only the states in the upper right quadrant of the plot are possible. Since the energy goes as $n_1^2 + n_2^2$, equal-energy surfaces for the ground state are defined by circles centered at the origin.

Suppose we now add $N$ electrons to the box. The ground state would consist of two electrons in each state (one with spin up and one with spin down), starting from the lowest-energy state

---

[6] The coefficient of the $\cos x$ term must equal zero because $\cos x$ is nonzero at $x = 0$. The restriction that $n$ be an integer ensures that the wave function vanishes at $x = L$.

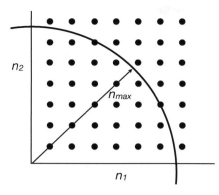

**Figure F.1** Distribution of energy states of a particle in a two-dimensional box. Equal-energy surfaces for the ground state are found along the curve with radius $n_{max}$.

and filling each state until the electrons have all been placed.[7] The maximum energy of a state in the system will be proportional to $n_{max}^2$. Electrons in this model system do not interact with each other nor with the environment. Thus, there is no potential energy and we can associate $E_{n_1,n_2,n_3}$ with the kinetic energy of the system. Thus, we can consider this system as an ideal gas of electrons. Since the density is uniform across the box, this system is often referred to as a *uniform, non-interacting, electron gas*.

The total kinetic energy will be the sum of the energies of all filled states, i.e., of all states with $n_1^2 + n_2^2 + n_3^2 < n_{max}^2$, which can be written as

$$K = 2 \frac{\hbar^2 \pi^2}{2mL^2} \sum_{n_1=1}^{n_1^2+n_2^2+n_3^2<n_{max}^2} \sum_{n_2=1} \sum_{n_3=1} \left( n_1^2 + n_2^2 + n_3^2 \right), \tag{F.18}$$

where the initial factor of 2 accounts for having two electrons per state. If $N$ is very big, then the separation between the states is small relative to $n_{max}$ and we can convert the sum to an integral, where the density of states is just unity. Equation (F.18) becomes

$$\begin{aligned} K &= \frac{\hbar^2 \pi^2}{mL^2} \int \int \int^{r^2 < n_{max}^2} r^2 dn_1 dn_2 dn_3 \\ &= \frac{\hbar^2 \pi^2}{mL^2} \times \frac{1}{8} \int_0^{n_{max}} r^2 \times 4\pi r^2 \, dr \,, \end{aligned} \tag{F.19}$$

where $r = \sqrt{n_1^2 + n_2^2 + n_3^2}$. The factor of 1/8 comes integrating over the full sphere when only one octant has permissible states (all $n$ must be positive). Evaluating Eq. (F.19) we find

$$K = \frac{\hbar^2 \pi^2}{mL^2} \times \frac{1}{8} \times 4\pi \left. \frac{r^5}{5} \right|_0^{n_{max}} = \frac{\hbar^2 \pi^3}{10mL^2} n_{max}^5 \,. \tag{F.20}$$

The number of filled states, which is just half the number of electrons, $N/2$, can be found by summing over the states in the same way we summed the energy. Following the same procedure

---

[7] From the Pauli exclusion principle discussed below in Appendix F.6.1, each state can have one electron with spin $+1/2$ and one with spin $-1/2$.

in which the sums are converted into integrals, we have

$$N = 2 \sum_{n_1=1} \sum_{n_2=1} \sum_{n_3=1}^{n_1^2+n_2^2+n_3^2 < n_{max}^2}$$

$$= 2 \times \frac{1}{8} \int_0^{n_{max}} 4\pi r^2 \, dr$$

$$= \frac{\pi}{3} n_{max}^3 \,. \tag{F.21}$$

The density of the electron gas is the number of electrons divided by the volume, $V = L^3$,

$$\rho = \frac{\pi}{3V} n_{max}^3 \,. \tag{F.22}$$

$n_{max}$ then is related to $\rho$ by

$$n_{max} = \left(\frac{3V\rho}{\pi}\right)^{1/3} = L\left(\frac{3\rho}{\pi}\right)^{1/3} \tag{F.23}$$

and the kinetic energy per volume is (using Eq. (F.20))

$$\mathcal{K} = \frac{K}{V} = \frac{1}{V} \frac{\hbar^2 \pi^3}{10mL^2} \left(\frac{3V\rho}{\pi}\right)^{5/3} = \frac{\hbar^2}{m} \frac{3}{10} (3\pi^2)^{2/3} \rho^{5/3} \,. \tag{F.24}$$

Equation (F.24) is the kinetic energy density (energy per volume) of a uniform, non-interacting, electron gas with electron density $\rho$, which is an important component in the Thomas-Fermi model in Chapter 4.

## F.5.2 Harmonic oscillator

The harmonic oscillator refers to a spring with the potential $U(x) = (1/2)kx^2$, where $k$ is the force constant.[8] To determine the quantum mechanical solution, we need to solve

$$E\psi(x) = \left(-\frac{\hbar^2}{2m} \frac{d^2}{dx^2} + \frac{1}{2}kx^2\right)\psi(x)\,. \tag{F.25}$$

The solution to this differential equation can be written in special functions called Hermite polynomials. For our purposes, it is important to know that the energy is again quantized and takes the form ($n \geq 0$)

$$E_n = \left(n + \frac{1}{2}\right)\hbar\omega\,, \tag{F.26}$$

where $\omega = \sqrt{k/m}$, the standard definition of frequency.

A very interesting outcome of the quantum solution is how the energy of the ground state in the quantum case differs from expectations from classical physics. The lowest energy of the classical harmonic oscillator is 0 – the spring is at rest in its equilibrium length. The lowest

---

[8] The classical solution to the harmonic oscillator is described in Appendix D.3.

energy in Eq. (F.26) is when $n = 0$, yet the energy is $E_0 = \hbar\omega/2$, which is a purely quantum-mechanical phenomenon. This energy is referred to as the *zero-point energy* and manifests itself in the specific heat of a low-temperature solid, which deviates from the classical result because of the zero-point energy of the lattice vibrations (phonons).

## F.5.3 The hydrogen atom

The foundation for our understanding of the electronic structure of atoms, molecules, and solids as well as the structure of the periodic table is the elegant solution to the hydrogen-like atom that consists of a nucleus (with a plus charge) and an electron that has a negative charge. This problem is thoroughly examined in most, if not all, introductory quantum mechanics texts, so we will not discuss the solution in any detail nor will we provide the solution.

The potential energy between an electron with charge $-e$ and a nucleus with charge $Ze$ is[9]

$$U(r) = -\frac{Ze^2}{r}, \tag{F.27}$$

where $r$ is the distance between the electron and the nucleus.

Because of the form of $U$, it is easiest to solve the Schrödinger equation in spherical-polar coordinates and then, by separating the variables, extract three separate differential equations that can be solved separately, yielding quantized solutions with three, related, quantum numbers

$$\text{total quantum number:} \quad n = 1, 2, 3, \ldots$$
$$\text{azimuthal quantum number:} \quad l = 0, 1, 2, \ldots, n - 1$$
$$\text{magnetic quantum number:} \quad m = -l, -l + 1, \ldots, 0, \ldots, l - 1, l \quad . \tag{F.28}$$

The energy of a hydrogen atom (to this approximation) depends only on the total quantum number $n$

$$E_n = -\frac{\mu Z^2 e^4}{2\hbar^2 n^2}. \tag{F.29}$$

The ground state for hydrogen ($Z = 1$) is $E_0 = -\mu e^4/2\hbar^2$, which has a magnitude of 13.6057 eV ($2.17990 \times 10^{-18}$ J). This unit of energy is called a Rydberg (Ry). Another unit of energy that is often used is the hartree, where 1 hartree = 2 Ry = 27.211385 eV.

The mean electron-nucleus distance in a hydrogen atom is found to be proportional to $a_o$, where

$$a_o = \frac{\hbar^2}{\mu e^2}. \tag{F.30}$$

In the simple Bohr model of an H atom, $a_o$ corresponds to the radius of the smallest orbit in the hydrogen atom. It is thus appropriately called the Bohr radius and has the value $a_o = 0.529$ Å.

---

[9] We are using cgs units in this equation. For MKS units, we would need to multiply this result by $1/4\pi\epsilon_o$, where $\epsilon_o$ is the permittivity constant. These equations are discussed in Appendix E.

A wave function for a single electron is called an *orbital*, a term that derives from the original Bohr model that spoke of electrons being in orbits around the nucleus. In the full quantum theory, there are no orbits, just the wave function $\psi$, from which we can determine the probability that an electron is located at a specific location. The states with $l = 0$ are called *s* orbitals and are always spherically symmetric, i.e., $\psi$ is a function of $r$ alone. Orbitals with $l = 1$ are called *p* orbitals and have directional character. For states with $l = 1$, $m$ can take the values $\pm 1$ and $0$. Thus there are three *p* orbitals. Similarly, the $l = 2$ states (the *d* orbitals), have $m = 0, \pm 1, \pm 2$ and so there are ten *d* orbitals, and so on. Drawings of these orbitals are given in most elementary quantum texts.

## F.6  ATOMS WITH MORE THAN ONE ELECTRON

To extend the calculations from the 1-electron H atom to multielectron systems, including the 2-electron He atom, is challenging because there are no exact solutions. Since electrons repel each other, their motion must be *correlated* so that they stay apart from each other. The 1-electron functions derived for the H atom cannot describe that correlated motion. The decrease in the energy when the correlated motion is included is called the *correlation energy*. As is discussed in Chapter 4, determining the correlation energy is essential when describing the energetics of multielectron systems. However, while the 1-electron hydrogen orbitals are not strictly accurate for He, they are approximately correct and we can (conveniently) use them to describe the basic properties of its electronic structure. Indeed, we can speak very freely of $1s$, $2s$, $2p$, ... orbitals for He and more complex atoms, even though the exact forms of their wave functions are not known.

Because the solutions are approximately correct for multielectron atoms, the quantum solution to the hydrogen atom can form the basis for our understanding of the properties of the elements. For multielectron systems, the degeneracy between orbitals with the same $n$ but different $l$ values is removed, which leads to the structure we see in the periodic table. There are many subtleties in this picture, including the relative energies of the different orbitals across the periodic table, which we will not discuss.

### F.6.1  Electron spin and the Pauli exclusion principle

So far we have neglected a very important feature of an electron – the *electron spin*. Each electron has an intrinsic angular momentum, or spin, specified by a quantum number $m_s$, where $m_s = +\frac{1}{2}$ or $-\frac{1}{2}$. Thus we have four quantum numbers to consider $n$, $l$, $m$, and $m_s$. An important result is the *Pauli exclusion principle*, which states that no electrons can exist in the same quantum state. Thus, in a hydrogen atom, for any given state with $n$, $l$, and $m$, we can have two electrons, one with spin $+\frac{1}{2}$ and one with spin $-\frac{1}{2}$.

The spin of the electron leads to a magnetic moment that can couple with the magnetic moment of the orbital angular moment. Since the orbital momentum depends on the orbital (i.e., $s$, $p$, $d$, ...), this *spin-orbit coupling* removes the degeneracy between the orbitals, which leads to splittings that have been observed in the hydrogen electronic spectra. We note that this energy is generally quite small.

We can include the spin explicitly in the wave function, for example as

$$\Psi(\mathbf{r}, \mathfrak{s}) = \psi(\mathbf{r})\mathfrak{s},$$ (F.31)

where $\mathfrak{s}$ is the spin wave function, i.e., $\mathfrak{s} = \alpha$ (spin up) or $\beta$ (spin down).

### F.6.2 Indistinguishability and antisymmetric wave functions

We cannot identify one electron versus another – they are indistinguishable particles. We thus have the conundrum of developing wave functions that meet the requirements of the Pauli exclusion principle yet that maintain the indistinguishability of the electrons. One approach is to have a wave function for a system of electrons that is *antisymmetric* for an exchange of the spatial and spin coordinates of any pair of electrons. By antisymmetric, we mean that when two electrons are exchanged for each other, $\psi \rightarrow -\psi$, which satisfies the Pauli principle that the states are not the same but also maintains the indistinguishability, since $|\psi^2|$ is unchanged.

Consider two electrons 1 and 2 in a hydrogen-like atom with states given by $n_1, l_1, m_1, m_{s_1}$ and $n_2, l_2, m_2, m_{s_2}$. Also suppose that we can use solutions from the 1-electron hydrogen atom (i.e., that $n$, $l$, and $m$ are still good quantum numbers).[10] The antisymmetric wave function would take the form

$$\psi = \psi_{n_1,l_1,m_1,m_{s_1}}(1)\psi_{n_2,l_2,m_2,m_{s_2}}(2) - \psi_{n_1,l_1,m_1,m_{s_1}}(2)\psi_{n_2,l_2,m_2,m_{s_2}}(1).$$ (F.32)

When 1 and 2 are exchanged $\psi \rightarrow -\psi$. Note that if all quantum numbers were the same, then $\psi = 0$, which is not allowed. Suppose that we have $N$ electrons in a system whose individual wave functions are denoted by $\psi_i$, where $i$ could indicate the four quantum numbers of a hydrogen-like atom. Slater pointed out that we can represent an antisymmetric wave function as the determinant

$$\psi(1, 2, 3, \ldots, N) = \frac{1}{\sqrt{N}} \begin{vmatrix} \psi_a(1) & \psi_a(2) & \ldots & \psi_a(N) \\ \psi_b(1) & \psi_b(2) & \ldots & \psi_b(N) \\ \vdots & \vdots & \vdots & \vdots \\ \psi_N(1) & \psi_N(2) & \ldots & \psi_N(N) \end{vmatrix}.$$ (F.33)

### F.6.3 The exchange energy

An important ramification of the requirement for antisymmetric wave functions is a difference in behavior of pairs of spins with the same sign and pairs of spins with opposite sign. This difference in behavior leads to an energy that is purely quantum mechanical in nature, called the *exchange energy*.

---

[10] These solutions are not strictly correct if we have more than one electron (e.g., for He) – the electron-electron interactions modify the wave functions.

Consider the free electron gas from Appendix F.5.1 for two electrons with *opposite* spins. The antisymmetric wave function formed from the plane waves[11] $e^{\mathbf{k}\cdot\mathbf{r}}$ and $e^{\mathbf{k}'\cdot\mathbf{r}}$ is

$$
\begin{aligned}
\Psi_{\uparrow\downarrow} &= \frac{1}{\sqrt{2}V}\begin{vmatrix} e^{i\mathbf{k}\cdot\mathbf{r}_1}\alpha(1) & e^{i\mathbf{k}\cdot\mathbf{r}_2}\alpha(2) \\ e^{i\mathbf{k}'\cdot\mathbf{r}_1}\beta(1) & e^{i\mathbf{k}'\cdot\mathbf{r}_2}\beta(2) \end{vmatrix} \\
&= \frac{1}{\sqrt{2}V}\left[ e^{i(\mathbf{k}\cdot\mathbf{r}_1+\mathbf{k}'\cdot\mathbf{r}_2)}\alpha(1)\beta(2) - e^{i(\mathbf{k}'\cdot\mathbf{r}_1+\mathbf{k}\cdot\mathbf{r}_2)}\alpha(2)\beta(1) \right],
\end{aligned}
\tag{F.34}
$$

where the spin variables are $\alpha$ and $\beta$, indicating spin up and spin down, respectively. Look at the structure of $\Psi_{\uparrow\downarrow}$. In the first term, electron 1 has spin up and electron 2 has spin down, while in the second term, 1 has spin down and 2 has spin up. As noted above, both terms are needed to ensure that the wave function is antisymmetric.

The probability of finding the electron at a point with given spin is

$$
\begin{aligned}
|\Psi_{\uparrow\downarrow}|^2 = \frac{1}{2V^2}\Big[ &\alpha^2(1)\beta^2(2) + \alpha^2(2)\beta^2(1) \\
&+ 2\mathbb{R}\Big\{ e^{i\mathbf{k}\cdot(\mathbf{r}_1-\mathbf{r}_2)}e^{i\mathbf{k}'\cdot(\mathbf{r}_2-\mathbf{r}_1)}\alpha(1)\alpha(2)\beta(1)\beta(2) \Big\} \Big],
\end{aligned}
\tag{F.35}
$$

where $\mathbb{R}$ indicates the real part of the following expression. Integrating over the spin variables (which we indicate by $\int d\mathfrak{s}$) will give the spatial probability distribution, which is[12]

$$
P_{\uparrow\downarrow} = \int |\Psi_{\uparrow\downarrow}|^2 d\mathfrak{s}_1 d\mathfrak{s}_2 = \frac{1}{2V^2}(1 + 1 + 0) = \frac{1}{V^2},
\tag{F.36}
$$

which is a constant independent of position, which is what one would expect for a uniform electron gas.

Now suppose the spins have the same sign, in which case

$$
\begin{aligned}
\Psi_{\uparrow\uparrow} &= \frac{1}{\sqrt{2}V}\begin{vmatrix} e^{i\mathbf{k}\cdot\mathbf{r}_1}\alpha(1) & e^{i\mathbf{k}\cdot\mathbf{r}_2}\alpha(2) \\ e^{i\mathbf{k}'\cdot\mathbf{r}_1}\alpha(1) & e^{i\mathbf{k}'\cdot\mathbf{r}_2}\alpha(2) \end{vmatrix} \\
&= \frac{1}{\sqrt{2}V}\left[ e^{i(\mathbf{k}\cdot\mathbf{r}_1+\mathbf{k}'\cdot\mathbf{r}_2)} - e^{i(\mathbf{k}'\cdot\mathbf{r}_1+\mathbf{k}\cdot\mathbf{r}_2)} \right]\alpha(1)\alpha(2).
\end{aligned}
\tag{F.37}
$$

Squaring the wave function and integrating over the spins we again find the spatial probability, which in this case is

$$
\int |\Psi_{\uparrow\uparrow}|^2 d\mathfrak{s}_1 d\mathfrak{s}_2 = \frac{1}{V^2}\left\{ 1 - \mathbb{R}e^{i(\mathbf{k}'-\mathbf{k})\cdot(\mathbf{r}_1-\mathbf{r}_2)} \right\}.
\tag{F.38}
$$

This result is really quite fascinating – it says that the probability distribution for a *noninteracting* electron gas is **not** uniform, i.e., it depends on the positions $\mathbf{r}_1$ and $\mathbf{r}_2$. There is a correlation between the motions of electrons with parallel spins that arises from the requirement of having

---

[11] As noted in Appendix F.5.1, the wave function for a free electron could be written as either a plane wave or in terms of sin and cos. Here, the plane waves are more convenient.

[12] While we write an integral over spin variables, since they are discrete the integral really represents a sum. To evaluate the probability, we need the relations $\int \alpha^2(1)d\mathfrak{s} = 1 = \int \beta^2(1)d\mathfrak{s}$ while $\int \alpha(1)\beta(1)d\mathfrak{s} = 0$.

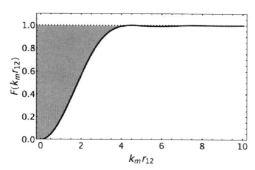

**Figure F.2** The probability distribution for electrons with parallel spins in a non-interacting electron gas ($\mathcal{F}(k_m r_{12})$ from Eq. (F.39)) plotted versus $k_m r_{12}$. The shaded area is the deviation of the probability from uniformity and shows a "hole" near $k_m r_{12} = 0$ that indicates the like-signed electrons avoid each other. It is referred to as the Fermi hole.

antisymmetric wave functions. Note that in Eq. (F.38), when $\mathbf{r}_1 = \mathbf{r}_2$, the probability vanishes, so that electrons with parallel spins are not found at the same point.

Now consider a system with a large number of electrons, $N$. The probability distribution for electrons with parallel spins is found by averaging over all the occupied orbitals, just as we did for the kinetic energy of an electron gas in Appendix F.5.1. In this case, the probability, $P_{\uparrow\uparrow}$, is

$$
\begin{aligned}
P_{\uparrow\uparrow}(r_{12}) &= \frac{\sum_{\mathbf{k}}^{k_m} \sum_{\mathbf{k'}}^{k_m} \int |\Psi_{\uparrow\uparrow}|^2 d\sigma_1 d\sigma_2}{\sum_{\mathbf{k}}^{k_m} \sum_{\mathbf{k'}}^{k_m} 1} \\
&= \frac{1}{V^2} \frac{\sum_{\mathbf{k}}^{k_m} \sum_{\mathbf{k'}}^{k_m} \left\{ 1 - \mathbb{R}e^{i(\mathbf{k'}-\mathbf{k})\cdot(\mathbf{r}_1-\mathbf{r}_2)} \right\}}{\sum_{\mathbf{k}}^{k_m} \sum_{\mathbf{k'}}^{k_m} 1} \\
&= \frac{1}{V^2} \frac{\int^{|\mathbf{k}|<k_m} d\mathbf{k} \int^{|\mathbf{k'}|<k_m} d\mathbf{k'} \left\{ 1 - \mathbb{R}e^{i(\mathbf{k'}-\mathbf{k})\cdot(\mathbf{r}_1-\mathbf{r}_2)} \right\}}{\int^{|\mathbf{k}|<k_m} d\mathbf{k} \int^{|\mathbf{k'}|<k_m} d\mathbf{k'}} \\
&= \frac{1}{V^2} \left\{ 1 - 9 \left( \frac{\sin(k_m r_{12}) - k_m r_{12} \cos(k_m r_{12})}{(k_m r_{12})^3} \right)^2 \right\} \\
&= \frac{1}{V^2} \mathcal{F}(k_m r_{12}),
\end{aligned}
\tag{F.39}
$$

where $k_m$ indicates the highest occupied orbital, we converted the sums to integrals (as we did in Appendix F.5.1), and integrated. $r_{12} = |\mathbf{r}_1 - \mathbf{r}_2|$ is the magnitude of the interelectronic separation. We plot $\mathcal{F}(k_m r_{12})$ in Figure F.2. The first term in $\mathcal{F}$ is equivalent to the result from electrons with opposite spins in Eq. (F.36). Thus, the second term corrects for the extra repulsion between electrons with parallel spins, creating a "hole" in the probability near $k_m r_{12} = 0$, which is referred to as the Fermi hole and arises from the antisymmetric nature of electrons.

The net electrostatic (Coulomb) energy for an electron gas is the average electrostatic repulsion between the electrons based on the above probabilities, i.e.,

$$
J_{quant} = \int (P_{\uparrow\downarrow}(r_{12}) + 2P_{\uparrow\uparrow}(r_{12})) \frac{e^2}{r_{12}} d\mathbf{r}_1 d\mathbf{r}_2 ,
\tag{F.40}
$$

where $e$ is the electron charge and the factor of 2 before $P_{\uparrow\uparrow}$ accounts for the electrons being either both up or both down.

The classical Coulomb energy (Eq. (E.10)) of an electron gas would be based on a uniform distribution of the electrons, without the Fermi hole in Figure F.2. The change in energy that arises from the presence of the Fermi hole is referred to as the *exchange energy*. Subtracting the classical result from $J_{quant}$, we find (after many manipulations described in detail elsewhere [251])

$$
\begin{aligned}
E_{ex} &= -\frac{9N^2e^2}{V}\int_0^\infty \frac{1}{r}\left(\frac{\sin(k_m r_{12}) - k_m r_{12}\cos(k_m r_{12})}{(k_m r_{12})^3}\right)^2 4\pi r^2 dr \\
&= -\frac{9\pi N^2 e^2}{V k_m^2},
\end{aligned}
\tag{F.41}
$$

where we have integrated by parts and expressed the answer in terms of the known integrals.[13] The exchange energy density is $E_{ex}$ divided by the volume $V$ and is, writing the electron density as $\rho = N/V$,

$$
\mathcal{E}_{ex} = \frac{E_{ex}}{V} = -\frac{9\pi e^2}{k_m^2}\rho^2.
\tag{F.42}
$$

The value for $k_m$ is related to $n_{max}$ in Figure F.1 by

$$
k_m = \frac{2\pi n_{max}}{L},
\tag{F.43}
$$

which in turn is related to $\rho$ in Eq. (F.23), giving

$$
k_m = 2\pi\left(\frac{3}{\pi}\right)^{2/3}\rho^{1/3}.
\tag{F.44}
$$

Putting Eq. (F.44) into Eq. (F.43) we have that the total exchange energy density is

$$
\mathcal{E}_{ex} = -\frac{3e^2}{4}\left(\frac{3}{\pi}\right)^{1/3}\rho^{4/3}.
\tag{F.45}
$$

## F.7 EIGENVALUES AND EIGENVECTORS

There are very few problems that can be solved exactly in quantum mechanics. Typically we use a variational approach, in which we parameterize the wave function and then vary the parameters until the minimum calculated energy is found. For example, suppose we have a *trial wave function* $\psi_t$ that depends on some set of parameters. The average energy can be determined as shown in Eq. (F.9),

$$
E_t = \frac{\langle\psi_t|\mathcal{H}|\psi_t\rangle}{\langle\psi_t|\psi_t\rangle},
\tag{F.46}
$$

where $E$ is the energy associated with the Hamiltonian $\mathcal{H}$. The subscript $t$ on $E$ indicates that this is the energy of the specific trial wave function $\psi_t$.

---

[13] $\mathrm{si}(x) = -\int_x^\infty \sin(t)/t\, dt$ and $\mathrm{ci}(x) = -\int_x^\infty \cos(t)/t\, dt$.

The variational theory states that the energy calculated based on an approximate wave function is an *upper bound* to the true energy $E$, i.e., that

$$E \leq E_t \,. \tag{F.47}$$

If the trial wave function is the correct solution to the Schrödinger equation, then the true energy is obtained. Any parameters in the wave function can be varied to find the lowest value of the energy, which will be the best estimate for the correct function and energy based on this trial function. For convenience we will drop the subscript $t$ from the rest of this discussion.

A common practice is to approximate the wave function as a linear set of functions, typically solutions to a similar, known, problem. For example, for a set of $m$ functions

$$\psi = \sum_{i=1}^{m} c_i \phi_i \,, \tag{F.48}$$

where the set of functions $\{\phi_i\}$ is called a *basis set*. The goal is to determine the set of coefficients $\{c_i\}$ that minimize the energy. The accuracy, of course, will depend on the quality of the basis set chosen to approximate the system, with the assumption that the closer the basis set functions are to the correct wave function, the lower the energy. As an example, consider the solution for a multielectron atom. A typical basis set would be made up of the functions that describe the hydrogenic orbitals.

From Eq. (F.46), we have

$$E \sum_{i=1}^{m} \sum_{j=1}^{m} c_i c_j S_{ij} = \sum_{i=1}^{m} \sum_{j=1}^{m} c_i c_j H_{ij} \,, \tag{F.49}$$

where the Hamiltonian integrals are

$$H_{ij} = \int \phi_i^*(\mathbf{r}) \mathcal{H}(\mathbf{r}) \phi_j(\mathbf{r}) d\mathbf{r} = \langle \phi_i | \mathcal{H} | \phi_j \rangle \,, \tag{F.50}$$

and the *overlap* integrals are

$$S_{ij} = \int \phi_i^*(\mathbf{r}) \phi_j(\mathbf{r}) d\mathbf{r} = \langle \phi_i | \phi_j \rangle \,. \tag{F.51}$$

Minimizing the energy with respect to the parameters leads to the set of equations

$$\sum_{i=1}^{m} \sum_{j=1}^{m} c_i \left( H_{ij} - E S_{ij} \right) = 0 \qquad j = 1, 2, \ldots m \,, \tag{F.52}$$

which is just a set of $m$ simultaneous homogeneous linear equations whose solution is found in standard ways. If we consider $H_{ij}$ and $S_{ij}$ as an element of a matrix $\mathbf{H}$ and $\mathbf{S}$, respectively, then the set of equations becomes $\mathbf{H} - E\mathbf{S} = 0$. The eigenvalues of the matrix are the energies of the eigenstates $E_i$ and the eigenvectors the coefficients $\{c\}_i$ for those states. Note that it is often easiest if one assumes a basis set whose functions are orthogonal and normalized, i.e, so that $S_{ij} = 1$ if $i = j$ and 0 otherwise. This matrix approach is the basis for most methods used to solve complex quantum mechanical problems. As noted, the key to obtaining a high-quality result is having a good choice of basis functions.

## F.8  MULTIELECTRON SYSTEMS

This book is focused on materials, which by their nature have many nuclei and many electrons. The Hamiltonian for such a system is made up of a number of parts. First, consider a system with $M$ nuclei and $N$ electrons. There will be a kinetic energy term for each electron, whose total contribution to the Hamiltonian is

$$\sum_{i=1}^{N} -\frac{\hbar^2}{2m_i}\nabla_i^2 \, . \tag{F.53}$$

There will also be an interaction term between each electron and each nucleus, which we write as

$$\sum_{i=1}^{N} v(\mathbf{r}_i) \, , \tag{F.54}$$

where the potential acting on electron $i$ from the $M$ nuclei is

$$v(\mathbf{r}_i) = -\sum_{\alpha=1}^{M} \frac{Z_\alpha e^2}{r_{i\alpha}} \, . \tag{F.55}$$

Finally, there is a term from the electrons interacting with each other through the standard electrostatic interaction, which we can write in many ways,[14] for example

$$\sum_{i}\sum_{j>i} \frac{e^2}{r_{ij}} \, . \tag{F.56}$$

Putting it all together, the net Hamiltonian is

$$\mathcal{H} = \sum_{i=1}^{N} \left( -\frac{\hbar^2}{2m_i}\nabla_i^2 + v(\mathbf{r}_i) \right) + \sum_{i}\sum_{j>i} \frac{e^2}{r_{ij}} \, . \tag{F.57}$$

This form of the Hamiltonian assumes that the motion of the electrons is so fast relative to the nuclei that the nuclei can be considered as fixed – it ignores any coupling between electronic and nuclear motion. This assumption is called the *Born-Oppenheimer approximation*. With that assumption, if the nuclear positions are fixed, then their interactions with the electrons, given in Eq. (F.55), act like an external, fixed, potential.

It is common to rewrite Eq. (F.57) using the atomic units introduced in Appendix F.5.3, in which the energy is in hartree, the distance is in bohr, the mass unit is the electron mass ($m_e$) and the unit of charge is the electron charge $e$. In this case, the Hamiltonian becomes

$$\mathcal{H} = \sum_{i=1}^{N} \left( -\frac{1}{2}\nabla_i^2 + v(\mathbf{r}_i) \right) + \sum_{i}\sum_{j>i} \frac{1}{r_{ij}} \tag{F.58}$$

---

[14] See discussion about Eq. (3.5).

and

$$v(\mathbf{r}_i) = -\sum_{\alpha=1}^{M} \frac{Z_\alpha}{r_{i\alpha}}. \tag{F.59}$$

## F.9   QUANTUM MECHANICS OF PERIODIC SYSTEMS

We will be primarily interested in the behavior of electrons in a crystal. Since the structure of the crystal lattice (i.e., the nuclei) is periodic, then the external potential $v(\mathbf{r})$ in the Hamiltonian (Eq. (F.55)) will be periodic, and thus the electronic structure will be periodic.

The basis for creating a periodic solution is Bloch's theorem. We start by identifying $\psi_{\mathbf{k}}(\mathbf{r})$ as solutions to the Schrödinger equation for a potential $v(\mathbf{r})$, where $\mathbf{k}$ is a vector indicating the three quantum numbers for the states.

Bloch's theorem states that all solutions $\psi_{\mathbf{k}}(\mathbf{r})$ for a periodic potential $v(\mathbf{r})$ have to meet in a continuous way throughout space. One way to write this conditions is that

$$\psi_{\mathbf{k}}(\mathbf{r}) = u_{\mathbf{k}}(\mathbf{r})e^{i\mathbf{k}\cdot\mathbf{r}}, \tag{F.60}$$

where $u_{\mathbf{k}}(\mathbf{r})$ is periodic on the lattice, i.e.,

$$u_{\mathbf{k}}(\mathbf{r}+\mathbf{R}) = u_{\mathbf{k}}(\mathbf{r}), \tag{F.61}$$

for every $\mathbf{R}$ in the direct lattice. Eq. (F.60) and Eq. (F.61) imply that

$$\psi_{\mathbf{k}}(\mathbf{r}+\mathbf{R}) = e^{i\mathbf{k}\cdot\mathbf{R}}\psi_{\mathbf{k}}(\mathbf{r}). \tag{F.62}$$

Equation (F.62) is a statement of Bloch's theorem, which is the basis for the electronic structure methods for crystals discussed in Chapter 4. The proof of Bloch's theorem is straightforward and available from a variety of sources. We emphasize that Eq. (F.62) imposes a stringent condition on any solution to the Schrödinger equation for any periodic potential.

## F.10   SUMMARY

This brief introduction to quantum mechanics is certainly not exhaustive and barely scratches the surface. The important results are that particles are described with wave functions $\psi(\mathbf{r})$, the probability of finding a particle at $\mathbf{r}$ is $\psi^*(\mathbf{r})\psi(\mathbf{r})$, and the fundamental basis for quantum mechanics is the Schrödinger equation. We have not provided much in the way of detail and the reader is encouraged to explore one (or more) of the many fine texts on quantum mechanics listed in the Suggested reading section.

## Suggested reading

There are many good books available on quantum mechanics.

- You might want to start with a physical chemistry text; for example *Schaum's Outline of Physical Chemistry* [293], *Physical Chemistry: A Molecular Approach*, by Donald A. McQuarrie and John D. Simon [224], provide general discussions.
- While a bit old-fashioned, the classic *Introduction to Quantum Mechanics with Applications to Chemistry* by Linus Pauling and E. Bright Wilson [253], is worth a look as well.
- For those a bit more adventurous, seek a more physics-based treatment. There are a large number of titles to choose from.

# Statistical thermodynamics and kinetics

In this appendix, we introduce the basic concepts and assumptions of statistical thermodynamics. We discuss time averages and introduce the concept of an ensemble average. We present the important ensembles used in atomistic simulations and show their connection to thermodynamics through a function called the partition function. Finally, we summarize some important results often used in atomistic simulations. We also introduce some basic concepts from kinetics.

## G.1 BASIC THERMODYNAMIC QUANTITIES

In Table G.1, we define the thermodynamic quantities used in this text. Quantities such as $E$, $H$, and $A$ are *extensive*, i.e., they are additive and scale linearly with the size of the system. In contrast, quantities like $T$, $P$ and $\mu$ are *intensive* - they do not scale with system size. Note that we use the notation $A$ for the Helmholtz free energy, which in some communities is denoted by $F$. Note also that the symbol $U$ always indicates a potential energy, while $K$ indicates the kinetic energy.

The thermodynamic quantities in Table G.1 can be related to each other through the standard differential forms.[1]

---

[1] Assuming a system with $r$ types of particles,

$$dA = -SdT - PdV + \sum_{i=1}^{r} \mu_i dn_i \tag{G.1}$$

$$dG = -SdT + VdP + \sum_{i=1}^{r} \mu_i dn_i$$

$$dH = TdS + VdP + \sum_{i=1}^{r} \mu_i dn_i$$

$$dE = TdS - PdV + \sum_{i=1}^{r} \mu_i dn_i$$

From these equations we can read off such relationships as $(\partial A/\partial T)_{N,V} = -S$, using the definition of the total derivative, which, if $f = f(x, y)$, is $df = (\partial f/\partial y)_x dy + (\partial f/\partial x)_y dy$. We can also use the result that if $dz = adx + bdy$, then $(\partial a/\partial y)_x = (\partial b/\partial x)_y$ to derive a series of relations between the partial derivatives. For example, from the result for $dG$, we have $(\partial S/\partial P)_{T,N} = (\partial V/\partial T)_{P,N}$. The equations connecting the partial derivatives are called Maxwell's relations.

Table G.1 **Definition of common thermodynamic symbols**

| | |
|---|---|
| $N$ = number of particles | $E = U + K$ = internal energy |
| $V$ = volume | $S$ = entropy |
| $P$ = pressure | $H = E + PV$ = enthalpy |
| $T$ = temperature | $A = E - TS$ = Helmholtz free energy |
| $\mu$ = chemical potential | $G = E - TS + PV$ = Gibbs free energy |

## G.2 INTRODUCTION TO STATISTICAL THERMODYNAMICS

In Chapter 5, we discuss the cohesive energy of materials and models for the interactions between atoms in a number of types of systems. Everything we discuss in that chapter, however, is true only for systems at a temperature of 0 K. To apply these concepts to materials at finite (i.e., greater than zero) temperatures requires invoking results from a field of study called statistical mechanics. Statistical mechanics gives an understanding of how the macroscopic properties of a system are related to its microscopic variables. For example, statistical mechanics shows how the $10^{23}$ degrees of freedom of the atoms in a bulk system reduce to the few parameters (e.g., pressure, volume, and temperature) that describe the thermodynamics, or macroscopic behavior, of that system. There are strictly two kinds of studies in statistical mechanics. Statistical thermodynamics describes the behavior of systems in equilibrium and provides the connection to thermodynamics. Nonequilibrium statistical mechanics focuses on systems away from equilibrium. The latter topic is beyond the scope of this book and will not be discussed.

A principal focus of statistical thermodynamics is the calculation of average quantities and the connection of these averages to what we see in nature. Statistical thermodynamics also provides a framework for understanding the role of fluctuations and their connection to phase stability, heat flow, etc.

There are many books on statistical thermodynamics (see, for example, those listed in the bibliography at the end of the chapter). In this appendix, we summarize some of the concepts that are important to properly use basic simulation techniques such as the Monte Carlo and molecular dynamic methods.

## G.3 MACROSTATES VERSUS MICROSTATES

The thermodynamic state of a system is referred to as a *macrostate*, which refers to the set of macroscopic properties that define the state. The macrostate categorizes the overall, bulk behavior of a system and can be described by the thermodynamic *constraints* that operate on it. For example, the thermodynamic properties of a system with a constant number of particles

$N$, a constant volume $V$, and a constant temperature $T$ will be different than one in which the pressure $P$ is held constant and the volume allowed to vary – it is the difference between a system described by the Helmholtz free energy $A(N, V, T)$ or the Gibbs free energy $G(N, P, T)$.

The *microstate* of a system is the instantaneous value of its internal variables. For example, if a system has $N$ atoms, the microstate at a specific time $t_o$ is the set of atomic coordinates and momenta of all the atoms at $t_o$. The microstates will change with time as the system evolves.

Statistical mechanics tells us how to connect averages over the microstates to the thermodynamics of the macrostates. It turns out that there are two ways we can think about making that connection, by time averages or by ensemble averages.

## G.4 PHASE SPACE AND TIME AVERAGES

Suppose we have a system with $N$ atoms, where $N$ is very large. If the motion of those atoms obeys classical mechanics, then at a time $t$ the atoms have a specific set of positions and momenta. The time-dependent properties of the system are completely characterized by the $6N$ quantities known as the mechanical degrees of freedom – the $3N$ coordinates of the positions $(\mathbf{r}_1, \mathbf{r}_2, \mathbf{r}_3, \ldots, \mathbf{r}_N)$ and the $3N$ coordinates of the momenta $(\mathbf{p}_1, \mathbf{p}_2, \mathbf{p}_3, \ldots, \mathbf{p}_N)$, i.e., by its instantaneous microstate.[2] These positions and momenta are coordinates in a $6N$-dimensional space, called *phase space*. At any point in time, the system is characterized by a point in phase space, i.e., the set of $3N$ positions and $3N$ momenta at that time. As time progresses, the system will move through phase space as the atoms acquire new positions and momenta. We call the locus of points traced in that motion through phase space the *trajectory* of the motion.[3]

We can designate a point in phase space at time $t$ as $(\mathbf{r}^N(t), \mathbf{p}^N(t))$, where the notation $\mathbf{r}^N(t)$ means the set of $N$ position vectors at time $t$, $\mathbf{r}^N(t) = \{\mathbf{r}_1(t), \mathbf{r}_2(t), \ldots, \mathbf{r}_N(t)\}$, while $\mathbf{p}^N(t)$ is a set of momentum vectors, $\mathbf{p}^N(t) = \{\mathbf{p}_1(t), \mathbf{p}_2(t), \ldots, \mathbf{p}_N(t)\}$. We could try to explicitly follow the change in these variables over time. However, knowing the instantaneous state of a system does not tell us much – it is too complicated for us to consider for even relatively small values of $N$.

We can start to see the complexity of phase space in Figure G.1. In Figure G.1a we show the phase space of the simple one-dimensional harmonic oscillator of Appendix D.3, where the velocity $v_x(t)$ is plotted versus the position $x(t)$ over time. It is not very interesting. There are only two degrees of freedom in the problem and the total energy $E$, which is the sum of kinetic energy (a function of $v_x^2$) and the potential energy (a function of $x^2$) is a constant – the motion of the oscillator is such that the kinetic energy and potential energy exchange over time to keep the energy fixed. Thus, the velocity and position are coupled and phase space is simple. In Figure G.1b, we plot $v_x(t)$ versus $x(t)$ for one atom chosen from a three-dimensional molecular

---

[2] There are three degrees of freedom per coordinate $(x, y, z)$ and three per each momentum $(p_x, p_y, p_z)$.

[3] Technically, we can remove three degrees of freedom by ignoring the motion of the center of mass of the particles. With $N$ very large, that is irrelevant.

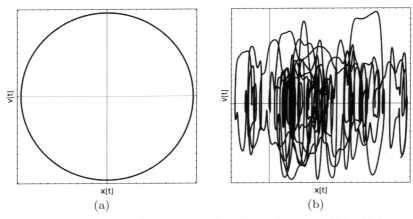

**Figure G.1** A simple view of phase space. Plotted are $v_x(t)$ versus $x(t)$ for (a) the one-dimensional harmonic oscillator of Appendix D.3 and (b) one coordinate of one atom over a sub-femtosecond time span from a three-dimensional molecular dynamics simulation.

dynamics simulation of many atoms interacting with a Lennard-Jones potential. Shown is a sub-femtosecond part of its trajectory. There is no obvious coupling between $x$ and $v_x$. We can tell some things, for example that the atom seems to be moving relative to a fixed position in space, but in most ways the figure is not particularly informative. Of course, this plot does not tell the whole story for even this atom. We have shown only the $x$ and $v_x$ components. We really should be showing the trajectory as a function of all three directions and all three velocities. Such a plot would require a six-dimensional space,[4] which we cannot visualize. That still would not represent the system, however, because the motion of any atom is heavily influenced by the motions of the other atoms, with kinetic and potential energies being exchanged between them through collisions and longer-ranged interactions. Thus, we need the full $6N$-dimensional space to represent the system. Since we have difficulties in extracting information from a two-dimensional phase space, understanding the $6N$-dimensional space will clearly be hopeless.

The goal is to connect the microstates to the macrostates by means of quantities averaged over the microstates. The most intuitive idea is that the value we observe for the macrostate in an experiment would be equal to the average of that quantity over time. For example, the observed total energy $E_{obs}$ would be a *time average* of the value of $E(t)$ over some time period $t_o$,

$$E_{obs} = \langle E \rangle = \frac{1}{t_o} \int_0^{t_o} E(t')dt' , \qquad (G.2)$$

where $E$ is a continuous variable of time. $E(t)$ is the value of $E$ determined at the point in phase space at time $t$, i.e., $E(t) = K(\mathbf{p}^N(t)) + U(\mathbf{r}^N(t))$. What is a sufficient time $t_o$ for a time average that corresponds to a macroscale observable? $t_o$ must be long enough so that the system can visit sufficient points in the parts of phase space accessible to it.

[4] A six-dimensional space would be a plot with six axes: $x$, $y$, $z$, $v_x$, $v_y$, and $v_z$.

# G.5  ENSEMBLES

There is another way to calculate average quantities.[5] When we performed the time average, we evaluated $E$ at each configuration visited in phase space along the trajectory. Suppose that along that trajectory, the system visited (or became arbitrarily close to) configurations more than once. We could imagine keeping track of the number of times the system took on a specific configuration. The average could then be determined by the value of, for example, $E_\alpha$, at each distinct configuration ($\alpha$) multiplied by the number of times the system was in that configuration $n_\alpha$. To find the average, we would just need to divide by the number of configurations $N_{config}$,

$$\langle E \rangle = \frac{1}{N_{config}} \sum_{\alpha=1}^{N_{config}} n_\alpha E_\alpha = \sum_{\alpha=1}^{N_{config}} \rho_\alpha E_\alpha \,. \tag{G.3}$$

In the second part of Eq. (G.3), we introduced the *probability density*, $\rho_\alpha$, which is the fraction of all possible states that are in configuration $\alpha$. For continuous systems, the sum in Eq. (G.3) becomes an integral. An important property of the probability density is that it is normalized, i.e.,

$$\sum_{\alpha=1}^{N_{config}} \rho_\alpha = 1 \,. \tag{G.4}$$

Instead of following a single system over a long period of time, imagine that a very large number of identical systems were created and allowed to evolve independently – an *ensemble* of systems. While these systems are identical, they will be in different states. The probability density of being in a given state can be found by counting the number of times that state is represented in the ensemble of systems and dividing by the number of systems. Averages are then found as in Eq. (G.3). This way of looking at averaging comes originally from the work of J. Williard Gibbs and is called an ensemble average.

The probability densities for an ensemble of systems will depend on the constraints put upon the system. For most of the systems we will consider, the constraints will be that certain thermodynamic quantities are fixed. Thus, we will sometimes write the ensemble average in the form $\langle \rangle_\beta$, where $\beta$ lists what constraints we have placed on the system.

## G.5.1  Weighting functions and probability

Consider the probability of being in a state, $\rho_\alpha$. At this point, we do not have a way to specify what factors will influence that probability, for example, how it might depend on thermodynamic variables such as temperature. We can, however, introduce a function that takes on that role, with the hope that we can eventually specify what that function is. We will call this function

---

[5] This section uses results from the discussion of probability in Appendix C.4.

$w_\alpha$ for state $\alpha$. Because $w_\alpha$ will specify how thermodynamic quantities weight the probability of a state relative to another state, it is called a weighting function.

$\rho_\alpha$ is thus assumed to be of the form $\rho_\alpha \propto w_\alpha$. However, we want the probabilities to be normalized, as discussed in Appendix C.4, so we can write[6]

$$\rho_\alpha = \frac{w_\alpha}{\sum_\alpha w_\alpha} = \frac{1}{Q} w_\alpha \,, \tag{G.5}$$

where the sum in the denominator is over all states. The normalization function,

$$Q = \sum_\alpha w_\alpha \,, \tag{G.6}$$

is called the *partition function* and is of fundamental importance in statistical mechanics, providing a critical link to thermodynamics. An average of a quantity $B$ is given by

$$\langle B \rangle = \frac{1}{Q} \sum_\alpha w_\alpha B_\alpha \,. \tag{G.7}$$

The properties of an ensemble of systems depend on the constraints and each set of constraints corresponds to a different ensemble. As we shall see, each ensemble can be characterized by a different weighting function $w_\alpha$ and thus a different partition function $Q$.

In this book, we will deal with many classical systems with continuous potentials, which are characterized by a phase space with positions and momenta as the fundamental coordinates. These are continuous variables, so we must replace the sums over states in the above equations by integrals. Since we can in principle have all possible positions and momenta, we must integrate over all coordinates and over all momenta. For such systems, the sum over states takes the form for an $N$ particle system [223]

$$\sum_\alpha f_\alpha \rightarrow \frac{1}{N! h^{3N}} \int f(\mathbf{r}^N, \mathbf{p}^N) d\mathbf{r}^N d\mathbf{p}^N \,. \tag{G.8}$$

The $N!$ accounts for the fact that the particles are indistinguishable, i.e., since we cannot tell which particle is which, there are $N!$ ways we could distribute them. The $h^{3N}$ connects our classical system to quantum mechanics and is required to yield the correct results for the entropy of an ideal gas.

## G.5.2 Ergodicity

We have assumed that the two ways of taking averages, by averaging either over time or over an ensemble of systems, are equivalent. Systems for which this equivalence holds are said to be *ergodic*. Ergodicity is not an easy concept and proving that a system is ergodic is not often possible. However, it is reasonable to assume that the large systems we will consider in this text are ergodic.

---

[6] $\sum_\alpha \rho_\alpha = \sum_\alpha w_\alpha / \sum_\alpha w_\alpha = 1$.

## G.5.3 Canonical ensemble (*NVT*)

In the laboratory, it is not uncommon to consider systems in which a set number of atoms of a material $N$ is put in a container with a fixed volume $V$ and then held at a constant temperature $T$. The free energy associated with systems with constant $NVT$ is the Helmholtz free energy $A$. Systems with these constraints are referred to as being in the *canonical ensemble*, which is sometimes referred to as the *NVT* ensemble in reference to the fixed thermodynamic variables. As stated, the canonical ensemble is characterized by having a fixed number of particles in a fixed volume at constant temperature (*NVT*). All other thermodynamic quantities will fluctuate (in equilibrium) around their average values. As is discussed in Chapter 7, this is the standard ensemble of most Monte Carlo (MC) simulations.

The weighting function describes how thermodynamics influences the probability and, for the canonical ensemble, is given by[7]

$$w_\alpha = e^{-E_\alpha/k_B T} , \tag{G.9}$$

where $E_\alpha$ is the energy of state $\alpha$ and $k_B$ is the Boltzmann constant ($k_B = R/N_A$, the gas constant divided by Avogadro's number). We come across this form of weighting function very often. It is referred to as the Boltzmann factor. The factor $1/k_B T$ occurs so frequently that it is often designated by $\beta = 1/k_B T$. We will not use that form in this text, preferring to explicitly include $T$.

The canonical partition function of Eq. (G.6) is given by

$$Q_{NVT} = \sum_\alpha e^{-E_\alpha/k_B T} . \tag{G.10}$$

Evaluating $Q_{NVT}$ requires a sum over the Boltzmann factors for all states in the system. In most cases there are far too many possible states and we cannot evaluate $Q$ directly. Thus, while we know the weighting function $w_{NVT}$ for each state, from Eq. (G.5) we do not know the actual probability density $\rho_{NVT}$ for each state. Suppose, however, that we consider two microstates, $\alpha$ and $\alpha^*$. We can evaluate the *relative* probability of the states by

$$\frac{\rho_{\alpha^*}}{\rho_\alpha} = \frac{e^{-\beta E_{\alpha^*}}}{Q} \frac{Q}{e^{-\beta E_\alpha}} = e^{-\beta(E_{\alpha^*} - E_\alpha)} . \tag{G.11}$$

It is very important to note that *the relative probability between states does not depend on an evaluation of the partition function*. Being able to determine the relative probability between states is the basis of the Metropolis Monte Carlo method described in Chapter 7.

From Eq. (G.7), averages in the canonical ensemble take the form,

$$\langle E \rangle = \frac{1}{Q_{NVT}} \sum_\alpha e^{-E_\alpha/k_B T} E_\alpha , \tag{G.12}$$

where we have shown the average energy as an example.

---

[7] A nice proof of this is given in [62].

The Helmholtz free energy, $A$, is related to the canonical partition function by [223]

$$A = -k_B T \ln Q_{NVT} . \tag{G.13}$$

From thermodynamics, $(\partial (A/T)/\partial (1/T))_{N,V} = E$. It is reassuring to note that taking the derivative of $A$ with respect to $1/T$ in Eq. (G.13), using the definition of $Q_{NVT}$ in Eq. (G.10), recovers the expression for the average energy in Eq. (G.12).[8]

For continuous systems described by positions and momenta, the partition function takes the form

$$Q_{NVT} = \frac{1}{N!h^{3N}} \iint e^{-\mathcal{H}(\mathbf{r}^N, \mathbf{p}^N)/k_B T} d\mathbf{r}^N d\mathbf{p}^N , \tag{G.14}$$

where the value of the Hamiltonian function $\mathcal{H}(\mathbf{r}^N, \mathbf{p}^N) = K(\mathbf{p}^N) + U(\mathbf{r}^N)$ is the total energy [223]. In Eq. (G.14), the limits of integration for the positional coordinates are such that they lie in the volume $V$. The components of the momenta can take on any values between $-\infty$ and $\infty$.[9]

If the potential depends only on the positions,[10] the integrals over momenta and positions are separable into a set of integrals over the momenta multiplied by a set of integrals over the positions. We can then evaluate the integrals over the momenta separately and, it turns out, analytically[11] and the partition function becomes

$$Q_{NVT} = \frac{1}{N! \Lambda^{3N}} \int e^{-U(\mathbf{r}^N)/k_B T} d\mathbf{r}^N , \tag{G.15}$$

where $\Lambda = h/\sqrt{2\pi m k_B T}$ is the *thermal de Broglie wavelength*. The momenta thus do not appear directly in $Q_{NVT}$. For quantities that depend only on position, for example, the potential energy, averages take the form

$$\langle U \rangle = \frac{1}{Z_{NVT}} \int e^{-U(\mathbf{r}^N)/k_B T} U(\mathbf{r}^N) d\mathbf{r}^N , \tag{G.16}$$

where the *configurational integral* is

$$Z_{NVT} = \int e^{-U(\mathbf{r}^N)/k_B T} d\mathbf{r}^N . \tag{G.17}$$

Note that the de Broglie wavelength cancels out in Eq. (G.16).

We can also evaluate the averages of quantities that depend only on the momenta. In this case, the integrals over the coordinates cancel out of the problem. For example, the average

---

[8] A useful exercise is to show that this statement is true.

[9] From Eq. (D.7), the kinetic energy for each atom takes the form $K(p_x, p_y, p_z) = (p_x^2 + p_y^2 + p_z^2)/2m$.

[10] For example, if there are no velocity-dependent terms, such as those that arise from a frictional force as discussed in Chapter 13.

[11] The integrals over each coordinate of the momenta for each atom are $\int_{-\infty}^{\infty} \exp(-(p_x^2/2m)/k_B T) dp_x$. These are just Gaussian integrals of the form $\int_{-\infty}^{\infty} \exp(-ax^2) dx = \sqrt{\pi/a}$. Thus $\int_{-\infty}^{\infty} \exp(-(p_x^2/2m)/k_B T) dp_x = \sqrt{2\pi m k_B T}$ for each coordinate. For each atom we have three identical integrals multiplying each other, one for $p_x$, one for $p_y$, and one for $p_z$. Thus, the contribution from each atom is $(2\pi m k_B T)^{3/2}$. If all the atoms are the same, then we have a product of $N$ such integrals, so the net contribution of $K$ to the partition function is $(2\pi m k_B T)^{3N/2}$.

kinetic energy, starting from Eq. (G.14) and Eq. (G.12), is

$$\langle K \rangle = \frac{\int \exp(-K(\mathbf{p}^N)/k_B T) K(\mathbf{p}^N) d\mathbf{p}^N}{\int \exp(-K(\mathbf{p}^N)/k_B T) d\mathbf{p}^N} . \tag{G.18}$$

These integrals are easy to evaluate (though tedious) using the basic approach described in footnote 11 and we find that the average kinetic energy is related to temperature by

$$\langle K \rangle = \frac{3}{2} N k_B T . \tag{G.19}$$

Equation (G.19) is a very important result and an example of the *equipartition theorem*, which states that in thermal equilibrium, energy is equally partitioned over its various forms.[12] Equation (G.19) is employed in simulations (e.g., molecular dynamics) to determine the temperature from the average kinetic energy.

From the virial theorem in classical mechanics [223] we can define an expression for an instantaneous "pressure" function, $\mathcal{P}$, whose average is the pressure of the system $P$, i.e., $\langle \mathcal{P} \rangle = P$

$$\mathcal{P} = \frac{N}{V} k_B T - \frac{1}{3V} \left\langle \sum_{i=1}^{N} \mathbf{r}_i \cdot \nabla_i U \right\rangle. \tag{G.20}$$

For a potential energy (Chapter 5) described by a sum over central-force potentials $\phi(r)$ that depend only on the distance $r_{ij}$ between particles $i$ and $j$, the pressure equation is[13]

$$\mathcal{P} = \frac{N}{V} k_B T - \frac{1}{3V} \left\langle \sum_{i=1}^{N} \sum_{j>i}^{N} r_{ij} \frac{d\phi}{dr_{ij}} \right\rangle. \tag{G.21}$$

## G.5.4 Maxwell-Boltzmann distribution

We can derive from the canonical partition function another very important result – the distribution of speeds of particles in an ideal gas in thermal equilibrium. This distribution is called the Maxwell-Boltzmann distribution after its originators, James Clark Maxwell and Ludwig Boltzmann. For an atom, it takes the form

$$f_P(p_x, p_y, p_z) = \left( \frac{1}{2\pi m k_B T} \right)^{3/2} e^{-(p_x^2 + p_y^2 + p_z^2)/2m k_B T} , \tag{G.22}$$

where $m$ is the mass and $T$ is the temperature. Note that the Maxwell-Boltzmann distribution is the familiar Gaussian distribution (called the normal distribution in statistics).

---

[12] More specifically, the equipartition theory states that each degree of freedom whose contribution to the energy is a function of the square of its coordinate contributes $1/2 k_B T$ to the average energy. Thus, the average kinetic energy, which is proportional to $v_x^2 + v_y^2 + v_z^2$, has the value of $3/2k_B T$ for each atom. A harmonic oscillator, with $U(x) = 1/2k(x^2 + y^2 + z^2)$, would have an average energy of $1/2k_B T$ for each coordinate.

[13] The pressure equation for the embedded-atom model (a many-body potential for metals) is given in Eq. (6.19).

### G.5.5 Microcanonical ensemble (*NVE*)

The microcanonical ensemble is characterized by a constant number of particles, constant volume, and constant total energy (*NVE*). In Chapter 6 we discuss that this is the ensemble of standard molecular dynamics simulations. The weighting function for this ensemble is given by

$$w_{NVE} = \delta\big(\mathcal{H}(\mathbf{r}^N, \mathbf{p}^N) - E\big), \tag{G.23}$$

where $\delta(x)$ is the Dirac delta function of Appendix C.5.1. What this function says is that all states for which the total energy is the prescribed value, $E$, are equally probable, but that all states for which the energy is not that value are forbidden, i.e., the system is constrained such that the Hamiltonian takes on a constant value. Only those momenta and positions consistent with this constraint will be accessible to this system.

The partition function in the microcanonical ensemble is just the integral of the weighting function over all states

$$Q_{NVE} = \frac{1}{N!h^{3N}} \int \delta\big(\mathcal{H}(\mathbf{r}^N, \mathbf{p}^N) - E\big) d\mathbf{r}^N d\mathbf{p}^N. \tag{G.24}$$

The connection to thermodynamics arises from the relation of the entropy $S$ to the partition function

$$S = k_B \ln Q_{NVE}, \tag{G.25}$$

where $k_B$ is Boltzmann's constant. This is perhaps the most fundamental of all ensembles, in that it establishes the idea that entropy is a measure of the randomness of the system. A particularly nice description of how to prove this connection is given in Callen's book on thermodynamics [55].

The kinetic, $K$, and potential, $U$, energies can change so long as the total energy, $E = K + U$, is constant. At equilibrium, the kinetic and potential energies will fluctuate around their average values. The temperature in this ensemble is not constant, as it is in the canonical ensemble. At equilibrium, the temperature will fluctuate around its average value, as well. However, the only definition we have for the temperature is the relation in Eq. (G.19) relating the average temperature to the average kinetic energy, as discussed further in Chapter 6. The pressure in this ensemble also fluctuates around some average value as it does in the canonical ensemble (and any other ensemble in which $V$ is fixed).

### G.5.6 Canonical and microcanonical ensembles compared

To simplify the discussion, suppose a system consists of three identical "atoms" (1, 2, and 3) and that each atom has only $m$ possible energy levels, as pictured in Figure G.2 for $m = 6$. At any time, each atom will be in one of the energy levels, indicated by $n$. For convenience of future discussions, we will take the energy of the $n$th level of an atom as $\epsilon_n = n\epsilon_o$. The total energy is $E = E_1 + E_2 + E_3$, where the energy of the $i$th atom is $E_i = n_i\epsilon_o$. In terms of the energy levels, $E = \epsilon_o(n_1 + n_2 + n_3)$. The microstate of this system can be indicated by $\{E_1, E_2, E_3\}$ or, equivalently, by $\{n_1, n_2, n_3\}$. The connection to the macrostate will be through something we can measure, which in this case will be the average of the total energy $\langle E \rangle$.

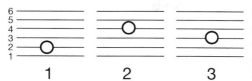

**Figure G.2** A schematic system with three identical atoms, each with six possible energy levels $n$ with energy $\epsilon_n = n\epsilon_o$. The microstate is the instantaneous set of the configurations of the atoms, which in this case is indicated by their energy $\{E_1, E_2, E_3\}$.

Suppose the instantaneous state of the system in Figure G.2 was measured at $M$ separate times, creating a list of the values of the microstate $\{n_1, n_2, n_3\}$. Counting the number of times that each atom was in the $n$th energy level and then dividing by $M$ creates the probability density $\rho_n$ of an atom being in the $n$th level. The average energy for each atom is $\rho_n$ for that atom times the energy of the $n$th level, which for this simple example of a set of identical atoms is just $\rho_n \epsilon_n$. The average total energy is $\langle E \rangle = N\rho_n\epsilon_n$, where $N$ is the number of atoms.

Suppose the system was in the microcanonical ensemble. The instantaneous energy of the system, $E = E_1 + E_2 + E_3 = \epsilon_o(n_1 + n_2 + n_3)$, would be constant. If $E = 9\epsilon_o$, then $\{2, 4, 3\}$ and $\{3, 3, 3\}$ would be allowable microstates,[14] while $\{3, 3, 4\}$ would not be. There is equal probability of being in any of the allowable microstates, i.e., $\{2, 4, 3\}$ is as equally probable as $\{3, 3, 3\}$. Note that the number of accessible microstates rapidly expands with increasing energy, which is consistent with our ideas of entropy (cf. Eq. (G.25)).

If the system in Figure G.2 were in the canonical ensemble, then all microstates would be possible, i.e., it is possible to have any value of $n$ in each of the three states. The *probability* of each microstate would not, however, be identical. Indeed, the probability of microstate $\alpha$ is $\exp(-\beta E_\alpha)/Q$, where $E_\alpha$ is the energy of the set of $\{n_1, n_2, n_3\}$ for state $\alpha$ and $Q = \sum_\alpha \exp(-\beta E_\alpha)$. As noted above, while all states are possible, the number that are *probable* is a relatively small subset of all the possible states. The most probable microstates represent a balance between the number of states, which increases with energy, and the Boltzmann factor $\exp(-\beta E_\alpha)$, which decreases with energy.

## G.5.7 Other ensembles

Most systems in the laboratory have a constant pressure $P$ and temperature $T$. Neither the canonical ($NVT$) nor the microcanonical ensemble ($NVE$) ensemble is best to describe such a system and another is needed. That system is designated by ($NPT$), in which the total number of particles, the pressure, and the temperature are fixed, and is called the *isobaric-isothermal ensemble*. In this ensemble, the total energy and the volume are not fixed and at equilibrium each fluctuates around some mean. This ensemble is especially useful when studying phase transitions, during which there are generally volume changes at constant $P$.

The partition function of the $NPT$ ensemble is given by

$$Q_{NPT} = \frac{1}{V_o N! h^{3N}} \int dV \int e^{-(\mathcal{H}(\mathbf{r}^N, \mathbf{p}^N) + PV)/k_B T} d\mathbf{r}^N \mathbf{p}^N, \tag{G.26}$$

---

[14] $n_1 + n_2 + n_3 = 9$.

where $V_o$ is an unimportant constant that defines some basic unit of volume. The connection to thermodynamics is through the relation

$$G = -k_B T \ln Q_{NPT} , \qquad (G.27)$$

where $G$ is the Gibbs free energy.

In the grand canonical ($\mu V T$) ensemble, the chemical potential $\mu$ is fixed. Thus, the number of particles fluctuates, along with the pressure and energy. This ensemble is very useful for the study of alloy properties, though effective implementation can be quite subtle. The partition function for a one-component system is

$$Q_{\mu V T} = \sum_{N=0}^{\infty} e^{(\mu N / k_B T)} Q_{NVT} , \qquad (G.28)$$

where we note that since particles are discrete, we use sums, not integrals, over the number of atoms. The connection to thermodynamics is through the relation

$$PV = -k_B T \ln Q_{\mu V T} . \qquad (G.29)$$

## G.6  FLUCTUATIONS

Unless they are fixed by thermodynamic constraints, in equilibrium all thermodynamic quantities fluctuate around an average value. Suppose, for example, we had a variable $F$, which could be energy or temperature or whatever. The standard deviation, $\sigma_F$, of $F$ is a measure of the fluctuations of $F$ and is given by

$$\sigma_F^2 = \langle (F - \langle F \rangle)^2 \rangle = \langle F^2 \rangle - \langle F \rangle^2 = \langle (\delta F)^2 \rangle . \qquad (G.30)$$

From thermodynamics, the derivative of $E$ with respect to $T$ at constant volume is defined as the heat capacity,

$$C_V = (\partial E / \partial T)_{N,V} . \qquad (G.31)$$

Consider the definition of the average energy $\langle E \rangle$ in the canonical ensemble from Eq. (G.12).[15] Taking the derivative with respect to $T$ and rearranging, we find[16]

$$C_V = \frac{1}{k_B T^2} \left( \langle E^2 \rangle - \langle E \rangle^2 \right) = \frac{1}{k_B T^2} \langle \left[ E - \langle E \rangle \right]^2 \rangle = \frac{1}{k_B T^2} \sigma_E^2 , \qquad (G.32)$$

---

[15] We will suppress the reference to $NVT$ in the averages in this section as we have established that we are using the canonical ensemble.

[16] $\partial \langle E \rangle / \partial T = \partial [f/Q]/\partial T$, where $f = \sum_n E_n \exp(-E_n/(k_B T))$ and $Q = \sum_n E_n \exp(-E_n/(k_B T))$. Using the chain rule, $\partial [f/Q]/\partial T = (1/Q)\partial [f]/\partial T + f \partial [1/Q]/\partial T$. In the first term, we need to evaluate $\partial \exp(-E_n/(k_B T))/\partial T = (E_n/(k_B T^2)) \exp(-E_n/(k_B T))$, so the first term is $\langle E^2 \rangle/(k_B T^2)$. The second term is $f \partial [1/Q] = -(f/Q^2)\partial Q/\partial T = -(1/(k_B T^2))(f/Q)^2 = -\langle E \rangle^2/(k_B T^2)$.

where $\sigma_E$ is the standard deviation of $E$. Equation (G.32) shows that the heat capacity, the heat energy required to increase the temperature of a substance at constant volume, is related to the mean-square deviation of the instantaneous fluctuations of the energy from its mean value, i.e.,

$$\sigma_E^2 = k_B T^2 C_V . \tag{G.33}$$

Equation (G.33) is a remarkable result. It shows that the size of the spontaneous fluctuations of the energy is related to the rate at which energy changes in response to changes in the temperature. It also enables us to estimate the relative width of the distribution in energy for a system.

If we assume that $E$ is normally distributed, its probability density will be a Gaussian of the form

$$f_E(E) = \frac{1}{\sqrt{2\pi}\,\sigma_E} e^{-(E - \langle E \rangle)^2 / (2\sigma_E^2)} . \tag{G.34}$$

From standard statistics, the relative root-mean-square (rms) value of the distribution is defined as

$$\frac{\sqrt{[E - \langle E \rangle]^2}}{\langle E \rangle} = \frac{\sqrt{k_B T^2 C_V}}{\langle E \rangle} \sim O\left(\frac{1}{\sqrt{N}}\right) , \tag{G.35}$$

which represents the relative width of the distribution. Since $C_V$ is an intensive quantity, it is proportional to $N$, the number of atoms in the system, as is $\langle E \rangle$. Thus the relative width of the distribution goes as $1/\sqrt{N}$, which becomes vanishingly small as $N$ approaches the Avogadro's number of atoms in bulk systems – the distribution becomes effectively a Dirac delta function.

We can write $\sigma_E^2 = \gamma N$, where $\gamma$ is a constant that includes the $k_B T^2$ term. $E$ is also extensive, so write $E - \langle E \rangle = N(E_n - \langle E_n \rangle)$, where $E_n$ is the energy per atom (an intensive variable). Rewriting the probability distribution we find

$$f_E(E) = \frac{N}{\sqrt{2\pi\gamma N}} e^{-N(E_n - \langle E_n \rangle)^2 / (2\gamma)} , \tag{G.36}$$

where we include the factor of $N$ to ensure that it is normalized. We plot $f_E(E)$ in Figure G.3 for a series of values of $N$. In this figure, we used values of the heat capacity per atom for copper and scaled the energy values by dividing by the cohesive energy. Note that by 2000 atoms, the distribution is highly peaked. As $N$ increases it would become sharper and sharper. Figure G.3 may look similar to Figure C.6 in Appendix C.5.1. It should, since Eq. (G.34) is the same function used to represent a delta function in Eq. (C.54).

The important point of this section is that it explains why the thermodynamic variables in the laboratory seem like constants even though they may not be fixed by the thermodynamic constraints. $E$ does fluctuate around its mean, but we will never measure those fluctuations for bulk systems because the relative deviation from the mean is immeasurably tiny. For small systems, however, such as in computer simulations or, perhaps, for very small nanosystems, these fluctuations can be seen.

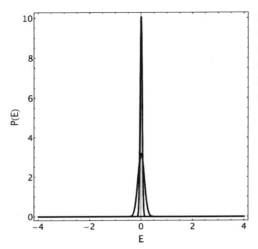

**Figure G.3** Schematic figure of distribution of the values of $E$ around its mean for different values of the number of atoms $N$. Constants were determined based on estimates for copper. The broadest peak had $N = 1$, the middle peak had $N = 10$, and the sharp peak had $N = 2000$, as described in the text.

### G.6.1 Equivalence of ensembles

In the canonical ensemble, $N$, $V$, and $T$ are fixed and all other thermodynamic quantities fluctuate around some average value. In the microcanonical ensemble, $N$, $V$, and $E$ are fixed. The two ensembles are complementary – the two variables that differ make an intensive-extensive pair. In thermodynamics, variables occur in such intensive-extensive pairs, e.g., $(T, E)$, $(P, V)$, and $(\mu, N)$.

Suppose we consider two ensembles characterized by the intensive variable $f$ and the extensive variable $F$, respectively. Now suppose we are interested in the average value of some quantity $B$. From thermodynamics, we must have that $\langle B \rangle_f = \langle B \rangle_F$ if $F = \langle F \rangle_f$. For finite-sized systems, $\langle B \rangle_f = \langle B \rangle_F + C$, where $C$ is a correction term that is of order $\mathcal{O}(1/N)$ [5]. The corrections die off rapidly with system size, but for small systems, such as in most simulations, they can make a difference of a few percent [5]. Thus, a molecular dynamics calculation in the microcanonical ensemble ($NVE$) and a Monte Carlo calculation in the canonical ensemble ($NVT$) may give values for qualities such as the pressure that differ by a small amount, strictly because of the difference in ensembles.

## G.7 CORRELATION FUNCTIONS

It is standard in statistical analysis to measure the correlation between variables, i.e., the degree to which the response of one variable is influenced by, or coupled to, another. In statistical mechanics, these are described by *correlation functions*. Suppose we have two quantities $A$ and $B$. Using the standard statistical definition, the correlation between those two quantities is given by

$$c_{AB} = \frac{\langle (A - \langle A \rangle)(B - \langle B \rangle) \rangle}{\sigma_A \sigma_B} , \qquad (G.37)$$

where the standard deviations are $\sigma_A = \sqrt{\langle A - \langle A \rangle \rangle^2}$, with a similar term for $B$. The correlation function $c_{AB}$ as defined in Eq. (G.37) is normalized such that if the values of $A$ and $B$ are perfectly correlated, for example if $A = B$, then $c_{AB} = 1$. If $A$ and $B$ are independent and not correlated at all, then $c_{AB} = 0$.

## G.7.1 Time correlation functions

An important class of correlation functions involve correlations between one time $t$ and another time $t'$. The correlation between $A(t)$ and $B(t')$ is called a *time correlation function* and has the form

$$c_{AB}(t) = \frac{\langle (A(t) - \langle A \rangle)(B(0) - \langle B \rangle) \rangle}{\sigma_A \sigma_B} , \tag{G.38}$$

where $t'$ has been set to zero since only the time difference matters. An *autocorrelation function* describes a correlation between the value of a quantity at one time and its value at another, i.e.,

$$c_{AA}(t) = \frac{\langle (A(t) - \langle A \rangle)(A(0) - \langle A \rangle) \rangle}{\sigma_A^2} . \tag{G.39}$$

The most important example of a time correlation function is arguably the *velocity autocorrelation function* (the correlation of the velocity with itself), which is defined as

$$c_{vv}(t) = \frac{\langle \mathbf{v}(t) \cdot \mathbf{v}(0) \rangle}{\langle v^2 \rangle} . \tag{G.40}$$

The average in Eq. (G.40) is over all the atoms. Since $\mathbf{v}$ is a vector, its average is zero. $\langle v^2 \rangle$ in the denominator of Eq. (G.40) is a scalar, which can be related to temperature by remembering that $\langle K \rangle = m \langle v^2 \rangle / 2 = 3k_B T / 2$, so that $\langle v^2 \rangle = 3k_B T / m$. Thus,

$$c_{vv}(t) = \frac{m}{3k_B T} \langle \mathbf{v}(t) \cdot \mathbf{v}(0) \rangle . \tag{G.41}$$

At $t = 0$, $c_{vv}(0) = 1$. At long times, as $t \to \infty$, $c_{vv}(t) \to 0$ – there is a total loss of correlation between the velocity of an atom at $t = 0$ and its velocity at $t = \infty$.

Consider what $c_{vv}(t)$ measures. It is the average of the dot product of an atom's velocity at $t = 0$ and its velocity at a later time $t$. The dot product is $\langle \mathbf{v}(t) \cdot \mathbf{v}(0) \rangle = \langle v(t)v(0) \cos \theta(t) \rangle$, where $\theta(t)$ is the angle between the velocities at time $t$ and time $t = 0$, $v(0)$ and $v(t)$ are the speeds (positive quantities) at the two times, and the average is over all the atoms in the system. We might expect that the direction of the motions of all the atoms would be very different over time, given that they suffer many different random collisions, and thus $c_{vv}(t)$ would go to zero rather quickly. In Figure G.4, however, we see rather complex behavior, with a long decay time to zero. These long-time tails have been the subject of a great deal of debate over the years. Note the structure in Figure G.4a for $T^* = 0.76$ (solid curve). The "well" region has $\langle v(t)v(0) \cos \theta_{v(0),v(t)} \rangle < 0$, which can only occur if $\theta_{v(0),v(t)} < 0$, which shows that the well corresponds to the atom's velocity having reversed itself, with the atom moving back towards its original position. This behavior is typically seen in dense systems at lower temperatures and corresponds to a "backscattering" event caused by collisions with neighboring atoms.

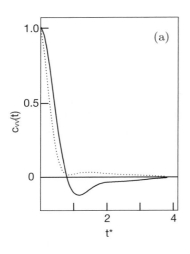 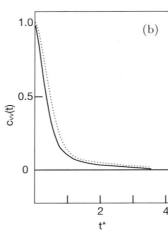

**Figure G.4** Velocity autocorrelation function for the Lennard-Jones fluid. (a) $\rho^* = 0.85$ with the solid curve at $T^* = 0.76$ and the dotted curve with $T^* = 4.70$. (b) $\rho^* = 0.65$ with the solid curve at $T^* = 1.43$ and the dotted curve with $T^* = 5.09$. The Lennard-Jones potential is defined in Eq. (6.21) and the reduced units in Table 6.1 in Section 6.2. Figures adapted from [197] and [139].

The diffusion constant is related to the velocity autocorrelation function by [223]

$$D = \lim_{t \to \infty} \frac{k_B T}{m} \int_0^t c_{vv}(t')dt' = \lim_{t \to \infty} \frac{1}{3} \int_0^t \langle \mathbf{v}_i(t') \cdot \mathbf{v}_i(0) \rangle dt' . \tag{G.42}$$

Equation (G.42) is an example of a *Green-Kubo* relation, relating a macroscopic transport coefficient (e.g., $D$) to integrals over time correlation functions. Other Green-Kubo relations are available, for example, for the shear viscosity and thermal conductivity. In Chapter 6 we discuss how to use the molecular dynamics methods to evaluate $c_{vv}$ and similar functions.

Comparing Eq. (G.42) with the relation for the diffusion constant at long times in Eq. (2.2),

$$D = \frac{1}{6t} \langle \mathbf{r}^2 \rangle , \tag{G.43}$$

where $\langle \mathbf{r}^2 \rangle$ is the mean square displacement. Equating the two equations implies

$$\langle \mathbf{r}^2 \rangle = 2t \int_0^t \langle \mathbf{v}_i(t') \cdot \mathbf{v}_i(0) \rangle dt' . \tag{G.44}$$

It is straightforward to prove Eq. (G.44) starting from $\mathbf{r}(t) = \int_0^t \mathbf{v}(\tau)d\tau$,[17] thus establishing that Eq. (G.43) is equivalent to Eq. (G.42).

---

[17] $\mathbf{r}(t) = \int_0^t \mathbf{v}(\tau)d\tau$, so

$$\langle \mathbf{r}^2 \rangle = \left\langle \int_0^t \mathbf{v}(\tau_1)d\tau_1 \cdot \int_0^t \mathbf{v}(\tau_2)d\tau_2 \right\rangle$$

$$= \int_0^t d\tau_1 \int_0^t d\tau_2 \langle \mathbf{v}(\tau_1) \cdot \mathbf{v}(\tau_2) \rangle$$

$$= \int_0^t d\tau_1 \int_0^{\tau_1} d\tau_2 \langle \mathbf{v}(\tau_1) \cdot \mathbf{v}(\tau_2) \rangle + \int_0^t d\tau_2 \int_0^{\tau_2} d\tau_1 \langle \mathbf{v}(\tau_1) \cdot \mathbf{v}(\tau_2) \rangle$$

$$= 2 \int_0^t d\tau_1 \int_0^{\tau_1} d\tau_2 \langle \mathbf{v}(\tau_1) \cdot \mathbf{v}(\tau_2) \rangle$$

## G.7.2 Spatial correlation functions

Another important correlation is that between the positions of particles. In a solid, for every atom vibrating around its lattice site, its neighbors are all vibrating around their lattice positions at positions and directions relative to that site. In a fluid, where particles diffuse readily, there are no set positions, but particles are located relative to each other in an average sense. There are correlation functions that differentiate between those types of correlations, the most useful of which being the *pair distribution function*. The pair distribution function is written as $g_2(\mathbf{r}_i, \mathbf{r}_j)$, where the indices $i$ and $j$ indicate the atoms. Since we will find an average, the specific indices are, of course, arbitrary. This function gives the probability of finding a pair of atoms a certain distance apart in a material. Since it depends only on distance, it is often written as $g_2(r)$, or more simply, as $g(r)$, which is the notation adopted here. A method to calculate $g(r)$ from distributions of atomic positions is discussed in Appendix 6.9.3.

As defined, $g(r)$ does not depend on the direction in the lattice, but just yields information about the distances. For example, while we may not know where exactly on the sphere each atom may lie, the average number of neighbors in a sphere of radius $r_o$ around a central atom is

$$n = 4\pi \int_0^{r_o} r^2 g(r) dr .$$ (G.45)

A typical $g(r)$ for a liquid is given in Figure G.5a. Note that there is a large nearest-neighbor peak and an oscillatory pattern at longer distances that asymptotes to a value of 1. For a solid, as shown in Figure G.5b, $g(r)$ has peaks centered at each neighbor distance. For an $fcc$ structure, for example, there would be peaks at $r_{nn}$, $\sqrt{2}r_{nn}$, $\sqrt{3}r_{nn}$, and so on, where $r_{nn}$ is the nearest-neighbor distance, as shown in Figure G.5b. Unlike $g(r)$ for the fluid, the peaks in the solid drop essentially to zero in between the peaks. Integrating $g(r)$ over each peak (using Eq. (G.45) with the appropriate limits) will give the number of neighbors at that distance. For neighbor shells in an $fcc$ structure, these would be 12, 6, 24, . . .

Given that $g(r)$ gives the average distance between atoms in a material, it should not be surprising that the average potential energy and average pressure for a system described by central-force pair potentials can be written in terms of the radial distribution function. These relations are [62]

$$\langle U \rangle = 2\pi \frac{N^2}{V} \int_0^\infty r^2 \phi(r) g(r) dr$$ (G.46)

$$= 2 \int_0^t d\tau_1 \int_0^{\tau_1} d\tau_2 \langle \mathbf{v}(0) \cdot \mathbf{v}(\tau_2 - \tau_1) \rangle$$

$$= 2 \int_0^t d\tau_1 \int_0^t d\tau_2 \langle \mathbf{v}(0) \cdot \mathbf{v}(\tau) \rangle$$

$$= 2t \int_0^t \langle \mathbf{v}(0) \cdot \mathbf{v}(\tau) \rangle d\tau$$

In the second line, the order of the averaging was rearranged. In the third line, we broke the integral into two, equivalent, integrals in which one variable was always kept less than the other. In the fifth line, we recognized that correlations only depend on the time difference between two variables, not on the absolute time. In the sixth line, we did a change of variables to $\tau = \tau_2 - \tau_1$.

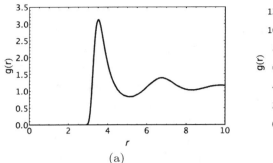
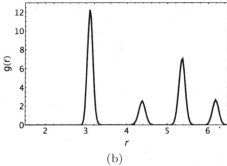

**Figure G.5** Typical radial distribution functions: (a) for a liquid and (b) for a low temperature solid in an *f cc* structure.

and

$$\langle P \rangle = N k_B T - \frac{2\pi}{3} \frac{N^2}{V} \int_0^\infty r^2 \left( r \frac{\partial^2 \phi(r)}{\partial r} \right) g(r) dr .$$ (G.47)

For systems whose interactions are not radially symmetric (e.g., molecules have shape, which is reflected in their interactions), the simple radial distribution function given here does not provide enough information about the structure of the material to be particularly useful. There are a number of ways to represent distribution functions for those materials [129].

There are many other measures of structure that we could define. For example, a translational order parameter can be written as

$$\rho(\mathbf{k}) = \frac{1}{N} \sum_{i=1}^N \cos(\mathbf{k} \cdot \mathbf{r}_i) ,$$ (G.48)

where $\mathbf{k}$ is a selected reciprocal lattice vector, and can be used to monitor changes in structure. For an *fcc* lattice, for example, $\mathbf{k} = (2\pi/a_o)(-1, 1, -1)$, where $a_o$ is the unit cell length, not the simulation cell. For a perfect *fcc* solid $\rho(\mathbf{k}) = 1$ while the average of $\rho(\mathbf{k}) = 0$ for a liquid. As we see in Chapter 6, monitoring $\rho(\mathbf{k})$ is an excellent way to probe the phase of the system.

## G.8 KINETIC RATE THEORY

Kinetic rate theory describes how systems evolve with time. More specifically, rate theory describes systems that evolve over time via discrete events, which are generally *activated* processes that depend on jumps over barriers from one basin of attraction to another. In this section we review the basic theory behind rate processes – transition-state theory.

### G.8.1 Basics

We often discuss rates in the context of chemical kinetics, the reaction of molecules to form new molecules. Many other types of processes share a key feature with chemical reactions – their dynamics can be characterized by infrequent events. An infrequent event is one in which

**Figure G.6** Typical topology of a system manifesting infrequent events, in this case an atom jumping out of one potential well to another. The atom will vibrate in its well until it has sufficient velocity in the right direction to escape. Adapted from [333].

the particles in the system jump from one state to another quickly relative to the long times in which they are localized in essentially the same place. For example, an atom diffusing in a solid generally is located at a specific site in the lattice, vibrating around its average position. Diffusion occurs in those rare cases when atoms jump to other sites. Our goal is to understand the rates at which these jumps occur.

## G.8.2 Potential surfaces and reaction coordinates

In Figure G.6, we show a schematic view of the potential energy surface of an atom as it sits in a site in a solid. This figure is a contour plot, with the heavy lines representing lines of constant energy. In the middle is a basin in which an atom vibrates around its mean position, while at the right and left are barriers between the original basin and adjacent ones. The motion of the atom is shown as the light line. The atom spends most of its time moving around the first basin. Every once in a while, an atom develops sufficient energy in a mode that points toward an adjacent basin and it moves over the barrier and through the dividing surface (the vertical lines) at the ridge top between the basins. The trajectory between basins is referred to as the *transition coordinate* or, because of rate theory's origin in chemistry, the *reaction coordinate*. The atom may cross the dividing surface to the other basin or it may recross back into its original basin. The key to rate theory is that the time for jumping to another site is much shorter than the time the atom vibrates in its basin.

Rare-event processes all have this basic topology – barriers between basins such that the crossing of barriers is infrequent. The goal of rate theory is to describe how long, on average, an atom will sit in its well before making a jump to a new location.

In Figure G.7 we look at the potential surface between two basins in three dimensions. We can see the minimum activation barrier is at a *saddle point*, by which we mean that the potential surface is shaped like a saddle, with a minimum-energy path between basins that has a negative curvature. The potential has a positive curvature in the direction perpendicular to the lowest-energy path. The shape is thus similar to that of a saddle used to ride horses.

We can see the shape of the potential surface a bit easier by plotting the energy along specific paths in Figure G.7. In Figure G.8a we show the change in energy along the coordinate connecting two basins. The figure shows the potential energy change $\Delta U$ from going from state $A$ to state $B$, which contributes to the overall driving force for the reaction, $\Delta G$.

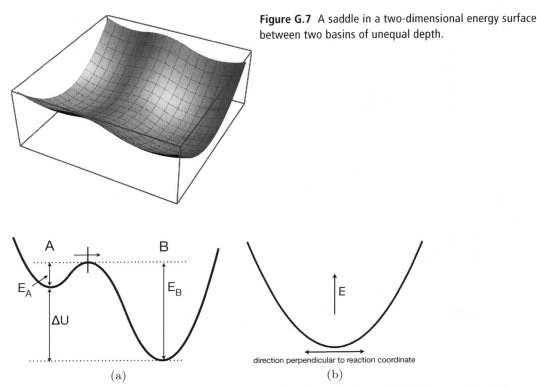

**Figure G.7** A saddle in a two-dimensional energy surface between two basins of unequal depth.

**Figure G.8** The energy along paths on the potential surface shown in Figure G.7. (a) Cross-sectional view along the minimum-energy reaction pathway between the two basins. (b) The energy in the direction perpendicular to the reaction coordinate along the path shown in (a).

As an atom moves between the neighboring basins, there is an energy barrier between the wells. The magnitude of the energy barrier for $A \rightarrow B$, $\Delta E_A$, is called the activation energy of the process. The position at the top of the barrier is called the transition state, and is indicated by the line at the top of the barrier (the dividing surface of Figure G.6). The activation for $B \rightarrow A$ is $\Delta E_B$. The activation energy is the minimum energy to cross the barrier, i.e., it is the difference in energy at the transition point and the bottom of the well of the basin that the atom started in.

The shape of the potential surface along the transition coordinate shows a negative curvature ($\partial^2 U/\partial X^2 < 0$, where $X$ represents the reaction coordinate) at the transition point. In directions perpendicular to the reaction coordinate, $Y$, however, the atom sits in a well at the transition point ($\partial^2 U/\partial Y^2 > 0$), as shown schematically in Figure G.8b.

### G.8.3 Chemical reaction rates

Consider the basic chemical reaction

$$A + B \underset{k_f}{\overset{k_r}{\rightleftharpoons}} C \,,$$ 
(G.49)

in which $A$, $B$, and $C$ represent atoms or molecules. $k_f$ is the temperature-dependent *rate constant* for the forward reaction, $A + B \rightarrow C$, while $k_r$ is the rate constant for the reverse reaction, $C \rightarrow A + B$. The rate constants are measures of the speed of the reaction and appear in the rate equation for the production of the reactants. For example, the change in concentration of $C$ for the forward reaction is

$$\frac{d[C]}{dt} = k_f[A][B] , \tag{G.50}$$

where [ ] indicates concentration. The units of $k$ depend on the type of process being described. Typically $k$ is measured in units of $s^{-1}$. For gas-phase reactions, however, rate constants are often measured in some combination of moles, volume, and seconds, depending on the order of the reaction.[18]

Experimentally it has been found that the temperature dependence of $k_f$ can be well described in many cases by the Arrhenius equation

$$k_f = Ae^{-E_a/RT} , \tag{G.51}$$

where $R$ is the gas constant. $A$ is the pre-exponential (or frequency factor) and $E_a$ is the activation energy for the reaction. Both $A$ and $E_a$ are usually assumed to be constants for the reaction [155]. The Arrhenius relation describes well not just the temperature dependence of chemical reaction rates but also most processes whose rate is determined by infrequent events. For some reactions, the equation for the rate constant takes the form

$$k_f = AT^n e^{-E_a/RT} , \tag{G.52}$$

where $n$ is an additional constant. The parameters $A$, $n$, and $E_a$ are usually based on experiment and tabulated.[19]

A reverse reaction such as $B \rightarrow A$ in Figure G.8a would typically have a different activation energy than the forward reaction and thus a different rate. If $\Delta G = G_B - G_A$ is the free energy difference between $A$ and $B$, then from Figure G.8a $\Delta G < 0$ and the activation barrier in the reverse direction would be $E_a - \Delta G = E_a - (\Delta H - T\Delta S)$, where the enthalpy change on reaction is $\Delta H$ and the entropy change $\Delta S$. The rate constant for the reverse reaction is related to the forward rate constant by [140]

$$K_{eq} = \frac{k_f}{k_r} = e^{-\Delta G/RT} . \tag{G.53}$$

Thermochemical quantities are tabulated for standard states of 1 atmosphere, so care must be taken in Eq. (G.53) if other units (e.g., concentration) are used [140].

---

[18] The order of a reaction with respect to a specific reacting species is the power to which the concentration of that species is raised in the rate equation. For example, if a reaction like $A + 2B \rightarrow C$ had a rate equation $d[C]/dt = k[A][B]^2$. The order of the reaction for $A$ is 1 and that for $B$ is 2, with a total order of reaction of $1 + 2 = 3$.

[19] The constants can be determined from experiments on the reaction rates as a function of temperature by plotting $\log(k(T))$ versus $1/T$. If Eq. (G.51) holds, it would be a straight line with slope $E_a/R$ and intercept of $\log(A)$.

## G.8.4 Transition-state theory

A commonly used approach to understanding rates is through transition-state theory (TST). The output of TST is the rate constant for the crossing from one state $A$ across a barrier to another state $B$. It may be the rate of a reaction, with states representing molecules, or it could be a diffusion event, in which case the states are atoms in different sites. There are a number of ways to arrive at the equations of TST. We will use one based on simple ideas from equilibrium thermodynamics. Note that TST ignores all quantum effects (such as tunneling).

Transition-state theory is based on the idea that a new state is formed at the transition point of a reaction, i.e., at the saddle point shown in Figure G.7. From the TST point of view, we would rewrite the chemical reaction in Eq. (G.49) as

$$A + B \rightleftharpoons (AB)^{\ddagger} \rightarrow C, \qquad (G.54)$$

where the superscript $(AB)^{\ddagger}$ indicates the transition state. Moreover, it is assumed that the transition state is in equilibrium with the system and the reactants, which allows us to make use of results from thermodynamics.

Since the reactants are in equilibrium with the transition state, we can write the equilibrium constant of the reaction to form the transition state as [155]

$$K^{\ddagger} = \frac{[AB]^{\ddagger}}{[A][B]}, \qquad (G.55)$$

where $[AB]^{\ddagger}$ is the concentration of the transition state.[20] Thus, $[AB]^{\ddagger}$ is

$$[AB]^{\ddagger} = K^{\ddagger}[A][B]. \qquad (G.56)$$

The reaction rate will be proportional to the product of the concentration of the transition state and the frequency of the transition state decomposing into the product. We will discuss that frequency below, but for now we will call the frequency $\nu$ and have

$$\text{reaction rate} = \nu[AB]^{\ddagger} = \nu K^{\ddagger}[A][B]. \qquad (G.57)$$

However, from standard thermodynamics, the equilibrium constant is given by

$$K^{\ddagger} = e^{-\Delta G^{\ddagger}/RT}, \qquad (G.58)$$

where $\Delta G^{\ddagger}$ is the free energy difference between the transition state and the reactants. Thus, from Eq. (G.50) we can write the rate constant $k$ as

$$k = \nu K^{\ddagger} = \nu e^{-\Delta G^{\ddagger}/RT} = \nu e^{\Delta S^{\ddagger}/R} e^{-\Delta H^{\ddagger}/RT}. \qquad (G.59)$$

Thus, we can identify $\Delta H^{\ddagger}$ with the energy $E_A$ in Figure G.8a.

---

[20] It is important to remember that thermodynamics does not describe individual events but rather an ensemble of events. Thus, we cannot consider just one transition state but rather a concentration of transition states.

It remains to determine a value for the frequency of decomposition from the transition state. There are a number of ways to think about what that frequency is. In transition-state theory, it is the frequency of the specific vibration in the transition state that leads directly to the formation of the products. In a diffusion problem, that might be the specific vibration that leads from the transition state to another atom site.

We can estimate that frequency by remembering the assumption that everything is in equilibrium. From quantum mechanics, we can associate an energy $h\nu$ to the vibrational mode, where $h$ is Planck's constant. From statistical mechanics, remembering the equipartition theorem from Appendix G.5.3, we can associate an energy of $1/2\,k_B T$ to each of the coordinates of that vibration in thermal equilibrium. Examining Figure G.8a, we see that the frequency of vibrations directly along the reaction coordinate is imaginary, while the frequencies in the two directions perpendicular to the reaction coordinate are real; thus we can associate a total thermal energy of $k_B T$ to the vibrational mode.[21] Equating the two expressions, we can estimate $\nu$ as

$$\nu = \frac{k_B T}{h} \,. \tag{G.60}$$

There is a problem with the argument in Eq. (G.60). Not all vibrations will actually lead to the product, and thus one typically adds a correction factor $\kappa$ to account for the fraction of vibrations that do not decompose. The transition-state theory rate constant then becomes

$$k = \kappa \frac{k_B T}{h} e^{\Delta S^{\ddagger}/R} e^{-\Delta H^{\ddagger}/RT} \,. \tag{G.61}$$

The parameters in Eq. (G.61) can be determined from fitting to experimental data.

There are far more elegant ways to derive transition-state rate constants involving results from statistical mechanics. They are even more illuminating of the basic physics of activated processes. However, for our purposes, this simple view seems sufficient.

Transition-state theory is very useful as a guide for understanding the basic physics of infrequent events. It is not, however, the complete story. It is based on assumptions that may not be true in all cases, especially that the transition state is in equilibrium. However, it is an excellent way to directly understand how the potential surface is related to reaction rates.

### G.8.5 Harmonic transition-state theory

A particularly simple approximation to transition-state theory assumes that we have identified the saddle point of the reactive pathway on the potential surface. If we assume that the potential surface can be approximated locally in the regions near the wells as a harmonic potential, then the rate constant can be written simply as

$$k^{HTST} = \nu_o e^{-\Delta E_A/k_B T} \,, \tag{G.62}$$

---

[21] From Eq. (D.16), the frequency of a vibration is $\nu = \sqrt{k/m}/(2\pi)$, where $k = \partial^2 U/\partial x_{\alpha}^2$ for each coordinate $x_{\alpha}$. Along the reaction coordinate in Figure G.8a, $\partial^2 U/\partial x_{\alpha}^2 < 0$, so $k < 0$ and $\nu$ is imaginary.

where, from the statistical mechanical analysis, the pre-exponential factor is

$$v_o = \frac{\prod_{i=1}^{3N} v_i^{min}}{\prod_{i=1}^{3N-1} v_i^{sad}}.$$

(G.63)

The $\{v_i^{min}\}$ are the vibrational frequencies (for all atoms in the system) when the particle is at the minimum of the well of state $A$ and $\{v_i^{sad}\}$ are the non-imaginary frequencies at the saddle point. There are only $3N - 1$ of these since the frequency associated with the reaction coordinate is imaginary, as it involves a negative curvature as shown in Figure G.8a, as discussed in footnote 21. These frequencies can be found from the matrix of second derivatives of the potential. If there are no recrossings and the potential surface is harmonic, then Eq. (G.62) is the exact expression for the rate. For most systems, $k^{HTST}$ is a reasonable approximation to the true rate. The harmonic approximation in sometimes called the Vineyard equation [323].

## G.9   SUMMARY

In this appendix we review many of the basic ideas behind the statistical basis of thermodynamics. We introduce phase space and the contrasting view of thermodynamics presented by the idea of ensembles. The basic ensembles of simulations are described and important results are derived. Finally, correlation functions are introduced and the basic ones used in simulations are described.

### Suggested reading

There are many books on statistical thermodynamics. Some I recommend include:

- David Chandler, *Introduction to Modern Statistical Mechanics* [62]. An excellent and concise introduction to statistical mechanics.
- H. B. Callen, *Thermodynamics and an Introduction to Thermostatistics* [55]. I find this a delightful book, with an interesting perspective.
- McQuarrie's book *Statistical Mechanics* is a good place to start [223].
- Brief backgrounds are given in D. Frenkel and B. Smit, *Understanding Molecular Simulations* [109], and M. P. Allen and D. J. Tildesley, *Computer Simulation of Liquids* [5].
- Kinetic rate theory is discussed in many sources. A very nice, but somewhat advanced, text with a materials perspective is *Kinetics of Materials*, by Balluffi *et al.* [21]. A more introductory presentation is available in any physical chemistry text, e.g., [224].

# H Linear elasticity

Suppose we have an object under some prescribed load (i.e., an applied force). We can describe how the object will deform in response to that load with the results of *elasticity* theory. For the purposes of this textbook, we will restrict the discussion to the regime of small displacements, in which we can use a linear version of elasticity theory. The fundamental assumptions of *linear elasticity* are that (1) the displacements (strains) are small and (2) there are linear relationships between the strains and their associated stresses (we define stress and strain hereinafter). The assumption of linear elasticity is reasonable for many applications and is used extensively in structural analysis.

We note that there is a further restriction to linear elasticity. The applied stress must be low enough so that yielding does not occur, i.e., so that the material does not undergo permanent deformation. Consider a thin metal rod, for example. If one applies a small force to the rod, it deforms but springs back to its original state when the force is removed. If you keep increasing the force, eventually the rod bends and does not return to the original state when the force is removed. That deformation is caused by the movement of linear defects called dislocations, which are described in Appendix B.5. We also describe a basic model of plastic deformation in terms of dislocation motion in that section. Later in this chapter, in Appendix H.5, we discuss the relationship between elastic and plastic strain.

## H.1 STRESS AND STRAIN

The *stress* is defined as the force acting on an area of an object. The stress clearly has directionality – a force acting in one direction will deform an object differently from a force acting in another direction. We could indicate that directionality in a number of ways. Generally, since we usually use a Cartesian coordinate system, we follow the notation indicated in Figure H.1a. In that figure, we consider a small cube of material, with sides $\delta x$, $\delta y$, and $\delta z$. Imagine forces acting on the front plane (i.e., defined by a constant value of $x$), which could be in the $x$, $y$, or $z$ directions. By convention, the force per area on that plane is defined as the stress and is indicated by $\sigma_{x\alpha}$, where $\alpha = x$, $y$, or $z$ depending on the direction of the force. The stress normal to the front plane is $\sigma_{xx}$, while $\sigma_{xy}$ and $\sigma_{xz}$ are *shearing* stresses. For stresses on the plane perpendicular to the $y$ axis, the stresses take the form $\sigma_{y\alpha}$ and similarly for stresses on the plane perpendicular to the $z$ axis. Thus, there are nine possible stress components.

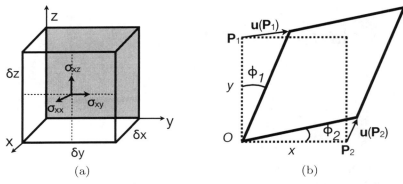

**Figure H.1** (a) Components of the stress tensor. The components of the stress acting on the front plane are $\sigma_{xx}$, $\sigma_{xy}$, and $\sigma_{xz}$. (b) Shear displacements $u_1$ and $u_2$ and angles $\phi_1$ and $\phi_2$ for a sheared volume element.

The stress is thus an object called a second-rank tensor, which we denote by $\boldsymbol{\sigma}$.[1] We can represent that tensor in a matrix form as

$$\boldsymbol{\sigma} = \begin{pmatrix} \sigma_{11} & \sigma_{12} & \sigma_{13} \\ \sigma_{21} & \sigma_{22} & \sigma_{23} \\ \sigma_{31} & \sigma_{32} & \sigma_{33} \end{pmatrix}. \tag{H.1}$$

If there are no torques, as is usually the case, then

$$\sigma_{ij} = \sigma_{ji}. \tag{H.2}$$

If an object is acted upon by a sufficiently small stress, it will deform elastically, by which we mean that there is no permanent change in the shape of the object and that it returns to its original shape when the stress is removed. Consider first a simple, one-dimensional, example of a rod that initially has one end at $x_1$ and the other at $x_2$, with a length $\ell_o = x_2 - x_1$. Suppose that under a stress, the ends of the rod are moved to $x_1 \to x_1'$ and $x_2 \to x_2'$, with a new length $\ell = x_2' - x_1'$. The *strain* is defined as

$$\epsilon = \frac{\ell - \ell_o}{\ell_o}. \tag{H.3}$$

The strain is a dimensionless number measuring the relative size of the elastic deformation of the object.

Define a *displacement* function, $u(x)$, as

$$u(x) = x' - x. \tag{H.4}$$

$u(x)$ is a continuous function that maps the old shape into the new shape. Then,

$$x_1' = x_1 + u(x_1) \quad \text{and} \quad x_2' = x_2 + u(x_2) \tag{H.5}$$

[1] For more details on vectors and tensors, please see Appendix C.1.

and

$$\ell = x_2' - x_1' = u(x_2) - u(x_1) + \ell_o \tag{H.6}$$

with the strain becoming

$$\epsilon = \frac{u(x_2) - u(x_1) + \ell_o - \ell_o}{\ell_o} = \frac{u(x_2) - u(x_1)}{\ell_o}, \tag{H.7}$$

or

$$u(x_2) = u(x_1) + \epsilon \, \ell_o . \tag{H.8}$$

However, $u(x)$ is a continuous function and if we assume that the displacements are slowly varying, then we can write an expansion for $u(x_2)$ relative to $u(x_1)$,

$$u(x_2) = u(x_1 + \ell_o) \approx u(x_1) + \left( \frac{du(x)}{dx} \right)_{\ell_o=0} \ell_o . \tag{H.9}$$

Equating Eq. (H.9) with Eq. (H.8) we have that the strain is

$$\epsilon = \frac{du(x)}{dx} . \tag{H.10}$$

Now consider the more complicated situation shown in Figure H.1b. The undistorted material is defined by $\mathbf{P}_1 = (0, y_1)$ and $\mathbf{P}_2 = (x_2, 0)$. The displacements are $u(\mathbf{P}_1) = (u_x(\mathbf{P}_1), u_y(\mathbf{P}_1))$ and $u(\mathbf{P}_2) = (u_x(\mathbf{P}_2), u_y(\mathbf{P}_2))$. From basic trigonometry, the angles after distortion are

$$\phi_1 = \tan^{-1} \left( \frac{u_x(\mathbf{P}_1)}{y_1 + u_y(\mathbf{P}_1)} \right) \tag{H.11a}$$

$$\phi_2 = \tan^{-1} \left( \frac{u_y(\mathbf{P}_2)}{x_2 + u_x(\mathbf{P}_2)} \right) \tag{H.11b}$$

We can approximate the equations in Eq. (H.11) by taking advantage of the assumption that all the distortions are small. Using similar logic as in Eq. (H.9), we can expand $u_y(\mathbf{P}_2)$ in terms of $x_2$ as

$$u_y(\mathbf{P}_2) \approx \left( \frac{\partial u_y(\mathbf{P}_2)}{\partial x_2} \right) x_2 . \tag{H.12}$$

Assuming that the distortions are small relative to the initial size of distorted region,

$$\frac{1}{x_2 + u_x(\mathbf{P}_2)} \approx \frac{1}{x_2}, \tag{H.13}$$

and so (using similar logic for the $\phi_1$ term)

$$\phi_1 = \tan^{-1} \left( \frac{(\partial u_x(\mathbf{P}_1)/\partial y_1)y_1}{y_1} \right) = \tan^{-1} \left( \frac{\partial u_x}{\partial y_1} \right) \approx \left( \frac{\partial u_x}{\partial y} \right) \tag{H.14a}$$

$$\phi_2 = \tan^{-1} \left( \frac{(\partial u_y(\mathbf{P}_2)/\partial x_2)x_2}{x_2} \right) = \tan^{-1} \left( \frac{\partial u_y}{\partial x_2} \right) \approx \left( \frac{\partial u_y}{\partial x} \right) \tag{H.14b}$$

where in the last part of the equation we take advantage of the fact that for small angles, $\tan^{-1}\theta \approx \theta$. We take as the strain the average of the angles and so have (in general)[2]

$$\epsilon_{ij} = \frac{1}{2}\left(\frac{\partial u_i}{\partial x_j} + \frac{\partial u_j}{\partial x_i}\right). \tag{H.15}$$

As for the stress, $\epsilon_{ij} = \epsilon_{ji}$. Please remember that in Figure H.1b, it is assumed that the strains are very small and thus that the angles are small as well.

## H.2 ELASTIC CONSTANTS

In linear elasticity, the stresses are linearly proportional to the strains, with the proportionality constants being called the *elastic constants* $\{c_{ijkl}\}$, which are equivalent to the Hooke's law description of a spring. The complication is that each strain $\epsilon_{kl}$ is coupled potentially to all possible stresses, $\sigma_{ij}$. Thus, there are 27 possible elastic constants, which are denoted by $c_{ijkl}$, and the stresses are related to the strains by

$$\sigma_{ij} = \sum_{k=1}^{3}\sum_{l=1}^{3} c_{ijkl}\epsilon_{kl} = c_{ijkl}\epsilon_{kl}. \tag{H.16}$$

In Eq. (H.16), in the second part of the expression we use the Einstein notation, in which repeated indices are summed.[3]

The strain energy *density* of a linear elastic system is[4]

$$w = \frac{1}{2}\sigma_{ij}\epsilon_{ij} = \frac{1}{2}c_{ijkl}\epsilon_{kl}\epsilon_{ij}. \tag{H.17}$$

The strain energy is the integral of the strain energy density, $w$, over a volume of material

$$W = \int w\, dV = \frac{1}{2}\int c_{ijkl}\epsilon_{kl}\epsilon_{ij}\, dV. \tag{H.18}$$

## H.3 ENGINEERING STRESS AND STRAIN

There are times when it is more convenient to employ strains with a slightly different definition,

$$\epsilon_i = \epsilon_{ii} = \frac{\partial u_i}{\partial x_i}$$
$$\gamma_k = \gamma_{ij} = \left(\frac{\partial u_i}{\partial x_j} + \frac{\partial u_j}{\partial x_i}\right) \quad i \neq j \neq k. \tag{H.19}$$

Since this definition is often favored by engineers, we shall refer to it as engineering strain, by which we do *not* mean the non-differential strain measured in an experiment. Note that the

---

[2] We use the notation that $i$ and $j$ refer to the coordinates, in which $x_1 = x$, $x_2 = y$, $x_3 = z$.
[3] For example, $a_i b_i = \sum_{i=1}^{3} a_i b_i$.
[4] Compare again with the simple harmonic oscillator in Appendix D.3.

shear strains $\gamma$ are defined as twice the strains defined in Eq. (H.15). The stress is often given by the notation $\sigma_i = \sigma_{ii}$ and $\tau_k = \sigma_{ij}$ for $i \neq j \neq k$.

The engineering stress can be written in terms of the engineering strain for cubic crystals as

$$
\sigma = \begin{pmatrix} \sigma_x \\ \sigma_y \\ \sigma_z \\ \tau_x \\ \tau_y \\ \tau_z \end{pmatrix} = \begin{pmatrix} C_{11} & C_{12} & \dots & C_{16} \\ C_{21} & C_{22} & \dots & C_{26} \\ C_{31} & C_{32} & \dots & C_{36} \\ C_{41} & C_{42} & \dots & C_{46} \\ C_{51} & C_{52} & \dots & C_{56} \\ C_{61} & C_{62} & \dots & C_{66} \end{pmatrix} \begin{pmatrix} \epsilon_x \\ \epsilon_y \\ \epsilon_z \\ \gamma_x \\ \gamma_y \\ \gamma_z \end{pmatrix} = \mathbf{C}\epsilon , \tag{H.20}
$$

where the engineering elastic constants are related to the constants in Eq. (H.16) by $C_{mn} = c_{ijkl}$ with the numbering

$$
\begin{array}{llllllll}
ij & \text{or} & kl & \quad 11 & 22 & 33 & 23 & 31 & 12 \\
m & \text{or} & n & \quad 1 & 2 & 3 & 4 & 5 & 6 ,
\end{array} \tag{H.21}
$$

where we have used symmetry to reduce the numbers of elastic constants. The notation $\mathbf{C}$ indicates the $6 \times 6$ matrix of constants. We have written the six independent strain and stress terms as vectors, not matrices.

With the notation in Eq. (H.20), the elastic strain energy becomes

$$
W = \int w \, dV = \frac{1}{2} \int C_{ij}\epsilon_i\epsilon_j \, dV . \tag{H.22}
$$

# H.4 ISOTROPIC SOLIDS

For an *isotropic* solid all properties are independent of direction. In that case, only two elastic constants are independent. The elastic constant matrix becomes

$$
\mathbf{C} = \begin{pmatrix} \lambda + 2\mu & \lambda & \lambda & 0 & 0 & 0 \\ \lambda & \lambda + 2\mu & \lambda & 0 & 0 & 0 \\ \lambda & \lambda & \lambda + 2\mu & 0 & 0 & 0 \\ 0 & 0 & 0 & \mu & 0 & 0 \\ 0 & 0 & 0 & 0 & \mu & 0 \\ 0 & 0 & 0 & 0 & 0 & \mu \end{pmatrix} . \tag{H.23}
$$

In Eq. (H.23), we introduced the common materials constants:

$$
\begin{aligned}
\mu &= C_{44} = \frac{1}{2}(C_{11} - C_{12}) \\
\lambda &= C_{12} \\
\lambda + 2\mu &= C_{11}
\end{aligned} \tag{H.24}
$$

where $\mu$ is the *shear modulus* and $\lambda$ is the *Lamé constant*.

Applying Eq. (H.20), the stress in the isotropic case can be written as

$$\sigma_{11} = (\lambda + 2\mu)\epsilon_{11} + \lambda\epsilon_{22} + \lambda\epsilon_{33} \qquad \sigma_{23} = 2\mu\epsilon_{23} \tag{H.25}$$
$$\sigma_{22} = \lambda\epsilon_{11} + (\lambda + 2\mu)\epsilon_{22} + \lambda\epsilon_{33} \qquad \sigma_{31} = 2\mu\epsilon_{31}$$
$$\sigma_{33} = \lambda\epsilon_{11} + \lambda\epsilon_{22} + (\lambda + 2\mu)\epsilon_{33} \qquad \sigma_{12} = 2\mu\epsilon_{12}\,,$$

with the standard definition of stress and strain and

$$\sigma_x = (\lambda + 2\mu)\epsilon_x + \lambda\epsilon_y + \lambda\epsilon_z \qquad \tau_x = \mu\tau_x \tag{H.26}$$
$$\sigma_y = \lambda\epsilon_x + (\lambda + 2\mu)\epsilon_y + \lambda\epsilon_z \qquad \tau_y = \mu\gamma_y$$
$$\sigma_z = \lambda\epsilon_z + \lambda\epsilon_y + (\lambda + 2\mu)\epsilon_z \qquad \tau_z = \mu\gamma_z$$

with engineering stress and strain.

We can, of course, write the strain in terms of the stress. From Eq. (H.25) we have

$$\epsilon_{11} = \frac{1}{E}\left(\sigma_{11} - \nu(\sigma_{22} + \sigma_{33})\right) \qquad \epsilon_{23} = \frac{\sigma_{23}}{2\mu} \tag{H.27}$$
$$\epsilon_{22} = \frac{1}{E}\left(\sigma_{22} - \nu(\sigma_{11} + \sigma_{33})\right) \qquad \epsilon_{31} = \frac{\sigma_{31}}{2\mu}$$
$$\epsilon_{33} = \frac{1}{E}\left(\sigma_{33} - \nu(\sigma_{11} + \sigma_{22})\right) \qquad \epsilon_{12} = \frac{\sigma_{12}}{2\mu}\,,$$

New materials constants were introduced in Eq. (H.27),

$$E = \frac{\mu(3\lambda + 2\mu)}{\mu + \lambda} = \frac{9\mu B}{3B + \mu} = 2\mu(1 + \nu)$$
$$\nu = \frac{\lambda}{2(\mu + \lambda)} = \frac{3B - 2\mu}{2(3B + \mu)} \tag{H.28}$$
$$K = \frac{3}{3\lambda + 2\mu} = \frac{1}{B}$$

where $E$ is *Young's modulus*, $\nu$ is *Poisson's ratio*, and $K$ is the *compressibility* (the inverse of the *bulk modulus B*).

Some additional relationships between the constants are

$$\mu = \frac{E}{2(1 + \nu)}$$
$$\lambda = \frac{\nu E}{(1 + \nu)(1 - 2\nu)} = \frac{2\nu\mu}{1 - 2\nu}\,. \tag{H.29}$$

If a solid consists of atoms that interact only through central-force potentials (a function of distance between atoms only) and the atoms sit in centers of symmetry in the crystal, then additional relationships exist between the elastic constants, called the *Cauchy relations* [207]. The most useful of the Cauchy relations is probably[5]

$$C_{12} = C_{44} \quad \text{or} \quad \frac{C_{12}}{C_{44}} = 1\,. \tag{H.30}$$

---

[5] The expression for systems under finite pressure $P$ is $c_{44} - c_{12} + 2P = 0$ [176].

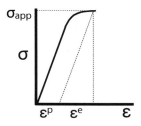

**Figure H.2** Schematic stress-strain curve showing plastic, $\epsilon^p$, and elastic, $\epsilon^e$, strains as a function of an applied stress, $\sigma_{app}$. The total strain is the sum of the elastic and plastic strains.

The usefulness of Eq. (H.30) is that any deviation of $C_{12}/C_{44}$ from 1 indicates important information about the intermolecular forces of elastic bodies.

From Eq. (H.24) and Eq. (H.30) it is tempting to assume that $\nu = 1/4$ for a crystal that interacts only through central forces.[6] However, this assumption neglects the fact that crystals have crystallography and so cannot actually be isotropic. For example, detailed atomistic simulations on *fcc* crystals that interact with the (central-force) Lennard-Jones potential show that the Poisson's ratio for deformation along the [100] direction is $\nu_{100} = 0.347$ [267]. Only when they averaged the properties over many crystallographic orientations did the authors find that $\nu = 1/4$, as expected for purely isotropic materials.

## H.5 PLASTIC STRAIN

When an object is subject to a sufficient stress, called the *yield stress*, permanent deformations are induced, such that the system does not return to its original state after the stress is removed. This permanent deformation is referred to as *plastic deformation* or *plasticity*.

In Figure H.2 we show a schematic view of a stress-strain curve that includes yielding. As the stress is increased from zero, the strain initially increases elastically, with a linear slope given by the appropriate elastic constant. At the yield stress, the material starts to yield, introducing a permanent deformation. Suppose the stress is increased until the applied stress, $\sigma_{app}$, is beyond the yield point. If we unload the sample (i.e., remove the stress), the system relaxes elastically along a line with the same slope as in the loading stage, determined by the elastic constants. However, since the sample was deformed plastically, it does not relax back to the starting point. The amount of strain associated with the plastic deformation is the *plastic strain $\epsilon^p$*.

To be more explicit, suppose in Figure H.2 we are loading an isotropic material in simple shear. We thus have only one stress component, say $\sigma_{12}$. From Eq. (H.27), $\sigma = 2\mu\epsilon$, where $\mu$ is the shear modulus. The total strain is thus just

$$\epsilon = \epsilon^p + \epsilon^e , \tag{H.31}$$

with $\epsilon^e = \sigma_{app}/2\mu$.

---

[6] From Eq. (H.24), $\mu = C_{44}$ and $\lambda = C_{12}$, thus the Cauchy relation in Eq. (H.30) is equivalent to $\mu = \lambda$. From Eq. (H.29), $\lambda = (2\nu/(1 - 2\nu))\mu$, or $2\nu/(1 - 2\nu) = 1$.

In Appendix B.5.1 we show that it is the movement of curvilinear defects called dislocations that leads to plastic deformation. A simple expression relating plastic strain to the dislocation density and the average displacement of each dislocation is given.

## Suggested reading

- A useful, simple, introduction to linear elasticity is given in *Theory of Elasticity*, third edition, by Stephen Timoshenko [314].

# 1 Introduction to computation

This chapter introduces some basic concepts used in the computations in this text. It is not a programming guide, as each software system has its own language and defined functions. We discuss some general methodologies that are common among programming languages and that crop up in a number of the methods in this text, for example the calculation of random numbers. We also discuss a few numerical methods. Specific implementations of the various methods are presented online at http://www.cambridge.org/lesar.

## 1.1 SOME BASIC CONCEPTS

Computers are discrete and thus all problems, whether discrete or continuous in space or time, must be converted to discrete methods on a computer. The requirement of having discrete methods presents challenges and guides the development of most of the models seen in this text. Some methods, such as molecular dynamics in Chapter 6, may be continuous in one dimension (space), but are solved with discrete time steps. Others, such as the Potts model of grain growth in Chapter 10, are discrete in both space and time.

## 1.2 RANDOM-NUMBER GENERATORS

A common need in essentially all of the methods discussed in this text is for random numbers. It is in many ways odd to discuss random numbers when talking about computers, which are precise and anything but random. Algorithms have been developed, however, that yield series of numbers that *look* random, at least relative to certain statistical measures of randomness. These algorithms are generally referred to as *random-number generators*. The challenge is that generators are not all of the same quality. The good news is that random-number routines in common software frameworks, such as Mathematica® or MATLAB®, are of good quality. A good discussion of the quality of random-number sequences is given in [234].

Consider the following sequence of numbers,

$$I_{i+1} = aI_i \pmod{m},\qquad(I.1)$$

where $I_1$ is some number that serves as the start of the random number sequence, $a$ and $m$ are constants, and the function $\mathrm{mod}(x, m)$ returns the remainder of $x/m$.[1] Park and Miller [250] propose

$$a = 7^5 = 16807 \quad m = 2^{31} - 1 = 2147483647 \tag{I.2}$$

based on extensive testing. The result of the sequence in Eq. (I.1) is a long string of positive integers whose maximum value is $m$ and whose properties satisfy many statistical tests of randomness. We can convert that sequence into a sequence of numbers between 0 and 1 by dividing the sequence by $m$,

$$F = \{I_i\}/m . \tag{I.3}$$

Note that all numbers in the sequence $F$ are $0 < F_i < 1$.

To generate a sequence of numbers with Eq. (I.1), a starting number must be chosen, which we can call $I_0$. This number is referred to as the *seed* of the random-number generator. For example, suppose we choose $I_0 = 362355$. The first five numbers in the sequence would be

$$F = \{1795133191, 805784434, 787104256, 361965072, 1873276800\}$$
$$\text{and} \tag{I.4}$$
$$F/m = \{0.835924, 0.375223, 0.366524, 0.168553, 0.872312\} .$$

Note that once the seed is chosen, the sequence of numbers is the same, a point that will be returned to below.

Our assertion is that the sequence $\Re = F/m$ is a sequence of numbers that has the same statistical properties as a sequence of real random numbers.[2] Without giving a complete analysis, we can start by showing a plot of the $F/m$ sequence for 10 000 numbers, which is shown in Figure I.1. The mean of the 10 000 numbers in that sequence is 0.498661 with a standard deviation of 0.2889. The numbers certainly look "random" in Figure I.1 and the mean and standard deviation are consistent with what is expected for a random sequence. Indeed, this simple generator passes all of the standard statistical tests for randomness and has seen a great deal of successful use [264].

This simple procedure for generating a sequence of random numbers is essentially what all modern random-number generators do: they start with some seed, and generate a long sequence of numbers whose properties satisfy statistical tests of randomness. Most modern random-number generators are more sophisticated than that in Eq. (I.1), but all have the same basic features.

Two sequences of numbers that start from the same seed are identical. Thus, it is important to use a different seed to generate a new sequence of random numbers. One approach is to tie

---

[1] $\mathrm{mod}(3,4)=3$, $\mathrm{mod}(8,4)=0$, $\mathrm{mod}(7,4)=3$, etc.

[2] Since we know that the numbers are not strictly random, they are sometimes referred to as "pseudorandom". We will not make that distinction here.

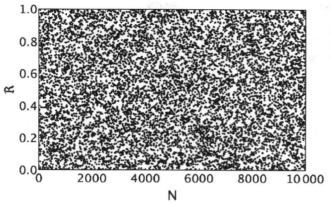

**Figure I.1** Results from a simple random-number generator. The points indicate the value and are plotted as a function of the order they appear in the random sequence.

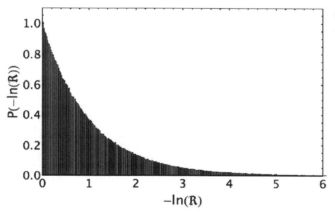

**Figure I.2** Probability distribution of $-\ln(\Re)$.

that seed to the time in seconds that has passed since a given start time. Since the seed changes as time changes, one is ensured of having a different seed each time.

As written, we have a set of random numbers, $\{\Re\}$, on the range $(0,1)$. It is simple to change the range of the random numbers. For example, if $\Re$ is a random number on $(0,1)$, $2\Re - 1$ is a random number on $(-1,1)$. Suppose we define a function called `ceiling(x)` that rounds $x$ to the next highest integer.[3] Then, if $\Re$ is a random number between 0 and 1, `ceiling(`$n\Re$`)` is a random integer between 1 and $n$.

In the discussion in Section 9.2 of assigning time to a kinetic Monte Carlo step, we make use of the statistical properties of the log of random numbers, specifically, the expression $-\ln(\Re)$, where $\Re$ is a random number on $(0,1)$. In Figure I.2 we show the probability distribution of a list of 1 million values of $-\ln(\Re)$. Note that while it is highly peaked near the origin, the mean and standard deviation of this distribution are equal to 1.

---

[3] Many programming languages, including various software platforms, have functions that do what `ceiling(x)` does.

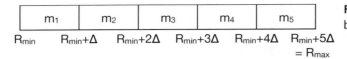

**Figure I.3** Schematic view of the binning procedure for $n_{bin} = 5$.

## I.2.1 Random numbers from a normal distribution

A *normal* (or Gaussian) distribution has a general form

$$\rho(x) = \frac{1}{\sigma\sqrt{2\pi}} e^{-(x-\langle x \rangle)^2/2\sigma^2} \, , \tag{I.5}$$

where the mean is $\langle x \rangle$ and the standard deviation is $\sigma$.

Box and Muller [40] show that a random variable $x$ can be generated from this distribution using

$$x = \langle x \rangle + \sigma \tau \, , \tag{I.6}$$

where $\tau$ is a random number generated from a normal distribution with zero mean and unit variance. $\tau$ can be calculated from

$$\tau_1 = (-2 \ln \Re_1)^{1/2} \cos(2\pi \Re_2) \qquad \tau_2 = (-2 \ln \Re_1)^{1/2} \sin(2\pi \Re_2) \, , \tag{I.7}$$

where $\Re_1$ and $\Re_2$ are random numbers on (0,1). Either or both of $\tau_1$ and $\tau_2$ could be used to generate the correct distribution.

## I.3 BINNING

We often want to calculate probability distributions of quantities that are continuous functions. As noted above, computers are not continuous, so we need to seek a way to represent the continuous probability distribution in a discrete way. The basic approach is to *bin* the values to create a histogram from which a normalized probability distribution can be obtained.

As an example, consider the calculation of the end-to-end probability distribution from the discussion of the random walk model in Chapter 2. $\mathcal{P}(R_n)$ in Eq. (2.16) tells us the likelihood of finding that the end-to-end distance (in a jump sequence of length $n$) is $R_n$.

Suppose we have performed $m$ random-walk simulations and thus have $m$ values for $R_n$ for each of the $n$ jumps in the sequence. We can create a discrete representation of $\mathcal{P}(R_n)$ for a specific $R_n$ by dividing the $m$ values into *bins* that each represent a finite range of values of $R_n$.

Consider a set of $m$ values of $R_n$ for a specific value of $n$, as shown in Figure I.3 for $n_{bin} = 5$. Suppose the maximum value from that list is $R_{max}$ and the minimum value is $R_{min}$. If we break the data into $n_{bin}$ equally spaced bins, then the bin width is

$$\Delta = \frac{R_n^{max} - R_n^{min}}{n_{bin}} \, . \tag{I.8}$$

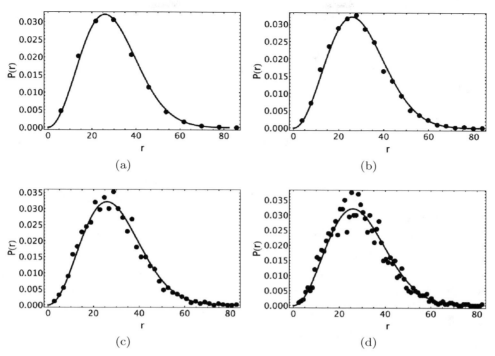

**Figure I.4** Calculated end-to-end probability distribution based on $m = 2000$ simulations at the end of a $n = 1000$ random walk on a square lattice with $a = 1$. The solid curve in each figure is the analytical result from Eq. (2.15) and Figure 2.3b. (a) $n_{bin} = 10$, (b) $n_{bin} = 20$, (c) $n_{bin} = 30$, (d) $n_{bin} = 80$.

We now need to count how many of the values of $R_n$ lie within each bin, i.e., how many values lie in the range $R_n^{min}$ to $R_n^{min} + \Delta$, how many lie in the range $R_n^{min} + \Delta$ to $R_n^{min} + 2\Delta$, and so on. Identifying the $i$th bin as the region $R_n^{min} + (i - 1)\Delta < R_n < R_n^{min} + i\Delta$, we need to determine $m_i$, the number of values of $R_n$ that lie in the $i$th bin. The probability of a value for $R_n$ lying within the $i$th bin is then just (from Eq. (C.38))

$$\mathcal{P}_i = \frac{m_i}{m}. \tag{I.9}$$

In Figure I.4 we show the calculated end-to-end probability distribution for $m = 2000$ simulations at the end of a $n = 1000$ step random walk on a simple cubic lattice in three dimensions with the lattice parameter $a = 1$. Based on the same data set, we show in that figure the dependence of the calculated probability distribution on the number of bins, $n_{bin}$, used in the binning process and compare the binned values with the analytic result from Eq. (2.15). We see that for $n_{bin} = 10$, the statistical deviation of the calculated result and the exact answer is excellent, while for $n_{bin} = 20$ through $n_{bin} = 80$ the results show increasing statistical variation from the exact value. This variation is common, and easily understood. There are $m = 2000$ total values of the end-to-end distance at $n = 1000$. Thus, on average, there are $m/n_{bin}$ values in each bin.[4]

---

[4] Of course, there are many more values at the peak of the distribution than at the tails.

At $n_{bin} = 10$, there are on average 200 points per bin, while for $n_{bin} = 10$ there are only 25 points and we expect much greater statistical fluctuations for a smaller number of data points. There is a tradeoff whenever one calculates probability distributions in this way: more bins mean a finer, more continuous, result, yet will require many more simulations to lower the statistical uncertainties.

## I.4 NUMERICAL DERIVATIVES

A common task in computational modeling is the calculation of derivatives on a lattice, for example in the phase-field method of Chapter 12. The phase-field method involves continuous order parameters that are solved on a grid, usually of uniform spacing. In the Allen-Cahn equation of Eq. (12.6), for example, we need to evaluate the Laplacian, $d^2\phi/dx^2 + d^2\phi/dy^2 + d^2\phi/dz^2$. For the Cahn-Hillard equation in Eq. (12.9), the fourth-order derivatives of $\phi$, $\nabla^2\nabla^2\phi$, are needed. To implement these equations on a grid requires approximate forms for these derivatives.

### I.4.1 One-dimensional derivatives

There are quite a number of approximations available in the literature for derivatives evaluated on a grid. For example, suppose we know the values of a function, $\phi$, at an equally spaced series of points in one dimension $\{\ldots, x_{i-3}, x_{i-2}, x_{i-1}, x_i, x_{i+1}, x_{i+2}, x_{i+3}, \ldots\}$. The derivative $d\phi_i/dx$ is easy to evaluate with the *central difference* formula

$$\frac{d\phi_i}{dx} = \frac{\phi_{i+1} - \phi_{i-1}}{2a}, \tag{I.10}$$

where $a$ is the grid spacing. Note that the derivative at point $i$ depends on the values at points $i - 1$ and $i + 1$.

We could evaluate the second derivative by taking a central difference of the first derivatives, i.e.,

$$\frac{d^2\phi_i}{dx^2} = \frac{\frac{d\phi_{i+1}}{dx} - \frac{d\phi_{i-1}}{dx}}{2a} = \frac{\frac{\phi_{i+2}-\phi_i}{2a} - \frac{\phi_i-\phi_{i-2}}{2a}}{2a} = \frac{(\phi_{i+2} + \phi_{i-2} - 2\phi_i)}{4a^2}. \tag{I.11}$$

This expression is generally not that accurate, as it involves the values at $x_i$ and $x_{i\pm2}$ and does not employ the values at $x_{i\pm1}$. It is often referred to as a *non-compact* form. A more accurate expression is the compact form, which involves a central finite difference using the derivatives at $x_{i\pm1/2}$, e.g., $d\phi_{i+1/2}/dx = (\phi_{i+1} - \phi_i)/a$ and $d\phi_{i-1/2}/dx = (\phi_i - \phi_{i-1})/a$, which leads to

$$\frac{d^2\phi_i}{dx^2} = \frac{\frac{d\phi_{i+1/2}}{dx} - \frac{d\phi_{i-1/2}}{dx}}{a} = \frac{\frac{\phi_{i+1}-\phi_i}{a} - \frac{\phi_i-\phi_{i-1}}{a}}{a} = \frac{\phi_{i+1} + \phi_{i-1} - 2\phi_i}{a^2}. \tag{I.12}$$

The values for the first and second derivatives for the system in Figure 12.2 are shown in Figure I.5 evaluated with the values for $\phi$ at $t = 0$.

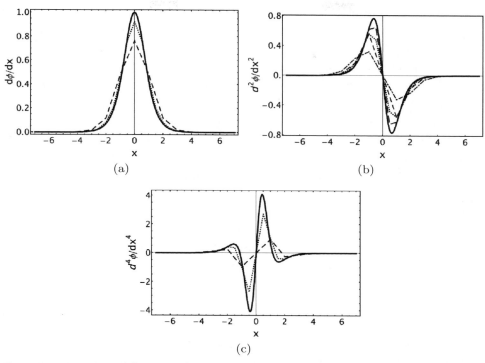

**Figure I.5** Comparison of discrete and exact derivatives of $\phi = \tanh(\gamma x)$, shown as the solid black lines. (a) $d\phi/dx$ from Eq. (I.10): dashed line with a grid size of 1 and dotted line with a grid size of 1/2. (b) $d^2\phi/dx^2$ from the non-compact form in Eq. (I.11): dot-dashed line with a grid size of 1 and dotted line with a grid size of 1/2; $d^2\phi/dx^2$ from the compact form in Eq. (I.12): small-dashed line with a grid size of 1 and large-dashed line with a grid size of 1/2. (c) $d^4\phi/dx^4$ from Eq. (I.13): dashed line with a grid size of 1 and dotted line with a grid size of 1/2.

Using the form for the second derivative of a function yields an approximation for the fourth derivative

$$\frac{d^4\phi_i}{dx^4} = \frac{\phi_{i+2} + \phi_{i-2} - 4(\phi_{i+1} + \phi_{i-1}) + 6\phi_i}{a^4}. \tag{I.13}$$

In Figure I.5, we show a comparison of various methods for evaluating one-dimensional derivatives of $\phi(x) = \tanh(\gamma x)$, where $x$ is given in units of $1/\gamma$. We also show the effect of mesh size on the accuracy of the derivatives. Specifically, in Figure I.5a is shown the first derivative, $d\phi/dx$. The solid curve is the exact result, the long dashes are with a grid of $a = \Delta x = 1/\gamma$ using Eq. (I.10), and the dotted line is for a grid with $a = 0.5/\gamma$.

In Figure I.5b, the Laplacian ($d^2\phi/dx^2$) is shown. The solid curve is the exact result, the long dashes are with a grid of $a = 1/\gamma$ using the non-compact expression in Eq. (I.11), the short dashes are for a grid with $a = 1/\gamma$ using the compact expression of Eq. (I.12), and the heavy solid curve is for a grid with $a = 0.5/\gamma$ using the compact expression.

In Figure I.5d we show the fourth derivative, $d^4\phi/dx^4$. The solid curve is the exact result, the long dashes are with a grid of $a = 1/\gamma$ using Eq. (I.13), and the short dashes are for a grid with $a = 0.5/\gamma$.

Consider the derivative $d\phi/dx$ in Figure I.5a. The numerical derivative with a mesh size of $a = 1/\gamma$ is not an unreasonable approximation, though the error is significantly reduced for a mesh size of $a = 0.5/\gamma$.

In Figure I.5b, we compare the results using the non-compact and compact expressions for $d^2\phi/dx^2$ for a mesh size of $a = 1/\gamma$. The compact expression is a much better representation of the derivative. We also show the results using the compact expression with a mesh size of $a = 0.5/\gamma$, which shows much better agreement with the exact derivative.

In Figure I.5d the results for $d^4\phi/dx^4$ are shown, which shows similar mesh size dependence. Not unexpectedly, we see that the results are better for smaller mesh sizes. Given that the width of an interface goes as $\sim 1/\gamma$, where $\gamma$ depends on the model parameters as in Eq. (12.20), the choice of mesh depends on the model and the desired accuracy. As in almost all modeling, the best thing to do is to try different parameters and meshes and see if the results converge.

## I.4.2 Two-dimensional derivatives

The basic formulae given above can be extended to more than one dimension. Consider, for example, an evaluation of the Laplacian for a two-dimensional simulation. The form would depend on the mesh used in the problem (the expression for a square grid would be different than that used on a triangular grid, for example).

Consider a square grid of points $\{x_i, y_i\}$. For convenience, we number the nearest-neighbor sites around $i$ (north, south, east and west) as $i + 1$ to $i + 4$ and the next nearest-neighbors (along the diagonals) as $i + 5$ to $i + 8$. The simplest form for evaluating $\nabla^2\phi = d^2\phi/dx^2 + d^2\phi/dy^2$ would be to use the compact expression from Eq. (I.12) for the nearest neighbors,

$$\frac{d^2\phi}{dx^2}(i) = \frac{\phi_{i+1} + \phi_{i+3} - 2\phi_i}{a^2} \qquad \frac{d^2\phi}{dy^2}(i) = \frac{\phi_{i+2} + \phi_{i+4} - 2\phi_i}{a^2}, \qquad (I.14)$$

which can be written as

$$\nabla^2\phi(i) = \frac{1}{a^2} \sum_{j=1}^{4} \left(\phi(x_{i+j}, y_{i+j}) - \phi(x_i, y_i)\right), \qquad (I.15)$$

Of course, we could also use the next-nearest neighbors to determine the Laplacian, which would lead to

$$\nabla^2\phi(i) = \frac{1}{(\sqrt{2}a)^2} \sum_{j=5}^{8} \left(\phi(x_{i+j}, y_{i+j}) - \phi(x_i, y_i)\right). \qquad (I.16)$$

Note the factor of $\sqrt{2}$ to account for the distance to the next-nearest neighbor.

In a recent study of a phase-field model of grain growth, Tikare *et al.* used a simple average of the nearest-neighbor and next nearest neighbor expressions [311], i.e.,

$$\nabla^2 \phi(i) = \frac{1}{2a^2} \left( \sum_{j=1}^{4} \left( \phi(x_{i+j}, y_{i+j}) - \phi(x_i, y_i) \right) + \frac{1}{2} \sum_{j=5}^{8} \left( \phi(x_{i+j}, y_{i+j}) - \phi(x_i, y_i) \right) \right),$$

(I.17)

where the first sum is over the nearest neighbors and the second over the next-nearest neighbors.

Similar equations to those in Appendix I.4.1 and Appendix I.4.1 can be derived in three dimensions. For example, Eq. (I.14) could be extended to the nearest neighbors on a three-dimensional lattice, and so on.

## I.5    SUMMARY

We presented a few numerical methods used throughout this text, including the generation of random numbers, the binning of continuous variables, and the evaluation of derivatives on a grid. The presentation is geared towards being able to use the methods used in this text and should not be taken as definitive or rigorous.

### Suggested reading

There are many books on computation.

- An excellent choice for most methods is the *Numerical Recipes* books by Press *et al.* They are available in a number of different computer languages, for example in [264].
- Many of these topics are discussed in texts geared towards the software platforms used in the online exercises. For example, *An Introduction to Scientific Computing: Twelve Computational Projects Solved with MATLAB* [76] and, while a bit dated, *Computer Science with Mathematica*® [213].

# REFERENCES

[1] Abell, G. C. 1985. Empirical chemical pseudopotential theory of molecular and metallic bonding. *Physical Review B*, **31**, 6184–6196.

[2] Alder, B. F., and Wainwright, T. E. 1957. Phase transition for a hard sphere system. *Journal of Chemical Physics*, **27**, 1208–1209.

[3] Alder, B. J., and Wainwright, T. E. 1959. Studies in molecular dynamics: 1. General method. *Journal of Chemical Physics*, **31**, 459–466.

[4] Alinaghian, P., Gumbsch, P., Skinner, A. J., and Pettifor, D. G. 1993. Bond order potentials: a study of s- and sp-valent systems. *Journal of the Physics of Condensed Matter*, **5**, 5795–5810.

[5] Allen, M. P., and Tildesley, D. J. 1987. *Computer Simulation of Liquids*. Oxford: Oxford University Press.

[6] Allen, S. M., and Cahn, J. W. 1979. Microscopic theory for antiphase boundary motion and its application to antiphase domain coarsening. *Acta Metallurgica*, **27**, 1085–1095.

[7] Allison, J., Li, M., Wolverton, C., and Su, X. M. 2006. Virtual aluminum castings: an industrial application of ICME. *JOM*, **58**(11), 28–35.

[8] Almohaisen, F. M., and Abbod, M. F. 2010. Grain growth simulation using cellular automata. *Advanced Materials Research*, **89–91**, 17–22.

[9] Anderson, M. P., Srolovitz, D. J., Grest, G. S., and Sahni, P. S. 1984. Computer simulation of grain growth: 1. Kinetics. *Acta Metallurgica*, **32**, 783–791.

[10] Arfken, G. B., and Weber, H.-J. 2001. *Mathematical Methods for Physicists*. New York: Academic Press.

[11] Ashbaugh, H. S, Patel, H. A., Kumar, S. K., and Garde, S. 2005. Mesoscale model of polymer melt structure: Self-consistent mapping of molecular correlations to coarse-grained potentials. *Journal of Chemical Physics*, **122**, 104908.

[12] Ashby, M. F. 1992. Physical modelling of materials problems. *Materials Science and Technology*, **8**, 102–111.

[13] Ashby, M. F. 2011. *Materials Selection in Mechanical Design*. 4 edn. Oxford, UK: Butterworth-Heinemann.

[14] Ashby, M. F., and Johnson, K. 2009. *Materials and Design: The Art and Science of Material Selection in Product Design*. 2 edn. Oxford, UK: Butterworth-Heinemann.

[15] Ashcroft, N. W., and Mermin, N. D. 1976. *Solid State Physics*. Monterey, CA, USA: Brooks Cole.

[16] Aziz, R. A., and Chen, H. H. 1977. An accurate intermolecular potential for argon. *Journal of Chemical Physics*, **67**, 5719–5726.

[17] Backman, D. G., Wei, D. Y., Whitis, D. D., Buczek, M. B., Finnigan, P. M., and Gao, D. M. 2006. ICME at GE: accelerating the insertion of new materials and processes. *JOM*, **58**(11), 36–41.

[18] Baettig, P., Schelle, C. F., LeSar, R., Waghmare, U.V., and Spaldin, N. A. 2005. Theoretical prediction of new high-performance lead-free piezoelectrics. *Chemistry of Materials*, **17**, 1376–1380.

[19] Bales, G. S., and Chrzan, D. C. 1994. Dynamics of irreversible island growth during submonolayer epltaxy. *Physical Review B*, **50**, 6057–6067.

[20] Baletto, F., and Ferrando, R. 2005. Structural properties of nanoclusters: energetic, thermodynamic, and kinetic effects. *Reviews of Modern Physics*, **77**, 371–423.

[21] Balluffi, R. W., Allen, S. M., and Carter, W. C. 2005. *Kinetics of Materials*. New York: Wiley-Interscience.

[22] Basanta, D., Bentley, P. J., Miodownik, M. A., and Holm, E. A. 2003. Evolving cellular automata to grow microstructures. Pages 1–10 of: Ryan, C., Soule, T., Keijzer, M., Tsang, E. P. K., Poli, R., and Costa, E. (eds), *EuroGP 2003*, vol. LNCS 2610. Berlin: Springer-Verlag.

[23] Baskes, M. I. 1987. Application of the embedded-atom method to covalent materials: a semiempirical potential for silicon. *Physical Review Letters*, **59**, 2666–2669.

[24] Baskes, M. I. 1999. Many-body effects in fcc metals: a Lennard–Jones embedded-atom potential. *Physical Review Letters*, **83**, 2592–2595.

[25] Battaile, C. C., and Srolovitz, D. J. 2002. Kinetic Monte Carlo simulation of chemical vapor deposition. *Annual Review of Materials Research*, **32**, 297–319.

[26] Battaile, C. C., Srolovitz, D. J., and Butler, J. E. 1997. A kinetic Monte Carlo method for the atomic-scale simulation of chemical vapor deposition: application to diamond. *Journal of Applied Physics*, **82**, 6293–6300.

[27] Battaile, C. C., Srolovitz, D. J., Oleinik, I. I., Pettifor, D. G., Sutton, A. P., Harris, S. J., and Butler, J. E. 1999. Etching effects during the chemical vapor deposition of (100) diamond. *Journal of Chemical Physics*, **111**, 4291–4299.

[28] Becquart, C. S., and Domain, C. 2011. Modeling microstructure and irradiation effects. *Metallurgical and Materials Transactions A.*, **42**, 852–870.

[29] Bellucci, D., Cannillo, V., and Sola, A. 2010. Monte Carlo simulation of microstructure evolution in biphasic-systems. *Ceramics international*, **36**, 1983–1988.

[30] Belonoshko, A. B. 1994. Molecular dynamics of silica at high pressures – equation of state, structure, and phase transitions. *Geochimica et Cosmochimica Acta*, **58**, 1557–1566.

[31] Bercegeay, C., and Bernard, S. 2005. First-principles equations of state and elastic properties of seven metals. *Physical Review B*, **72**, 214101.

[32] Bernardes, N. 1958. Theory of solid Ne, A, Kr, and Xe at 0 K. *Physical Review*, **112**, 1534–1539.

[33] Berry, J., Elder, K. R., and Grant, M. 2008. Melting at dislocations and grain boundaries: a phase field crystal study. *Physical Review B*, **77**, 224114.

[34] Biner, S. B., and Morris, J. R. 2002. A two-dimensional discrete dislocation simulation of the effect of grain size on strengthening behaviour. *Modelling and Simulation in Materials Science and Engineering*, **10**, 617–635.

[35] Bishop, G. H., Harrison, R. J., Kwok, T., and Yip, Sidney. 1982. Computer molecular dynamics studies of grain boundary structures. I. Observations of coupled sliding and migration in a three dimensional simulation. *Journal of Applied Physics*, **53**, 5596–5608.

[36] Boal, D. 2002. *Mechanics of the Cell*. Cambridge, UK: Cambridge University Press.

[37] Boettinger, W. J., Warren, J. A., Beckermann, C., and Karma, A. 2002. Phase-field simulation of solidification. *Annual Review of Materials Research*, **32**, 163–194.

[38] Bortz, A. B., Kalos, M. H., and Lebowitz, J. L. 1975. A new algorithm for Monte Carlo simulation of Ising spin systems. *Journal of Computational Physics*, **17**, 10–18.

[39] Bos, C., Mecozzi, M. G., and Sietsma, J. 2010. A microstructure model for recrystallisation and phase transformation during the dual-phase steel annealing cycle. *Computational Materials Science*, **48**, 692–699.

[40] Box, G. E. P., and Muller, M. E. 1958. A note on the generation of random normal deviates. *The Annals of Mathematical Statistics*, **29**, 610–611.

[41] Brenner, D. W. 1990. Empirical potential for hydrocarbons for use in simulating the chemical vapor deposition of diamond films. *Physical Review B*, **42**, 9458–9471.

[42] Brenner, D. W. 1996. Chemical dynamics and bond-order potentials. *Materials Research Society Bulletin*, **21**(2), 36–41.

[43] Broughton, J. Q., and Gilmer, G. H. 1983. Molecular dynamics investigation of the crystal fluid interface. 1. Bulk Properties. *Journal of Chemical Physics*, **79**, 5095–5104. (Note that the sign of the coefficient $C_4$ is incorrect in Eq. 2.)

[44] Buchelnikov, V. D., and Sokolovskiy, V. V. 2011. Magnetocaloric effect in NiMnX (X = Ga, In, Sn, Sb) Heusler alloys. *The Physics of Metals and Metallography*, **112**, 633–665.

[45] Buckingham, A. D. 1959. Molecular quadrupole moments. *Quarterly Review of the Chemical Society*, **13**, 183–214.

[46] Buckingham, A. D. 1975. Intermolecular forces. *Philosophical Transactions of the Royal Society of London A*, **272**, 5–12.

[47] Buckingham, A. D., and Utting, B. D. 1970. Intermolecular forces. *Annual Review of Physical Chemistry*, **21**, 287–316.

[48] Buehler, M. J. 2010. Colloquium: Failure of molecules, bones, and the Earth itself. *Reviews of Modern Physics*, **82**, 1459–1487.

[49] Bulatov, V. V., and Cai, W. 2006. *Computer Simulations of Dislocations*. New York: Oxford.

[50] Burke, J. E., and Turnbull, D. 1952. Recrystallization and grain growth. *Progress in Metal Physics*, **3**, 220–292.

[51] Caflisch, R. E. 1998. Monte Carlo and quasi-Monte Carlo methods. *Acta Numerica*, **7**, 1–49.

[52] Cahn, J. W., and Hilliard, J. E. 1958. Free energy of a nonuniform system. I. Interfacial free energy. *Journal of Chemical Physics*, **28**, 258–267.

[53] Cahn, J. W., and Hilliard, J. E. 1959. Free energy of a nonuniform system. III. Nucleation in a two-component incompressible fluid. *Journal of Chemical Physics*, **31**, 688–699.

[54] Cai, W., Bulatov, V. V., Chang, J., Li, J. P., and Yip, S. 2003. Periodic image effects in dislocation modelling. *Philosophical Magazine*, **83**, 539–567.

[55] Callen, H. B. 1985. *Thermodynamics and an Introduction to Thermostatistics*. New York: Wiley.

[56] Car, R., and Parrinello, M. 1985. Unified approach for molecular dynamics and density-functional theory. *Physical Review Letters*, **55**, 2471–2474.

[57] Catlow, C. R. A. 2005. Energy minimization techniques in materials modeling. Pages 547–564 of: Yip, S. (ed), *Handbook of Materials Modeling*, vol. A. Dordrecht: Springer.

[58] Ceder, G., Morgan, D., Fischer, C., Tibbetts, K., and Curtarolo, S. 2006. Data-mining-driven quantum mechanics for the prediction of structure. *Materials Research Society Bulletin*, **31**(12), 981–985.

[59] Ceperley, D., and Alder, B. 1980. Ground state of the electron gas by a stochastic method. *Physical Review Letters*, **45**, 566–569.

[60] Ceperley, D., and Alder, B. 1986. Quantum Monte Carlo. *Science*, **231**, 555–560.

[61] Chan, P. Y., Tsekenis, G., Dantzig, J., Dahmen, K. A., and Goldenfeld, N. 2010. Plasticity and dislocation dynamics in a phase field crystal model. *Physical Review Letters*, **105**, 015502.

[62] Chandler, D. 1987. *Introduction to Modern Statistical Mechanics*. Oxford: Oxford University Press.

[63] Chen, J., Im, W., and Brooks III, C. L. 2006. Balancing solvation and intramolecular interactions: toward a consistent generalized Born force field. *Journal of the American Chemical Society*, **128**, 3728–3736.

[64] Chen, L.-Q. 2002. Phase-field models for microstructure evolution. *Annual Review of Materials Research*, **32**, 113–140.

[65] Chen, L.-Q., and Hu, S. 2004. Phase-field method applied to strain-dominated microstructure evolution during solid-state phase transformations. Pages 271–296 of: Raabe, D., Roters, F., Barlat, F., and Chen, L.-Q. (eds), *Continuum Scale Simulations of Engineering Materials*. Weinheim: Wiley-VCH.

[66] Cherry, M., and Imwinkelried, E. 2006. How we can improve the reliability of fingerprint identification. *Judicature*, **90**, 55–57.

[67] Cho, J., Terry, S. G., Levi, C. G., and LeSar, R. 2005. A kinetic Monte Carlo simulation of thin film growth by physical vapor deposition under substrate rotation. *Materials Science and Engineering A*, **391**, 390–401.

[68] Chopard, B., and Droz, M. 1998. *Cellular Automata Modeling of Physical Systems*. Cambridge, UK: Cambridge University Press.

[69] Ciacchi, L. C., and Payne, M. C. 2004. The entry pathway of $O_2$ into human ferritin. *Chemical Physics Letters*, **390**, 491–495.

[70] Clementi, E, and Roetti, E. 1974. Roothan–Hartree–Fock atomic wave functions. *Atomic Data and Nuclear Data Tables*, **14**, 177–478.

[71] Coffey, W. T., Kalmykov, Yu. P., and Waldron, J. T. 1996. *The Langevin Equation: With Applications in Physics, Chemistry, and Electrical Engineering*. Singapore: World Scientific.

[72] Committee, Integrated Computational Materials Engineering. 2008. *Integrated Computational Materials Engineering: A Transformational Discipline for Improved Competitiveness and National Security*. Tech. rept. National Research Council.

[73] Conway, J. H. 1970. *Game of Life*. This famous model was never published by the author!

[74] Cranford, S., and Buehler., M. J. 2011. Coarse-graining parameterization and multiscale simulation of hierarchical systems: Part I: Theory and Model Formulation. Pages 13–34 of: Derosa, P., and Cagin, T. (eds), *Multiscale Modeling: From Atoms to Devices*. Boca Raton, FL, USA: CRC Press.

[75] Curtarolo, S., Morgan, D., and Ceder, G. 2005. Accuracy of *ab initio* methods in predicting the crystal structures of metals: A review of 80 binary alloys. *Computer Coupling of Phase Diagrams and Thermochemistry*, **29**, 163–211.

[76] Danaila, I., Joly, P., Kaber, S. M., and Postel, M. 2007. *An Introduction to Scientific Computing: Twelve Computational Projects Solved with MATLAB*. New York: Springer.

[77] Darden, T., York, D., and Pederson, L. 1993. Particle mesh Ewald: an N log(N) method for Ewald sums in large systems. *Journal of Chemical Physics*, **98**, 10089–10092.

[78] Daw, M. S., and Baskes, M. I. 1983. Semiempirical, quantum mechanical calculation of hydrogen embrittlement in metals. *Physical Review Letters*, **50**, 1285–1288.

[79] Daw, M. S., and Baskes, M. I. 1984. Embedded-atom method: derivation and application to impurities, surfaces, and other defects in metals. *Physical Review B*, **29**, 6443–6453.

[80] Daw, M. S., Foiles, S. M., and Baskes, M. I. 1993. The embedded-atom method: a review of theory and applications. *Materials Science Reports*, **9**, 251–310.

[81] De Miguel, E., Rull, L. F., Chalam, M. K., and Gubbins, K. E. 1991. Liquid crystal phase diagram of the Gay–Berne fluid. *Molecular Physics*, **74**, 405–424.

[82] de Wit, R. 1960. The continuum theory of stationary dislocations. *Solid State Physics*, **10**, 249–292.

[83] Déprés, C., Robertson, C. F., and Fivel, M. C. 2004. Low-strain fatigue in AISI 316L steel surface grains: a three-dimensional discrete dislocation dynamics modelling of the early cycles I. Dislocation microstructures and mechanical behaviour. *Philosophical Magazine*, **84**, 2257–2275.

[84] Déprés, C., Robertson, C. F., and Fivel, M. C. 2006. Low-strain fatigue in 316L steel surface grains: a three dimension discrete dislocation dynamics modelling of the early cycles. Part 2: Persistent slip markings and micro-crack nucleation. *Philosophical Magazine*, **86**, 79–97.

[85] Déprés, C., Fivel, M., and Tabourot, L. 2008. A dislocation-based model for low-amplitude fatigue behaviour of face-centred cubic single crystals. *Scripta Materialia*, **58**, 10861089.

[86] Devincre, B., and Kubin, L. 2010. Scale transitions in crystal plasticity by dislocation dynamics simulations. *Comptes Rendus Physique*, **11**, 274–284.

[87] Devincre, B., and Kubin, L. P. 1997. Mesoscopic simulations of dislocations and plasticity. *Materials Science and Engineering A*, **234–236**, 8–14.

[88] Devincre, B., Kubin, L. P., and Hoc, T. 2006. Physical analyses of crystal plasticity by DD simulations. *Scripta Materialia*, **54**, 741–746.

[89] Devincre, B., Hoc, T., and Kubin, L. 2008. Dislocation mean free paths and strain hardening of crystals. *Science*, **320**, 1745–1748.

[90] Dick, B. G., and Overhauser, A. W. 1958. Theory of dielectric constants of alkali halide crystals. *Physical Review*, **112**, 90–103.

[91] Ding, J., Carver, T. J., and Windle, A. H. 2001. Self-assembled structures of block copolymers in selective solvents reproduced by lattice Monte Carlo simulation. *Computational and Theoretical Polymer Science*, **11**, 483–490.

[92] Dirac, P. A. M. 1930. A note on the exchange phenomena in the Thomas atom. *Proceedings of the Cambridge Philosophical Society*, **26**, 376–385.

[93] Doi, M. 1995. *Introduction to Polymer Physics*. Oxford, UK: Oxford University Press.

[94] Doi, M., and Edwards, S. F. 1986. *The Theory of Polymer Dynamics*. Oxford, UK: Oxford University Press.

[95] Duff, N., and Peters, B. 2009. Nucleation in a Potts lattice gas model of crystallization from solution. *Journal of Chemical Physics*, **131**, 184101.

[96] El-Awady, J. A., Biner, S. B., and Ghoniem, N. M. 2008. A self-consistent boundary element, parametric dislocation dynamics formulation of plastic flow in finite volumes. *Journal of the Mechanics and Physics of Solids*, **56**, 20192035.

[97] Elder, K. R., and Grant, M. 2004. Modeling elastic and plastic deformations in nonequilibrium processing using phase field crystals. *Physical Review E*, **70**, 051605.

[98] Elder, K. R., Thornton, K., and Hoyt, J. J. 2011. The Kirkendall effect in the phase field crystal model. *Philosophical Magazine*, **91**, 151–164.

[99] Fermi, E. 1927. Un Metodo Statistice per la Determinazione di Alcune Proprieta dell'Atomo. *Rendiconti Accademia Lincei*, **6**, 602–607.

[100] Ferris, K. F., Peurrung, L. M., Loni, M., and Marder, J. 2007. Materials informatics: fast track to new materials. *Advanced Materials and Processes*, **165**, 50–51.

[101] Finnis, M. W., and Sinclair, J. E. 1984. A simple empirical N-body potential for transition metals. *Philosophical Magazine A*, **50**, 45–55.

[102] Fiolhais, C., Nogueira, F., and Marques, M. A. L. (eds). 2003. *A Primer in Density Functional Theory*. Lecture Notes in Physics, vol. 620. New York, NY: Springer.

[103] Fischer, C. C., Tibbetts, K. J., Morgan, D., and Ceder, G. 2006. Predicting crystal structure by merging data mining with quantum mechanics. *Nature Materials*, **5**, 641–646.

[104] Flory, P. J. 1953. *Principles of Polymer Chemistry*. Ithaca, NY: Cornell University Press. Chapter 10.

[105] Foiles, S. M. 1985. Calculation of the surface segregation of Ni–Cu alloys with the use of the embedded-atom method. *Physical Review B*, **32**, 7685–7693.

[106] Foiles, S. M. 1996. Embedded-atom and related methods for modeling metallic systems. *Materials Research Society Bulletin*, **21**(2), 24–28.

[107] Foreman, A. J. E., and Makin, M. J. 1966. Dislocation motion though a random array of obstacles. *Philosophical Magazine*, **14**, 911–924.

[108] Foreman, A. J. E., and Makin, M. J. 1967. Dislocation motion though a random array of obstacles. *Canadian Journal of Physics*, **45**, 511–517.

[109] Frenkel, D., and Smit, B. 2002. *Understanding Molecular Simulations*. Second edn. San Diego, CA: Elsevier Academic Press.

[110] Frisch, U., Hasslacher, B., and Pomeau, Y. 1986. Lattice-gas automata for the Navier-Stokes equation. *Physical Review Letters*, **56**, 1505–1508.

[111] Frost, H. J., and Thompson, C. V. 1988. Computer simulation of microstructural evolution in thin films. *Journal of Electronic Materials*, **17**, 447–458.

[112] Gandin, Ch.-A., and Rappaz, M. 1994. A coupled finite element-cellular automaton method for the prediction of dendritic grain structures in solidification processes. *Acta Metallurgica et Materialia*, **42**, 2233–2246.

[113] Gardner, M. 1985. *Wheels, Life, and Other Mathematical Amusements*. New York: W.H. Freeman & Company.

[114] Gaskell, D. R. 2008. *Introduction to the Thermodynamics of Materials*. Fifth edn. London: Taylor & Francis.

[115] Gater, A. J. P., Brown, S. G. R., Baker, L., and Fourlaris, G. 2011. A fast Bortz–Kalos–Lebowitz implementation of the kinetic Monte Carlo technique for the prediction of strain ageing. *Scripta Materialia*, **65**, 400–403.

[116] Gay, J. G., and Berne, B. J. 1981. Modification of the overlap potential to mimic a linear site-site potential. *Journal of Physical Chemistry*, **74**, 3316–3319.

[117] Gaylord, R. J., and Nishidate, K. 1996. *Modeling Nature: Cellular Automata Simulations with Mathematica*. Santa Clara, CA: TELOS.

[118] Germann, T. C., and Kadau, K. 2008. Trillion-atom molecular dynamics becomes a reality. *International Journal of Modern Physics C*, **9**, 1315–1319.

[119] Ghoniem, N. M., and Sun, L. Z. 1999. Fast-sum method for the elastic field off three-dimensional dislocation ensembles. *Physical Review B*, **60**, 128–140.

[120] Ginstead, C. M., and Snell, J. L. 1997. *Introduction to Probability*. 2 edn. Providence, RI, USA: American Mathematical Society. Available as a free download at http://www.dartmouth.edu/~chance/teaching_aids/books_articles/probability_book/book.html.

[121] Ginzburg, V. L., and Landau, L. D. 1950. On the theory of superconductivity. *Zh. Eksp. Teor. Fiz.*, **20**, 1064. translation in L. D. Landau, Collected Papers (Pergamon, Oxford, 1965), p. 546.

[122] Glazier, J. A., and Graner, F. 1993. Simulation of the differential adhesion driven rearrangement of biological cells. *Physical Review E*, **47**, 2128–2154.

[123] Glicksman, M. E. 1999. *Diffusion in Solids: Field Theory, Solid-State Principles, and Applications*. New York: Wiley-Interscience.

[124] Glotzer, S. C., and Paul, W. 2002. Molecular and mesoscale simulation methods for polymer materials. *Annual Review of Materials Research*, **32**, 401–436.

[125] Goetz, R. L., and Seetharaman, V. 1998. Modeling dynamic recrystallization using cellular automata. *Scripta Materialia*, **38**, 405–413.

[126] Goldstein, H., Poole, C. P., and Safko, J. L. 2001. *Classical Mechanics* (third edition). New York: Addison-Wesley.

[127] Goringe, C. M., Bowler, D. R., and Hernandez, E. 1997. Tight-binding modelling of materials. *Reports on Progress in Physics*, **60**, 1447–1512.

[128] Graner, F., and Glazier, J. A. 1992. Simulation of biological cell sorting using a two-dimensional extended Potts model. *Physical Review Letters*, **69**, 2013–2016.

[129] Gray, C. G., and Gubbins, K. E. 1985. *Theory of Molecular Fluids*. Oxford: Oxford University Press.

[130] Greengard, L., and Rokhlin, V. 1987. A fast algorithm for particle simulations. *Journal of Computational Physics*, **73**, 325–348.

[131] Greengard, L. F. 2003. *The Rapid Evaluation of Potential Fields in Particle Systems*. Cambridge, MA: MIT Press.

[132] Guillot, B. 2002. A reappraisal of what we have learnt during three decades of computer simulation on water. *Journal of Molecular Liquids*, **101**, 219–260.

[133] Gulluoglu, A. N., Srolovitz, D. J., LeSar, R., and Lomdahl, P. S. 1989. Dislocation distributions in two dimensions. *Scripta Metallurgica*, **23**, 1347–1352.

[134] Gurney, J. A., Rietman, E.A., Marcus, M. A., and Andrews, M. P. 1999. Mapping the rule table of a 2-D probabilistic cellular automaton to the chemical physics of etching and deposition. *Journal of Alloys and Compounds*, **290**, 216–229.

[135] Haile, J. M. 1997. *Molecular Dynamics Simulation: Elementary Methods*. New York: Wiley-Interscience.

[136] Haire, K. R., Carver, T. J., and Windle, A. H. 2001. A Monte Carlo lattice model for chain diffusion in dense polymer systems and its interlocking with molecular dynamics simulation. *Computational and Theoretical Polymer Science*, **11**, 17–28.

[137] Hand, D., Mannila, H., and Smyth, P. 2001. *Principles of Data Mining*. Cambridge, Massachusetts, USA: MIT Press.

[138] Hansen, J.-P. 1986. Molecular-dynamics simulation of Coulomb systems in two and three dimensions. Pages 89–129 of: Ciccotti, G., and Hoover, W.G. (eds), *Molecular Dynamics Simulation of Statistical-Mechanical Systems*. Amsterdam: North-Holland.

[139] Hansen, J.-P., and McDonald, I. 1986. *Theory of Simple Liquids*. London: Academic Press.

[140] Harris, S. J., and Goodwin, D. G. 1993. Growth on the reconstructed diamond (100) surface. *Journal of Physical Chemistry*, **97**, 23–28.

[141] Harun, A., Holm, E. A., Clode, M. P., and Miodownik, M. A. 2006. On computer simulation methods to model Zener pinning. *Acta Materialia*, **54**, 3261–3273.

[142] Hassold, G. N., and Holm, E. A. 1993. A fast serial algorithm for the finite temperature quenched Potts model. *Computers in Physics*, **7**, 97–107.

[143] Hesselbarth, H. W., and Göbel, I. R. 1991. Simulation of recrystallization by cellular automata. *Acta Metallurgica et Materialia*, **39**, 2135–2143.

[144] Hill, J.-R., Freeman, C. M., and Subramanian, L. 2000. Use of force fields in materials modeling. Pages 141–216 of: Lipkowitz, K. B., and Boyd, D. B. (eds), *Reviews in Computational Chemistry*, vol. 16. New York: Wiley-VCH.

[145] Hill, N. A. 2000. Why are there so few magnetic ferroelectrics? *Journal of Physical Chemistry B*, **29**, 6694–6709.

[146] Hirschfelder, J. O., Curtiss, C. F., and Bird, R. B. 1964. *Molecular Theory of Gases and Liquids*. Second edn. New York: Wiley. See section 13.3a.

[147] Hirth, J. P., and Lothe, J. 1992. *Theory of Dislocations*. Malabar, Florida: Kreiger Publishing.

[148] Hirth, J. P., Zbib, H. M., and Lothe, J. 1998. Forces on high velocity dislocations. *Modelling and Simulation in Materials Science and Engineering*, **6**, 165–169.

[149] Hohenberg, P., and Kohn, W. 1964. Inhomogeneous electron gas. *Physical Review B*, **136**, B864–B871.

[150] Holm, E. A. 1992. Modeling microstructural evolution in single-phase, composite and two-phase polycrystals. Ph.D. thesis, University of Michigan. Available from Proquest Microfilm.

[151] Holm, E. A., and Battaile, C. C. 2001. The computer simulation of microstructural evolution. *JOM*, **53**(9), 20–23.

[152] Holm, E. A., Glazier, J. A., Srolovitz, D. J., and Grest, G. S. 1991. Effects of lattice anisotropy and temperature on domain growth in the two-dimensional Potts model. *Physical Review A*, **43**, 2662–2668.

[153] Holm, E. A., Miodownik, M. A., and Rollett, A. D. 2003. On abnormal subgrain growth and the origin of recrystallization nuclei. *Acta Materialia*, **51**, 27012716.

[154] Hoover, W. G. 1985. Canonical dynamics: equilibrium phase-space distributions. *Physical Review A*, **31**, 1695–1697.

[155] House, J. E. 2007. *Principles of Chemical Kinetics*. Second edn. Amsterdam: Academic Press.

[156] Howard, J. 2001. *Mechanics of Motor Proteins and the Cytoskeleton*. Sunderland, MA, USA: Sinauer Associates.

[157] Hull, D., and Bacon, D. J. 2001. *Introduction to Dislocations*. Fourth edn. Oxford, UK: Butterworth Heinemann.

[158] Huo, S. 1995. *get title*. Ph.D. thesis, Kansas State University.

[159] Israelachvili, J. 1992. *Intermolecular and Surface Forces*. London: Academic Press.

[160] Jones, G. W., and Chapman, S. J. 2012. Modeling growth in biological materials. *SIAM Review*, **54**, 52–118.

[161] Jones, J. E., and Ingham, A. E. 1925. On the calculation of certain crystal potential constants, and on the cubic crystal of least potential energy. *Proceedings of the Royal Society of London. Series A*, **107**, 636–653.

[162] Jorgensen, W. L., Chandrasekhar, J., Madura, J. D., Impey, R. W., and Klein, M. L. 1983. Comparison of simple potential functions for simulating liquid water. *Journal of Chemical Physics*, **79**, 926–935.

[163] Kammer, D., and Voorhees, P. W. 2008. Analysis of complex microstructures: Serial sectioning and phase-field simulations. *Materials Research Society Bulletin*, **33**(6), 603–610.

[164] Kansuwan, P., and Rickman, J. M. 2007. Role of segregating impurities in grain-boundary diffusion. *Journal of Chemical Physics*, **126**, 094707.

[165] Kawasaki, K. 1966. Diffusion constants near the critical point for time-dependent Ising models. I. *Physical Review*, **145**, 224–230.

[166] Kawasaki, K., Nagai, T., and Nakashima, K. 1989. Vertex models for two-dimensional grain growth. *Philosophical Magazine B*, **60**, 399–421.

[167] Kaxiras, E. 2003. *Atomic and Electronic Structure of Solids*. Cambridge, UK: Cambridge University Press.

[168] Keeler, G .J., and Batchelder, D. N. 1970. Measurement of the elastic constants of argon from 3 to 77 K. *Journal of Physics C*, **3**, 510–522.

[169] Kim, S. G., Horstemeyer, M. F., Baskes, M. I., *et al.* 2009. Semi-empirical potential methods for atomistic simulations of metals and their construction procedures. *Journal of Engineering Materials and Technology*, **131**, 041210.

[170] Kim, Y. S., and Gordon, R. G. 1974. Unified theory for the intermolecular forces between closed shell atoms and ions. *Journal of Chemical Physics*, **61**, 1–16.

[171] Kocks, U. F., and Mecking, H. 2003. Physics and phenomenology of strain hardening: The FCC case. *Progress in Material Science*, **48**, 171–273.

[172] Kofke, D. A. 2011. *CE 530 Molecular Simulation*. http://www.eng.buffalo.edu/~kofke/ce530/Lectures/Lecture12/sld001.htm.

[173] Kohn, W., and Sham, L. J. 1965. Self-consistent equations including exchange and correlation effects. *Physical Review*, **140**, A1133–A1138.

[174] Kollman, P. A., Massova, I., Reyes, C., *et al.* 2000. Calculating structures and free energies of complex molecules: Combining molecular mechanics and continuum models. *Accounts of Chemical Research*, **33**, 889–897.

[175] Körner, C., Attar, E. and Heinl, P. 2011. Mesoscopic simulation of selective beam melting processes. *Journal of Materials Processing Technology*, **211**, 978–987.

[176] Korpiun, P., and Lüscher, E. 1977. Use of force fields in materials modeling. Page 729 of: Venables, J. A., and Klein, M. L. (eds), *Rare Gas Solids*, vol. 2. London: Academic Press.

[177] Koslowski, M., Cuitiño, A. M., and Ortiz, M. 2002. A phase-field theory of dislocation dynamics, strain hardening and hysteresis in ductile single crystals. *Journal of the Mechanics and Physics of Solids*, **50**, 2597–2635.

[178] Koslowski, M., Thomson, R., and LeSar, R. 2004a. Avalanches and scaling in plastic deformation. *Physical Review Letters*, **93**, 125502.

[179] Koslowski, M., Thomson, R., and LeSar, R. 2004b. Dislocation structures and the deformation of materials. *Physical Review Letters*, **93**, 265503.

[180] Kratky, O., and Porod, G. 1949. X-ray investigation of chain-molecules in solution. *Recueil des Travaux Chimiques des Pays-Bas*, **68**, 1106–1122.

[181] Kremer, K. 2003. Computer simulations for macromolecular science. *Macromolecular Chemistry and Physics*, **204**, 257–264.

[182] Kroc, J. 2002. Application of cellular automata simulations to modelling of dynamic recrystallization. Pages 773–782 of: Sloot, P. M. A., Tan, C. J. K., Dongarra, J. J., and Hoekstra, A. G. (eds), *Computational Science – ICCS 2002*, vol. LNCS 2329. Berlin: Springer-Verlag.

[183] Kubelka, J., Hofrichter, J., and Eaton, W. A. 2004. The protein folding "speed limit". *Current Opinion in Structural Biology*, **14**, 76–88.

[184] Kubin, L. P., Canova, G., Condat, M., Devincre, B., Pontikis, V., and Bréchet, Y. 1992. Dislocation microstructures and plastic flow: A 3D simulation. *Solid State Phenomena*, **23–24**, 455–472.

[185] Kubin, L. P., Devincre, B., and Thierry, H. 2009. The deformation stage II of face-centered cubic crystals: Fifty years of investigations. *International Journal of Materials Research*, **100**, 1411–1419.

[186] Kundin, J., Raabe, D., and Emmerich, H. 2011. A phase-field model for incoherent martensitic transformations including plastic accommodation processes in the austenite. *Journal of the Mechanics and Physics of Solids*, **59**, 2082–2102.

[187] Lal, M. 1969. Monte Carlo computer simulations of chain molecules I. *Molecular Physics*, **17**, 57–64.

[188] Landau, D. P., and Binder, K. 2000. *A Guide to Monte Carlo Simulations in Statistical Physics*. Cambridge: Cambridge University Press.

[189] Langevin, P. 1908. Sur la théorie du mouvement brownien. *Comptes Rendus de l'Académie des Sciences*, **146**, 530–533. English translation in American Journal of Physics 65, 1079–1081 (1997).

[190] Leach, A. R. 2001. *Molecular Modelling: Principles and Applications*. Second edn. Harlow, UK: Prentice-Hall.

[191] Lee, C. T., Yang, W. T., and Parr, R. G. 1988. Development of the Colle-Salvetti correlation-energy formula into a functional of the electron density. *Physical Review B*, **37**, 785–789.

[192] Lenosky, T. J., Kress, J. D., Kwon, I., *et al.* 1997. Highly optimized tight-binding model of silicon. *Physical Review B*, **55**, 1528–1544.

[193] Lépinoux, J., and Kubin, L. P. 1987. The dynamic organization of dislocation structures: a simulation. *Scripta Metallurgica*, **21**, 833–838.

[194] Lépinoux, J., Weygand, D., and Verdier, M. 2010. Modeling grain growth and related phenomena with vertex dynamics. *Comptes Rendus Physique*, **11**, 265–273.

[195] LeSar, R., and Etters, R. D. 1988. Character of the $\alpha$-$\beta$ phase transition in solid oxygen. *Physical Review B*, **37**, 5364–5370.

[196] LeSar, R., and Rickman, J. M. 1996. Finite-temperature properties of materials from analytical statistical mechanics. *Philosophical Magazine B*, **73**, 627–639.

[197] Levesque, D., and Verlet, L. 1970. Computer "experiments" on classical fluids. III. Time-dependent self-correlation functions. *Physical Review A*, **2**, 2514–2528.

[198] Lewis, A. C., and Geltmacher, A. B. 2006. Image-based modeling of the response of experimental 3D microstructures to mechanical loading. *Scripta Materialia*, **55**, 81–85.

[199] Lewis, A. C., Suh, C., Stukowski, M., Geltmacher, A. B., Spanos, G., and Rajan, K. 2006. Quantitative analysis and feature recognition in 3-D microstructural data sets. *JOM*, **58**(12), 52–56.

[200] Lewis, A. C., Suh, C., Stukowski, M., Geltmacher, A. B., Rajan, K., and Spanos, G. 2008. Tracking correlations between mechanical response and microstructure in three-dimensional reconstructions of a commercial stainless steel. *Scripta Materialia*, **58**(7), 575–578.

[201] Lin, K., and Chrzan, D. C. 1999. Kinetic Monte Carlo simulation of dislocation dynamics. *Physical Review B*, **60**, 3799–3805.

[202] Lin, Y., Ma, H., Matthews, C. W., *et al.* 2012. Experimental and theoretical studies on a high pressure monoclinic phase of ammonia borane. *Journal of Physical Chemistry C*, **116**, 2172–2178.

[203] Lishchuk, S. V., Akid, R., Worden, K., and Michalski, J. 2011. A cellular automaton model for predicting intergranular corrosion. *Corrosion Science*, **53**, 2518–2526.

[204] Liu, Y., Baudin, T., and Penelle, R. 1996. Simulation of normal grain growth by cellular automata. *Scripta Materialia*, **34**, 1679–1683.

[205] Liu, Z.-K., Chen, L.-Q., and Rajan, K. 2006. Linking length scales via materials informatics. *JOM*, **58**(11), 42–50.

[206] Lothe, J., Indenbom, V. L., and Chamrov, V. A. 1982. Elastic field and self-force of dislocations emerging at free surfaces of an anisotropic half-space. *Physica Status Solidi (b)*, **111**, 671–677.

[207] Love, A. E. H. 1944. *A Treatise on the Mathematical Theory of Elasticity*. 4 edn. New York: Dover.

[208] Lukas, H., Fries, S. G., and Sundman, B. 2007. *Computational Thermodynamics: The CALPHAD Method*. Cambridge, UK: Cambridge University Press.

[209] MacKerell, A. D. *et al.* 1998. All-atom empirical potential for molecular modeling and dynamics studies of proteins. *Journal of Physical Chemistry B*, **102**, 35863616.

[210] MacPherson, R. D., and Srolovitz, D. J. 2007. The von Neumann relation generalized to coarsening of three-dimensional microstructures. *Nature*, **446**, 1053–1055.

[211] Madec, R., Devincre, B., and Kubin, L. 2004. On the use of periodic boundary conditions in dislocation dynamics simulations. *Solid Mechanics and Its Applications*, **115**, 35–44.

[212] Madras, N., and Sokal, A. D. 1988. The pivot algorithm: a highly efficient Monte Carlo method for the self-avoiding walk. *Journal of Statistical Physics*, **50**, 109–186.

[213] Maeder, R. E. 2000. *Computer Science with Mathematica*. Cambridge, UK: Cambridge University Press.

[214] Mahoney, M. W., and Jorgensen, W. L. 2000. A five-site model for liquid water and the reproduction of the density anomaly by rigid, nonpolarizable potential functions. *Journal of Chemical Physics*, **112**, 8910–8922.

[215] Malcherek, T. 2011. The ferroelectric properties of $Cd_2Nb_2O_7$: a Monte Carlo simulation study. *Journal of Applied Crystallography*, **44**, 585–594.

[216] Malvern, L. E. 1969. *Introduction to the Mechanics of a Continuous Medium*. London: Prentice-Hall.

[217] March, N. H. 1975. *Self-Consistent Fields in Atoms*. Oxford, UK: Pergamon Press.

[218] Martin, J. W., Doherty, R. D, and Cantor, B. 1997. *Stability of Microstructure in Metallic Systems*. Second edn. Cambridge, UK: Cambridge University Press.

[219] Martin, R. M. 2008. *Electronic Structure: Basic Theory and Practical Methods*. Cambridge, UK: Cambridge University Press.

[220] Martínez, E., Marian, J., Arsenlis, A., Victoria, M., and Perlado, J. M. 2008. Atomistically informed dislocation dynamics in fcc crystals. *Journal of the Mechanics and Physics of Solids*, **56**, 869–895.

[221] Massalski, T. B. 1996. Structure of Stability of Alloys. Chap. 3, pages 135–204 of: Cahn, R. W., and Haasen, P. (eds), *Physical Metallurgy*, vol. 1. Amsterdam: North-Holland.

[222] McDowell, D. L., and Olson, G. B. 2008. Concurrent design of hierarchical materials and structures. *Scientific Modeling and Simulation*, **15**, 207–240.

[223] McQuarrie, D. A. 1976. *Statistical Mechanics*. New York: Harper and Row.

[224] McQuarrie, D. A., and Sion, J. D. 1997. *Physical Chemistry: A Molecular Approach*. Mill Valley, CA, USA: University Science Books.

[225] Mendelev, M. I., and Bokstein, B. S. 2010. Molecular dynamics study of self-diffusion in Zr. *Philosophical Magazine*, **90**, 637–654.

[226] Merriam-Webster. 2011. *Dictionary*. http://www.merriam-webster.com/dictionary.

[227] Metropolis, N., and Ulam, S. 1949. The Monte Carlo method. *Journal of the American Statistical Association*, **44**, 335–341.

[228] Metropolis, N., Rosenbluth, A. W., Rosenbluth, M. N., Teller, A. H., and Teller, E. 1953. Equation of state calculations by fast computing machines. *Journal of Chemical Physics*, **21**, 1087–1092.

[229] Meyers, M. A., Mishra, A., and Benson, D. J. 2006. Mechanical properties of nanocrystalline materials. *Progress in Materials Science*, **51**, 427–556.

[230] Miguel, M. C., Vespignani, A., Zapperi, S., Weiss, J., and Grasso, J. R. 2001. Intermittent dislocation flow in viscoplastic deformation. *Nature*, **410**, 667–671.

[231] Miodownik, M., Holm, E. A., and Hassold, G. N. 2000. Highly parallel computer simulations of particle pinning: Zener vindicated. *Scripta Materialia*, **42**, 1173–1177.

[232] Miodownik, M. A. 2002. A review of microstructural computer models used to simulate grain growth and recrystallisation in aluminium alloys. *Journal of Light Metals*, **2**, 125–135.

[233] Mora, L. A. B., Gottstein, G., and Shvindlerman, L. S. 2008. Three-dimensional grain growth: Analytical approaches and computer simulations. *Acta Materialia*, **56**, 5915–5926.

[234] Morokoff, W. J., and Caflisch, R. E. 1994. Quasi-random sequences and their discrepancies. *SIAM Journal on Scientific Computing*, **15**, 1251–1279.

[235] Müller-Plathe, F. 2002. Coarse-graining in polymer simulation: from the atomistic to the mesoscopic scale and back. *ChemPhysChem*, **3**, 754–769.

[236] Nada, H., and van der Eerden, J. P. J. M. 2003. An intermolecular potential model for the simulation of ice and water near the melting point: A six-site model of $H_2O$. *Journal of Chemical Physics*, **118**, 7401–7413.

[237] Najafabadi, R., and Yip, S. 1983. Observation of finite-temperature Bain transformation (f.c.c. $\leftrightarrow$ b.c.c.) in Monte Carlo simulation of iron. *Scripta Metallurgica*, **17**, 1199–1204.

[238] Neumann, R., and Handy, N. C. 1997. Higher-order gradient corrections for exchange-correlation functionals. *Chemical Physics Letters*, **266**, 16–22.

[239] Nielsen, S. O., Lopez, C. F., Srinivas, G., and Klein, M. L. 2004. Coarse grain models and the computer simulation of soft materials. *Journal of the Physics of Condensed Matter*, **16**, R481–512.

[240] Nogueira, F., Castro, A., and Marques, M. A. L. 2003. A tutorial on density functional theory. *Lecture Notes in Physics*, **620**, 218256.

[241] Nose, S. 1984. A molecular dynamics method for simulations in the canonical ensemble. *Molecular Physics*, **52**, 255–268.

[242] Olson, G. B. 1997. Computational design of hierarchically structured materials. *Science*, **277**, 1237–1242.

[243] Onsager, L. 1944. Crystal statistics. I. A two-dimensional model with an order–disorder transition. *Physical Review*, **65**, 117–149.

[244] Onuchic, J. N., and Wolynes, P. G. 2004. Theory of protein folding. *Current Opinion in Structural Biology*, **14**, 70–75.

[245] Onuchic, J. N., LutheySchulten, Z., and Wolynes, P. G. 1997. Theory of protein folding: the energy landscape perspective. *Annual Review of Physical Chemistry*, **48**, 545–600.

[246] Oono, Y., and Puri, S. 1987. Computationally efficient modeling of ordering of quenched phases. *Physical Review Letters*, **58**, 836–839.

[247] Orowan, E. 1934. Zur Kristallplastizität III: Über die Mechanismus des Gleitvorganges. *Zeitschrift für Physik*, **89**, 634–659.

[248] Packard, N. H., and Wolfram, S. 1985. Two-dimensional cellular automata. *Journal of Statistical Physics*, **38**, 901–946.

[249] Pao, C.-W., Foiles, S. M., Webb III, E. B., Srolovitz, D. J., and Floro, J. A. 2009. Atomistic simulations of stress and microstructure evolution during polycrystalline Ni film growth. *Physical Review B*, **79**, 224113.

[250] Park, S. K., and Miller, K. W. 1988. Random number generators: good ones are hard to find. *Communications of the ACM*, **31**, 1192–1201.

[251] Parr, R. G., and Yang, W. 1989. *Density-Functional Theory of Atoms and Molecules*. Oxford, UK: Oxford University Press.

[252] Parrinello, M., and Rahman, A. 1981. Polymorphic transitions in single crystals: A new molecular dynamics method. *Journal of Applied Physics*, **52**, 7182–7190.

[253] Pauling, L., and Wilson Jr, E. B. 1935. *Introduction to Quantum Mechanics with Applications to Chemistry*. New York: Dover.

[254] Payne, M. C., Teter, M. P., Allan, D. C., Arias, T. A., and Joannopolis, J. D. 1992. Iterative minimization techniques for ab initio total-energy calculations: molecular dynamics and conjugate gradients. *Reviews of Modern Physics*, **64**, 1045–1097.

[255] Peach, M., and Koehler, J. S. 1950. The forces exerted on dislocations and the stress fields produced by them. *Physical Review*, **80**, 436–439.

[256] Perdew, J. P, and Wang, Y. 1992. Accurate and simple analytic representation of the electron-gas correlation energy. *Physical Review B*, **45**, 13244–13249.

[257] Perdew, J. P, and Zunger, A. 1981. Self-interaction correction to density-functional approximations for many-electron systems. *Physical Review B*, **23**, 5048–5079.

[258] Perdew, J. P, Burke, K., and Ernzerhof, M. 1996. Generalized gradient approximation made simple. *Physical Review Letters*, **77**, 3865–3868.

[259] Pettifor, D. G., and Oleynik, I. I. 2004. Interatomic bond-order potentials and structural prediction. *Progress in Material Science*, **49**, 285–312.

[260] Phillips, R. 2001. *Crystals, Defects, and Microstructures: Modeling Across Scales*. Cambridge, UK: Cambridge University Press.

[261] Polanyi, M. 1934. Über eine Art Gitterstrung, die einen Kristall plastisch machen könnte. *Zeitschrift für Physik*, **89**, 660–664.

[262] Ponder, J. W., and Case, D. A. 2003. Force fields for protein simulations. *Advances in Protein Chemistry*, **66**, 27–85.

[263] Porter, D. A., and Easterling, K. E. 1992. *Phase Transformations in Metals and Alloys*. Second edn. Boca Raton, FL, USA: CRC Press.

[264] Press, W. H., Teukolsky, S. A., Vetterling, W. T., and Flannery, B. P. 1992. *Numerical Recipes in Fortran 77*. Cambridge, UK: Cambridge University Press.

[265] Project, Paraview. 2011. *ParaView – Open Source Visualization Software*. http://www.paraview.org/.

[266] Püschl, W. 2002. Models for dislocation cross-slip in close-packed crystal structures: a critical review. *Progress in Material Science*, **47**, 415–461.

[267] Quesnel, D. J., Rimai, D. S., and DeMejo, L.P. 1993. Elastic compliances and stiffnesses of the fcc Lennard–Jones solid. *Physical Review B*, **48**, 6795–6807.

[268] Raabe, D. 2002. Cellular automata in materials science with particular reference to recrystallization simulation. *Annual Review of Materials Research*, **32**, 53–76.

[269] Raabe, D. 2004a. Cellular, lattice gas, and Boltzmann automata. Pages 57–76 of: Raabe, D., Roters, F., Barlat, F., and Chen, L.-Q. (eds), *Continuum Scale Simulations of Engineering Materials*. Weinheim: Wiley-VCH.

[270] Raabe, D. 2004b. Overview of the lattice Boltzmann method for nano- and microscale fluid dynamics in materials science and engineering. *Modelling and Simulation in Materials Science and Engineering*, **12**, R13–R46.

[271] Rajan, K. 2008. Combinatorial material sciences: experimental strategies for accelerated knowledge discovery. *Annual Review of Materials Research*, **38**, 299–322.

[272] Rappaz, M., and Gandin, Ch.-A. 1993. Probabilistic modeling of microstructure formation in solidification processes. *Acta Metallurgica et Materialia*, **41**, 345–360.

[273] Read, W. T. 1953. *Dislocations in Crystals*. New York: McGraw Hill. Available online at http://ia700308.us.archive.org/31/items/dislocationsincr032720mbp/dislocationsincr032720mbp.pdf.

[274] Rittner, J. D., and Seidman, D. N. 1996. ⟨110⟩ symmetric tilt grain-boundary structures in fcc metals with low stacking-fault energies. *Physical Review B*, **54**, 6999–7015.

[275] Rodrigues, P. C. R, and Fernandes, F. M. S. S. 2007. Phase diagrams of alkali halides using two interaction models: a molecular dynamics and free energy study. *Journal of Chemical Physics*, **126**, 024503.

[276] Rollett, A. D., Srolovitz, D. J., and Anderson, M. P. 1989. Simulation and theory of abnormal grain growth – anisotropic grain boundary energies and mobilities. *Acta Metallurgica*, **37**, 1227–1240.

[277] Rose, J. H., Smith, J. R., Guinea, F., and Ferrante, J. 1984. Universal features of the equation of state of metals. *Physical Review B*, **29**, 2963–2969.

[278] Rothman, D. H., and Zaleski, S. 1994. Lattice-gas models of phase separation: Interfaces, phase transitions, and multiphase flow. *Reviews of Modern Physics*, **66**, 1417–1479.

[279] Rountree, C. L., Kalia, R. K., Lidorikis, E., Nakano, A., Van Brutzel, L., and Vashishta, P. 2002. Atomistic aspects of crack propagation in brittle materials: Multimillion atom molecular dynamics simulations. *Annual Review of Materials Research*, **32**, 377–400.

[280] Ryckaert, J.-P., Ciccotti, G., and Berendsen, H. J. C. 1977. Numerical integration of the cartesian equations of motion of a system with constraints: Molecular dynamics of n-alkanes. *Journal of Computational Physics*, **23**, 327–341.

[281] Sauzay, M., and Kubin, L. P. 2011. Scaling laws for dislocation microstructures in monotonic and cyclic deformation of fcc metals. *Progress in Materials Science*, **56**, 725–784.

[282] Scheraga, H. A., Khalili, M., and Liwo, A. 2007. Protein-folding dynamics: Overview of molecular simulation techniques. *Annual Review of Physical Chemistry*, **58**, 57–83.

[283] Schiferl, S. K., and Wallace, D. C. 1985. Statistical errors in molecular dynamics simulations. *Journal of Chemical Physics*, **83**, 5203–5209.

[284] Seidman, D. N. 2002. Subnanoscale studies of segregation at grain boundaries: Simulations and experiments. *Annual Review of Materials Research*, **32**, 235–269.

[285] Sept, D., and MacKintosh, F. C. 2010. Microtubule elasticity: Connecting all-atom simulations with continuum mechanics. *Physical Review Letters*, **104**, 018101.

[286] Shan, T.-R., Devine, B. D., Hawkins, J. M., Asthagiri, A., Phillpot, S. R., and Sinnott, S. B. 2010. Second-generation charge-optimized many-body potential for $Si/SiO_2$ and amorphous silica. *Physical Review B*, **82**, 235302.

[287] Shewmon, P. 1989. *Diffusion in Solids*. Warrendale, PA: TMS.

[288] Shin, C. S., Robertson, C. F., and Fivel, M. C. 2007. Fatigue in precipitation hardened materials: a three-dimensional discrete dislocation dynamics modelling of the early cycles. *Philosophical Magazine*, **87**, 3657–3669.

[289] Shirinifard, A., Gens, J. S., Zaitlen, B. L., Pop?awski, N. J., Swat, M., and Glazier, J. A. 2009. 3D multi-cell simulation of tumor growth and angiogenesis. *PLoS ONE*, **4**, e7190.

[290] Singh, D. J. 1994. *Planewaves, Pseudopotentials, and the LAPW Method*. Boston: Kluwer Academic Publishers.

[291] Soper, A. K. 1996. Empirical potential Monte Carlo simulation of fluid structure. *Chemical Physics*, **202**, 295–306.

[292] Sørensen, M. R., and Voter, A. F. 2000. Temperature-accelerated dynamics for simulation of infrequent events. *Journal of Chemical Physics*, **112**, 9599–9606.

[293] Spiegel, M. R. 1988. *Schaum's Outline of Physical Chemistry*. Second edn. Boston: McGraw-Hill.

[294] Srolovitz, D. J., Anderson, M. P., Sahni, P. S., and Grest, G. S. 1984a. Computer simulation of grain growth: 2. Grain-size distribution. *Acta Metallurgica*, **32**, 793–802.

[295] Srolovitz, D. J., Anderson, M. P., Grest, G. S., and Sahni, P. S. 1984b. Computer simulation of grain growth: 3. Influence of a particle distribution. *Acta Metallurgica*, **32**, 1429–1438.

[296] Stillinger, F. H., and Webber, T. A. 1985. Computer simulation of local order in condensed phases of silicon. *Physical Review B*, **31**, 5262–5271.

[297] Suh, C. W., and Rajan, K. 2005. Virtual screening and QSAR formulations for crystal chemistry. *QSAR and Combinatorial Science*, **24**(February), 114–119.

[298] Sutton, A. P. 1993. *Electronic Structure of Materials*. New York: Oxford.

[299] Sutton, A. P., Godwin, P. D., and Horsfield, A. P. 1996. Tight-binding theory and computational materials synthesis. *Materials Research Society Bulletin*, **21**(2), 42–28.

[300] Svyetlichnyy, D. S. 2010. Modelling of the microstructure: from classical cellular automata approach to the frontal one. *Computational Materials Science*, **50**, 92–97.

[301] Swaminarayan, S., LeSar, R., Lomdahl, P. S., and Beazley, D. 1998. Short-range dislocation interactions using molecular dynamics: annihilation of screw dislocations. *Journal of Materials Research*, **13**, 3478–3484.

[302] Swendsen, R. H., and Wang, J.-S. 1986. Replica Monte Carlo simulation of spin glasses. *Physical Review Letters*, **57**, 2607–2609.

[303] Swope, W. C., Andersen, H. C., Berens, P. H., and Wilson, K. R. 1982. A computer simulation method for the calculation of equilibrium constants for the formation of physical clusters of molecules: application to small water clusters. *Journal of Chemical Physics*, **76**, 637–649.

[304] Syha, M., and Weygand, D. 2010. A generalized vertex dynamics model for grain growth in three dimensions. *Modelling and Simulation in Materials Science and Engineering*, **18**, 015010.

[305] Taleb, A., and Stafiej, J. 2011. Numerical simulation of the effect of grain size on corrosion processes: surface roughness oscillation and cluster detachment. *Corrosion Science*, **53**, 2508–2513.

[306] Taylor, G. I. 1934. The mechanism of plastic deformation of crystals. Part I: Theoretical. *Proceedings of the Royal Society of London. Series A*, **145**, 362–387.

[307] Terentyev, D., Zhurkin, E. E., and Bonny, G. 2012. Emission of full and partial dislocations from a crack in BCC and FCC metals: an atomistic study. *Computational Materials Science*, **55**, 313–321.

[308] Tersoff, J. 1988. New empirical approach for the structure and energy of covalent systems. *Physical Review B*, **37**, 6991–7000.

[309] Thomas, L. H. 1927. The Calculation of Atomic Fields. *Proceedings of the Cambridge Philosophical Society*, **23**, 542–548.

[310] Thornton, K., and Poulsen, H. F. 2008. Three-diemensional materials science: an intersection of three-dimensional reconstructions and simulations. *Materials Research Society Bulletin*, **33**(6), 587–595.

[311] Tikare, V., Holm, E. A., Fan, D., and Chen, L.-Q. 1998. Comparison of phase-field and Potts models for coarsening processes. *Acta Materialia*, **47**, 363–371.

[312] Tikare, V., Braginsky, M., Bouvard, D., and Vagnon, A. 2010. Numerical simulation of microstructural evolution during sintering at the mesoscale in a 3D powder compact. *Computational Materials Science*, **48**, 317–325.

[313] Tilocca, A. 2008. Short- and medium-range structure of multicomponent bioactive glasses and melts: an assessment of the performances of shell-model and rigid-ion potentials. *Journal of Chemical Physics*, **129**, 084504.

[314] Timoshenko, S. 1970. *Theory of Elasticity*. New York: McGraw-Hill.

[315] Tsuchiya, T., and Tsuchiya, J. 2011. Prediction of a hexagonal $SiO_2$ phase affecting stabilities of $MgSiO_3$ and $CaSiO_3$ at multimegabar pressures. *Proceeding of the National Academy of Science, USA*, **108**, 1252–1255.

[316] Uberuaga, B. P., Bacorisen, D., Smith, R., Ball, J. A., Grimes, R. W., Voter, A. F., and Sickafus, K. E. 2007. Defect kinetics in spinels: long-time simulations of $MgAl_2O_4$, $MgGa_2O_4$, and $MgIn_2O_4$. *Physical Review B*, **75**, 104116.

[317] Upmanyu, M., Srolovitz, D. J., Shvindlerman, L. S., and Gottstein, G. 1999. Misorientation dependence of intrinsic grain boundary mobility: simulation and experiment. *Acta Materialia*, **47**, 3901–3914.

[318] van der Giessen, E., and A. Needleman, A. 2002. Micromechanics simulation of fracture. *Annual Review of Materials Research*, **32**, 141–162.

[319] van Duin, A. C. T., Dasgupta, S., Lorant, F., and Goddard, W. A. 2001. ReaxFF: A reactive force field for hydrocarbons. *Journal of Physical Chemistry A*, **105**, 9396–9409.

[320] Vanderzande, C. 1998. *Lattice Models of Polymers*. New York, NY: Cambridge University Press.

[321] Verlet, L. 1967. Computer "experiments" on classical fluids. I. Thermodynamical properties of Lennard–Jones molecules. *Physical Review*, **159**, 98–103.

[322] Verlet, L. 1968. Computer "experiments" on classical fluids. II. Equilibrium correlation functions. *Physical Review*, **165**, 201–214.

[323] Vineyard, G. H. 1957. Frequency factors and isotope effects in solid state rate processes. *Journal of Physics and Chemistry of Solids*, **3**, 121–127.

[324] Volterra, V. 2005. *Theory of Functionals and of Integral and Integro-Differential Equations*. New York: Dover.

[325] von Appen, J., Dronskowski, R., and Hack, K. 2004. A theoretical search for intermetallic compounds and solution phases in the binary system Sn/Zn. *Journal of Alloys and Compounds*, **379**, 110–116.

[326] von Neumann, J. 1952. Discussion: shape of metal grains. Pages 108–110 of: Herring, C. (ed), *Metal Interfaces*. Cleveland, OH: American Society for Metals.

[327] Voter, A. F. 1986. Classically exact overlayer dynamics: diffusion of rhodium clusters on Rh(100). *Physical Review B*, **34**, 6819–6829.

[328] Voter, A. F. 1994. The embedded-atom method. Pages 77–89 of: Westbrook, J. H., and Fleischer, R. L. (eds), *Intermetallic Compounds: Vol. 1, Principles*. New York: Wiley.

[329] Voter, A. F. 1996. Interatomic Potentials for Atomistic Simulations. *Materials Research Society Bulletin*, **21**(2). The entire issue is devoted to potentials and has good articles on almost all types of potentials discussed in this text. These articles have many references.

[330] Voter, A. F. 1997a. Hyperdynamics: accelerated molecular dynamics of infrequent events. *Physical Review Letters*, **78**, 3908–3911.

[331] Voter, A. F. 1997b. A method for accelerating the molecular dynamics simulation of infrequent events. *Journal of Chemical Physics*, **106**, 4665–4677.

[332] Voter, A. F., and Chen, S. P. 1986. Accurate interatomic potentials for Ni, Al, and $Ni_3Al$. *Materials Research Society Symposium Proceedings*, **82**, 175–180.

[333] Voter, A. F., Montelenti, F., and Germann, T. C. 2002. Extending the time scale in atomistic simulation of materials. *Annual Review of Materials Research*, **32**, 321–346.

[334] Vukcevic, M. R. 1972. The elastic properties of cubic crystals with covalent and partially covalent bonds. *Physica Status Solidi (b)*, **50**, 545–552.

[335] Wales, D. J. 2003. *Energy Landscapes*. Cambridge, UK: Cambridge University Press.

[336] Wallin, M., Curtin, W. A., Ristinmaa, M., and Needleman, A. 2008. Multi-scale plasticity modeling: Coupled discrete dislocation and continuum crystal plasticity. *Journal of the Mechanics and Physics of Solids*, **56**, 3167–3180.

[337] Wallqvist, A., and Mountain, R. D. 2007. Molecular models of water: derivation and description,. *Reviews in Computational Chemistry*, **13**, 183–247.

[338] Wang, H. Y., and LeSar, R. 1995. O(N) algorithm for dislocation dynamics. *Philosophical Magazine A*, **71**, 149–164.

[339] Wang, H. Y., and LeSar, R. 1996. An efficient fast-multipole algorithm based on an expansion in the solid harmonics. *Journal of Chemical Physics*, **104**, 4173–4179.

[340] Wang, H. Y., LeSar, R., and Rickman, J. M. 1998. Analysis of dislocation microstructures: impact of force truncation and slip systems. *Philosophical Magazine*, **78**, 1195–1213.

[341] Wang, J.-S., and Swendsen, R. H. 1990. Cluster Monte Carlo simulations. *Physica A*, **167**, 565–579.

[342] Wang, Y., and Li, J. 2010. Phase field modeling of defects and deformation. *Acta Materialia*, **58**, 1212–1235.

[343] Wang, Y., Srolovitz, D. J., Rickman, J. M., and LeSar, R. 2003. Dislocation motion in the presence of diffusing solutes: a computer simulation study. *Acta Metallurgica*, **51**, 1199–1210.

[344] Wang, Y. F., Rickman, J. M., and Chou, Y. T. 1996. Monte Carlo and analytical modeling of the effects of grain boundaries on diffusion kinetics. *Acta Materialia*, **44**, 2505–2513.

[345] Wang, Yu. U, Jin, Y. M., and Khachaturyan, G. 2005. Dislocation dynamics – phase field. Pages 2287–2305 of: Yip, S. (ed), *Handbook of Materials Modeling*. Dordrecht: Springer.

[346] Wang, Z., Ghoniem, N. M., and LeSar, R. 2004. Multipole representation of the elastic field of dislocation ensembles. *Physical Review B*, **69**, 174102.

[347] Wang, Z., Ghoniem, N. M., Swaminarayan, S., and LeSar, R. 2006. A parallel algorithm for 3D dislocation dynamics. *Journal of Computational Physics*, **219**, 608–621.

[348] Wang, Z. Q., Beyerlein, I. J., and LeSar, R. 2007. Dislocation motion in high-strain-rate deformation. *Philosophical Magazine*, **87**(16), 2263–2279.

[349] Wang, Z. Q., Beyerlein, I. J., and LeSar, R. 2008. Slip band formation and mobile density generation in high rate deformation of single fcc crystals. *Philosophical Magazine*, **88**, 1321–1343.

[350] Wang, Z. Q., Beyerlein, I. J., and LeSar, R. 2009. Plastic anisotropy in fcc single crystals in high rate deformation. *International Journal of Plasticity*, **25**, 26–48.

[351] Weertman, J., and Weertman, J. R. 1992. *Elementary Dislocation Theory*. New York: Oxford University Press.

[352] Williams, P. L., and Mishin, Y. 2009. Thermodynamics of grain boundary premelting in alloys. II. Atomistic simulation. *Acta Materialia*, **57**, 3786–3794.

[353] Wilson, E. B., Decius, J. C., and Cross, P. C. 1955. *Molecular Vibrations*. New York: Dover Publications, Inc.

[354] Wolf, D. 2005a. Introduction: Modeling crystal interfaces. Pages 1925–1930 of: Yip, S. (ed), *Handbook of Materials Modeling*, vol. A. Dordrecht: Springer.

[355] Wolf, D. 2005b. Structure and energy of grain boundaries. Pages 1953–1983 of: Yip, S. (ed), *Handbook of Materials Modeling*, vol. A. Dordrecht: Springer.

[356] Wolf, D., Keblinski, P., Phillpot, S. R., and Eggebrecht, J. 1999. Exact method for the simulation of Coulombic systems by spherically truncated, pairwise $r^{-1}$ summation. *Journal of Chemical Physics*, **110**, 8254–8282.

[357] Wolff, U. 1989. Collective Monte Carlo updating for spin systems. *Physical Review Letters*, **62**, 361–364.

[358] Wolfram, S. (ed). 1986. *Theory and Applications of Cellular Automata*. Singapore: World Scientific. especially the outline and Chapter 1.1.

[359] Wolfram, S. 2002. *A New Kind of Science*. Champaign, IL: Wolfram Media.

[360] Woodward, C., Trinkle, D. R., L. G. Hector, Jr., and Olmsted, D. L. 2008. Prediction of dislocation cores in aluminum from density functional theory. *Physical Review Letters*, **100**, 045507.

[361] Wulf, W. M. 1998. The image of engineering. *Issues in Science and Technology Online*, **Winter**. http://www.issues.org/15.2/wulf.htm.

[362] Yamakov, V., Wolf, D., Phillpot, S. R., and Gleiter, H. 2002. Grain-boundary diffusion creep in nanocrystalline palladium by molecular-dynamics simulation. *Acta Materialia*, **50**, 61–73.

[363] Yip, S. 2005. Atomistic calculations for structure-property correlations. Pages 1931–1951 of: Yip, S. (ed), *Handbook of Materials Modeling*, vol. A. Dordrecht: Springer.

[364] Yu, J., Sinnott, S. B., and Phillpot, S. R. 2007. Charge-optimized many body potential for the Si/SiO$_2$ System. *Physical Review B*, **75**, 085311.

[365] Zacate, M. O., Grimes, R. W., Lee, P. D., LeClair, S. R., and Jackson, A. G. 1999. Cellular automata model for the evolution of inert gas monolayers on a calcium (111) surface. *Modelling and Simulation in Materials Science and Engineering*, **7**, 355–367.

[366] Zacharapolous, N., Srolovitz, D. J., and LeSar, R. 1997. Dynamic simulation of dislocation microstructures in Mode III cracking. *Acta Materialia*, **45**, 3745–3763.

[367] Zacharapolous, N., Srolovitz, D. J., and LeSar, R. 2003. Discrete dislocation simulations of the development of a continuum plastic zone ahead of a Mode III crack. *Journal of the Mechanics and Physics of Solids*, **51**, 695–713.

[368] Zbib, H. M., Rhee, M., and Hirth, J. P. 1998. On plastic deformation and the dynamics of 3D dislocations. *International Journal of Mechanical Sciences*, **40**, 113–127.

[369] Zhang, H. Z., Liu, L. M., and Wang, S. Q. 2007. First-principles study of the tensile and fracture of the Al/TiN interface. *Computational Materials Science*, **38**, 800–806.

[370] Zhou, C. Z., and LeSar, R. 2012a. Dislocation dynamics simulations of Bauschinger effects in metallic thin films. *Computational Materials Science*, **54**, 350–355.

[371] Zhou, C. Z., and LeSar, R. 2012b. Dislocation dynamics simulations of plasticity in polycrystalline thin films. *International Journal of Plasticity*, **30–31**, 185–201.

[372] Zhou, C. Z., Biner, S. B., and LeSar, R. 2010a. Discrete dislocation dynamics simulations of plasticity at small scales. *Acta Materialia*, **58**, 1565–1577.

[373] Zhou, C. Z., Biner, S. B., and LeSar, R. 2010b. Simulations of the effect of surface coatings on plasticity at small scales. *Scripta Materialia*, **63**, 1096–1099.

[374] Zhou, C. Z., Beyerlein, I. J., and LeSar, R. 2011a. Plastic deformation mechanisms of fcc single crystals at small scales. *Acta Materialia*, **59**, 7673–7682.

[375] Zhou, N., Shen, C., Mills, M. J., Li, J., and Wang, Y. 2011b. Modeling displacive-diffusional coupled dislocation shearing of $\gamma'$ precipitates in Ni-base superalloys. *Acta Materialia*, **59**, 3484–3497.

[376] Zwanzig, R. 2001. *Nonequilibrium Statistical Mechanics*. New York: Oxford University Press.

# INDEX

Printed in the United States
By Bookmasters